Heredity and Society

Heredity and Society

Readings in Social Genetics

Second Edition

Edited by

Adela S. Baer

San Diego State University

Macmillan Publishing Co., Inc.
New York

Collier Macmillan Publishers
London

Macmillan Publishing Co., Inc.
866 Third Avenue, New York, New York 10022

Collier Macmillan Canada, Ltd.

Library of Congress Cataloging in Publication Data

Baer, Adela S   (date)   comp.
    Heredity and society.

    1. Human genetics—Social aspects—Addresses, essays, lectures. I. Title.
QH438.7.B3   1977   301.24'3   76-7397
ISBN 0-02-305160-4

Printing: 1 2 3 4 5 6 7 8   Year: 7 8 9 0 1 2 3

For I. Michael Lerner, teacher, and for Eugene M. Swenson and Golda M. Swenson, supporters

preface

This set of readings was collected for the purpose of informing undergraduate and graduate students of biology and allied fields about the genetic aspects of contemporary social problems. Too often in the arena of public affairs one hears that the world's problems have a single master cause (ideology yesterday, overpopulation today). This book is designed to help correct any such misconception in the academic or public mind about the biological bases of society's predicaments. Ecology, medicine, and agriculture notwithstanding, genetics is germane to the dimensions and difficulties of human welfare. Moreover, the borderline between scientific disciplines, such as genetics and medicine, is often a fertile ground for new concepts and discoveries. This book furnishes many vistas of interactions between genetics and allied fields, some applied and some not— interactions that enrich insights on both sides.

The traditional concerns of genetics are presented in this book in a way that is convenient for exploring biosocial problems. The history of the development of genetical thought is explored in the first section. These reflections on genetics in terms of its social setting in Europe, Russia, and the United States illuminate the problems of prejudice and social policy that have befallen genetics. In turn, Dobzhansky shows us how our political imagery and compartmentalization express a widespread misconception of the biological nature of the human species.

Basic problems in understanding gene expression are highlighted by consideration of hybrid vigor in agricultural crops, the activation of mutagens by liver extracts in the system developed by Bruce Ames, the seemingly capricious appearance of schizophrenia in families, the genetic wear-and-tear involved in aging, and the relevance of IQ scores to human intelligence. Genetic recombination is not expounded directly in the book but is closely related to, or underlies, what is said about biopolitical fallacies on genetics and eugenics, about the importance of hybrid corn and new wheat varieties in the world economy, and about the genetic basis of human behavior and intelligence.

Mutation is the primary concern of the section on genetic hazards in the environment, but it is also dealt with here in papers on the genetic engineering of new crops, on the aging process, on medical genetics, and on planned or unplanned genetic engineering of human beings.

The genetic structure of populations, as well as natural and artificial selection, is taken up in discussions of environmental mutagens, plant breeding advances, and the evolution of human aggression, intelligence, disease resistance, and other attributes. Family planning, genetic counseling practices, abortion—all discussed here—also bear on the question of the genetic structure of the human population. Other practical concerns in population and quantitative genetics are discussed in various readings.

In this edition special emphasis is put on the social parameters of medical genetics. This field is currently expanding at a dramatic rate and, more than any others, raises pressing questions of social goals, social costs, and emotional consequences. These issues are investigated both in the section on medical genetics and in the readings by Kirk, Gary, and Fuhrmann in other sections.

The future of human evolution is the focus of the last section in the book, but the premise of any significant human future is based on ability to cope with current problems of political misjudgment, food production, environmental pollution, and not the least, population growth. Thus the final readings in the book depend on the resolution of issues raised in earlier sections. As Paul Tillich said about the experience of being, which is acutely relevant to a book on heredity and society, "In this region we do not *have* a question, we *are* the question."

Each section contains an introduction outlining the key ideas of, and controversies about, the chosen topics as well as a reference list for further study. In selecting articles for this book I have chosen readable, concise presentations of ideas and facts over lengthier, perhaps more comprehensive, reports. In some areas a wealth of lively articles was available, yet only one could be included.

I am grateful to numerous colleagues, students, and friends for their encouragement, suggestions, and other aid on this project. It is a pleasure to acknowledge the particular assistance on the first edition of Fran Whitfield, William J. Libby, Frederick J. deSerres, Herman W. Lewis, Frank J. Ratty, Kenneth M. Taylor, Wayne Daugherty, Quinton B. Welch, and my daughters, Susan and Nicoli Baer. For the second edition the assistance of Jean Passino, John Jenkins, Robert M. Kitchen, and Mary Fitzgerald is gratefully acknowledged. I am especially indebted to the authors and publishers who graciously gave their permission to have their articles reprinted in the book.

A. S. B.

contributors

(Senior authors in the case of multiple authorship)

A. C. Allison, *National Institute for Medical Research, Mill Hill, London, England.*

Bruce N. Ames, *Department of Biochemistry, University of California, Berkeley.*

Francisco J. Ayala, *Department of Genetics, University of California, Davis.*

M. B. Baird, *Masonic Medical Research Laboratory, Utica, New York.*

Peter S. Carlson, *Department of Crop and Soil Sciences, Michigan State University, Lansing.*

E. W. Caspari, *Department of Biology, University of Rochester, New York.*

C. T. Chetsanga, *Department of Natural Sciences, University of Michigan, Dearborn.*

A. Comfort, *Institute for Higher Studies, 2311 Garden Street, Santa Barbara, California.*

Margaret J. Corey, *P. O. Box 111, Nackawic, New Brunswick, Canada.*

James F. Crow, *Department of Genetics, University of Wisconsin, Madison.*

Beryl L. Crowe, *The Evergreen State College, Olympia, Washington.*

The late Theodosius Dobzhansky, *Department of Genetics, University of California, Davis.*

Alan Ducatman, *Department of Community Medicine, Mount Sinai School of Medicine of the City University of New York.*

The late L. C. Dunn, *Columbia University, New York.*

Daniel Dykhuizen, *Department of Genetics, Research School of Biological Sciences, The Australian National University, Canberra.*

L. Erlenmeyer-Kimling, *Department of Medical Genetics, New York State Psychiatric Institute, New York, New York.*

Amitai Etzioni, *Center for Policy Research, Inc., 475 Riverside Drive, New York, New York.*

John Fletcher, *Interfaith Metropolitan Theological Education, Inc., 1419 V Street NW, Washington, D.C.*

F. C. Fraser, *Department of Biology, McGill University, Montreal, Quebec, Canada.*

Walter Fuhrmann, *Institut für Humangenetik der Universität, Giessen, Germany.*

Lawrence E. Gary, *Institute for Urban Affairs and Research, Howard University, Washington, D.C.*

Marshall F. Gilula, *Life Energies Research Institute, 3240 Gifford Lane, Coconut Grove, Florida.*

Richard A. Goldsby, *Department of Chemistry, University of Maryland, College Park.*

The late Richard B. Goldschmidt, *Zoology Department, University of California, Berkeley.*

John B. Graham, *Department of Pathology, University of North Carolina Medical School, Chapel Hill.*

Garrett Hardin, *Department of Biological Sciences, University of California, Santa Barbara.*

Jack R. Harlan, *Agronomy Department, University of Illinois, Urbana.*

Leonard L. Heston, *Department of Psychiatry, University of Minnesota, Minneapolis.*

R. Holliday, *National Institute for Medical Research, Mill Hill, London, England.*

Arthur R. Jensen, *Institute of Human Learning, University of California, Berkeley.*

Dudley Kirk, *Food Research Institute, Stanford University, Stanford, California.*

Richard C. Lewontin, *Biological Laboratories, Harvard University, Cambridge, Massachusetts.*

The late W. Farnsworth Loomis, *Department of Biochemistry, Brandeis University, Waltham, Massachusetts.*

Paul C. Mangelsdorf, *510 Caswell Road, Chapel Hill, North Carolina.*

Ei Matsunaga, *National Institute of Genetics, Mishima, Japan.*

Robert W. Miller, *Epidemiology Branch, National Cancer Institute, National Institute of Health, Bethesda, Maryland.*

James V. Neel, *Department of Human Genetics, University of Michigan Medical School, Ann Arbor, Michigan.*

Louis P. Reitz, *Crop Research Division, Agricultural Research Service, U.S. Department of Agriculture, Beltsville, Maryland.*

Robert L. Sinsheimer, *Division of Biology, California Institute of Technology, Pasadena, California.*

Staffan Skerfving, *National Institute of Public Health, Solnavägenl, Solna, Sweden.*

Curt Stern, *Zoology Department, University of California, Berkeley.*

D. R. Stoltz, *Research Laboratories, Food and Drug Directorate, Ottawa, Ontario, Canada.*

contents

Introduction

Introduction

Every age is characterized by critical problems. As people become aware that a change in the natural or social environment is interfering with the "normal" course of living, they identify a problem and anxiously, or even desperately, strive to solve it.

Today we are probably more aware of critical social issues than in the past—global communication networks and high levels of literacy and education facilitate this awareness. In truth, today we probably have more problems than during any previous age. To the recurring crises brought on by local wars, plagues, pestilence, and other natural disasters, we have added the possibility of annihilation of the human species through nuclear or chemical-biological warfare or through pollution-caused destruction of the food chain on which all life depends. In addition, there is the unprecedented possibility of inadvertent bungling of human evolution itself.

Many of our present critical problems are traceable to the growth of technology, and thus of science (Brown, 1971; Fuller, 1972). Technical advances in medicine, agriculture, transportation, and so on, have produced an enormous population, much longer lived than formerly, which (particularly in the United States) voraciously consumes natural resources. Of course, the blatant paradox is that technology has produced the present problems by surmounting previous inefficiency or ineffectiveness in medicine, agriculture, and other realms. And although technology can solve the currently pressing problems of population growth and its associated ills of pollution and misuse of natural resources, the problems of the future have not been seriously investigated, let alone anticipated. Most of us are willing to leave that to the next generation. Our hope is to be able to explore the current issues in as comprehensive a manner as possible. Thus, this book seeks to determine the genetic implications, for human and other organisms, of current social problems and practices—thereby to move toward and better discern a "science of mankind."

The book explicates humanity's continual problems in food production by describing the input of genetic science into improved plant and animal breeding. The readings summarize the triumph of hybrid corn and go on to the somewhat tarnished green revolution in grain production.

1

Possibilities for agricultural improvement by genetic engineering are also discussed.

Human problems of survival in terms of surmounting disease are discussed in the book from a variety of perspectives. Food additives and other environmental chemicals, with their attendant dangers in increasing mutations, are considered. The contrived decline in infections and congenital diseases and the resulting increase in viability and longevity are scrutinized in genetic and evolutionary terms.

The focus on the peculiar institution of war is in terms of the genetic effects of atomic radiation and chemicals. Biological weapons are not discussed in the book because little nonsecret information on them is available. War as a cloak for genocide, as was espoused by Hitler, is discussed.

Sociological aspects of genetics are interwoven among other topics in the book: Abortion and other methods of birth control are discussed from a genetic point of view, trends in genetic counseling and screening are highlighted, intelligence is given a forum, and proposed genetic change by engineering at the DNA level is presented and debated from diverse viewpoints.

Some of these topics are merely timely; others are timeless. The advances in science generate changes in civilization which are largely unpredictable, while at the same time the laws of nature, of which genetic relations form some of the most elegant in biology, endure and persist. Genetics is becoming increasingly a component of medical and agricultural education and will soon become a major concern in psychology and the social sciences. These trends are all to the good. It is also important for biologists who do not classify themselves as geneticists (such as ecotacticians) to appreciate more fully the genetic aspects of biological phenomena, not only in their academic studies, but also in their philosophical or political ruminations on human beings and the natural world.

The generations now living, but particularly the young, will have to make critical decisions on the fate of mankind. It is well that these decisions be based as fully as possible on a sound evaluation of scientific knowledge, not the least of which is genetic knowledge. We are just now emerging from the Dark Ages in terms of ecological awareness. Genetic awareness is not yet that far along. However, we must keep in mind that even with imaginative decisions based on humanistic and scientific understanding, humanity will not attain a genetic paradise in the classical sense. The best of all possible human existences on earth is one of sustained genetic diversity which permits evolutionary adaptation to unpredictable environmental change (Dobzhansky, 1962).

For the reader only casually exposed to genetics heretofore, most introductory biology texts explain Mendelian and chromosomal genetics as well as mutation in a concise fashion (e.g., Baer et al., 1971). Lengthier treatments of genetics are given by Lerner and Libby (1976) and Strickberger (1976), among others.

Bibliography

Baer, A. *The Genetic Perspective.* Philadelphia: Saunders, 1977.

Baer, A., W. Hazen, D. Jameson, and W. Sloan. *Central Concepts of Biology.* New York: Macmillan, 1971.

Bodmer, W. F., and L. L. Cavalli-Sforza. *Genetics, Evolution, and Man.* San Francisco: Freeman, 1976.

Brown, M., ed. *The Social Responsibility of the Scientist.* New York: Free Press, 1971.

Dobzhansky, T. *Mankind Evolving.* New Haven, Conn.: Yale University Press, 1962.

Dunn, L. C. *Heredity and Evolution in Human Populations*, rev. ed. New York: Atheneum, 1968.

Fuller, W., ed. *The Biological Revolution: Social Good or Social Evil?* Garden City, N.Y.: Doubleday Anchor, 1972.

Jenkins, J. *Genetics.* Boston: Houghton Mifflin, 1975.

Lerner, I. M., and W. J. Libby. *Heredity, Evolution, and Society*, 2nd ed. San Francisco: Freeman, 1976.

Stern, C. *Principles of Human Genetics*, 3rd ed. San Francisco: Freeman, 1973.

Strickberger, M. W. *Genetics*, 2nd ed. New York: Macmillan, 1976.

Sutton, H. E. *An Introduction to Human Genetics*, 2nd ed New York: Holt, Rinehart and Winston, 1975.

one

Genetics and the Social Structure

Genetics has experienced problems of acceptance in the society of scientists and problems of understanding and application in society as a whole. The first social setting that genetics encountered was the world of traditional biology. As shown in Goldschmidt's paper here, genetics equilibrated with that world only gradually. The young science then found itself embroiled in other worlds: those of politics and philosophy.

Two great social tragedies involving genetics occurred in Germany and Russia. Russia experienced the rampant antigeneticism of Lysenko from the 1930s through the 1950s. Lysenko espoused Stalin's ideology and became Stalin's expert on the improvement of farm productivity, a pressing problem. He denounced genetics as being anti-Marxist, according to Stalin's interpretation, and trumpeted the wonders of scientifically questionable measures to improve agriculture. He thus prevented any farm improvement by means of genetic techniques, and what is more, persecuted the biologists who opposed him. Few Mendelian geneticists survived the Lysenko era in Russia, as Caspari and Marshak report in this section. Other sources of information on the political problems of genetics and other sciences in Russia include Joravsky (1970), Lerner (1972), Lerner and Libby (1976), Sakharov (1968), Medvedev (1969), and Wallace (1972). The last two chapters of Medvedev's book are of particular interest. One describes the beneficial changes that have occurred in Russia since the deposition of Lysenko in 1965. The other analyzes how such pseudoscientific doctrines come about and may become politically entrenched—an object lesson for every society. Commentary on Medvedev's elucidation of Lysenkoism is given by Hechter (1970), and the story of this book's publication in the United States is given in Lerner's foreword to the book.

The inhuman events involving genetics in Germany during the Hitler dictatorship are discussed by Dunn in this section. The events are woven by Dunn into the social fabric of the eugenics movement from 1900 onward. Hitler was obsessed with the idea of race purification. Inept "scientific" doctrines for genetic improvement offered him a rationale for the sterilization (in 1933) and later the gas chamber murder (in 1939) of the allegedly unfit: mental patients, those with deformities, the feebleminded, Jews, and others. Over 200,000 people were sterilized; at least 6 million, mostly Jews, were exterminated (see Tenenbaum

1956, for details). Haller (1963, 1968), Ludmerer (1972), and others have helped elucidate the decay of eugenics into racism that occurred in the United States during the first half of this century.

In reference to both Lysenko in Russia and Hitler in Germany, we see the tendency in the public mind to view plants, animals, and human beings in terms of ideal types—optimal phenotypes. As Dobzhansky points out in the final paper in this section, such typological thinking is erroneous, widespread, and not yet in danger of dying out. Through Dobzhansky's description of the work of Charles Darwin and Gregor Mendel, we see that pluralistic, populational thinking puts us much nearer the evolutionary truth of nature than does typological thinking. In culmination, Dobzhansky skillfully demonstrates how these two types of thinking about the biological world influence our political views, whether liberal or conservative, particularly with regard to racism.

One often hears that science is, or is not, politically and socially neutral. What are the key ideas presented in this section that bear on this question?

Bibliography

Haldane, J. B. S. *Heredity and Politics*. New York: Norton, 1938. A brilliant attack on the eugenics movement that had wide impact.

Haller, J. S., Jr. *Outcasts from Evolution: Scientific Attitudes of Racial Inferiority 1859–1900*. Urbana, Ill.: University of Illinois Press, 1971.

Haller, M. H. *Eugenics. Hereditarian Attitudes in American Thought*. New Brunswick, N.J.: Rutgers University Press, 1963.

Haller, M. H. Social science and genetics: A historical perspective. In *Genetics*, Biology and Behavior Series, D. C. Glass, ed. New York: Rockefeller University Press, 1968, pp. 215–225.

Harmsen, H. The German sterilization act of 1933. *Eugenics Rev.*, **46**:227–232, 1955. A historical review.

Hechter, D. Biology and politics. *Bull. Atomic Sci.*, Apr. 1970, pp. 54–56.

Joravsky, D. *The Lysenko Affair*. Cambridge, Mass.: Harvard University Press, 1970.

Lerner, I. M. Dialectical materialism and Soviet science. *Quart. Rev. Biol.*, **47**:313–316, 1972.

Lerner, I. M., and W. J. Libby. *Heredity, Evolution, and Society*, 2nd ed. San Francisco: Freeman, 1976.

Ludmerer, K. M. *Genetics and American Society: A Historical Appraisal.* Baltimore, Md.: Johns Hopkins University Press, 1972.

McClearn, G. E., and J. C. DeFries. *Introduction to Behavioral Genetics.* San Francisco: Freeman, 1973. Chapter 11 discusses the eugenics movement and the racism resulting from it in the United States, including the relevant court decisions.

Medvedev, Z. A. *The Rise and Fall of T. D. Lysenko.* Translated by I. M. Lerner. New York: Columbia University Press, 1969.

Sakharov, A. D. *Progress, Coexistence, and Intellectual Freedom.* New York: Norton, 1968.

Slater, E. German eugenics in practice. *Eugenics Rev.,* **27**:285–295, 1936.

Tenenbaum, J. *Race and Reich: The Story of an Epoch.* New York: Twayne, 1956.

Wallace, B. *Essays in Social Biology, Vol* **2**. *Genetics, Evolution, Race, Radiation Biology.* Englewood Cliffs, N.J.: Prentice-Hall, 1972. Contains a chapter of take-home lessons from Lysenkoism.

1   Fifty Years of Genetics

Richard B. Goldschmidt

When the baby was born fifty years ago whose development into a giant some of us have witnessed, no proud parents hailed the new addition to the biological family, no relatives surrounding the crib heaped praise and presents upon the child. Quite the contrary: Everybody looked askance at the odd creature which did not seem at all to fit into the respectable family. How could such an unimportant thing as purple and white pea flowers impart decisive knowledge on heredity, which everybody knew, was a phenomenon embracing the entire organism? Who could believe that the playing with numbers and symbols could lead to anything but oddities? Everybody was convinced that the offspring of two different parents was halfbreed and their offspring a quarterbreed. There was just no room for the strange story of the pea flowers. It seems incredible to us now how few zoologists and botanists realized that a turning point in the study of heredity had arrived and how fewer still were sufficiently impressed as to start, at once, work on those lines.

There were other obstacles to quick success. Among the rediscoverers of Mendel's laws, the already most famous one, de Vries, was more interested in the aberrant behavior of his Oenothera " mutants " than in the simple facts of Mendelian heredity, which were only a side issue with him and which he subsequently played down. Tschermak was engaged in agricultural and horticultural plant breeding, which did not lend him much authority with zoologists and botanists, as things were. Correns, who had gone farthest in the new line of work and followed it up vigorously, happened to be a difficult and queer personality, not the man to impress and convince his reluctant fellow botanists. He worked at the time in the laboratory of the great plant physiologist, Pfeffer. When he asked his boss for a small plot of land in the Botanical Garden in which to continue his experiments, Pfeffer refused, saying that this was not scientific botany but just gardening, which could not be done in his institute. Thus, the important results which Correns accomplished in the following years, working his own garden plot with no help but his wife's, remained largely unknown to his fellow botanists. He had no students or collaborators and grew to middle age before the hostile profession gave him a small professorship.

In view of such a situation it was a lucky event for the young science that a man of unusual energy, endowed with strong fists and a fighting spirit, William Bateson, had already embarked upon what was to become genetical work, again in the face of hostile university colleagues, but backed by a committee of the Royal Society. Thus, with a group of enthusiastic collaborators, he was at once prepared to start Mendelian experiments with both animals and plants. He was soon followed by a few men in different countries, some of whom had also to pay for entering an

Excerpt from " Fifty years of genetics," *Amer. Naturalist*, **86**:313–340, 1950. This address was prepared to be read at the opening of the Golden Jubilee Meeting of the Genetics Society of America held at Columbus, Ohio, Sept. 11–14, 1950. Reprinted with the permission of The University of Chicago Press.

unpopular line of work. Cuénot in France quickly became one of the founding fathers, which did him little good in the eyes of the scientific powers-that-be in Paris. Toyama in Japan started his pioneer work on silkworms with the result that ten years later I found him past middle age and still an assistant professor. Standfuss and Lang in Switzerland and Johannsen in Denmark, Baur in Germany, Castle, Davenport and Shull in this country and Nilsson-Ehle in Sweden, soon joined the small group, of whom the last named had to wage a hard uphill fight to find success. After the middle of the first decade, genetics began to become more or less respectable and at the end of the first decade (i.e., 1910) was established as an up-and-coming branch of biology.

... Boveri and Sutton had shown in the first years of Mendelism that the new laws are completely explained by the location of the unit factors, later called genes, in the chromosomes and by the maneuvers of maternal and paternal chromosomes in the meiotic divisions. At about the same time it was realized that the enigmatic accessory chromosomes control the distribution of the sexes among the offspring. Thus, as early as three years after the rediscovery of Mendel, genetics might have assumed her proper place in biology.

That this took much longer is due, at least in part, to the unfortunate circumstance that Bateson, the man who exercised the greatest influence upon the development of genetics in its childhood, just did not believe in chromosomes (and probably did not know much about them). He actually authored the slogan that genetics must do its work without taking notice of eventual carriers of Mendelian unit factors. Johannsen, the greatest influence in the following years, concurred in this opinion. The leading continental geneticist, Correns, did not exactly share these views and was even willing to accept the role of the chromosomes. However, in a queer kind of jealousy, he always insisted that the decisive facts were the genetical ones which might find interesting but unnecessary confirmation from cytology. Thus, early genetics worked exclusively with the symbolic formulae without the benefit of visualizing their workings in terms of chromosomes. For some reason, which is difficult to understand, even many good biologists were unable to handle this symbolism and therefore derided it. Strange as it sounds, Hans Spemann once told me after a genetical lecture that he never was able or would be able to work those notations. Still, at the end of the first decade when I communicated to a meeting of the German zoologists a simple case of dominant sex-linked heredity which I had found in the nun moth, I was reprimanded by my elders for presenting what nobody in the audience could possibly understand. It is rather fortunate for the development of genetics that (at that time) those who were unable to understand the new science did not have the backing of authoritarian rulers who might have been willing to declare their ignorance of the law of the land.

In spite of all these handicaps to clear sailing and the small size of the crew steering the genetics ship through sometimes turbulent waters, the accomplishments in the first decade laid the solid basis for the subsequent phenomenal rise of genetics. The universal occurrence of Mendelian heredity was demonstrated in plants, animals and man for all types of visible characters which lent themselves to analysis. ... The first links with neighboring sciences were forged when human traits were found to Mendelize, and both medicine and anthropology started, hesitatingly, to incorporate the new knowledge. Eugenics began to enter its first rather primitive Mendelian phase. The discovery that characters of crop plants and domestic animals showed Mendelian behavior started the genetical work in agriculture. . . .

After this diversion into the childhood diseases of genetics, we may finish the appraisal of the Gargantuan infancy of our science by stating that after a ten-year struggle for a place in the sun the still-small group of geneticists felt that they could proclaim genetics as an established science. The private conferences which Bateson had convoked from time to time were enlarged in 1911 into the first International Conference on Genetics, held in Paris. It was wholly directed and financed by the seed firm of Vilmorin whose clever and energetic head had realized the supreme importance of genetics for plant breeding. Official French biology remained conspicuously absent, and few biologists from other countries attended. Johannsen, Cuénot and the Bateson group were the central figures, Baur and Nilsson-Ehle the oncoming young men. Though international, it was still a meeting of a small group regarded more or less as outsiders by their fellow biologists. However, the first genetic periodical had already been started in Germany, soon followed by one in England. Teaching of genetics had begun in a few universities. . . .

You may reproach me with spending so much time upon the childhood of genetics instead of concentrating on the problems of today. The reason is that too many of the younger generation, overwhelmed by the amount of recent work, do not realize the gigantic size of the contribution made by the pioneers in the short space of ten years (1900–1910) against serious odds and handicaps. They wrote the alphabet of genetics, without which no one working today could do his spelling. . . .

2   The Rise and Fall of Lysenko

E. W. Caspari and R. E. Marshak

Academician T. D. Lysenko is again in the news, probably for the last time. Within the last several months he has been removed as director of the Institute of Genetics of the Academy of Sciences of the U.S.S.R., and his entire institute is being completely reorganized along Western lines. Furthermore, special commissions will be set up to eradicate Lysenkoist doctrine from the curriculum and textbooks used in Soviet secondary schools and institutions of higher education. What is the significance of Lysenko's downfall and what does it reveal about the state of Soviet society and its present leadership? It is certainly no accident that Lysenko's exit has followed so closely the accession of Brezhnev and Kosygin to power in the Soviet Union.

From *Science*, **149** (July 16, 1965), 275–278. Copyright 1965 by the American Association for the Advancement of Science. Reprinted with permission of the authors and the publisher.

The fortunes of Lysenko have always been closely tied to the support which he secured from the top leadership of the Soviet Communist Party and government. Under Stalin, Lysenko became not only director of the Soviet Academy's Institute of Genetics but also president of the all-powerful Lenin Agricultural Academy. In this dual capacity Lysenko was in a position to crush his scientific opponents, and this he proceeded to do with great alacrity and thoroughness. When Khrushchev took over, the official attitude became one of "letting all flowers bloom," and Russian biologists formerly in disgrace were rehabilitated and allowed to resume their scientific work. Lysenko lost nominal control of the Agricultural Academy but remained director of the Academy's Genetics Institute and continued to exercise considerable influence in high government and party circles. Although Lysenko no longer dominated all Russian genetics, he still held Khrushchev's ear and received new honors from the Soviet government. The fall of Lysenko since Brezhnev and Kosygin came to power indicates that the new leaders have adopted a less emotional and more intellectual approach to science than Khrushchev did.

Not only does the decline in the fortunes of Lysenko reflect the continuing improvement in the scientific climate in the U.S.S.R. since Stalin's death, but his career is an object lesson on the harm which results from an attempt to impose an external dogmatism on science. There are numerous instances in history of such an attempt having been made—for example, Galileo's experience with the Inquisition in Rome in 1616, the Scopes trial in Tennessee in 1925, and others. But there are few cases in which the chief antagonist has been a scientist himself seeking to impose his dogmatic views on an entire scientific discipline in the name of a higher authority (in this case the Soviet Communist Party). Furthermore, the Lysenko affair has been marked by unusually brutal treatment of the opposition and by a conspicuous failure to attain the very objectives invoked to justify such treatment.

The Background

Just as the Fundamentalists' literal interpretation of the Bible led to the passage of the Tennessee law against the teaching of evolution and to the subsequent Scopes trial, a literal interpretation of Marx and Engels set the stage for the rise of Lysenko. It should be recalled that Marx and Engels published their major works before Mendelian genetics became known to the world (in 1900) and at a time when Lamarck's idea of the inheritance of acquired characteristics was still a respectable scientific theory. Even Darwin had assumed that the Lamarckian mechanism was of importance in evolution. The thesis that the environment can produce physiological and mental changes in man which can be passed on to later generations was embraced by Marx and Engels (great admirers of Darwin), who saw in such a process a happy device for hastening the achievement of a benevolent Communist society.

By 1925 the Lamarckian hypothesis was being abandoned in the West because all the evidence brought forth in its favor had turned out to be either invalid or faked. Moreover, in the West, Darwinism had been reinterpreted to exclude the Lamarckian factor in evolution. Not so in the Soviet Union. Since Marx and Engels were the prophets of the Russian Revolution, their fervent Darwinism (with its associated Lamarckian doctrine) provided the "scientific" rationale for dedicated Communists striving to mold Soviet man into a paragon of virtue, hard work, and social consciousness.

Several other factors contributed to the persistence of Lamarckian ideas in the Soviet Union. The great Pavlov, whose work on conditioned reflexes in animals had highlighted the importance of environmental influences in behavior, announced in 1923 that a conditioned reflex in mice was inherited and even advanced in succeeding generations. Pavlov's withdrawal of this claim when the evidence collapsed did not discourage the Lamarckian Marxists in the U.S.S.R.

A Russian scientist who more directly prepared the groundwork for Lysenko was the Russian horticulturist I. V. Michurin (the Russian counterpart to the American Burbank), who died in 1935. Michurin's extensive and, in many ways, successful empirical work in plant breeding was bolstered by theoretical views which were anti-Mendelian and, in a sense, pro-Lamarckian. Michurin believed (as Darwin did) that "heredity" is in some fashion diffused throughout the organism and can be modified by many types of environmental influences. This point of view was in the Lamarckian tradition and was taken over by Lysenko.

Another factor which contributed to Lysenko's rise was the Soviet Union's great need to improve its agriculture. In the Western world Mendelian genetics was being used to develop new strains of corn and other plants and to increase agricultural productivity. But the time scale was slow, because advances depended on patient breeding of the proper genetic material and laborious crossing experiments. The Soviet government was impatient and ready to back a scientist who promised rapid improvement of Soviet agriculture.

Attack on Mendelian Genetics

Thus the stage was set in the mid-1930's, at a time when Lamarckianism had been discredited in the Western world, for Lysenko's destructive attack on Mendelian genetics. At that time there were in the Soviet Union many very active and original geneticists. Some of them were working with problems of evolution and gene action in fruit flies. There was also an outstanding group of workers in plant genetics, under the leadership of Academician N. I. Vavilov, who were concerned with the study of the evolutionary origin of cultivated plants and the application of this knowledge to practical plant breeding. As a follower of Michurin and a devout Communist, Lysenko began a violent campaign against the Russian school of classical genetics.

Lysenko derided "Mendelian-Weismannian-Morganian" genetics as passive, "idealistic," and "metaphysical," in contrast to the active, "materialistic," and "empirical" character of his own physiological theory of heredity. He had no sympathy with Mendel's original abstract concept of the gene as a calculating unit designed to describe genetic experiments, nor with the later elaboration of the concept, in which the genes are viewed as autonomous units governing the inheritance of well-defined characters which could not be modified by environment. Lysenko ignored the fact that, by the mid-1930's, some rather concrete ideas about the molecular nature of the gene had been developed. He also claimed that

Mendelian genetics was anti-Darwinian, even though by 1930 modern evolutionists had succeeded in reconciling Darwin's theory of evolution with Mendelian genetics. Lysenko rejected outright the abundant experimental basis for the Mendelian mechanism of inheritance with the argument that the Mendelian rules are statistical in character. They therefore conflict, he asserted, with Marxist dialectics and cannot be regarded as " natural laws." This type of argument has been used at one time or another by Soviet scientists against Western work in quantum physics, relativity theory, theory of the chemical bond, and cybernetics.

Lysenko substituted for the Mendelian theory of heredity his own so-called "physiological" theory, following Michurin in the assumption that heredity is diffused through the whole organism, collected in the germ cells, and mixed in the course of fertilization; in this process, he argued, the weaker germ cell becomes assimilated by the stronger, in analogy to the assimilation of food in nutrition. Lysenko further proposed his theory of " development of plants by stages," according to which the successive stages in the development of a plant can be affected—that is, speeded up or slowed down—by environmental conditions such as temperature. If one of these stages is altered by environmental influences, the subsequent stages will also become changed and an organism with different physiological qualities, presumably better adapted to the environmental influence in question, will result. Lysenko finally claimed that these environmentally induced changes are transmitted to the progeny and thus result in better-adapted plant lines.

Agricultural Experiments

Lysenko's promise to improve agriculture was based on these ideas. Extensive experiments were carried out: early embryonic stages—that is, plant seeds—were subjected to treatment with strong environmental agents, like high and low temperatures, and the progeny of the treated plants were bred. Spectacular claims for the improvement of plant species by this method were made. But it appears that in fact these experiments did not measure up to expectations and that the slow " Western " method, based on quantitative classical genetics, is superior to a method based on exposing unselected material to altered environmental conditions. A particularly striking example of the effectiveness of Western methods is the successful use of hybrid corn in corn breeding—a technique which underlies the great productivity of the American corn-breeding industry. The method is based on the principle of heterosis, the production of higher yields of corn through the use of hybrid plants developed by crossing carefully inbred lines. Heterosis is well explained by classical genetics but is a complete mystery under Lysenkoist doctrine.

Another line of experimentation pursued by Lysenkoists was the production of " vegetative hybrids." If, as was claimed, heredity is diffused through the organism, it should be possible to produce hybrids by the union of two organisms, such as is accomplished in grafting one plant on another, just as well as through sexual combination. Grafting has been used for a long time in agricultural practice, but the Lysenkoists claimed that the graft hybrids propagate a combination of characters from the two species through future generations in the same way that sexual hybrids do. It is, of course, quite possible that in such grafts branches arise with tissues derived from both plants. Such "chimeras," also observed in the West, have, however, never been found to perpetuate hybrid characters in any experiments carried out outside of Russia.

The Lysenkoists attempted, furthermore, to apply this method of changing hereditary characters to animals. In this application the injection of blood from animals of one strain into animals of another was considered the equivalent of grafting in plants. It was, for instance, claimed that if white chickens are injected with blood from a colored strain, then in the offspring of these treated birds, partly and even fully colored animals appear with a certain frequency. Since spotting in animals is by no means infrequent and can readily be accounted for by lack of genetic purity, experiments of this type have to be carefully controlled in order to be convincing. The Lysenkoists did not supply sufficient information to permit an evaluation of their results.

Lysenko the Victor

The dubious character of Lysenkoist research and the Lysenkoists' grandiose claims for success in plant breeding led, during the period 1936 to 1948, to a violent controversy between Lysenko and his followers, on the one hand, and the Western-oriented geneticists, under Vavilov, on the other. Finally, in 1948, at a meeting of the Lenin Agricultural Academy, Lysenko was declared the victor, and classical genetics was denounced as contrary to Darwinism, Michurinism, and dialectical materialism. Having won the official support of the Soviet Communist Party and of Stalin, Lysenko established control over Russian genetics and allied branches of biology. He proceeded to suppress research in classical genetics and to eliminate his opponents—by firing all of them from their jobs and having his most bitter enemies exiled to Siberia (Vavilov died in a Siberian labor camp in 1943). By the time of Stalin's death, in 1954, Lysenko had filled every position with one of his followers and had practically destroyed classical genetics in the Soviet Union.

The Khrushchev Period

When Khrushchev came to power in the Soviet Union, the conditions of scientific work began to improve greatly. Particularly in the mathematical and physical sciences, many of the ingredients of scientific freedom were restored to the Soviet researcher: the freedom to choose the subject of his investigations and to draw the conclusions to which they led without having to subject these to the arbitrary dictates of a superior power; the freedom to publish his scientific results and to engage in the usual forms of scientific criticism; the freedom to receive the voluminous Western scientific literature; and, to a lesser extent, the freedom to have personal contact with all scientists working in his own field. While these new freedoms were more grudgingly granted to the biological scientists, partly because of Lysenko's continued influence, the fact remains that Soviet biology began to lose its monolithic character. As mentioned earlier, Lysenko was deposed as president of the

Lenin Agricultural Academy, although he maintained his position as director of the Institute of Genetics in Moscow. More importantly, the surviving classical geneticists in Russia, under the leadership of N. P. Dubinin, were rehabilitated and allowed to resume non-Lysenkoist research in their laboratories.

The state of affairs in Russian genetics and biology during the Khrushchev period can aptly be termed a "cold war" situation. Lysenkoists and non-Lysenkoists were permitted to coexist in the Soviet Union, but there was very little communication between them. When it came to personal contacts with Western geneticists, the Lysenkoists still held the upper hand. Thus, when the 1958 International Genetics Congress was to be held in Montreal, 27 Soviet geneticists, both Lysenkoist and non-Lysenkoist, indicated an intention to attend the congress and submitted abstracts of papers to be delivered. About 2 weeks before the congress opened, many of the Soviet scientists informed the organizers of the Montreal congress that they would be unable to attend for a variety of reasons. Finally, a delegation of 11 persons, all of them Lysenkoists, appeared at the conference. Similarly, at the 1963 Genetics Congress in the Hague, the Russian delegation consisted exclusively of Lysenkoists.

It was symptomatic of the state of genetics in the Soviet Union in 1963 that one of the members of the Russian delegation at the Hague Congress stated in private conversation with Western geneticists that Lysenkoists still denied the existence of genes but were willing to accept the existence of DNA as hereditary material. This acknowledgement—that DNA is not an abstract "idealistic" concept but a real molecule playing an important role in heredity—heralded, more than any other single event until then, the beginning of the end of Lysenkoism in the Soviet Union. Such an admission by a Lysenkoist was, of course, a tribute to the remarkable developments which have taken place in Western genetics during the past couple of decades. Indeed, while Lysenko was imposing his out-dated mixture of Michurinism, Lamarckianism, and Marxism on Soviet research, the science of biology was entering a golden age in the Western world, primarily because of the fusion of genetics and biochemistry into the field which is now commonly called molecular biology.

Catching Up

The triumphs of molecular genetics have been so overwhelming in recent years that even members of the Lysenko school have started to work actively in the less controversial area of microbial genetics. But what is more to the point, non-Lysenkoist biologists in the Soviet Union have been trying to catch up on Western developments and to establish molecular biology as a field of its own. Unfortunately, the relatively small number of surviving classical geneticists and the weakness of Soviet biochemistry (weak for reasons of its own) have made this a slow process. If one examines recent Soviet journals of biology one finds a strange mixture of Lysenkoist papers and molecular-biology papers based on Western ideas and methods. Nothing very profound nor new has resulted from this Soviet work as yet, but this is not very surprising in view of the late start.

As distinguished members of the Soviet scientific community have become increasingly aware, during the past few years, of the impressive accomplishments of Western science, the full measure of the Lysenko disaster has permeated their consciousness. As a result, strong scientific pressures have been building up to curb Lysenkoism and to build up Western-type biology. About 3 years ago, while Khrushchev was still in power, the distinguished Soviet physicist Peter Kapitza spoke out against the intrusion of Marxist dialectics into science (with special reference to biology) and the harm which its uncritical acceptance had done to Soviet science. Kapitza's voice did not carry the full weight of the Soviet Academy of Sciences, but it was indicative of the fact that responsible Soviet scientists were becoming increasingly concerned about the extent to which Lysenkoism had damaged Soviet biology, the plant-breeding program, and agriculture in general.

It remained for the president of the Soviet Academy of Sciences, M. D. Keldysh, to deliver the *coup de grâce* several months ago when he announced the removal of Lysenko as director of the Genetics Institute, and stated: "The exclusive position held by Academician Lysenko must not continue. His theories must be submitted to free discussion and normal verification. If we create in biology the same normal scientific atmosphere that exists in other fields, we will exclude any possibility of repeating the bad situation we witnessed in the past." Keldysh's action and forthright statement suggest that the new political leadership in the Soviet Union will not permit Communist fanaticism to injure the best scientific interests of the Soviet state.

An Instructive Story

The rise and fall of Lysenkoism is a sad and instructive story. The rise of Lysenko was due to an unfortunate combination of circumstances: the existence of the philosophical dogma of the Soviet state—to wit, dialectical materialism—with its strong convictions concerning human heredity, the existence of a strong national tradition in empirical plant breeding, founded on the Lamarckian approach to genetics (Michurinism); the desire for rapid transformation of Soviet agriculture, and, finally, the presence of a powerful dictator, Stalin, able and willing to throw the full resources of his government behind a specific ideological position. To this potent brew was added an extraordinarily ambitious and ruthless scientific adventurer named Lysenko.

The final downfall of Lysenko can be attributed to a continuous relaxation of all these factors since the death of Stalin. The political and economic tenets of Marxism have been increasingly separated from dialectical materialism as the supreme arbiter of all scientific concepts and procedures; Michurinism has been placed in its proper historical perspective; it has been recognized that the neglect of classical genetics has been in good part responsible for the lack of productivity of Soviet, as compared to Western, agriculture; and finally Khrushchev and, to an even greater degree, Brezhnev and Kosygin have been more reluctant than Stalin was to use the authority of the government to decide questions of scientific doctrine.

17

The tragedy of Lysenkoism is that so much precious time has been lost for the biological sciences in the U.S.S.R. The consolation is that once the Soviet Union takes a major decision to develop a scientific area (as it did several years ago in mathematical economics and econometrics), lavish provision is made for laboratories and equipment, Western ideas are widely introduced into the educational system, and no effort is spared to attract talented persons into the new field. The recent removal of Lysenko implies unequivocally that such a major decision has been taken with regard to molecular biology and the biological sciences generally. We can only applaud this decision and state our earnest hope that Soviet biologists will soon take their rightful place on one of the great frontiers of modern science. And without much prescience we can predict that at the next International Genetics Congress, to be held in Tokyo in 1968, the non-Lysenkoists will be well represented in the Russian delegation!

3   Cross Currents in the History of Human Genetics

L. C. Dunn

There is, I believe, general agreement that interest and activity in human genetics has today reached a peak never before attained. The periodical literature of the last ten years and the reports of the increasingly frequent symposia and conferences devoted to genetic problems in man provide convincing evidence of this. It is also clear that interest in these problems is likely to increase greatly in the next years so that what we may be witnessing now is only the beginning of a kind of renaissance in which genetics in general stands a chance of being greatly enriched by research on man.

There are probably many reasons for this rather sudden spread of interest, but I think that now is not the best time to try to identify the specific causes and influences of the change. For me, at any rate, a more interesting question is why this period has been so long delayed. Why did human genetics develop so slowly? It is, after all, sixty years since the basic principle of heredity came to recognition. By 1915 the general architecture of the hereditary material was known (*The Mechanism of Mendelian Heredity*, by Morgan, Sturtevant, Muller and Bridges). Even if we date the definitive elucidation of the physical basis of heredity as late as from the publication of Morgan's *Theory of the Gene* in 1926, still that knowledge has been with us for 35 years.

Grateful acknowledgement is made to the Galton Laboratory, University College, London, for hospitality and library facilities and to Professor L. S. Penrose for helpful discussions during the writing of this paper.
From the *American Journal of Human Genetics*, **14** (1962), 1–13. Copyright 1962 by the American Society of Human Genetics. Reprinted with permission of the author and The University of Chicago Press.

I know the stock explanation of lack of progress used in those days when human geneticists were inclined to be apologetic: "You see, we can't experiment with man, and his generation time is long." True, no more can we do experimental breeding with him today, nor has his generation time decreased. Yet the rate of learning about human genetics has greatly increased today. Great progress has only recently been made in several fields in which essential steps opening them to investigation were taken long ago. The primary generalization of population genetics was adumbrated by Pearson in 1904 and clearly formulated by Hardy and by Weinberg in 1908, and its usefulness in human genetics demonstrated by Bernstein in 1924. By 1930 the groundwork of general theory in this field had been laid by Haldane, Fisher, Wright and others, but there has been a long lag period in the application of such methods to man. Many of the implications of Mendelian genetics for studying the transmission system, gene action, bio-chemical genetics and evolution in man were foreseen by Garrod in 1908 and some of them even in 1902 by Bateson. Yet they too have only recently been exploited in human genetics. Cytological study had, even in the 1920's, facilitated the resolution of genetical problems in other animals and in plants, yet did not begin to serve this function in human genetics until the mid-fifties.

I do not mean to say that new technical and analytical methods have not had important effects in facilitating progress. They certainly have. I do mean to say that methods and ideas already available were not, for many years, applied vigorously and with good results to the study of human genetics.

I have recently been re-examining the history of genetics in the formative period from 1900 to about 1930. I have gained the impression that influences which played on human genetics during that period had a good deal to do with delaying its progress in the next 20 years and have not yet ceased to operate. It was then caught up in the crosscurrents to which all studies of man are exposed. The effects of science on human life are always immanent, yet never so immediately apparent as when man himself is the object of inquiry. In the period of which I speak, his confidence as controller of his own destiny had been aroused by recent scientific discovery and by social and political conquests of new environments. Rapid translations of new knowledge into terms applicable to improvements of man's lot is at such times likely to take precedence over objective and skeptical evaluation of the facts, a danger of which cautious scientists have long been aware. The testing of hypotheses by factual observations and the construction of general theory, the normal methods of science, are certainly no less important when human beings are involved, yet one often finds these neglected in human genetics in that period. Progress in human genetics seemed to have been impeded less by lack of means than by lack of a clear scientific goal, and this at a time when the major problems of genetics were taking a clear form. The particulate nature of the transmission mechanism of heredity had focused attention on the means by which genetic elements reproduce and maintain their continuity with opportunity for change and evolution, and on the means by which genes control metabolism and development. But most observations on human heredity were not oriented in any clear way toward such problems. Matters of greater moment seemed to be the inheritance of "insanity," of "feeblemindedness" and other then vaguely defined mental ills, the effects of parental age or alcoholism or social status on the offspring, and similar studies pursued for immediate social ends.

An interesting comparison, which I shall not be able to pursue in detail, is that between such dominant interests of the period as those just cited, and the direction initiated by Garrod's paper of 1901 (*Lancet*, Nov. 30) on alcaptonuria and especially by his Croonian lectures of 1908. One reads today those lectures as published in successive issues of *Lancet* for that year with admiration for the depth and breadth of Garrod's scientific understanding of genetics and of evolution, and then turns with amazement to the reports of discussions on human heredity at the Royal Academy of Medicine which ran through five later issues of that same volume. Except for Garrod's strong supporter and genetical advisor, William Bateson, there is little evidence that the numerous participants in those debates realized what, in fact, the problems were. Karl Pearson, the director of the Galton Laboratory of National Eugenics, was reported to have said in the third debate: "His own view was that there was no truth in Mendelism at all." (*Lancet* 2, p. 1615). He insisted that he had been misreported, although two independent records confirmed the quoted statement (p. 1768), and that he had said that "Mendelism had not been demonstrated for any one character." (*Lancet* 2, p. 1708). But the main lesson we learn from the above is that Garrod's work had little effect until many years had passed; while those interested in the social applications proposed by eugenics largely dominated the field of human genetics.

It will, I think, be clear to anyone who examines the records of the period from 1900 to about the middle thirties that the manner in which the eugenics movement developed cast a long shadow over the growth of sound knowledge of human genetics. The ideals of eugenics as originally proposed by Galton in 1883 and restated in more concrete form in 1901 can hardly be held responsible for this, for they will appeal to most people as embodying a noble conception. But there grew up within the eugenics movement ambivalent attitudes through which it tended to become all things to all people, here a science, there a social movement, and in Germany an instrument, through the so-called eugenics laws of 1933, of the ferocious application of prejudice which seemed to many people to be the logical extension, the reductio ad absurdum, of ideas to be found in eugenic programs elsewhere. One effect of all this was to deflect attention from the essential scientific problems and to discourage persons interested in these from pursuing them with human material. It seems as though some perverse kind of Gresham's law might have been operating here, bad coin driving out the better.

A second cause of failure and delay in human genetics was the all too frequent relaxation of critical criteria and a lowering of standards which would not be tolerated in other branches of genetics. In course of time this, like the handicap imposed by eugenics, became less important in relation to the rising tide of good scientific work, both practical and theoretical, in human genetics. Signs of the change may be seen in Penrose's paper of 1932 and in Hogben's book of 1931 which contained a sharp attack on eugenics. As these changes went on, the name eugenics disappeared from several institutions and publications dealing with human genetics. On the other side, some of the eugenical organizations (like the American Eugenics Society) tended to assume a more responsible attitude toward the scientific facts underlying social applications and toward research in human genetics.

I think it can be shown, however, that neither of the chief defects seen in the adolescent period of human genetics has in fact disappeared today. Now while I suppose that the chief function of historical analysis is to gain views that are more satisfying, intellectually and esthetically, than those afforded by studying only the present state of knowledge, still it has its practical side as well, since we can hardly overlook lessons for the conduct of our lives in the present and the future.

In both of these respects the history of connections between eugenics and human genetics has a special relevance. The connections were very close, and were especially evident in the United States, where interest in both fields was widespread at the turn of the century. Human genetics was often treated as part of eugenics, or as it was often called, human betterment or race improvement. It was that part concerned with acquisition of knowledge of human heredity. The association tended to be maintained because both subjects were frequently pursued and often taught by the same persons. Those who had been attracted by the promise inherent in the newly discovered work of Mendel often added to their repertoire the results of earlier studies like those of Dugdale (1874), and others who had dealt with mental deficiency and criminality as social problems.

There were, however, a few whose position was most clearly stated later by Bateson (1919). "The eugenist and the geneticist will, I am convinced, work most effectively without organic connection, and though we have much in common, should not be brigaded together. Geneticists are not concerned with the betterment of the human race but with a problem in pure physiology, and I am a little afraid that the distinctness of our aims may be obscured." But in general the position in most countries was that implied by an index entry in the one serious attempt to trace from documents, the history of some of the important ideas about heredity. That was the essay of Alfred Barthelmess (1952) in which under the entry *Mensch* we find "sieh auch Eugenik."

The nature of the relations between eugenics and the study of human heredity was strongly influenced by three facts. The first is that the formulation of the problems and program of eugenics antedated the recognition of the particulate nature of heredity. Early work in eugenics was thus guided by a view of heredity which proved to be without general validity. The second is that eugenics achieved organized forms before genetics did. It thus became at the very least a part of the environment in which genetics grew up. The third is that stated by Bateson: they had different goals.

The development of my argument now requires a brief sketch of the history of eugenics. The best source book for this is Karl Pearson's great four volume biography of Galton (1914–1930).

We may formally date the beginnings of eugenics in its modern form from Galton's Huxley Lecture to the Royal Anthropological Institute (published in *Nature*, Nov. 1, 1901) on "The possible improvement of the human breed under existing conditions of law and sentiment." Galton's ideas on this subject had been adumbrated long before this time, first in a paper of 1865. By 1883 they had been given the name of eugenics but had not then attracted active attention. Nor did the seed sown before the anthropologists in 1901 appear to have taken root quickly in England (although it fared better in the United States) and it took further effort on Galton's part to get a fellowship in eugenics established at University College, London, in 1904. This led first to the organization of a Eugenics Record Office and then in 1907 to the Galton Laboratory of National Eugenics (both endowed by Galton) and in the next year to the organization of the Eugenics Education Society. One should note two coincidences of date: 1865 was also the year in which Mendel presented his results at Brünn; and 1900 was the year of the famous "rediscovery" of these. But there was no more connection between the ideas of Galton and Mendel in 1900–01 than there had been in 1865. It was chance of the same kind which gave both men the same year of birth—1822.

Galton's ideas concerning eugenics had been formed first after reading *On the Origin of Species*. The substitution of social control for natural selection in guiding human evolution was for Galton "the logical application of the doctrine of evolution to the human race," but the first ambivalence appeared when he added that the result of his study had been "to elicit the religious significance of the doctrine of evolution. It suggests an alteration in our mental attitudes and imposes a new moral duty." He had become convinced of the heritability of mental qualities through his studies first of *Hereditary Genius* (1869) and then of *Inquiries into Human Faculty* (1883), and had devised statistical methods for the study of inheritance which led him to his Law of Ancestral Heredity in 1897. His views on heredity were always based on this "law" which turned out to describe certain resemblances in graded or continuous characters between parents and offspring but of course provided no explanatory or general principle such as that discovered by Mendel. This is not to say that his eugenical proposals would have been invalidated by his acceptance of Mendel's principles. Those proposals were based primarily on the supposition that heredity was an important cause of differences in physical, mental, and moral qualities, and that was sufficient for his purposes.

Divergences soon appeared in England both among the supporters of eugenics and between these and the school which was shortly to call itself genetics, but was at first referred to as Mendelians. The internal cleavage in eugenics was that between the research and the propaganda interests, as represented by the Eugenics Laboratory and the Eugenics Education Society. The Laboratory resisted and resented interference with its primary function by the Society. "It will never do," wrote Galton to Karl Pearson (the director of the Eugenics Laboratory) on February 6, 1909, "to allow the Eugenics Education Society to anticipate and utilize the Eugenics Laboratory publications" (*Life*, Vol. IIIa, p. 371), and he reminded the Society of the "differences between the work of the two classes of publication." The founder of the movement saw quite clearly the distinction between research and propaganda, and in his last public lecture on "Probability, the Basis of Eugenics" (Oxford 1909) he came out for research as the immediate need, social application as the distant goal.

But dissension between Society and Laboratory continued and finally Galton was impelled in a letter to the *London Times* (Nov. 3, 1910) to make his position quite clear vis-à-vis the Eugenics Laboratory and the Eugenics Education Society.

"Permit me," he wrote, "as the founder of one and the honorary president of the other, to say that there is no other connection between them. . . . The Laboratory investigates without bias . . . large collections of such data as may throw light on many problems of eugenics. The business of the Society is to popularize the results." (*Life*, Vol. IIIa, p. 408). This cleavage, which was to reappear time and again as the movement grew, marked a separation, often not well defined, between those interested in science and those interested in social and political questions. The progress of genetics may not have been directly affected by such disagreements within the eugenics movement, but the occasional excesses of persons with political motivation revealed the source of danger which eventually broke into the open in Germany.

The other cleavage which became apparent at once in England was that between the Mendelians led by Bateson, and the followers of Galton, led first by Weldon and then by Karl Pearson, and known as the Biometricians. The verbal battles between these sharply opposed schools certainly did delay the development of both genetics and eugenics in Great Britain. Karl Pearson, the first director of the Eugenics Laboratory, and, after Galton, the leading eugenist, never recognized the importance of Mendel's principles upon which genetics was founded. As late as 1930 he could say (*Life of Galton*, Vol. III, p. 309) " during the last 25 years we seem scarcely nearer the exact knowledge of the laws of heredity; the farther we advance the more complex does the problem become." It was not that he (or Galton) failed to understand the primary principle of segregation, although he did not appreciate the relation of dominance to it. Indeed in 1904 Pearson foreshadowed an important extension of the principle of segregation by showing that Mendel's ratio $1DD : 2DR : 1RR$ tends to maintain itself indefinitely in random breeding populations of large size (cf. Wright 1959). Galton likewise applied the term "atomistic" to Mendel's system; but neither Galton nor Pearson nor their followers found their interest satisfied by the new principles of Mendel. The heredity in which they were interested could not (they thought) be studied in that way. What they thought important to understand was quantitative variation in human intellectual ability, and Mendelism they considered to be of no help at all. In fact at that stage it was not helpful. This of course is only to say that the purposes of the biometricians and eugenists differed from those of the protogeneticists. Purposes, like tastes, may not be fair game for scientific dispute, although neither side admitted that.

In general the alienation between the two schools was a local British affair. One aspect of it however involved the beginnings of the eugenics movement in the United States. There the ground had been prepared by studies like those of Dugdale (1874) on the Jukes family and of Alexander Graham Bell (1883) on deaf mutism. However in name and purposes the eugenics movement in the U.S.A. was clearly descended from the British one. It differed sharply from its parent in its attitude toward Mendelism. The first proponents of eugenics in the U.S.A., of which C. B. Davenport was the most active, were thoroughgoing Mendelians, and eugenists because they were Mendelians. In fact, Davenport might have been called a super-Mendelian. One has only to read his conclusions on the monofactorial inheritance of a violent temper or a wandering habit to realize this. The British eugenists correctly surmised that this attitude could (as in fact it did) bring the whole movement into disrepute. Dr. David Heron of the Galton Laboratory vigorously attacked in 1913 the first papers to come from Davenport's newly established Eugenics Record Office (founded 1910). Heron wrote (p. 5): "We have selected this rounded group of papers because they deal with a very pressing subject, that of mental defect, and in our opinion form a very apt illustration of the points just referred to, i.e., careless presentation of data, inaccurate methods of analysis, irresponsible expression of conclusions and rapid change of opinion. . . . The Mendelian conclusions drawn have no justification whatever." And further (p. 61): "The authors have in our opinion done a disservice to knowledge, struck a blow at careful Mendelian research, and committed a serious offense against the infant science of eugenics." Heron's criticism, it must be acknowledged, was more than merely another skirmish in the war being waged between the Biometricians and the Mendelians. In this case the point at issue was fundamental scientific method, and Davenport and his collaborators were at least guilty of a lack of caution from which the whole eugenics movement was to suffer. It was at this time, 1910–1915, that single gene interpretations began to be applied with great confidence (amounting in some cases to recklessness) to differences in mental ability and to mental diseases. The outstanding example was feeblemindedness, and on the basis of the first pedigrees published by Goddard in 1910 Davenport (1911 *Eugenics Record Office Bulletin* 1) adopted the hypothesis that mental deficiency in general was inherited as a Mendelian recessive. In this he was followed by many others, and eugenical programs and some legislation were based on this assumption. Stanley P. Davies, who reviewed the history of this period in 1923, called it "the alarmist period." The first fruits of new methods of mental testing were garnered rapidly and widely, and the overemphasis on bad heredity as the cause of mental deficiency and mental disease, and on restrictive or negative eugenics as the only possible cure of a social ill brought on its inevitable reaction. H. S. Jennings in 1925 attempted to restore common sense by his critical attack on the whole concept of unit characters and on the unreality of the either-or distinction between heredity and environment in the determination of human (or any other) characters. Raymond Pearl in 1928 said: "Orthodox eugenists are going contrary to the best established facts of genetical science and are in the long run doing their cause harm." One of the signs that the public image of eugenics had been affected by this and similar criticisms was revealed when G. K. Chesterton published in 1922 *Eugenics and Other Evils*. These essays are not the best example of Chesterton's wit and journalistic skill, but the main point was made sharply clear to his large audience. These essays, he said, had been accumulating since before the first World War, and he had thought the defeat of Germany would have rendered them obsolete. But, he said in his foreword: "It has gradually grown apparent, to my astounded gaze, that the ruling classes of England are still proceeding on the assumption that Prussia is a pattern for the whole world. For that reason, three years after the war with Prussia, I collect and publish these papers." The essence of his objections to eugenics is revealed in one sentence (p. 51): "Even if I were a eugenist I should not personally elect to waste my time locking up the feeble-minded. The people I should lock up would be the strong-minded." Although his criticisms were not always cogent, his suspicion of eugenics, race hygiene, and "scientific officialism" of the German type proved to have been well-founded.

In Germany the eugenics movement took the name Rassenhygiene from a book of that title published in 1895 by Alfred Ploetz who also founded in 1903 the chief German journal in this field, the *Archiv für Rassen-und-Gesellschaftsbiologie*. In an article in this journal in 1909 Galton agreed with the editor that Rassenhygiene and Eugenik were to be regarded as synonymous. Any misunderstanding on this score was removed in 1931 when the chief German society in this field, the Deutsche Gesellschaft für Rassenhygiene (founded in 1902) added "Eugenik" to its title. The direction in which Rassenhygiene led had become evident long before Hitler came to power; and the advent of the new laws for sterilization of the unfit and unwanted, and for the exclusion of Jews from the new state were greeted with editorial acclaim in the *Archiv*. The speed with which the first of these laws were prepared and promulgated within the first few months of 1933 is probably to be explained by the composition of the committee of experts which drafted them. This included Ploetz and his fellow eugenists Rüdin and Lenz and others who had worked in this field together with Heinrich Himmler. Frick, Hitler's Minister of the Interior, whose department was charged with the administration of the laws said upon their coming into force: " The fate of race-hygiene, of the Third Reich and the German people will in the future be indissolubly bound together " (*Arch. Rass. Ges. Biol.* Vol. 27, p. 451). The situation was made quite clear by von Verschuer in the introduction to his book *Leitfaden der Rassenhygiene*, published in 1941.

Es ist entscheidend für die Geschichte eines Volkes was der politische Führer von den Ergebnissen der Wissenschaft als wesentlich erkennt und zur Tat werden lässt. Die Geschichte unserer Wissenschaft ist aufs engste verknüpft mit der deutschen Geschichte der jüngste Vergangenheit. Der Führer des deutschen Reiches ist der erste Staatsmann der die Erkenntnisse der Erbiologie und Rassenhygiene zu einem leitenden Prinzip in der Staatsführung gemacht hat. (p. 11)

(Decisive for the history of a people in what the political leader recognizes as essential in the results of science and puts into effect. The history of our science is most intimately connected with German history of the most recent past. The leader of the German state is the first statesman who has wrought the results of genetics and race hygiene into a directing principle of public policy.)

This statement by a leading German human geneticist was made with some deliberation, for it appeared first in identical form in an article by von Verschuer in *Der Erbarzt* 1937, p. 97, and although it has been omitted in a recent edition of the above book (1959), the author has not to my knowledge publicly altered his position on enforced race hygiene. Although not all geneticists who remained in Germany thus accepted the eugenical and racial doctrines and practices of the Nazis, there is at least evidence that even the serious scientists among them underrated the dangers of the movement until it was too late. From this the melancholy historical lesson can be drawn that the social and political misuse to which genetics applied to man is peculiarly subject is influenced not only by those who support such misuse, but also by those who fail to point out, as teachers, the distinctions between true and false science.

In von Verschuer's book of 1941 Galton is acknowledged to be the modern founder of race hygiene as eugenics; but to Gobineau is given the greater credit of having first brought race into politics, thereby becoming the founder of political anthropology, a field in which the leading later exponents in Germany were Eugen Fischer and H. A. K. Günther. In contrast to the situation in Great Britain in which Galton had been unable to arouse the interest of anthropologists, the German eugenics movement had close connections with the kind of anthropology which was pursued by anthropometric methods. Since this was not guided by a theoretical rationale such as might have been supplied by population genetics, it fell the more quickly a victim to the pseudo-science of the promoters of the Aryan mythology. The chief research institute was the Kaiser Wilhelm Institute for Anthropology, Human Genetics, and Eugenics, of which Eugen Fischer was the director. Many members of this institute had become so politically involved with Nazism that after the defeat of Hitler's regime the institute was not continued by the West German State, thus fulfilling rather quickly Frick's prophecy of the interdependence of race hygiene and the Nazi state. It must be noted that in the debacle eugenics carried anthropology and human genetics down with it. There can be no doubt that in Germany, formerly a center of genetical research, the effect of its association with race hygiene was to delay for a generation the development of a science of human genetics.

In the United States, as in Britain, anthropologists in general did not respond to eugenical appeals. The kind of racialism which had become attached to eugenics was not calculated to appeal to persons whose profession it was to study and interpret human differences objectively and in socio-cultural as well as biological terms. Human genetics has today become a useful contributor to anthropology, mainly through gene frequency studies, and by the application of good objective methods generally untinged by racialism. However, there are still reminders of the uncritical use of what look like genetical methods applied to racial anthropology. What shall one say, for example, when three authors, after anthropometric examination of 44 Italian war orphans of whom the father was unknown but assumed to be "colored" draw sweeping conclusions concerning heterosis ("established with certainty"), inheritance of erythrocyte diameter ("very convincing") and other statements not supported by evidence. Yet these are statements made in 1960 by Luigi Gedda and his co-workers Serio and Mercuri in their recent book *Meticciato di Guerra*. R. R. Gates, who writes an introduction in English to this elaborate book refers to it as an important contribution to what he calls "racial genetics." Others will have greater difficulty in detecting any contribution to genetics, but may see in it, as I do, a reflection in 1960 of the uncritical naiveté of that early period of human genetics which delayed its progress. And the same year—1960—sees also the appearance of a new journal *Mankind Quarterly* devoted in part to racial anthropology of the above kind (again described as such by one of its editors—R. R. Gates) and embodying racist attitudes of the earlier period. Truly the past is not yet buried, and human genetics, in spite of its recent evidences of new life, is still exposed to old dangers.

Eugenics movements grew in many other countries in the period before and just following the first World War, but space will permit taking account of only one such development.

It may be regarded as an anomaly of history that in Russia eugenics did not appear in an organized form until after the revolution of 1917, and this notwithstanding the enunciation of ideas very similar to those of Galton by W. M. Florinsky in 1866. In 1919 a eugenics department was started in the Institute of Experimental Biology in Moscow under N. K. Koltzoff, and shortly thereafter a Eugenics Bureau began in Leningrad under J. A. Philiptschenko. By 1925 thirteen research articles on human genetics, *sensu stricto*, had been published from these institutes. The Russian Eugenics Society was founded in Moscow in 1920, with local branches in Leningrad, Saratov, and Odessa, and the *Russian Journal of Eugenics* under Koltzoff and Philiptschenko began in 1923 (cf. Koltzoff, 1925). The difficulties and later suppression of eugenics in the Soviet Union were foreshadowed by an event in connection with the publication in 1924 of Philiptschenko's book on eugenics. While it was in production at the government printing office, according to Weissenberg (1926) there was inserted in the introduction a statement to the effect that measures with important eugenical effects were the destruction of the bourgeoisie and the victory of the workers. What part the existence of eugenics there played in bringing about the suppression of genetics in the USSR is not clear to me. The first institute to be suppressed appears to have been that concerned with human genetics, the Gorky Institute for Medical Genetics, but this may have been incidental to the condemnation and execution of its director, S. G. Levit, as a "traitor."

Although the chief crosscurrents operating on human genetics were generated by persons pressing for social and political regulation of human breeding, frequently to the neglect of sound scientific method, others of less marked and definite character traced to relations, or lack of them, between genetics and medical research. Apart from lack of understanding of genetics on the part of physicians, there were frequent expressions of active lack of interest, since principles discovered in peas and exploited and extended by experiments with flies were not thought relevant to human beings. And if, physicians often said, a disease was inherited, that meant it couldn't be treated and knowledge about it was not likely to be useful. The gap due to mutual lack of appreciation and of common experience and training as between medical men and geneticists has shown some signs of narrowing, but is certainly far from being bridged. This would require further will and effort on both sides.

Nor can one conclude as yet that the confusion between the aim of eugenics and the facts of human heredity which Bateson pointed out has yet been cleared up. As eminent an acknowledged leader in genetics as Professor H. J. Muller has recently restated the adherence to ideas on controlled human breeding which he outlined in his book of 1935 *Out of the Night*. The central idea, eutelegenesis, had been developed by Herbert Brewer (1935) who probably was unaware of an earlier similar proposal by A. S. Serebrovsky (1929). The essential feature of the proposal was to utilize the sperm of men, chosen on the basis of achievement as superior, and by increasing through the use of artificial insemination the numbers of offspring of such superior sires to raise the average level of ability of the next generation. In his reiteration of

27

this proposal in 1959, Muller has refined and extended it. He now proposes to retain the whole genotype of such men (which the processes of meiosis would tend to break down) by multiplying samples of their diploid spermatogonial cells in tissue cultures and subsequently obtaining embryos from these by some form of ectogenesis. Even though the technical problems involved might in some future time be solved, several more important scientific ones would still remain. Such schemes assume that there is an ideal genotype for a human being. Plato could entertain such an ideal but can we do so after our experience of the variety of genotypes in successful populations? Can human cultures be maintained by an ideal genotype? Even though the proponents of eutelegenesis should admit that there might be several good kinds of human being, are there objective scientific criteria by which they might be selected? Even though choice of sperm or genotype donors were to rest with persons as benevolent and acute as I believe Professor Muller to be, selection would still be subject to changing tastes and ideals and thus to control by imposed power as in the Nazi state. Muller illustrates this in his own examples of the eminent men he might have chosen, for in his list of 1959 as compared with that of 1935, Lenin, Marx and Sun Yat Sen have been dropped and Einstein and Lincoln added.

To me such schemes seem to express the same sort of benevolent utopianism as did some of Galton's proposals of 60 years ago, but now they must be viewed in the light of some actual experience with them. Then as now they were backed by the prestige of men of deserved eminence in science, then of Galton, today of Muller, but this did not save the earlier programs from grave misuse and ultimate damage to both human society and science. In fact the high scientific standing of their proponents increases the dangers of uncritical acceptance of them as bases for social and political action, with the ever attendant risk of loss of public confidence in genetics as applied to man if or when their unsoundness becomes manifest.

Such considerations remind us of the dilemma which scientists face in their desire both to advance sound knowledge and to make it serve its essential social function. In the case of human genetics, I do not believe that the problems posed by the cohabitation of these two purposes are to be settled by divorce, as Bateson suggested. The problems posed by the continuing occurrence of diseases and defects ("Our load of mutations" Muller 1950) are real and they must be faced, both as biological and as social problems. Both sets of interests must be free to develop, and better together than separately for this is the condition under which common criteria for criticism and rigorous judgements, so badly needed in all fields affected by potential social applications, may be evolved.

If I have strayed somewhat from the limits of 1900–1930 that I had set for myself for a historical review, I suppose this is a reflection of my view that some of the cross currents operating in the earlier years of this century still play upon us. It seems to me that their influence in the first two or three decades was in part due to the lack of a clear vision of what studies of man have to contribute to the elucidation of general problems such as the mechanism of evolution and of gene action. The rise of population genetics and of physiological genetics have now turned attention to the rich source of material for these problems provided by human populations, and by the accumulated experience of medical and anthropological research. Recent discoveries, such as the identification of human genes controlling serological, biochemical, and developmental processes subject to the action of natural selection, should now give human genetics that orientation toward important biological problems which was not generally recognized in its early days. What seems to me to be most important, especially in its implications for the future, is the growing recognition of the logical unity of genetics, for its essential problems, being concerned with a system of elements having similar attributes in all forms of life, can be seen to transcend the special problems of the different categories of organisms. Human genetics, freed from the narrower bounds and conflicting purposes which hindered its early growth, seems clearly destined to play an important role in the advancement of the whole science of genetics of which it is a part. And that, in the long run, may constitute its best contribution to the satisfaction of human needs.

References

Bateson, B. 1928. *William Bateson, F. R. S. Naturalist.* Cambridge Univ. Press.

Bateson, W. 1919. Common sense in racial problems. The Galton Lecture. *Eugen. Rev.* 11.

Bateson, W., and Saunders, E. R. 1902. *Royal Society Evolution Reports 1902–1909.* Report 1, Dec. 17, 1901, *cf.* p. 133 for Bateson's interpretation of the observations of Garrod in *Lancet*, Nov. 30, 1901.

Barthelmess, A. 1952. *Vererbungswissenschaft.* Freiburg/München: Orbis Academicus.

Bernstein, F. 1924. Ergebnisse einer biostatistischen zusammenfassenden Betrachtung über die erblichen Blutstrukturen des Menschen. *Klin. Wschr.* 3: 1495–1497.

Brewer, H. 1935. Eutelegenesis. *Eugen. Rev.* 27: 121–126.

Chesterton, G. K. 1922. *Eugenics and other evils.* London.

Davenport, C. B. 1914. Reply to criticism of recent American work by Dr. Heron of the Galton Laboratory. *Eugenics Record Office, Cold Spring Harbor, N. Y., Bulletin 11.*

Davies, S. P. 1923. *The social control of the feeble minded.* Ph.D. Dissertation, Faculty of Political Science, Columbia Univ. Press.

Dugdale, R. L. 1874. *The Jukes; a study in crime, pauperism, disease and insanity.* New York.

Florinsky, W. M. 1866. *Uber die Vervollkommung und Entartung der Menschheit.* Petersburg.

Galton, F. For detailed references to the publications of Galton see Pearson (1914–1930).

Garrod, A. 1908. Croonian Lectures to the Royal Academy of Medicine. Inborn errors of metabolism. *Lancet* 2: 1–7; 73–79, 142–148; 214–220.

Hardy, G. H. 1908. Mendelian proportions in a mixed population. *Science* 28: 49–50.

Heron, D. 1913. Mendelism and the problem of mental defect—a criticism of recent American work. *Univ. of London, Publication of the Galton Laboratory, Questions of the Day and Fray, Number 7.*

Hogben, L. 1931. *Genetics principles in medicine and social science.* London: Williams & Norgate.

Hogben, L. 1933. *Nature and nurture.* London: George Allen & Unwin.

Jennings, H. S. 1925. *Prometheus, or biology and the advancement of man.* New York: E. P. Dutton & Co.

Joravsky, D. 1961. *Soviet Marxism and natural science. 1917–1932.* New York: Columbia Univ. Press.

Koltzoff, N. K. 1925. Die Rassenhygienische Bewegung in Russland. *Arch. Rassenb.* **17**: 96–99.

Muller, H. J. 1935. *Out of the night: a biologist's view of the future,* New York: Vanguard Press.

Muller, H. J. 1950. Our load of mutations. *Amer. J. Hum. Genet.* **2**: 111–176.

Muller, H. J. 1959. The guidance of human evolution. *Perspectives in Biol. and Med.* **3**: 1–43.

Pearl, R. 1928. Eugenics. *Proc. 5th Internat. Cong. Genet., Berlin* **1**: 260.

Pearson, K. 1904. On a generalized theory of alternative inheritance, with special reference to Mendel's laws. *Phil. Trans. Roy. Soc. London* A **203**: 53–86.

Pearson, K. 1914–1930. *The life, letters and labors of Francis Galton.* Vol. I, II, III, IIIa. Cambridge Univ. Press.

Penrose, L. S. 1932. On the interaction of heredity and environment in the study of human genetics (with special reference to Mongolian imbecility). *J. Genet.* **25**: 407.

Serebrovsky, A. A. 1929. *Antropogenetika Medikobiologcheskii Zhurnal* **5**: 3–19 (Russian, as cited by Joravsky, 1961.)

von Verschuer, O. 1959. *Genetik des Menschen.* Urban. Schwarzenberg.

Weinberg, W. 1908. Uber den Nachweis der Vererbung beim Menschen. *Jahreshefte Verein vaterl. Naturkunde Wüttemberg* **64**: 369–382 (*cf.* also Stern, C. 1943. The Hardy-Weinberg law. *Science* **97**: 137–138.)

Weissenberg, S. 1926. Theoretische und praktische Eugenik in Sowjet Russland. *Arch. Rassenb.* **18**: 81.

Wright, S. 1959. Physiological genetics, ecology of populations and natural selection. *Perspectives in Biol. and Med.* **3**: 107–151.

4   On Genetics and Politics

Theodosius Dobzhansky

There is a habit of thought perhaps as old as language itself that keeps getting in the way of our understanding of the history and nature of the processes of life. This is our tendency to think in terms of static types. Of course we must sort into categories the overwhelming diversity of phenomena we perceive and experience, and

From *Social Education,* **32** (Feb. 1968), 142–146. Copyright 1968 by the National Council for the Social Studies. Reprinted with permission of the National Council for the Social Studies and Theodosius Dobzhansky.

we do so in words like "man," "cat," or "dog." Such words do not refer to particular persons or animals but to abstract representatives of mankind, cat-kind, and dog-kind. Also, such words emphasize differences between "kinds" as if there were rigid boundaries; they give no hint of what the "kinds" may have in common. Nor do they take into account, or even suggest, the diversity within "kinds"—the diversity of persons, of cats, and of dogs. Moreover, individual persons and animals change and grow old.

Typological Thinking

Nevertheless, man has long sought for ways to unify as well as classify this diversity—since long before Bronowski said, beautifully, that "science is nothing else than the search to discover unity in the wild variety of nature—or more exactly, in the variety of our experience. Poetry, painting, the arts are in the same search, in Coleridge's phrase, for unity in variety. Each in its own way looks for likeness under the variety of human experience."

One tempting way to unify is to declare that the diversity and change are false appearances. Plato did so by his famous theory of ideas. He believed that God created eternal, unchangeable, and inconceivably beautiful prototypes, or ideas, of Man, Horse, and even such mundane and inanimate objects as Bed and Table; that individual persons, horses, beds, and tables are only pale shadows of their respective ideas; that acquiring wisdom means "seeing" the ideas where formerly one saw only their shadows. Aristotle assumed only one cosmic idea, which manifests itself in the visible world. He believed that nature's "purpose" was to realize the ideal form, and that all animals were variants of a single architectonic plan.

For more than 2,000 years—up to the time of Darwin—the organic world was viewed either as Plato or as Aristotle saw it, within the Western World. Either way was compatible with believing that living, and even nonliving, bodies constituted a "Great Chain of Being" that ranged from lesser to greater perfection. Leibniz in the seventeenth and Bonnet in the eighteenth century felt satisfied that the Chain is single and uninterrupted. Bonnet saw it starting with fire and "finer matters," and extending through air, water, minerals, corals, truffles, plants, sea anemones, birds, ostriches, bats, quadrupeds, monkeys, and so to man.

To Lamarck and to Geoffrey St. Hilaire in the nineteenth century, but not to their predecessors and contemporaries, the Great Chain implied evolution. Cuvier, dissenting, was of course absolutely right that there is no single plan of body structure common to all animals. Still, the hypothesis that living organisms are manifestations of a limited number of basic types or ideas of structure was useful because it inspired studies on comparative anatomy and classification. There is no doubt that such data, though collected for another purpose, did help to furnish a base for the theory of evolution.

Such are some of the pre-Darwinian roots of *typological thinking*. Their common quality is that they postulate, implicitly or explicitly, archetypes or ideal types.

31

Populational Thinking

The Darwin-Wallace theory of evolution shattered the basis of typological thinking over a century ago, but many biologists are still unaware of this profound implication. I agree with Ernst Mayr that "the replacement of typological thinking by populational thinking is perhaps the greatest conceptual revolution that has taken place in biology." My only cavil is that Mayr's "has taken place" is over-optimistic, though I do believe that this conceptual revolution is well under way. It has proceeded so slowly because the difference between typological and populational thinking is as subtle as it is profound, but also because habits of thought are at least as hard to change as any other habits. I will try to show this difference, and then to indicate how it is represented in and confuses so much of our sociological and political thinking.

Darwin characterized his great work *On the Origin of Species* as "one long argument." The heart of this argument was "that species are only strongly marked ... varieties, and that each species first existed as a variety." This amounts to erasing any sharp lines between "varieties" or "species" by in effect saying that there exist within a "species" the raw materials from which natural selection compounds new "species," which are in turn subject to further transformation into other new "species." The transformations are not a false front or an illusion; they are real novelties because they take a variety of directions. Thus neither the "varieties" nor the "species" can be diverse manifestations of Platonic archetypes or ideas or of an Aristotelian tendency, for they are changeable in different ways. To the typologist, who is eager to classify and pigeonhole, the presence of intermediate forms between "varieties," "species," "races" is a nuisance which one is tempted to shrug off whenever possible. But to the evolutionist borderline cases are a godsend.

Darwin's argument was limited to affirming that evolution *had* taken place. For roughly the next four decades, until about 1900, studies in comparative anatomy, embryology, and paleontology concentrated on discovering and examining further evidence to confirm the argument. Such evidence is now so well known that it is a part of elementary, college, and high school courses of biology.

Then for about 30 years after the turn of the century there was a rapid growth of knowledge of basic genetics, stemming largely from Gregor Mendel's rediscovered work of 1865 on transmission of characteristics from one generation to another. Around 1930 a mathematical theory was deduced which explains *what causes* evolution to take place. It took account of Darwin's argument that natural selection was the mechanism; it introduced the idea of adaptation to environment as the "pressure" that directed natural selection within a breeding population; and it incorporated the new hypothesis of gene mutations as explanation of the random appearance of entirely new characteristics within a population. Since then many scientists have demonstrated that the theory makes coordinated sense of experimental genetics, paleontology, zoological and botanical systematics, cytology, comparative morphology, embryology, anthropology, ecology, physiology, and biochemistry; and it has been strengthened by recent spectacular developments in molecular genetics.

This "synthetic" theory (so called because it synthesizes) demonstrates that in a non-uniform environment there is no such thing as a typical, average, or ideal type within sexually reproducing, outbreeding populations of a species such as man. It demonstrates that such a biological species is a population of unique individuals (except for identical twins), carriers of *diverse* genotypes. ("Genotype" stands for the inherited, and hence heritable, components of an individual; "phenotype" stands for the outwardly expressed characteristics.)

Now how great is the genetic diversity within a given species is only sketchily known at present. The genetic diversity within the human species in particular is still quite inadequately explored. Part of the problem relates to our discovering ever greater complexity in how "genes" work. We have tried to explain the effects by hypothesizing dominant, semi-dominant, incompletely recessive, and completely recessive genes; polygenes—combinations that operate together to determine a characteristic; supergenes—linked groups of genes which determine complexes of traits inherited as units; pleiotropism—an individual gene having an effect on several characteristics which do not have any obvious developmental interdependence. The simple Mendelian duality of dominant and recessive genes has long since proved to offer a very inadequate explanation of reality. No doubt all of these postulates must eventually be reduced to molecular terms. But for our purposes at the moment, the main thing we must guard against is thinking in terms of one "gene" for every "trait" and one "trait" for every "gene."

There is no doubt that the range of potential diversity, of potentially possible genotypes, within a species such as man is vastly greater than could ever be realized as actual genotypes. The number of genes in a human sex cell is not known, but it could hardly be less than the 10,000 estimated for *Drosophila*. If we assume only two variants, alleles, per gene (and many, or most, genes may have more than two alleles), this makes $2^{10,000}$ potentially possible genetically different gametes. (Each of the sex cells, the gametes, bears only one of the two sets of chromosomes of the organism.) Therefore the potentially possible zygotes (union of two sex cells, hence genotypes) would be $3^{10,000}$. Of course the actual genotypes realized are not just a random assortment of gene combinations, which such computations might imply, because of the complexity of how genes "work," which I alluded to just above. Still, it should be apparent that "potential" must exceed "actually expressed" by astronomical proportions.

The main point to grasp about populational thinking in terms of the synthetic theory of evolution is that a *range* of diversity within a species is necessary for a range of adaptability to varying environments. Typological thinking is plainly un-evolutionary or anti-evolutionary because in varying ways it reverts to postulating an ideal or optimal type; and an optimal type could only be relevant to such a completely static and uniform environment as could pertain to few species—certainly not to man, though perhaps to such as oysters.

If the bulk of the genetic diversity in a species is balanced, the optimal genotype becomes a will-o'-the-wisp. There could be no one "normal" genotype, but only arrays of genotypes which make their carriers able to live and reproduce successfully in the various environments the species inhabits. There would inevitably be ill-adapted, pathological variants at one extreme and exceptionally vigorous, out-standingly successful variants at the other extreme, the genetic elite. Then it is the diversity of genotypes and phenotypes in a population, including the diversity of behaviors, and the factors which maintain the diversity that should be studied, not any "typical" or even statistically average behavior.

It is undeniable that some mutant genes are deleterious in all existing environ-ments whether inheritance is homozygous or heterozygous. It is likewise undeniable that some kinds of heterozygous variants are maintained in a population by balancing selection even though inheritance of the homozygous condition may be lethal. The sickle-cell trait is perhaps one of the best-known examples of such a heterozygous condition being maintained. It appears that heterozygous inheritance of this "gene" confers somewhat superior resistance to malaria while homozygous inheritance often leads to an anemic condition that frequently kills.

The Confusion of Political and Sociological Genetics

How does this subtle difference between typological thinking and populational thinking find its way into politics and sociology? First I must point out that though the roots of populational thinking are less easily traceable than the roots of typological thinking, they can be found in the Christian view that each individual soul is as valuable as any other; that it is the individual person who commits sins; that we possess free wills and are accountable for our actions. But the formal philosophical codification of human individuality came only in 1785, in Kant's doctrine that every human being is an end in himself. However, this doctrine is implicit in the belief that every man has certain inalienable rights, which belief became popular during the Age of Enlightenment, and which serves as the foundation of democracy.

In general, typological thinking about man goes mostly with conservative political persuasions, and the view that every man is an end in himself goes with liberal. But consistency is not one of man's hallmarks. Strangely enough, conservatives tend to emphasize that different people are genetically different, while liberals tend to assert that every man is at birth virtually a *tabula rasa*, a blank slate, and that all behavioral differences between men are the result of upbringing and education.

It is no exaggeration to say that some version of this *tabula rasa* theory is enter-tained, explicitly or implicitly, by most social scientists and by some influential schools of psychology, especially those with psychoanalytical leanings. It is attractive because it seems to uphold an optimistic view of human nature: though a lot of people behave stupidly, wretchedly, and viciously, this is because they were badly brought up and were spoiled by corrupting influences of badly organized society; a better organized society, better care, guidance, and education would make everybody behave with goodness and reasonableness, because that capacity is the normal and universal endowment of every representative of the human species.

Does it not become apparent that the *tabula rasa* theory is but another form of typological thinking? In this case only a single type for the whole species *Homo sapiens* is proposed.

John Locke made the classical statement of the *tabula rasa* theory, but he did not claim anything so rash as that all infants are born alike. What he did claim, rather, is that there are no inborn "ideas." He wrote: "Let us then suppose the mind to be, as we say, white paper, void of all characters, without any ideas; how comes it to be furnished? ... All that are born into the world being surrounded with bodies that perpetually and diversely affect them, variety of ideas, whether care be taken about it or not, are imprinted on the minds of children." His expression of the *tabula rasa* theory had a great and abiding influence on the climate of ideas in which the civilized part of mankind lives. It was adopted by the thinkers of the Age of Enlightenment, from Voltaire and Helvetius, Rousseau and Condorcet, to Bentham and Jefferson. It has been, and continues to be, an important ingredient in the philosophy of democracy. But exaggeration and overextension of the theory—to the erroneous conclusion that people are innately uniform lumps of clay molded only by environmental conditions—that interpretation of Locke, I think it can be demonstrated, makes nonsense of democracy.

Here is another example of typological thinking as it has become, in this case, woven into conservative political and sociological thought. The current civil rights movement in the United States has prompted the racists to devise a curious quasi-scientific justification of their social and political biases. Their argument is that Negroes, compared to whites, have a smaller average brain size and lower average scores on various intelligence and achievement tests. Whether these alleged differences are real, and if real whether of environmental origin, need not concern us here. The point is that not even racists deny that all these measurements vary among whites as well as among Negroes. The variation ranges overlap so broadly that the large brains and the high scores in Negroes are decidedly above the white averages, and the small brains and the low scores in whites are decidedly below the Negro averages. Now, the typological reasoning goes about as follows. All Negroes are said to be but manifestations of the same Negro archetype, and all whites of the same white archetype, no matter what his brain size or his intelligence score may be, an individual is either a Negro or a white and should be treated accordingly. There seems to be some subtle implication here that whatever genes determine "Negroness" and "whiteness"—"polygenes," "supergenes," or "pleiotropic"—also determine the brain size.

The populational, and in this case liberal, approach is simply that an individual is a unique genotype (except, again, for identical twins)—a unique combination of a huge number of different characteristics which makes him different from any other person, be that a member of his family, clan, race, or mankind. Beyond the universal rights of all human beings, a person ought to be evaluated on his own individual merits.

The populational approach does not invoke misguided extension of the *tabula rasa* theory, it does not assume that human nature is uniform, thereby reducing individuality to the status of a veneer applied by infant-rearing practices and circumstances of a person's biography. That is starting with a valid premise of Locke's *tabula rasa* theory but going on to an erroneous conclusion. The valid premise that population thinking accepts is that the behavioral development of *Homo sapiens* is

35

remarkably malleable by external circumstances. The geneticist is constantly forced to remind his colleagues, especially those in the social sciences, that what is inherited is not this or that particular phenotypic "trait" or "character" but a genotypic potentiality for an organism's developmental response to its environment. Given a certain genotype and a certain sequence of environmental situations, the development follows a certain path. The carriers of other genetic endowments in the same environmental sequence might well develop differently. But also, a given genotype might well develop phenotypically along different paths in different environments. In most abbreviated terms, the observed phenotypic variance has both a genetic and an environmental component.

How "stabilized" or how "plastic" any phenotypic expression might be varies from species to species, but in general what is necessary for survival and reproduction is rather "fixed." For instance, with very few exceptions, human babies are born with two eyes, a four-chambered heart, a suckling instinct, physiological mechanisms for maintaining constant body temperature, etc. However, sometimes stabilization is disadvantageous. If a *Drosophila* larva is given ample food, it develops into a fly of a certain size. If food is scarce, although above a certain minimum, the starving larva does not die, but pupates and gives an adult insect of a diminutive size. Also, the number of eggs deposited by a *Drosophila* female on abundant food and at favorable temperatures may easily be ten times greater than with scarce food and an unfavorable temperature. Evidently fixed body size and number of eggs produced would be disadvantageous. Natural selection has operated on the genotype to destabilize these characteristics and make them contingent on the environment in which an individual finds itself. It may be that the increased body size among many populations of human beings over the last few centuries (and particularly the last few generations) represents a similar mechanism. (Note the small size of medieval armor, and also the evidence from clothing and shoe manufacturers that nowadays people tend to be bigger than their grandparents, and even parents.)

Now the paramount adaptive characteristic of man is his educability—his capacity to adjust his behavior to circumstances in the light of experience—and that educability is a universal property of all nonpathological individuals. All individuals which belong to the adaptive norm of our species have this capacity of educability through symbolic thinking and communication by symbols, language. It is perhaps justified to say that human evolution has stabilized this capacity, although consequent overt, observable behavior is not stabilized. Locke went no further than that in his original *tabula rasa* theory, although it is easy to see how it could be exaggerated into the illusion that man at birth is a complete *tabula rasa* as far as his prospective behavioral development is concerned.

In reality, this educability goes hand in hand with the populational thinking which argues that genetic diversity enhances the adaptability of our species in particular. Any human society has diverse functions, and as civilization has developed and continues to develop, the diversity of vocations has increased and continues to increase enormously. The division of labor in a primitive society is distributed chiefly between sexes and among different age groups. In a civilized society it is distributed among castes and ultimately, individuals. The trend of cultural evolution is obviously not toward making everybody have identical occupations, but toward a more and more differentiated occupational structure. What would be the most adaptive response to this trend? Certainly nothing that would encourage genetic uniformity. Although the genetically guaranteed educability of our species makes most individuals trainable for most occupations, it is highly probable that individuals have more genetic adaptability to some occupations than to others. Although almost everybody could become, if brought up and properly trained, a fairly competent farmer, or a craftsman of some sort, or a solider, sailor, tradesman, teacher, or priest, certain ones would be more easily trainable to be soldiers and others to be teachers, for instance. And it is even more probable that only a relatively few individuals would have the genetic wherewithal for certain highly specialized professions, such as musician or singer or poet, or for high achievement in sports or wisdom or leadership. To argue that only environmental circumstances and training determine a person's behavior makes a travesty of democratic notions of individual choice, responsibility, and freedom.

Human educability is traditionally emphasized by those who espouse liberal political views, while genetic differences are harped upon by conservatives. As pointed out above, this is sheer confusion. The main tenet of liberalism is, it seems to me, that every human being, every individual of the species *Homo sapiens*, is entitled to equal opportunity to achieve the realization of his own particular potentialities, at least so far as compatible with realization of the potentialities of other people. Because human beings are individuals and not "types," because they are all different, equality of opportunity is necessary. A class or caste society leads unavoidably to misplacement of talents. The biological justification of equality of opportunity is that a society should minimize the loss of valuable human resources, as well as the personal misery resulting from misplaced abilities, and thus enhance its total adaptiveness to variable environments.

two   # Genetics and Agriculture

5. Hybrid Corn: Its Genetic Basis and Its Significance
in Human Affairs
Paul C. Mangelsdorf [1951]

6. New Wheats and Social Progress
Louis P. Reitz [1970]

7. Our Vanishing Genetic Resources
Jack R. Harlan [1975]

8. Plant Cell Cultures: Genetic Aspects of Crop
Improvement
Peter S. Carlson and Joseph C. Polacco [1975]

Breeding plants and animals for human uses, including food, began in antiquity (Heiser, 1973). Farmers and herders simply allowed only the "best" phenotypes to reproduce. Today this is still the case worldwide. The only large reservoir of animals and plants for human use in which selective breeding is not extensively practiced is the aquatic environment, especially the oceans.

The pioneer achievement of applied genetics is that of hybrid corn, although this was disputed in Russia until recently (see paper by Caspari and Marshak in Section One). Mangelsdorf tells the dramatic story of hybrid corn here (see also Griliches, 1960; Mangelsdorf, 1974). As Mangelsdorf predicted in 1951, the overwhelming problem in agriculture today is that 400 million human beings in the world are poorly nourished (in calories) and many of them are also deficient in protein intake. This segment of humankind comprises mainly the poor of southern Asia (Poleman, 1975) although poor people everywhere suffer from hunger. The millions of malnourished exist despite the fact that yields in crops and livestock have been enormously increased in the twentieth century by the aid of genetics and allied fields.

The papers by Reitz and by Carlson and Polacco presented in this section illuminate the current hopes and plans of genetic "medicine" for agriculture. Reitz describes the delicate interdependence of crop genetic improvement and fertilizers, irrigation, pest control, and other factors. He also puts the panacea of the "green revolution" in proper perspective: Crop improvement must be coupled with population control to prevent further starvation (see also Harris, 1973). Current trends in plant breeding are also reported by Chase (1968), Harpstead (1971), Hightower (1974), Levins (1974), Sigurbjornsson (1971), and Wade (1974a, b, c). The relationships between increased food production and population increase have been reviewed by Paddock (1970) and Poleman (1975). The relationships between food production and socioeconomic organization have been reviewed by Walters (1975).

The paper by Carlson and Polacco discusses a form of genetic engineering for new crop plants based on test-tube research. Although such endeavors are in their infancy, they offer a faint hope on the horizon for important breakthroughs in the food problem. Other imaginative schemes are in various stages of development: An artificially hybridized grain crop, triticale, is already widely planted

(Hulse and Spurgeon, 1974), and the test-tube designing of cereal crops that are fertilizer-independent is on the drawing board (Hardy and Havelka, 1975).

A recent advance in plant genetics is the development of male-sterile strains of corn, wheat, and sorghum to prevent unwanted self-fertilization—this formerly required emasculation by hand—and thus permit the production of a desired hybrid by wind pollination by another variety (Curtis and Johnston, 1969). However, this advance, like seemingly all human schemes, is flawed. For example, corn plants carrying the male sterility factor have been severely attacked by a new virulent strain of fungus that causes blight (rotting leaves, stalk, and ears). In 1970 the blight spread from Florida to Texas and Minnesota. Over 80 per cent of hybrid corn grown in the United States in 1970 contained this male sterility factor and was thus highly susceptible to the blight (Tatum, 1971; Wade, 1972). Luckily, the blight caused only a 15 per cent reduction in the U.S. crop in 1970. Blight-resistant strains of corn have been developed, of course, but calamity could occur for any of our major crops if new mutant strains of parasites go on the rampage. As Reitz suggests in his paper, more tolerance of genetic diversity in modern farm crops is urgently needed to prevent such epidemics. The paper by Harlan in this section explains the importance of conserving genetic variability in agricultural species, both to preserve disease-resistant genes and for other reasons (see also Frankel, 1975; Miller, 1973).

It is important to mention that the proliferation of the use of pesticides and herbicides in agriculture (see Irving, 1970) and warfare (Neidlands et al., 1972) has profound repercussions on both the deliberate victims (insects, weeds) and the accidental ones (other plants and animals and the citizenry). A U.S. Government projection of 1967 said the non-Western world should increase pesticide usage 600 per cent by 1985 to feed itself adequately (Paddock, 1970). Fortunately, this flood of chemicals has not taken place, largely because of the sustained public outcry of the last few years and, more recently, because oil, a necessary part of the pesticide manufacturing process, is becoming scarce and expensive (Djerassi et al., 1974; Ennis et al., 1975). In addition, genetics is now contributing detailed plans to control pest populations (Smith and von Borstel, 1972).

The readings in this section deal with plant genetics, not because animal genetics is intrinsically less interesting

(Rendel, 1970; Lerner, 1973; Francis, 1970), but because plants, being at the lowest level of food chains, are a more abundant source of energy for human survival. But, parenthetically, the recent development of artificial insemination procedures for animal breeding which now permit the stored semen of a single bull to father thousands of calves— even bull semen stored ten years has been used successfully —has fueled the imagination of H. J. Muller and others who wish to use the stored semen of select human donors to improve the quality of human genes (see paper by Dunn in Section One).

Bibliography

Brewbaker, J. L. *Agricultural Genetics*. Englewood Cliffs, N.J.: Prentice-Hall, 1964.

Brown, J. W. Native American contributions to science, engineering, and medicine. *Science*, **189**: 38–40, 1975.

Chase, S. S. The vanishing plant breeder: Who will expand our food supply? *Bull. Atomic Sci.*, Dec. 1968, pp. 10–13.

Curtis, B. C., and D. R. Johnston. Hybrid wheat. *Sci. Amer.*, **220**: 21–29, 1969.

Djerassi, C., C. Shih-Coleman, and J. Diekman. Insect control of the future: Operational and policy aspects. *Science*, **186**: 596–607, 1974.

Eckholm, E. P. The deterioration of mountain environments. *Science*, **189**: 764–770, 1975.

Ennis, Jr., W. B., W. M. Dowler, and W. Klassen. Crop protection to increase food supples. *Science*, **188**: 593–598, 1975.

Francis, J. Breeding cattle for the tropics. *Nature*, **227**: 557–560, 1970.

Frankel, O. H. Genetic conservation: Our evolutionary responsibility. *Genetics*, **78**: 53–65, 1975.

Greenland, D. J. Bringing the green revolution to the shifting cultivator. *Science*, **190**: 841–844, 1975.

Griliches, Z. Hybrid corn and the economics of innovation. *Science*, **132**: 275–280, 1960.

Hardy, R. W. F., and U. D. Havelka. Nitrogen fixation research: A key to world food? *Science*, **188**: 633–643, 1975. Discusses the possibility of transferring bacterial nitrogen-fixing genes to crop plants with the aim of reducing their dependence on nitrogen fertilizer.

Harpstead, D. D. High-lysine corn. *Sci. Amer.*, **225**: 34–42, 1971.

Harris, M. The withering green revolution. *Natural History*, Mar. 1973, pp. 20–23.

Heiser, C. B. *Seed to Civilization: The Story of Man's Food.* San Francisco: Freeman, 1973.

Hightower, J. *Hard Tomatoes, Hard Times.* Washington, D.C.: Agribusiness Accountability Project, 1972. Criticizes American agriculture for emphasizing machine-handling convenience in crop selection rather than the interest of the consumer.

Hulse, J. H., and D. Spurgeon. Triticale. *Sci. Amer.,* **231**:72–80, 1974.

Irving, G. W., Jr. Agricultural pest control and the environment. *Science,* **168**:1419–1424, 1970.

Lerner, I. M. Heterosis and the future of animal improvement. *Perspectives Bio. Med.,* **16**: 581–589, 1973.

Lerner, I. M., and H. P. Donald. *Modern Developments in Animal Breeding.* New York: Academic Press, 1966.

Levins, R. Genetics and hunger. *Genetics,* **78**:67–76, 1974.

Mangelsdorf, P. C. *Corn: Its Origin, Evolution, and Improvement.* Cambridge, Mass.: Belknap (Harvard University Press), 1974.

Mayer, A., and J. Mayer. Agriculture, the island empire. *Daedalus,* **103**:83–95, 1974.

Miller, J. Genetic erosion: Crop plants threatened by governmental neglect. *Science,* **182**:1231–1233, 1973.

Moore, P. D. Evolution of tolerance to pollution. *Nature,* **258**:13–14, 1975.

Mulvihill, J. J. Congenital and genetic disease in domestic animals. *Science,* **176**: 132–137, 1972.

Neidlands, J. B., G. H. Orians, E. W. Pfeiffer, A. Vennema, and A. H. Westing. *Harvest of Death: Chemical Warfare in Vietnam and Cambodia.* New York: Free Press, 1972.

Paddock, W. C. How green is the green revolution? *BioScience,* **20**:897–902, 1970.

Poleman, T. T. World food: A perspective. *Science,* **188**: 510–518, 1975.

Rendel, J. Conservation of animal genetic resources. *Science Journal,* **6**:49–55, 1970. Discusses genetic improvement of farm animals in underdeveloped countries.

Sigurbjornsson, B. Induced mutations in plants. *Sci. Amer.,* **224**:84–95, 1971.

Smith, H. H. Model systems for somatic cell plant genetics. *BioScience,* **24**:269–276, 1974.

Smith, R. H., and R. C. von Borstel. Genetic control of insect populations. *Science,* **178**:1164–1174, 1972.

Tatum, L. A. The Southern corn leaf blight epidemic. *Science,* **171**: 1113–1116, 1971.

Walters, H. Difficult issues underlying food problems. *Science,* **188**:524–530, 1975. Discusses social, economic, and political problems in world food production.

Wade, N. A message from corn blight: The dangers of uniformity. *Science*, **177**:678–679, 1972.

Wade, N. Green revolution: Creators still quite hopeful on world food. *Science*, **185**:844–845, 1974a.

Wade, N. Green revolution (I): A just technology, often unjust in use. *Science*, **186**: 1093–1096, 1974b.

Wade, N. Green revolution (II): Problems of adapting a western technology. *Science*, **186**:1186–1192, 1974c.

5   Hybrid Corn: Its Genetic Basis and Its Significance in Human Affairs

Paul C. Mangelsdorf

Introduction

The spectacular advances which man has made during the past century in controlling his physical environment have been largely the product of research in the physical sciences. Man has been far more adept at understanding and manipulating the non-living components of the universe than he has in controlling the living. But revolutionary changes are imminent. Many scientists, including physicists and chemists, now predict that the second half of the twentieth century will belong, not to physics and chemistry but to the biological sciences. I believe that these predictions are quite sound.

The trend is already foreshadowed by the important advances which have been made during the past 25 years in the field of applied biology. The discovery of insulin, penicillin, streptomycin, cortisone and similar disease-destroying chemicals which are elaborated by living things are familiar to the man on the street. Less well known to the general public are the advances, perhaps even more far-reaching in their importance, which have been made in improving our basic food plants through the application of the principles of genetics. One of these—hybrid corn—is the subject of my paper. In my opinion hybrid corn is the most far-reaching development in applied biology of this quarter century. It has already affected more lives, I venture to guess, than any of the epoch-making discoveries in medical biology of the same period. Insulin and penicillin have saved thousands of lives in the past 25 years, but the new abundance of foodstuffs which hybrid corn has created has saved millions of lives in this period of the world's history. Hybrid corn may even prove, in future historical perspective, to have been one of the most important factors in saving our American culture and the European civilization from which it was born.

Two dramatic examples of the importance of hybrid corn are found in the role which it played during World War II and immediately thereafter. During the three war years—1942, 1943, and 1944—the American farmer, suffering from acute labor shortages on the one hand and from unfavorable weather on the other, produced 90 percent as much corn as he had during the previous four years of peace. Hybrid corn made it possible for him to step up, by approximately 20 percent, an already unprecedented production at a time when maximum production was desperately needed. Because of hybrid corn we not only suffered no real food shortages during the war but, on the contrary, we were able to ship food to our allies and to employ surplus grain as raw material for the manufacture of alcohol, synthetic rubber, explosives, and other materials of war.

From *Genetics in the 20th Century*, L. C. Dunn, ed., 1951, pp. 555–571. Copyright 1951 by the Macmillan Publishing Co., Inc. Reprinted with permission of the author and the publisher.

At the conclusion of the war our corn surplus immediately fulfilled another and more peaceful but no less important role. In the year ending June 30, 1947, the United States shipped to the hungry peoples of war-torn Europe 18 million long tons of food. Very little of this was corn, but food is food and 18 million long tons of food is the equivalent of 720 million bushels of corn. In this same year the increase in yield of the corn crop of the United States resulting from the use of hybrid corn is conservatively estimated to have been more than 800 million bushels.[1] Thus Europe's food deficits during the first post-war years were made up, with some to spare, by the surplus accruing from the use of hybrid corn. Western Europe became less receptive to communism because hybrid corn had made it possible for the New World to come to its aid in a time of great need. Thus the principles of heredity discovered by Gregor Mendel in 1865 and rediscovered in 1900 came to play an important, if not immediately obvious, part in stemming the tide of communism in Europe. Perhaps Russian antipathy to Mendel's laws and to modern genetic theory is not unfounded.

Methods of Producing Hybrid Corn

What is hybrid corn and how has it achieved such an epoch-making contribution to the world's food supply? In a sense all corn is hybrid corn since corn is a cross-fertilized species in which hybridization between individual plants and between varieties and races is constantly occurring. Corn has had a long history of repeated hybridization and the accompanying hybrid vigor or heterosis has played one of the major roles in its evolution under domestication. What the modern corn breeder has done is merely to exploit more fully than is possible in nature a tendency characteristic of the species.

As a geneticist concerned with the problem of evolution under domestication, I would (taking a leaf from Dobzhansky's book) define hybrid corn as a type of adaptive polymorphism in which natural selection, acting in a man-made environment, preserves certain chromosomes not primarily because of their intrinsic worth, but because they interact effectively with other chromosomes, similarly preserved, to produce a highly successful Mendelian population. As a practical corn breeder, I would define hybrid corn as any corn artificially hybridized in order to utilize the phenomenon of hybrid vigor or heterosis.

As most commonly used in the United States at the present time, hybrid corn involves as a first step the isolation of inbred strains. This is accomplished by self-pollination, an intensive form of inbreeding, which in the case of corn is accompanied

[1] It is difficult to determine precisely how much the American corn crop has been increased by the use of hybrid corn. McCall (1944) estimated that the increased yield in 1943 resulting from the planting of 49,428,000 acres of hybrid corn was 669,480,000 bushels, an average increase of approximately 13.5 bushels per acre. At this same rate the 1946 acreage of 61,614,000 acres would have resulted in an increase of 831,789,000 bushels. Data presented in this paper suggest that the increased yield accompanying the use of hybrid corn (but not necessarily resulting wholly from the superior productiveness of hybrid corn) is approximately 15 bushels per acre. At this rate the increased yield attributable to hybrid corn in 1946 was 924,210,000 bushels.

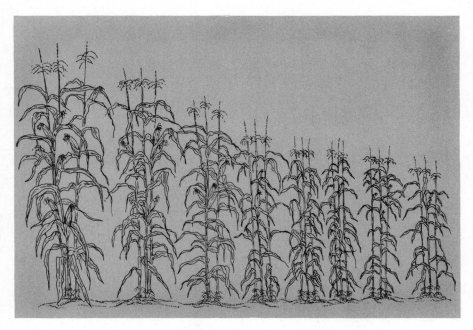

Figure 5-1.   The effect of inbreeding upon vigor in corn. The plants represent an F_1 hybrid and seven subsequent generations of self-pollination. Corn becomes homozygous and uniform after five or six generations of inbreeding.

by three principal effects: (1) the elimination of numerous deleterious recessives, (2) a marked loss in vigor and productiveness, (3) the establishment of homozygous strains of great uniformity. These uniform homozygous inbred strains usually yield less than half as much as the open-pollinated varieties from which they were derived. Their only value is as potential parents of productive hybrids (Figure 5-1).

When these unproductive inbred strains are crossed they give rise to vigorous hybrids, some of which are appreciably more prolific than the original open-pollinated corn from which they came and it is this phenomenon which is exploited in the modern commercial production of hybrid corn. The problem of hybrid seed production is substantially simplified by combining four inbred strains instead of two into a combination known as a double cross and this is now the common commercial procedure. It was the invention of the double-cross method by Jones in 1917 which made hybrid corn at once a practical reality instead of a future possibility. (See Figure 5-2.)

There are many variations of this prevailing double-cross method. Single crosses, the product of crossing two inbred strains, are employed when maximum uniformity is required as in the case of sweet corn used for canning. Three-way crosses produced by hybridizing a single cross with a third inbred strain are useful for testing the "combining ability" of new inbreds. This is also true of "top crosses" which are

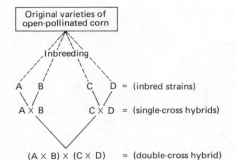

Original varieties of
open-pollinated corn

Inbreeding

A   B   C   D = (inbred strains)

A × B   C × D = (single-cross hybrids)

(A × B) × (C × D)   = (double-cross hybrid)

Figure 5-2. Diagram illustrating how commercial hybrid corn is usually produced.

hybrids of inbred strains and open-pollinated varieties. When two or more inbred strains are combined with an open-pollinated variety, the product becomes a "multiple top cross," a type of hybrid which has been very useful in the early stages of the corn-breeding program in Mexico. "Synthetic" varieties represent advanced generations of hybrids produced by combining a number of inbred strains, usually four or more. Other modifications of the common method have resulted from the discovery that inbred strains used in hybrids need not be inbred to the point of uniformity and homozygosity but can be used after one generation of self-pollination (Figure 5-3).

Figure 5-3. Ears illustrating how four uniform inbred strains are first combined into two single crosses and subsequently into a double cross. Single-crossed seed, since it is produced on inbred strains, is expensive. Double-crossed seed borne on vigorous single crosses is much cheaper to produce.

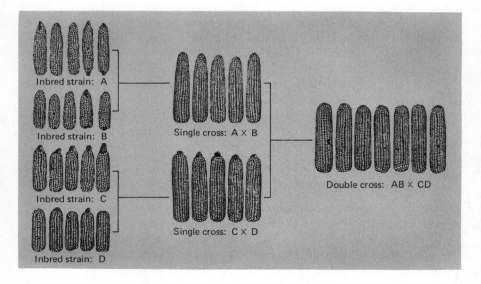

Inbred strain: A

Inbred strain: B

Single cross: A × B

Inbred strain: C

Single cross: C × D

Inbred strain: D

Double cross: AB × CD

The Genetic Basis of Hybrid Corn

The basis for all types of hybrid corn is the phenomenon known as hybrid vigor or heterosis which often accompanies the hybridization of unrelated varieties, races, or species. There is a divergence of opinion on how heterosis should be defined. In *Webster's International Dictionary* heterosis is, "the greater vigor or capacity for growth frequently displayed by crossed animals or plants as compared with those resulting from inbreeding." This is approximately the meaning which Shull (1948), who coined the term, intended that it should have. This is the definition to which the majority of geneticists would probably subscribe. To include as manifestations of "vigor" such unrelated characteristics as the capacity of insects and birds to produce more and better eggs or of excised roots to elaborate or utilize essential metabolites creates no fundamental conflict with this definition. Many geneticists, however, feel the need of more precise terms for particular types of heterosis. There is no reason why special terms with restricted definitions should not be coined to meet such needs.

On the question of the genetic basis of heretosis in corn there is even greater divergence of opinion than on its definition. There are at present two principal hypotheses. The one, foreshadowed by Bruce's speculation, first clearly expressed by Keeble and Pellew, and later elaborated upon by Jones in terms of the chromosome theory of heredity, holds that heterosis is caused by bringing together in the hybrid the dominant favorable genes of both parents. This theory says, in essence, that the hybrid is better than its parents because it exhibits the best qualities of both while concealing their defects. The second theory, actually the earlier, holds that heterozygosity *per se* is responsible for heterosis. Both Shull and East attributed heterosis to the physiological stimulus of heterozygosity. More recently East (1936) explained heterosis in terms of the complementary action of alleles at the same locus, a phenomenon to which Hull has given the term "overdominance" because the effect of the complementary action is supposedly greater than that resulting from dominance alone. This theory paraphrases an old axiom to say in essence that two alleles are usually better than one.

Fundamentally the two theories are not completely conflicting. Both involve orthodox genetic interpretations in terms of Mendelian principles. Both postulate the complementary action of genes. The principal difference between the theories is that one supposes the genes at different loci complement each other, while the other holds that different alleles at the same locus produce this effect. Since it is difficult, if not impossible, to distinguish between alleles and completely linked genes, there is at least one situation in which the two theories are identical. From the standpoint of the practical breeder the most important difference between the two theories is that if the first is correct, it should be possible to gain maximum improvement by accumulating the maximum number of dominant favorable genes in the homozygous condition since dominance is assumed to be usually incomplete. If the second theory is correct, maximum productiveness demands maximum heterozygosity. Perhaps it will be shown in the final analysis that both mechanisms are operating. Indeed, it

49

would be surprising if this were not true. In the meantime corn breeders continue to exploit the phenomenon of heterosis for the improvement of corn without being certain of its cause or in agreement on how to define it.

The History of Hybrid Corn

The history of hybrid corn is important in demonstrating how future advances in biology may be expected to occur. Most of the developments in science are the product of many men and many minds and in this respect hybrid corn is typical. Scientific achievements, like royalty and pure-bred livestock, often have long and complex genealogies. Lines of descent often trace back to more than one distinguished progenitor and the same progenitor sometimes appears more than once in the pedigree. Hybrid corn is no exception to this general rule. Two distinct lines of descent have converged to make hybrid corn an accomplished fact in our time and a third line has had a marked influence. If we trace these lines back to their recognizable sources they lead us to three famous biological scientists of the nineteenth century, Charles Darwin, Gregor Mendel, and Francis Galton.

One line of descent in the genealogy of hybrid corn, represented by the pioneering research of George H. Shull (1908, 1909), traces back through Johannsen, the Danish botanist, to Galton and from Galton to Darwin. Shull had no thought of improving the corn plant when he began his studies. His objective was to analyse the inheritance of quantitative characters and he chose corn as an appropriate experimental subject for this purpose. He inbred his lines to "fix" their characteristics and he crossed lines which had been thus inbred to study the inheritance of kernel-row-number.

That Shull should have been able from the limited experiments then completed and from unreplicated yield tests, not only to draw valid conclusions regarding the effects of inbreeding and crossbreeding, but also to design a new method of corn breeding based upon the exploitation of heterosis, is one of the most fortunate achievements of our time in the field of applied biology. Shull's idea of maintaining otherwise useless inbred lines of corn solely for the purpose of utilizing the heterosis resulting from their hybridization was revolutionary as a method of corn breeding. It is still the basic principle which underlies almost the entire hybrid corn enterprise. Hybrid corn did not come into being, however, until the first line of descent, represented by Shull, converged with the second line, represented by East, Hayes, and Jones, to produce a kind of genealogical "heterosis" whose results have been nothing short of explosive.

The second line of descent in the pedigree of hybrid corn, like the first, traces back to Charles Darwin. Darwin was interested in the effects of inbreeding and crossbreeding as factors in evolution and he conducted numerous experiments on self- and cross-fertilization in plants, including corn as one of his subjects. Darwin's results were known before their publication to Asa Gray with whom he carried on an extensive correspondence and who had visited him in England. One of Gray's students of that period was William Beal, who subsequently at Michigan State College wrote a review of Darwin's book and carried on experiments in corn with the direct purpose of utilizing hybrid vigor to increase yield, the first controlled experiments of their kind. Beal's work was the progenitor of the corn-breeding program at the University of Illinois under Holden, Shamel, Hopkins, and East (Holden, 1948) and this in turn was the parent of the long-term experiments at the Connecticut Agricultural Experiment Station on the effects of inbreeding and crossbreeding under East, Hayes, and Jones (Crabb, 1947). The Connecticut program reached fruition in 1917 in the invention of the double cross, a device for simplifying hybrid seed production, and in Jones' theory of heterosis which interpreted the phenomenon in terms of the chromosome theory of heredity and thus brought into the genealogy a third line of descent tracing back to Mendel through Morgan and his students. His theory undoubtedly gave as much stimulus as his double cross, if not more, to practical hybrid corn breeding. Historically, then, hybrid corn became transformed from a magnificent design into a practical reality when Jones' method of seed production made it feasible and his theory of heterosis made it plausible, a combination difficult for even the most conservative agronomist to resist. The impact of Jones' two contributions upon practical hybrid corn breeding is easily demonstrated by the sudden expansion of corn-breeding programs in various parts of the United States following the publication of his several papers (Mangelsdorf, 1948).

Edward Murray East is usually mentioned as one of the participants in the development of hybrid corn. East's part is not as specific as that of Shull and Jones. He made an important contribution, however, in initiating and maintaining the long-range experiments on inbreeding and crossbreeding which finally furnished two important keys to the problem of practical hybrid corn breeding. Without East's work we would not have had hybrid corn as soon as we did; indeed we might not have had it now, although it would inevitably have come sooner or later.

By 1933 hybrid corn was in commercial production on a substantial scale, and the United States Department of Agriculture began to gather statistics on it. By 1949, 77.6 percent of the total United States acreage was in hybrid form.

The roster of corn breeders who have produced this unprecedented accomplishment is a long one, including many distinguished names—Richey, Hayes, Wallace—and a score or more of others. Yet in the light of their huge accomplishment—65 million acres of hybrid corn in 1950—the number of men who have participated in the development is almost unbelievably small. Some 25 men, not more than 50 at the most, have played a major part in an agricultural revolution affecting more than 150 million Americans. One man in each six millions of the population has made an important contribution toward determining the prosperity and destiny of the entire group. These men have been catalyzing agents effective in concentrations of approximately 0.2 parts per million.

Significance of Hybrid Corn

The significance of hybrid corn in human affairs is manifold. It is an excellent example of scientific research at its best. It illustrates how the scholars of one generation reach new heights by almost literally standing upon the shoulders of those who have preceded them. It shows how painstaking experimentation and brilliant speculation complement each other to produce a near-miracle which neither would have accomplished alone. Hybrid corn illustrates especially well the necessity and importance of the free interplay of theory and practice which flourishes most successfully in a non-totalitarian society. Marxian ideology emphasizes the integration of theory and practice but modern exponents of Marx seem not to realize that this interaction cannot be achieved by edict or decree. It is something which, in the words of the popular song, "comes naturally" and it does so only in a free society where there is no neglect of the untrammeled search for truth on the one hand and no overemphasis on utilitarian aspects on the other.

The importance of the theoretical background would be difficult to overestimate. It would, for example, have been virtually impossible for so-called "practical" plant breeders, living in an authoritarian country and working under duress to improve the corn plant, ever to have discovered or invented hybrid corn. The reason is obvious. The first step in hybrid corn production is inbreeding and inbreeding leads not to immediate improvement but to a drastic reduction in yield. It would take a very brave corn breeder, indeed, to report to his superiors in the party hierarchy that he had, after two generations of inbreeding, succeeded in reducing yield by 50 per cent. Indeed, even in our own democracy it has not always been easy for corn breeders to defend this paradoxical procedure of "advancing backwards." I still have vivid memories of some of my own experiences in the early days of the corn-breeding program with which I was associated in Texas and I am sure that other corn breeders, like me, have been the target, if not of outright criticism, at least of good-natured ridicule at the hands of practical farmers.

Hybrid Corn and the Food Problem

But the real significance of hybrid corn lies, of course, in the contribution which it has made and can make to the world's food supply and here it can contribute more than is, at first glance, apparent.

Hybrid corn is much more than a method of increasing the productiveness of corn. In biological terms it is a kind of enzyme or catalyst which leavens the entire agricultural economy. In terms of anthropology it is a wedge which splits wide-open an entire cultural complex of long standing allowing a new cultural pattern to be formed.

The average acre yield of corn in the United States has increased from approximately 22 bushels per acre in the early 30's, when hybrid corn first began to be used commercially, to approximately 33 bushels in the late 40's when it occupied some 75 percent of the total corn acreage (Figure 5-4). An increase of 50 percent, when only three acres out of every four were planted to hybrid corn, suggests that if all were planted to hybrid corn the increase would be about $66\frac{2}{3}$ percent. Since increases of this magnitude are seldom met in controlled experiments, it follows that the use of hybrid corn has brought in its wake other improved agricultural practices including crop rotation, the use of fertilizers, and the growing of soil-

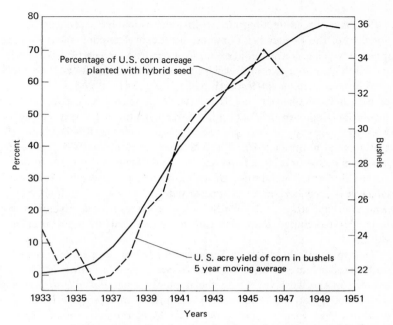

Figure 5-4. Graphs showing how the average yield of corn in the United States has risen in almost exact proportion to the increase in percentage of the total acreage planted to hybrid corn. With three acres out of every four now planted to hybrid corn, the average yield has increased from approximately 22 bushels per acre to approximately 33 bushels.

improvement crops. The successful utilization of hybrid corn has made the American farmer receptive to an entire complex of new and improved methods.

What has happened in the United States can be expected to happen eventually in other parts of the world where corn is an important food plant. Already corn production in Italy and other parts of Southern and Eastern Europe has been greatly increased by the importation of American hybrid corn seed which fortunately has proven to be well adapted. In 1949, 2000 tons of sixteen different American hybrids were planted in Italian fields on about 150,000 acres. In 1950, at least 3000 tons of hybrid seed were exported to Italy. Yields of the best American hybrids are substantially greater than those of the best native varieties. Increases of 25 to 30 percent are produced in experimental plots under controlled conditions, but the Italian farmer, like his American counterpart, reports much greater increases on his own farm.

In the countries of Latin America where corn is the basic cereal, new hybrids especially adapted to local conditions will usually have to be developed since hybrids from the United States are generally not adapted. Corn-breeding programs aimed at this objective are in progress in Mexico, Guatemala, Salvador, Costa Rica, Colombia, Venezuela, Brazil, Uruguay, Chile, and Argentina.

The corn-breeding program in Mexico which has been conducted under the auspices of the Rockefeller Foundation, in complete cooperation with the Government of Mexico, is a splendid example of exporting a technical skill for the benefit of a friendly neighbor without sacrifice to our own capital assets. The Mexican hybrid corn program, under the leadership of J. George Harar and the immediate direction of Edwin J. Wellhausen and Lewis M. Roberts, has succeeded in seven years in revolutionizing corn production in Mexico. Begun in 1943, it had by 1948 affected Mexico's corn crop so that, for the first time since the Revolution in 1912, Mexico had no need to import corn. The Mexican corn crop of 1950 is estimated to be the largest in its history. Here two men, in a population of some 22 million people, have shaped the nutritional destiny of the entire mass. The catalyst in this case has been effective in a concentration of approximately 0.1 parts per million. Since this represents approximately the same order of magnitude as the effectiveness of corn breeders in the United States, there is reason to believe that the figure may have a general validity.

To appreciate the full significance of what has been done in Mexico one must understand what corn means to the Mexican people. There is probably no other country in the world in which the population has become so dependent upon a single food plant. To millions of Mexicans, as to their ancestors before them for centuries past, corn is literally the staff of life; the daily bread which, eaten three times a day, 365 days a year, provides the fuel that keeps the human mechanism functioning. A Mexican laborer, when he can get it, will consume as much as two pounds of corn a day. When corn is plentiful the Mexican is happy and relatively prosperous. When it is scarce there is unrest and danger to stable government.

In Mexico, as in the United States, hybrid corn has brought other important innovations in its wake. A drastic change in his basic food plant has "softened up" the Mexican farmer, traditionally one of the world's most conservative individuals, making him receptive to other changes. As recently as 1943, for example, Mexico's most competent agricultural experts expressed the conviction that it would be impossible, because of Mexico's traditional reliance on corn, to introduce grain sorghums into Mexico in regions where rainfall is inadequate for corn and where the more drought-resistant sorghums might be expected to flourish. Today—seven years later, a short period in terms of agricultural progress—two agencies of the Mexican Government are each claiming the prerogative of distributing sorghum seed to the Mexican farmer. New wheat varieties resistant to rust, soy beans, sweet clover, and other new soil-improving crops are meeting the same warm reception that has greeted the introduction of hybrid corn. The entire agricultural enterprise of Mexico is in a state of flux as a consequence of the labors of a small band of well-trained geneticists.

What has been done in Mexico can presumably be done in other Latin-American countries where corn is the basic food plant. A second experiment under the auspices of the Rockefeller Foundation has been started in Colombia and it will soon be known whether the successful Mexican experiment can be duplicated in other countries. What can be done with corn in America can perhaps also be done with rice in China, with wheat in parts of India, and elsewhere in the world where the economy is strongly dependent upon a single crop. Experience in Mexico has shown that the most effective way to start an agricultural revolution is to improve the basic crop plant by the application of genetic principles. Soil improvement and the control of soil erosion are in the long run perhaps even more important objectives than crop improvement, but it is a drastic change in his basic crop plant which first brings new hope to the tradition-bound farmers of undeveloped countries and prepares the ground for other even more far-reaching changes.

Hybrid Corn and the Future

The problem of population and the food supply has now become one of the world's most acute and pressing problems and is directly or indirectly the cause of much of the world's present unrest. It is difficult for Americans, living in a land of agricultural surpluses, and taxed to maintain scarcity prices in the face of plenty, to realize that two-thirds of the world's people are, by modern nutritional standards, inadequately fed and are suffering from chronic malnutrition (Boyd-Orr, 1950). What can be done to solve the problem of the world's underfed people while the world's population is increasing at the unprecedented rate of 22 million people per year, a rate which promises to accelerate still further before it begins to decline?

There is no simple answer to this question. There are those, probably unduly pessimistic, who regard the problem as insoluble and who fear that the world must resign itself to the inexorable consequences of the Malthusian Law: ever-recurring famine and ever-increasing poverty. There are others, undoubtedly too optimistic or too sentimentally humanitarian, who believe that the United States alone can feed the world. They would have us continue indefinitely to produce and export surpluses to underfed peoples, not realizing that this measure is only a short-term palliative at best and that to export food is in reality to export a part of a permanent soil fertility. Between these two extremes lies the solution. The most hopeful circumstance at present lies in the fact that food is, after all, only one form of energy and man's greatest source of energy, solar radiation, is still almost untapped and virtually inexhaustible. The amount of solar radiation which falls upon all parts of the world exceeds by far in energy equivalents man's needs for food, fuel, and power. Man is an ingenious animal who has, sooner or later, always found a way to exploit to his own satisfaction or benefit the natural resources at his command. It is almost inconceivable that he will not also learn to utilize solar energy, especially since he has now in parts of the world accustomed himself to the luxury of unlimited power resulting from the exploitation of fossil fuels which represent the stored solar energy of the past. Probably he will also learn to control universally, as he already has in some parts of the world, the unrestricted growth of population which is the basic cause of the present problem. The great danger of this century is that he will not do so in time and that global chaos will prevail before he has succeeded either in harnessing solar energy for new methods of food production or in stabilizing his own reproductive rate. It is during this period that the more prosperous nations of the world must, in their own enlightened self-interest and not motivated merely by generosity or by sentimentalism, make every possible contribution toward developing the agricultural resources of the world. It is during this period that the improvement of our cultivated plants and domestic animals can make their most vital contribution. It is during this period that hybrid corn will be of even greater significance to human affairs than it has been in the past. Hybrid corn cannot solve the problem of the world's food supply, but combined with other effective methods it can surely help.

References

Boyd-Orr, John. 1950. The food problem. Sci. Amer. **183,** No. 2:11–15.

Brieger, F. G. 1950. The genetic basis of heterosis in maize. Genetics. **35**:420–445.

Cohen, I. Bernard. 1948. Science, servant of man. Boston. Pp. 176–203.

Crabb, A. Richard. 1947. The hybrid-corn makers. New Brunswick, N.J.

Crow, James F. 1948. Alternative hypotheses of hybrid vigor. Genetics. **33**:477–487.

Daniels, Farrington. 1950. Solar energy. *In* Centennial. Washington. Pp. 163-170.

Dobzhansky, Th. 1949. Observations and experiments on natural selection in Drosophila. Proc. 8th Int. Cong. Genetics, 210–224.

East, E. M. 1936. Heterosis. Genetics. **21**:375–397.

——— and Hayes, H. K. 1912. Heterozygosis in evolution and in plant breeding. U.S. Dept. Agr. Bull. **243**:1–58.

——— and Jones, D. F. 1919. Inbreeding and outbreeding. Philadelphia and London.

Holden, P. G. 1948. Corn breeding at the University of Illinois 1895 to 1900. Charlevoix, Michigan.

Johannsen, W. 1903. Ueber Erblichkeit in Populationen und in reinen Linien. Jena.

Jones, D. F. 1917. Dominance of linked factors as a means of accounting for heterosis. Genetics. **2**:466–479.

——— 1918. The effects of inbreeding and crossbreeding upon development. Conn. Agr. Expt. Sta. Bull. **207**:5–100.

Mangelsdorf, P. C. 1948. The history of hybrid corn. Jour. Heredity. **39**: 177–180.

McCall, M. A. 1944. Crop improvement, a weapon of war and an instrument of peace. Jour. Amer. Soc. Agron. **36**:717–725.

Richey, F. D. 1950. Corn breeding. Advances in Genetics. **3**:159–192.

Shull, G. H. 1908. The composition of a field of maize. Report Amer. Breeder's Assn. **4**:296–301.

——— 1909. A pure line method of corn breeding. Report Amer. Breeder's Assn. **5**:51–59.

——— 1948. What is "heterosis"? Genetics. **33**:439–446.

Sprague, G. F. 1946. The experimental basis for hybrid maize. Biol. Rev. **21**: 101–120.

Whaley, W. Gordon. 1944. Heterosis. Bot. Rev. **10**: 461–498.

6   New Wheats and Social Progress

Louis P. Reitz

Man's quest for food is everlasting. The supply of food can never be very far ahead of the need, because all food is perishable. Nor can man long survive when food is in short supply, because his body reserves are soon exhausted. We know this from very recent experience. In 1966, grounds for pessimism were easy to find: Our supplies of wheat were gone; India had had two droughts in succession; the U.S.S.R. had had two crop failures in 3 years; and Australia's wheat crop was poor (*1*). But the years since, while not entirely favorable, have brought supplies back to substantial amounts, worldwide, and reserve supplies have accumulated again in some places. Part of this turnaround is due to the new varieties of wheat, to slight expansion of acreage, and to the higher priority given wheat production in the assignment of resources, especially fertilizer, in the developing nations.

Jonathan Swift wrote approvingly of making "two blades of grass to grow where one grew before." This is about what the new wheats do when grown in association with good soil and crop management practices. Agricultural progress of this magnitude is having an impact on food supplies, markets, government farm programs, production practices, agribusiness, and population. Improved capability of producing the major cereal grains—wheat, rice, and corn, in particular—has given new hope of feeding the world's burgeoning population. If that hope could be realized without insurmountable side effects, this would surely lead to social progress. However, too much reliance on such agricultural development, good though it may be, will boomerang if other measures are not taken to balance the social pressures in every country, and may only alleviate the difficulties temporarily.

Wheat has been man's companion and food in Eurasia and North Africa for thousands of years. Prehistorians, in developing their time scale, have set wide limits for the first appearance of wheat on earth, and rightly so, because this event has repeatedly been set earlier and earlier.

Good archeological evidence shows that wheat was known in Neolithic times (*2*) and probably was a plant of great antiquity even then. Cereal pollen (not necessarily wheat) dating back to 10,000 B.C. is known (*3*). Principally, emmer was the species found associated with wheat husbandry of the 7th millennium (*4*).

What Is New?

Can there be anything new about a crop so old? We sometimes hear a similar question asked about man. The answer to both questions is "yes," in the long view. Both man and his food plants are subjected to stresses, and adjustments are made. In the press to obtain more food, wheat is grown now in selected sites in all continents, from latitude 60°N to 60°S, and from just below sea level to elevations of more than

3000 meters (10,000 feet). It exists in several thousand forms or varieties; nearly all of these may be broadly grouped into two species (*Triticum aestivum* L. and *T. durum* Desf.) having 21 and 14 pairs of chromosomes, respectively. Some varieties of recent origin have proved so highly productive and widely useful that they are being grown to the exclusion of other varieties over large areas.

What is really new about wheat boils down to one thing—higher grain yield per unit of land, popularly referred to as the "green revolution" (*4a*). To achieve this higher yield, modern breeders have manipulated several hundred known genes into desired combinations governing plant morphology, physiology, and resistance to disease, and uncounted other essential attributes, to create the varieties now associated with the green revolution (*5*).

Yields per unit of land have been increased dramatically in many countries in the last decade or two; they have been trebled in Mexico and doubled in the United States and in parts of India, Pakistan, Turkey, and many other countries. This is a consequence of (i) increased fertilization or plant food management, (ii) irrigation or better moisture management, (iii) control of pests, (iv) economic incentives, and (v) selection of responsive genotypes of wheat. New varieties undoubtedly provide a basis for obtaining higher yields, and they are a catalyst, making the other factors more interactive (*6*). These high-yielding wheats are not a single variety or genotype, as is erroneously implied in many reports. At least 50 varieties contribute to the worldwide green revolution.

The new wheats have been improved in three major ways. (i) They have shorter, stiffer straw than standard wheats; (ii) they have greater adaptability, hence are better suited to the environments where they are grown; and (iii) they are more resistant to diseases and insects. Most often talked about are the dwarf types, sometimes called semidwarfs to distinguish them from the extremely short (and worthless) freaks or dwarfs of purely genetic interest. The semidwarfs (Figure 6–1) grow from half to two-thirds the height of standard varieties (*7*). Of all the new wheats, the semidwarfs are the most spectacular. The Gaines variety, developed in the state of Washington, has established a new world record of 209 bushels per acre (*7*). The new Mexican varieties have brought hope for hunger-stricken people in a score of countries where it had seemed that population growth must inevitably bring famine (*6–9*). The availability of these varieties has made it possible for nations to plan and carry out practical programs for increasing their food supply without resorting to territorial expansion. These nations have increased the productivity of their own resources through adaptive research, efficiency, education, capital investment, and expanded trade (*10*). This has been a stimulating, new experience that can help to alleviate other problems. Unquestionably, such experience is giving man a new opportunity to improve the quality of his living and forestall famine (*11*). Whether he does so, or only increases the numbers who will die of starvation at a later date, is the decision he faces (*9, 12–16*). Some say it is already too late. Malthus' dire predictions have had to be reassessed several times. New technology, which led to development of the new wheats, has caused another revision in estimates of the timing of the disaster. But unless other parts of the equation are changed, the general concepts of Malthus surely will be proved right in the end.

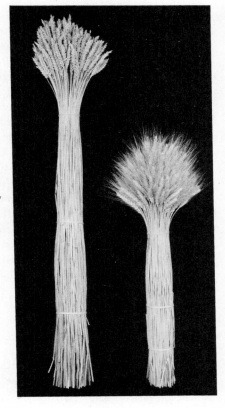

Figure 6-1. Semidwarf wheats, such as Gaines (right), are half to two-thirds the height of standard varieties. The sheaves pictured (about one-tenth actual size) show the full height minus a 5-centimeter stubble.

Genes from Japan

The details about the semidwarf wheats and how they were bred have been given in several accounts (see, for example, 6–8, 17–19). New varieties of this type are being released each year. The main genes for dwarfism trace to the Orient (8) especially to the Japanese variety Norin 10. Dwarf wheats appear to have a long history in Japan. They were first observed there by Americans as early as 1873 (20). No use of this information was made in the United States, and Japan's short wheats had to be rediscovered in 1946 (8). However, short types from Korea, China, and elsewhere were used earlier in the United States, in some cases with gratifying results. Short-stemmed mutants are readily induced, and a chemical spray (CCC), which must be applied to each crop, has been used to reduce the height of otherwise tall varieties (19). Italian breeders were among the first to use Japanese varieties in their wheat improvement program. Records of their crosses from 1911 onward reveal Japanese varieties in the pedigrees; from this work they derived early-maturing, short-stemmed varieties (8).

The semidwarfs first attracted attention in the United States soon after S. C. Salmon, a U.S. Department of Agriculture (USDA) agronomist, brought back

59

to the United States seed of Norin 10 and of 15 other varieties of this type in 1946 (8). These materials were distributed to U.S. breeders in the next 2 years. Beginning in 1949, a corps of workers at Washington State University, led by O. A. Vogel, USDA agronomist, utilized Norin 10 in a series of crosses with locally adapted varieties. From these crosses Vogel selected Gaines and Nugaines. Both varieties are short, have stiff straw, and, under good management conditions, give enormously high grain yields in the Pacific Northwest. They have winter habit of growth (that is, they require a period of cold to mature), hence they cannot be utilized where their requirements for cold during early growth are not met. They are not sufficiently resistant to the cereal rusts to avoid damage during epidemics. Early generation selections from crosses, and Norin 10, were made available in 1953 to N. E. Borlaug, a Rockefeller Foundation scientist in Mexico, and the crosses he obtained the next year mark the beginning of the new famous Mexican wheat varieties. Borlaug and his associates bred into these varieties dwarfing, spring habit, nonsensitivity to day length, early maturity, and a high degree of resistance to rust (6–8, 17). The most widely known and cultivated are the varieties Pitic 62, Lerma Rojo 64, Mexipak 65, Sonora 64, Indus 66, Penjamo 62, and Inia 66. But the pattern changes yearly as new and improved varieties appear. These varieties are now being used in at least 25 countries.

Essential Help

Other significant, yet little known, research developments took place at about the time of the semidwarf episode. These events are little known because they were diffuse, of varied character, slowly cumulative, and not widely publicized. Many people were involved, but there were no heroes. (I have no objection to heroes, but some events do not fit the pattern.) I refer to work done through people serving several agencies not yet mentioned here, especially the U.N. Food and Agriculture Organization (FAO), the Agency for International Development, many universities, the Ford Foundation, ministries of agriculture, and the numerous research people who published their findings in professional journals and research reports which disseminated valuable information. Three direct activities, among a score or more of importance in bringing the new wheats to the attention of many countries, are here singled out: (i) the work of the International Maize and Wheat Improvement Center (CIMMYT), through its research, testing, and training program (6, 21); (ii) the International Rust Nursery Program, with 40 countries cooperating (22); and (iii) the FAO Near-East Barley and Wheat Improvement Program, which has regularly (since 1952) introduced new germ plasm into its regional tests in a dozen countries. Finally, of most importance, are the farmers in all of these countries, who, at considerable risk, accepted the new varieties and the new technology (mainly higher rates of fertilization) needed to make them spectacularly productive.

Each country has its own success story (6, 17–19, 23), and many of them will never be told. Each is a little different from the others. Although the explanation given is usually simple (the use of new, high-yielding varieties), an interaction of many complex factors had to occur in order for grain yields to exceed, as they did last year, anything ever before recorded. Not the least of these factors were plentiful supplies of good seed and increased amounts of nitrogenous and phosphate fertilizers. The reduced cost of nitrogenous fertilizer has been significant in obtaining greater use of this fertilizer on wheat. Timely and proper sowing of seed, increased irrigation and water management (including drainage), weed and pest control, and prompt harvesting and handling of the crop were also required for success. A breakdown in any of these interlocking events reduced efficiency of production or nullified the benefits of the other practices. Where adapted to the environment, the new varieties have a yield potential from 30 to 100 percent higher than that of traditional local varieties. Therefore, they give increases in yield far beyond fertility levels formerly considered excessive for wheat. It was demonstrated that farmers will readily accept new practices when resources are available and new returns are substantial (1).

Government Policy and Payoff

The effect of government policy has been amply demonstrated in this experience. The constant support of the Mexican Ministry of Agriculture was crucial in the success of the Mexican cooperative program whereby Mexico increased its wheat production sixfold in 20 years. CIMMYT sprang from this program and formed the base of operations for helping other countries (17). There can be no doubt that the boldness and commitment of the Minister of Agriculture in Turkey to a wheat revolution was decisive. He shocked his own advisers in calling for, and obtaining, seed and fertilizer for 170,000 hectares the first year and 650,000 hectares (1,605,500 acres) the second in their accelerated program. In the affected areas, a threefold increase over the amounts grown with native varieties of seeds and with native agricultural practices was achieved (23).

The Indian Agricultural Research Institute, after 2 years of wide-scale testing of Mexican dwarf varieties of wheat, enlisted the help of the government of India in importing large quantities of two Mexican varieties. The government imported 250 tons of seed in 1965 and 18,000 tons in 1966, with the result that about 2 million hectares of dwarf wheats were harvested in 1968. "There is no parallel [up to 1968] for such rapid spread of new varieties in the world," commented M. S. Swaminathan, a member of the Institute (18). This program is credited with having broken a yield barrier in India that for 30 years had seemed impenetrable.

These and other importations and increases in the use of high yielding varieties of wheat have recently been summarized (24). Thirteen countries imported massive quantities of Mexican seed. West Pakistan imported 42,000 metric tons of seed for its 1968 crop, Turkey imported 22,100 tons for 1968, and India imported 18,000 tons for 1967. The combined area seeded to Mexican varieties in 1966 in these three countries was less than 8000 hectares; for 1969, in distinct contrast, it was estimated that 7 million hectares had been sown. Afghanistan, Nepal, Iran, Lebanon, Morocco, and Tunisia also produced sizable acreages of this type of wheat, bringing the total for Asia and North Africa to just under 8 million hectares (6, 24). In Europe and the Americas, at least another 2 million hectares of wheats derived from Norin 10 are grown. The result is that one or two genes for dwarfness and closely linked blocks of genes are present over large areas of the world. Since a relatively small number of varieties make up the major part of the seedlings, a large portion of the genetic makeup is nearly the same wherever the dwarf wheats are grown.

In 1967 and 1968 India produced 17 million tons of wheat, in contrast to a previous high of 12 million tons (18), and did so without an appreciable increase in total growing area. The estimate for 1969 is still higher (6). In West Pakistan almost half of the growing area was sown to the new wheats by 1969, and, with the total growing area increased by about 20 percent, production was almost double that of earlier years (6). In Turkey the new wheats contributed substantially to the 1968 and 1969 wheat crops, and projections indicate self-sufficiency by 1975 (23). Similar results on lesser acreages are evident in Iran, Afghanistan, Nepal, East Pakistan, Tunisia, and Morocco (1, 6).

In all of these estimates the contributions of many factors—the wheat genotype, the land selected, the fertilizer used, the amount of irrigation, and so on—are combined, and it is virtually impossible to separate them. In one study (1), the new varieties were roughly estimated to have added from 9.6 to 32 percent to normal wheat production in the Asian countries, the average being about 20 percent. Extending these known responses to larger acreages is hazardous, as yields will depend additionally upon factors such as future expansion and improvement of irrigation systems; future prices of grain; cost of fertilizer and other inputs; damage from pests; and government policies.

What Next?

Never before have so few genes been responsible for filling so many mouths with food. The genetic base of the new Mexican-type wheats is, unavoidably, relatively narrow. Breeders urgently need to broaden that base as a safeguard against catastrophic events that might follow the appearance of a new pathogenic culture of rust or some other disease. All cereal-growing regions where rust is a constant threat have seen resistance in their crops overcome by new, virulent organisms. There is no assurance that this will not happen again, and it could occur more often than it has in the past, requiring replacement varieties every few years.

When new varieties are grown on large acreages, the old varieties may quickly become extinct. This diminishes the diversity of germ plasm in nature. The U.S. Department of Agriculture has been aware of this process and has built a living herbarium of world germ plasm comprising over 20,000 accessions of wheat. The collection dates back to 1897, when the U.S. government organized the Seed and Plant Introduction Office. In harmony with activities with FAO, the International Biological Program, and other programs, a new effort is being made to obtain seed of indigenous wheats in those areas where introduction of new varieties threatens to make them extinct.

Conventional Varieties Improved

I have emphasized the semidwarf wheats of the Mexican and Gaines type because they give the most dramatic results. This does an injustice to breeders of improved conventional varieties. These varieties, while not semidwarf, have greater straw strength, are somewhat shorter, and yield more grain than older varieties. They are disease-resistant and, in some cases, notably insect-resistant. Salmon *et al.* (*25*) assessed a half century of wheat improvement in the United States and estimated that the better varieties available in 1950 yielded 40 percent more grain than the varieties in use in 1900. The advances occurred mostly in the last decade of the 50-year period and were evident in all regions of the country. More recent but similar evidence has been presented for various regions (*26*). Many of the improved conventional varieties are of value primarily because of their resistance to disease or insects and show little merit over other varieties when these hazards are not present, or when the resistance has been overcome by the pathogen or pest. Prior to 1944 the United States had not harvested a single crop of wheat that exceeded 1 billion bushels, whereas since 1944 all but seven crops have exceeded this amount. The latest three averaged 1.5 billion bushels. The acreage has declined slightly, and the average yield has about doubled (*27*). For a long time, yields in northern Europe have been the highest in the world (*27*). As in other examples presented here, use of improved varieties was only part of the reason for the increased yields. Of great importance were soil management, the use of fertilizer, the more timely tillage made possible by improvements in power machinery, and the use of supplemental chemicals for controlling diseases, insects, and weeds.

Summary

Will the upward trend in all food production, so dramatically exemplified by the new wheats, be adequate to meet the needs of the growing population? Yes, for a while. No one knows for how long (*14*). The prophets of doom will undeniably be proved right in the long run unless their basic assumptions are nullified by concrete acts, and soon. At some point in time, either a zero population growth must be achieved or vast new sources of food must be developed, and purchasing power increased. There is nothing on the research horizon to reject "a prodigious need for mankind to practice human husbandry" (*12*). Our waste products have reached levels that cause major concern, and it may well be that both agricultural and social advancement will be halted by the demands dictated by population growth and the by-products of what now passes for progress but also brings environmental un-balance (*15*). At least, life will be different, and it may be catastrophic (*16*, *28*). The "Three Ancients" (*29*) who helped plan and then, after a quarter of a century, reviewed the agricultural research and development work of the Rockefeller Foundation in developing nations concluded:

> We have discovered no magic formula for success in aid programs. We visualize no miracles and few easy solutions. But we do think that persistent use of science and common sense is the best guarantee of progress. . . . The fight

against ignorance and hunger is a tremendous undertaking and it will take tremendous efforts to win it.

The accumulation and use of research information for growing better and more nutritious crops needs to proceed in both the " have " and the " have not " countries. Among land-grown food crops, the cereal grains and large-seeded legumes are our greatest hope. Wheat is one of these, and it should be exploited to the limit. While wheat *production* is important, it is even more important to find ways to make wheat and other food crops available to remote, often very poor, people. Challenges remain; they are biological, environmental, economic, social, and international.

References

1. J. W. Willett, "The Impact of New Grain Varieties in Asia." *U.S. Dep. Agr. Publ. ERS-Foreign 175* (1969).
2. N. Jasny, "The Wheats of Classical Antiquity," *Johns Hopkins Univ. Stud. Hist. Polit. Sci. Ser. No. 62* (1944), p. 35.
3. K. V. Flannery, *Science* **147**, 1247 (1965).
4. H. Helbaek, "Plant collecting, dry-farming, and irrigation agriculture in prehistoric Deh Luran," in "Prehistory and Human Ecology: Deh Luran Plain," *Mem. Museum Anthropol. Univ. Mich. No. 1* (1969); *Science* **130**, 365 (1959).
4a. L. R. Brown, *Seeds of Change: The Green Revolution and Development in the 1970's* (Praeger, New York, 1970).
5. E. R. Ausemus, F. H. McNeal, J. W. Schmidt, in *Wheat and Wheat Improvement* (American Society of Agronomy, Madison, Wis., 1967), pp. 225–267.
6. *CIMMYT Rep.* (1968–69), pp. 57–101.
7. L. P. Reitz, in *U.S. Dep. Agr. Yearb.* (1968), pp. 236–239.
8. —— and S. C. Salmon, *Crop. Sci.* **8**, 686 (1968).
9. W. Paddock and P. Paddock, *Famine—1975?* (Little, Brown, Boston, 1967).
10. W. Weaver, "The President's Review" (Rockefeller Foundation, New York, 1958).
11. Q. M. West, "The revolution in agriculture: New hope for many nations," in *U.S. Dep. Agr. Yearb.* (1969), pp. 81–86.
12. W. E. Howard, *Bioscience* **19**, 779 (1969).
13. C. R. Wharton, *Foreign Affairs* **47**, 464 (1969).
14. "World Population and Food Supplies, 1980," *Amer. Soc. Agron. Spec. Publ. No. 6* (1965).
15. J. J. Spengler, *Science* **166**, 1234 (1969); P. R. Ehrlich and J. P. Holdren, *Bioscience* **19**, 1065 (1969).
16. F. S. L. Williamson, *Bioscience* **19**, 979 (1969).
17. N. E. Borlaug, *Phytopathology* **55**, 1088 (1965); ——, in *Proc. 3rd Int. Wheat Genet. Symp. Canberra* (1968), pp. 1–36.
18. M. S. Swaminathan, "Five Years of Research on Dwarf Wheats," *Publ. Indian Agr. Res. Inst. New Delhi* (1968).
19. "Cereal Improvement—Hybrid Varieties and Aspects of Culm Shortening," *Euphytica* **17**, Suppl. 1 (1968).

20. H. Capron, "Agriculture in Japan," *Rep. Commissioner Agr. for the Year 1873* (1874) [as cited by D. G. Dalrymple, *Bull. U.S. Dep. Agr. Foreign Agr. Serv.* (1969)].

21. "Preliminary Reports of the First Three Inter-American and the First Two Near East-American Spring Wheat Yield Nurseries," *CIMMYT Misc. Reps. 1–5* (ed. 2, 1966).

22. W. Q. Loegering and N. E. Borlaug, "Contribution of the International Spring Wheat Rust Nursery to Human Progress and International Good Will," *U.S. Dep. Agr. Publ. ARS 34–46* (1963).

23. L. M. Humphrey, "Mexican Wheat Comes to Turkey," *Publ. Food Agr. Div. USAID/Turkey, Ankara* (1969).

24. D. G. Dalrymple, "Imports and Plantings of High-Yielding Varieties of Wheat and Rice in the Less Developed Nations," *Bull. U.S. Dep. Agr. Foreign Agr. Serv.* (1969).

25. S. C. Salmon, O. R. Mathews, R. W. Leukel, *Advan. Agron.* **5**, 1 (1953).

26. L. P. Reitz, *Econ. Bot.* **8**, 251 (1954); ——— and S. C. Salmon, "Hard Red Winter Wheat Improvement in the Plains," *U.S. Dep. Agr. Tech. Bull 1192* (1959); V. A. Johnson, S. L. Shafer, J. W. Schmidt, *Crop Sci.* **8**, 187 (1968); F. H. McNeal and E. J. Koch, "Thirty Years of Testing Winter Wheats, *Triticum aestivum* L. em Thell., in the Western Wheat Region," *U.S. Dep. Agr. Publ. ARS 34–97* (1968); D. E. Anderson, *Farm Res.* (*N. Dak.*) **24**, 21 (1967).

27. "Agricultural Statistics" (*U.S. Dep. Agr. Publ.*) for 1969 and earlier years.

28. J. Platt, *Science* **166**, 1115 (1969).

29. E. C. Stakman, R. Bradfield, P. C. Mangelsdorf, *Campaigns against Hunger* (Belknap, Cambridge, Mass., 1967).

7   Our Vanishing Genetic Resources

Jack R. Harlan

All of the major food and fiber crops of the world are of ancient origin. The main sources of human nutrition today are contributed by such plants as wheat, rice, corn, sorghum, barley, potatoes, cassava, taro, yams, sweet potatoes, and grain legumes such as beans, soybeans, peanuts, peas, chickpeas, and so on. All of these plants were domesticated by Stone Age men some thousands of years ago and had become staples of the agricultural peoples of the world long before recorded history. We are not able to trace with certainty the genetic pathways that led to domestication, but we do know that these crops evolved for a long time under the guidance of man living in a subsistence agricultural economy. In the process of evolution, the domesticated forms often became strikingly different from their wild progenitors and generated enormous reserves of genetic variability.

Darwin opened his book *On the Origin of Species* with a discussion of variability of plants and animals under domestication. Genetic variability is the raw stuff of evolution, and he was struck by the range of morphological variation found in domesticated forms in contrast to their wild relatives. We are all familiar with the enormous differences among such breeds of dogs as Pekingese, dachshund, beagle, bulldog, Afghan, and Great Dane and how far removed they are in appearance from either wolves or any other wild species that could have been progenitor to domestic dogs. Similar ranges of diversity are seen in chickens, pigeons, cats, cattle, horses, and so on. Domestic plants exhibit the same phenomenon, especially among species that have been cultivated for a very long time and that have wide distributions. Genetic diversity is essential for evolution in nature and is, obviously, equally necessary for improvement by plant breeding.

Crop evolution through the millennia was shaped by complex interactions involving natural and artificial selection pressures and the alternate isolation of stocks followed by migrations and seed exchanges that brought the stocks into new environments and that permitted new hybridizations and recombinations of characteristics. Subsistence farmers of what we often call "primitive" agricultural societies have an intimate knowledge of their crops and a keen eye for variation. Artificial selection is often very intense, for the only forms to survive are those that man chooses to plant. The end products that emerged in primitive agricultural systems were variable, integrated, adapted populations called land races.

While land race populations are variable, diversity is far from random. They consist of mixtures of genotypes or genetic lines all of which are reasonably well adapted to the region in which they evolved but which differ in detail as to specific adaptations to particular conditions within the environment. They differ in reaction to diseases and pests, some lines being resistant or tolerant to certain races of pathogens and some to other races. This is a fairly effective defense against serious epiphytotics. Some components of the population are suceptible to prevalent pathogenic races, but not all, and no particular race of pathogen is likely to build up to epiphytotic proportions because there are always resistant plants in the population. Land races tend to be rather low yielding but dependable. They are adapted to the rather crude land preparation, seeding, weeding, and harvesting procedures of traditional agriculture. They are also adapted to low soil fertility; they are not very demanding, partly because they do not produce very much.

Land races have a certain genetic integrity. They are recognizable morphologically; farmers have names for them and different land races are understood to differ in adaptation to soil type, time of seeding, date of maturity, height, nutritive value, use, and other properties. Most important, they are genetically diverse. Such balanced populations—variable, in equilibrium with both environment and pathogens, and genetically dynamic—are our heritage from past generations of cultivators. They are the result of millennia of natural and artificial selections and are the basic resources upon which future plant breeding must depend.

In addition to variable land race populations, traditional agriculture generated enormous diversity in identifiable geographic regions called "centers of diversity" or "gene centers." Such centers are (or were) found on every continent, except

Australia where the native people did not cultivate plants. Wherever they are located they are always characterized by (i) very ancient agriculture, (ii) great ecological diversity (usually mountainous regions), and (iii) great human diversity in the sense of numerous culturally distinct tribes with complex interacting histories. Centers of diversity were first recognized and described by the great Russian agronomist and geneticist N. I. Vavilov in the 1920's and 1930's (1).

H. V. Harlan and M. L. Martini, concerned with genetic resources of barley, put it this way as early as 1936 (2):

> In the great laboratory of Asia, Europe, and Africa, unguided barley breeding has been going on for thousands of years. Types without number have arisen over an enormous area. The better ones have survived. Many of the surviving types are old. Spikes from Egyptian ruins can often be matched with ones still growing in the basins along the Nile. The Egypt of the Pyramids, however, is probably recent in the history of barley. In the hinterlands of Asia there were probably barley fields when man was young. The progenies of these fields with all their surviving variations constitute the world's priceless reservoir of germ plasm. It has waited through long centuries. Unfortunately, from the breeder's standpoint, it is now being imperiled. When new barleys replace those grown by the farmers of Ethiopia or Tibet, the world will have lost something irreplaceable.

That is the way it was before World War II. Genetic erosion was already well advanced in much of Europe, the United States, Canada, Japan, Australia, and New Zealand, where active plant breeding programs had been under way for some decades. But, the ancient reservoirs of germ plasm were still there in the more remote parts of the world and seemed to most people as inexhaustible as oil in Arabia. We could afford to squander our genetic resources because we never had much of our own, and we could always send collectors to such places as Turkey, Afghanistan, Ethiopia, India, Southeast Asia, China, Mexico, Colombia, and Peru and assemble all the diversity we could use. No one paid much attention to the prophetic warning of Harlan and Martini.

International Programs for Genetic Resource Conservation

After World War II, the picture began to change. Modern plant-breeding programs were established in many of the developing nations and often right in the midst of genetically rich centers of diversity. Some of the programs were successful, and new, uniform, high-yielding, modern varieties began to replace the old land races that had evolved over the millennia. The speed with which enormous crop diversity can be essentially wiped out is astonishing, and the slowness with which people have reacted to salvage of threatened genetic resources is dismaying (3).

Cries of alarm began to be sounded on the international scene about 15 years ago. A short chronology of events and actions associated with the Food and Agriculture Organization (FAO) of the United Nations is presented below.

1961. FAO convened a technical meeting on plant exploration and introduction. Among the recommendations was one to the effect that a panel of experts be appointed "to assist and advise the Director of the Plant Production and Protection Division in this field."

1962. A proposal for a Crop Research and Introduction Centre, Izmir (Turkey), was submitted to the United Nations Special Fund.

1963. The twelfth session of the FAO conference also recommended the establishment of a Panel of Experts on Plant Exploration and Introduction to advise FAO on these matters.

1964. The Crop Research and Introduction Centre, Izmir, became operative with U.N. Special Fund support. The Centre has collected, stored, and distributed germ plasm and now, under support of the Swedish government, is serving as a regional center for the Near East.

1965. The panel of experts was appointed.

1967. FAO and the International Biological Program (IBP) jointly sponsored a Technical Conference on Exploration, Utilization and Conservation of Plant Genetic Resources (4).

1968. A Crop Ecology and Genetic Resources Unit was established in the Plant Production and Protection Division, FAO.

1971. The Consultative Group on International Agricultural Research (CGIAR) was established under joint sponsorship of the World Bank, FAO, and U.N. Development Program (UNDP). Members include governments, private foundations, and regional development banks, and money is generated to support international agricultural research programs and institutes.

1971. A Technical Advisory Committee (TAC) was established to assist the CGIAR.

1972. Under joint sponsorship of TAC, FAO, and CGIAR a meeting was convened at Beltsville, Maryland, and a plan for a global network of Genetic Resources Centres was drawn up. Recommendations for location and funding were made and suggestions for international organization and coordination submitted to CGIAR through TAC.

1972. The U.N. Stockholm Conference on the Human Environment called for action on genetic resource conservation.

1973. A second FAO/IBP technical conference was convened in Rome (5).

1973. The CGIAR established a subcommittee on genetic resources.

1973. The International Board for Plant Genetic Resources (IBPGR) was established with a secretariat in FAO and financial resources provided by CGIAR, as recommended at the Beltsville meeting.

1974. Portions of the global strategy devised at Beltsville began to be funded through bilateral agreements with donor governments; for example, Sweden agreed to support the Izmir Centre for a time, and the Federal Republic of Germany agreed to support genetic resources centers in Ethiopia and Costa Rica. Other similar agreements have been or are being arranged.

Within the FAO structure, rather parallel developments took place with respect to forest genetic resources. Reports of technical conferences and meetings of the panel of experts, *Plant Introduction Newletter*, and *Forest Genetic Resources Information* are published by FAO.

It must be admitted that for all the organizational developments, and despite repeated and urgent pleas by the panel of experts, remarkably little collecting has been done to date. The Izmir Centre has been plagued with political, financial, administrative, and personnel problems from the start. It has managed to assemble a modest collection of some 10,000 accessions, and the long-term storage facilities now installed are excellent. The conception of the Izmir Centre is sound, and it is to be hoped that it will eventually perform the function for which it was established. FAO has conducted a few collecting expeditions and has given support to more, but the urgency of the situation demands much more vigorous action than has been generated so far.

The next few years, however, should show an increase in plant exploration. Funds should be available from the consultative group to support adequate exploration programs. For some regions it will probably be too late to salvage much.

It must also be admitted that much less would have been achieved without the dogged and determined insistence of Sir Otto Frankel of Australia (*4*). Through the years he has refused to abandon hope that serious action could, one day, be launched through an international cooperative program, and he has shaped most of the events described above.

Meanwhile, the international institutes, supported largely by CGIAR, have fared somewhat better. They each deal with one or a few crops and have usually understood that a part of the mission was to assemble and preserve germ plasm of the crops being developed. The world maize collection, for example, traces back to early international agricultural research sponsored by the Rockefeller Foundation in Mexico, Colombia, and elsewhere. A rather systematic effort was made to collect the races of maize, country by country, throughout Latin America. A major portion of the collection is maintained by the Centro Internacional de Mejoramento de Maíz y Trigo (CIMMYT) in Mexico, the Andean collection by the Instituto Colombiano Agrapucuerio (ICA) in Colombia, and the eastern South American collection is maintained at Piracicaba, Brazil. The maize collection appears to be in reasonably good shape, although some additional exploration is desirable.

The world rice collection has been growing rapidly in recent years through activities of the International Rice Research Institute (IRRI) in the Philippines. It is certainly not complete, but it is far better than it was 3 to 4 years ago. The Centro Internacional de Agricultura Tropical (CIAT) in Colombia is assembling cassava and beans. The International Institute of Tropical Agriculture (IITA) in Nigeria has been collecting cowpeas, pigeon peas, yams and other tropical tuber crops, and tropical vegetable species. The International Crops Research Institute for the Semi-Arid Tropics (ICRISAT), Hyderabad, India, has assumed responsiblity for world collections of sorghum, millets, chickpeas, and pigeon peas. The Centro Internacional de Papas (CIP) in Peru is starting to assemble potatoes for breeding work. All of these institutes are located in the tropics and should be able to maintain and rejuvenate collections of these crops much more efficiently than can be done in temperate countries.

National Programs: United States

The agriculture of the United States is an imported agriculture. Even crops domesticated by the American Indians—such as corn, potatoes, peanuts, cotton, tomatoes, and so on—originated in Latin America outside of the United States, and were introduced, some by Indians and some by Europeans. Because of our dependence on exotic germ plasm, the national government has sponsored collections and introductions from the beginning. As early as 1819, the Secretary of the Treasury issued a circular requesting Americans serving as consuls to send useful plant materials back to the United States. Formal plant exploration was conducted by the Office of the Patent Commissioner before 1862, when the U.S. Department of Agriculture (USDA) was created. In 1898 a Section of Seed and Plant Introduction was established in the USDA; and ever since, through various name changes and reorganizations, some unit within the department has been charged with responsibility for germ plasm assembly and maintenance (6).

A considerable impetus was given the plant introduction program by the Research and Marketing Act of 1946. Regional Plant Introduction Stations were established in the four administrative regions of the country. An Inter-Regional Potato Project was established in 1949 with a special station at Sturgeon Bay, Wisconsin, where exotic potato germ plasm could be grown and evaluated. A National Seed Storage Laboratory was built in Fort Collins, Colorado, and began operation in 1958. The primary objective of the laboratory is long-term seed storage, although research on the physiology of germination, dormancy, and longevity of seeds is also conducted.

Nearly 400,000 accessions have been introduced since 1898, but there has been substantial attrition over the years. The importance of maintenance was not at first generally realized, and much material was lost for one reason or another. Nevertheless, the present holdings of the USDA are very considerable and extremely important. The small grains (wheat, barley, oats, and rye) collection, for example, consists of more than 60,000 items, many of which could not possibly be replaced because they have disappeared from their original homelands. Substantial "world collections" of the major crops and many of the minor ones are being maintained at the Regional Plant Introduction Stations or through cooperative arrangements with other state and federal stations.

It would be nice to think that all the genetic diversity we will ever need is safely stored away in gene banks for future use. Unfortunately this is hardly the case. Some of our collections are large even when the numerous duplicates are accounted for, but none is really complete, and sources of diversity are drying up all over the world. We are particularly deficient in the wild and weedy relatives of our more important crops, and some geographic regions have been very poorly sampled. While the USDA has sponsored plant introduction work from the beginning, it has never been able to obtain enough support to systematically sample the world's germ plasm. The National Seed Storage Laboratory has received stepchild treatment with no increase in the operating budget for more than 15 years after establishment.

The southern corn leaf blight epidemic in 1970 aroused some activity in the area of crop vulnerability. A survey was commissioned by the National Academy of Sciences, resulting in a report on genetic vulnerability (7). It was found, not surprisingly, that not only corn but also every major crop we grow has a very narrow genetic base. The entire soybean industry, for example, traces back to six introductions from the same part of China. The leaf blight epidemic of 1970 came about because most of the hybrids produced had a common cytoplasm which conferred susceptibility to a particular race of the pathogen. We are just as vulnerable in sorghum where a cytoplasmic sterile system is used to produce hybrids. A crop-by-crop analysis reveals an extremely risky dependence on narrow genetic bases.

More than this, the number of crops we grow has been declining steadily. More and more people are being fed on fewer and fewer crops and these are becoming increasingly uniform, genetically.

After a series of meetings in Washington, an ad hoc committee drew up recommendations and presented them to the Agricultural Research Policy Advisory Committee (ARPAC) of the Agricultural Research Service. Among the recommendations was the establishment of a Genetic Resources Board at the national level which would, among other things, devise a national plan and program for systematic assembly, maintenance, evaluation, and utilization of plant genetic resources. It is to be hoped that a more systematic, coordinated, and effective program of genetic resource management can be generated for the country and that adequate financial support can be found. Approval for the board was obtained in January 1975.

National Programs: Other Countries

The U.S.S.R. probably has holdings of about the same magnitude as ours. No doubt, there is a good deal of duplication, yet they have arrays of collections that we do not have and we have materials they do not have. It would undoubtedly be of great mutual benefit if we could exchange collections and hold a complete set of duplicates in two different parts of the world. It would be a disaster if something should happen to either collection. Duplicate storage would be much safer.

National collections can be vulnerable. There is a heroic tale about the siege of Leningrad during World War II. People were dying of cold and starvation, reduced to eating rats, cats, dogs, dried glue from furniture joints and wall paper, or anything else that might prolong life. All this time, truckloads of edible seeds were in storage at the All-Union Institute of Plant Industry. The seeds were too precious to be sacrificed even at the cost of human life, and the collections survived for future use. We may pray that such a threat will never occur again, but prayer may not be adequate to save priceless genetic resources.

The Vavilovian emphasis on plant genetic resources persisted despite the long twilight of genetics under the political influence of T. D. Lysenko and Vavilov's tragic death as a result (8). The institute, which he directed for 20 years (1920–1940), was renamed the N. I. Vavilov All-Union Institute of Plant Industry (VIR) in 1968, just in time for the 75th anniversary of the organization in 1969 (9).

There may be some question as to how well theoriginal collections of the Vavilovian era have been maintained with respect to genetic authenticity, but there is no doubt that Soviet scientists are more collection minded than plant scientists elsewhere. Genetic resource management has been emphasized since 1920 and has become an integral part of the national agricultural development program. No doubt there are genes in Soviet collections that no longer exist anywhere else.

The Japanese, under the stimulus of H. Kihara, have also had strong genetic conservation programs, especially with certain crops. Expeditions have been sent to several centers of diversity over the decades and a national seed storage facility has been established at Hiratsuka. The University of Kyoto and the National Institute of Genetics, Mishima, have been especially active, although others have also participated.

Genetic resources centers with cold storage for long-term conservation have been established in a number of other countries. Some of the major ones include Brisbane, Australia; Prague, Czechoslovakia; Copenhagen, Denmark; Gatersleben, German Democratic Republic; Braunschweig-Volkenröde, German Federal Republic; New Delhi, India; Bari, Italy; Wageningen, Netherlands; and Warsaw, Poland. Others are being constructed or present facilities are being upgraded. Substantial holdings are being maintained in the United Kingdom, France, Sweden, Canada, and elsewhere. The necessity for genetic conservation is gradually being accepted throughout the world, but the urgency of salvage collection operations has yet to be generally appreciated.

A recent visit by a Plant Studies Delegation to the People's Republic of China revealed a somewhat ambiguous situation. The following observations may be pertinent. (i) China is, indeed, very rich in genetic diversity for many crops; (ii) Chinese scientists are not collection minded, and little effort is being made to conserve land races as they are replaced by modern varieties; (iii) the trend, at the moment, is to produce many species and varieties of fruits and vegetables, which tends to maintain diversity; and (iv) there is a strong emphasis on local self-sufficiency with respect to seed production at both the people's commune and production brigade levels which may tend to maintain variability at the national level. Overall, the picture is discouraging with respect to major crops. Two rice collections are being maintained, one for *japonica* and one for *indica* rices, but the ancient kaoliangs are disappearing from the Chinese sorghum belt, and the traditional millets are hanging on primarily in marginal dryland zones.

Altogether, a good deal has been done to collect genetic resources, and tentative, if unsystematic, steps have been taken to conserve much of it on a long-term basis. In view of the obvious limitations of our collections and in face of the current genetic "wipe out" of centers of diversity, it may be too little and too late. We continue to act as though we could always replenish our supplies of genetic diversity. Such is not the case. The time is approaching, and may not be far off, when essentially all the genetic resources of our major crops will be found either in the crops being grown in the field or in our gene banks. This will be a risky state of affairs and will demand a great deal more time and effort on genetic resource management than we have ever devoted to it in the past.

References and Notes

1. N. I. Vavilov, *Studies on the Origin of Cultivated Plants* (Institute of Applied Botany and Plant Breeding, Leningrad, 1926).
2. H. V. Harlan and M. L. Martini, *U.S. Department of Agriculture Yearbook of Agriculture, 1936* (Government Printing Office, Washington, D.C., 1936).
3. J. R. Harlan, *J. Environ. Qual.* 1, 212 (1972).
4. O. H. Frankel and E. Bennett, Eds., *Genetic Resources in Plants—Their Exploration and Conservation*, FAO/IBP (Blackwell, Oxford, 1970).
5. *Crop Genetic Resources for Today and Tomorrow* [J. G. Hawkes, Ed. (Cambridge Univ. Press, London, in press)] is an IBP synthesis volume and will include papers of the 1973 meeting.
6. Anonymous, *The National Program for Conservation of Crop Germ Plasm* (Univ. of Georgia Press, Athens, 1971).
7. *Genetic Vulnerability of Major Crops* (National Academy of Sciences, Washington, D.C., 1972).
8. Z. A. Medvedev, *The Rise and Fall of T. D. Lysenko*, I. M. Lerner, Transl. (Columbia Univ. Press, New York, 1969).
9. Anniversary volume of *Bulletin of Applied Botany, Genetics and Plant Breeding* (Leningrad, 1969), vol. 41, fasc. 1.

8   Plant Cell Cultures: Genetic Aspects of Crop Improvement

Peter S. Carlson and Joseph C. Polacco

Even the most optimistic agricultural experts foresee the need of quantum increases in agricultural production (*1*). Central among the obstacles that impede this goal of increased production is the real ceiling imposed by dwindling genetic variation in many crop species (*2*). Recent advances in plant cell and protoplast culture have drawn considerable attention because they form the basis of a novel technology to induce and recover agronomically desirable mutations, to make possible the rapid screening of naturally occurring variability, and to extend the range of plant hybridization beyond the bounds of sexual compatibility (*3, 4*). In this article, we shall attempt to evaluate current progress in this field and assess the potential for genetic applications of plant cell and protoplast culture both as techniques to expand the range of genetic variability and as tools for plant breeding.

The basic manipulations in vitro that are necessary for the genetic modification of plant cells are just now being developed, so that it is premature to judge the form of the final relationship between cell culture and crop improvement. As the disciplines of culture in vitro and plant breeding are juxtaposed in future research, new techniques will be developed that will permit greater utilization of cellular methods by the breeder. This article attempts to view the current relationships between

From *Science*, **188** (May 9, 1975), 622–625. Copyright 1975 by the American Association for the Advancement of Science. Reprinted with permission of the author and the publisher.

these two fields. We expect that the results and developments of future years will provide a more complete perspective. At present, the promise of plant cell culture technology appears to lie in two broad areas that are related to our extending the range of genetic variability—that is, parasexual hybridization and mutant selection.

Cell culture techniques offer a vehicle for the application of the methods of molecular genetics to higher plants in those few systems where the necessary manipulations in vitro have been defined (for example, tobacco, *Nicotiana tabacum*, and carrot, *Daucus carota*) (5). By employing single haploid or diploid cells as experimental organisms, it is possible to utilize procedures developed with microbial organisms to analyze and modify higher plants. It is now possible to grow large, relatively homogeneous populations of higher plant cells in a short period. This is accomplished, after inducing the cells to divide in a rapid unorganized fashion, by manipulating hormone levels in the culture medium. The resulting proliferation, termed a callus, can be dispersed and grown in liquid medium to generate a suspension culture consisting of single cells or small clumps of cells. Cells can be grown in chemically defined media, and defined mutants can be recovered by imposing chemical selection screens.

By enzymatically removing the walls from plant cells one can obtain large populations of protoplasts. These protoplasts can be induced to fuse with one another, and by this technique a number of interspecific and intergeneric heterokaryons and synkaryons have been produced. In some systems it is possible to stimulate the regeneration of single cells and of protoplasts into entire plants. This is normally accomplished by manipulating hormone levels in the culture medium.

In general, plantlets develop from undifferentiated cells along one of two pathways. With some species, cells in suspension culture can be stimulated to undergo organization directly into a plantlet through stages which resemble normal embryogenesis. With other species, cells from suspension cultures are allowed to proliferate prior to the application of the hormonal stimuli for shoot and root regeneration; hence, organ differentiation occurs within a large cell mass. Since whole plants can be recovered from cultured cells, a standard Mendelian analysis can be performed upon variants recovered in vitro. Because of their properties, plant cells can be manipulated as microbes; the traits they acquire in culture can also be analyzed genetically and physiologically in the whole plant.

Components of Cellular Technology

The specific manipulations in vitro that are of importance to the plant breeder can be summarized as follows.

CELLULAR CLONING AND THE RAPID PROPAGATION OF GENOTYPES. In numerous species of higher plants, large, genetically uniform populations of individuals can be produced from daughter cells ultimately derived from a single cell (6). Such cloning, which could prove important for increasing a superior genotype, can be accomplished with only a minimum of technology. Rapid cellular proliferation followed by plant regeneration from this tissue are the two required manipulations for inducing cloning in vitro. These two manipulations are generally routine in a wide variety of crop species, and have been accomplished even when regeneration from long-term undifferentiated cell cultures has proved impossible. Cellular cloning is currently being utilized in several crop species (7).

INDUCTION OF HAPLOID TISSUE FROM ANTHER AND POLLEN CULTURE. Haploid plantlets and callus tissue can be obtained in a number of genera by the techniques of anther and pollen culture (8). Within the immature pollen of cultured anthers, the vegetative cell can undergo division to generate an embryoid or an undifferentiated callus mass with the gametic number of chromosomes. Haploid cell lines are of importance because they permit direct selection for mutant phenotypes, and because they can readily be induced to form diploid tissue that yields isogenic lines. Anther-derived haploid plantlets are currently being used in only a few breeding programs (9) because they have not been routinely and widely produced among agronomically important crops (8).

EXTENDING THE RANGE OF GENETIC VARIABILITY BY WAY OF INDUCED MUTATIONS. Defined mutant types can be generated and recovered from cells of higher plants cultured in vitro (3). Cell cultures offer a microbial-like genetic system because it is possible to examine large numbers of individuals under known conditions, and it is often possible to work with haploid cell lines. Furthermore, individual plant cells have limited metabolic reserves, and their ability to grow on a completely defined, simple media allows selection for a wide range of variants. A number of mutant selection techniques, taken directly from microbial work, have been successfully utilized with plant cells. The direct application of microbial mutant selective procedures to higher plant cells may not be satisfactory for the plant breeder's needs. Microbial mutants have been recovered in order to analyze the biochemical and genetic organization of these cells. The aim of the plant breeder is not so much to analyze processes as to utilize them in maximizing yields. Novel selective schemes will be needed to accomplish this objective.

EXTENDING THE RANGE OF GENETIC VARIABILITY BY WAY OF SOMATIC HYBRIDIZATION AND PROTOPLAST MANIPULATION. A hybridization breeding program in vitro offers the possibility of producing hybrid plants containing genetic information from sexually incompatible species. Hybrid individuals can be produced by utilizing methods that do not involve standard sexual mechanisms (10). The isolation of viable protoplasts that undergo proliferation by cell division has now been reported in a number of plant genera (11). Furthermore, the fusion of protoplasts from different plant taxa can routinely be induced to occur with a high frequency (12). However, the rapid selection of hybrid clones in vitro awaits the development of adequate selective markers for the recognition of hybrid cells. Regeneration of hybrid cells into entire plants also presents a major problem. Protoplasts can be stimulated to take up macromolecules, viruses, organelles, and whole bacteria (3, 4). This property, when more completely characterized, may permit the introduction of large blocks of genetic information.

TRANSFER OF SPECIFIC GENETIC INFORMATION. The transformation of genetic information is a common means of genetically modifying bacteria. If this process could be routinely accomplished in higher plants, it might well provide an important tool for the breeder. Although there are claims for the occurrence of transformation-like events in higher plants (13), the literature is not clear and no reproducible systems are known that would enable one to examine this phenomenon.

CURRENT TECHNICAL LIMITATIONS. At present, the techniques we have described can be utilized with only a few species of higher plants, and in no species are these techniques routine procedures. The most obvious limitation to the manipulation of plant cells in vitro is our inability to achieve the regeneration of whole plants from cell cultures of cultivars of the important crop species. While there is no reason to suspect that this problem is more than technical, genetic variability induced in vitro cannot be utilized by the breeder unless whole plants (and gametes) can be recovered. An equally severe limitation of cellular technology stems from our limited knowledge of plant physiological and biochemical processes. The recovery of plant cells with genetic variation in vitro is dependent upon our being able to recognize distinct cellular phenomena. By further research we should be able to gain some insight into the molecular and cellular mechanisms underlying agronomically important traits. The physiological components of these whole-plant characters must then be duplicated in vitro. Selection schemes for recovering variants in processes unique to higher plants must also be developed. These are only several of the problems that limit the applicability of cell culture techniques to plant breeding. They serve to illustrate the need for further research that will enable us to more adequately define the role that cellular techniques will play in plant breeding practices.

Do These Techniques Offer Unique Solutions to Agronomic Problems?

There are a number of superficial situations in which an attempt to duplicate a natural stress in vitro might result in a cell culture yielding an agronomically interesting variant. Populations of mutagenized cells or of hybrid cells could be exposed to the agent of stress, and survivors could be recovered as potentially interesting individuals. Relatively straightforward selective procedures might be designed to recover, for example, cells tolerant to toxic amounts of specific ions, including hydrogen ions; to pollutants or herbicides; or to extremes of temperature or water stress; or to recover cells that can better utilize available nutrients. Similar schemes might be developed for recovering variants that are resistant to certain diseases or insensitive to pathogens, or that display an improved nutritional quality. It is not yet clear whether such variants would be expressed as positive agronomic changes in the whole plants.

We describe below two examples of situations in which cell cultures might be particularly useful for plant breeders. We think that many other examples of the applicability of cell culture techniques to a wide range of agronomic problems could be found. They represent a different perspective and approach from that normally taken by a plant breeder.

Nutritional Improvement

The protein quality of plant seeds is not of optimal nutritive value for humans. A high proportion of seed protein has been characterized as storage protein and is devoid of any known enzymatic activity. In general, cereal grains are nutritionally poor in lysine, threonine, or tryptophan (14). Soybeans and the other legumes are nutritionally limiting in methionine (14).

Plant breeders following traditional methodologies are actively engaged in selecting grain and legume varieties with more nutritionally balanced seed protein. The opaque-2 and floury-2 mutants of maize, first named for their altered macroscopic endosperm properties, were found to contain greatly enhanced amounts of protein-bound lysine (15) as a result of alterations in the relative amounts of lysine-poor and lysine-rich endosperm protein (16). Several workers have examined seeds from 9000 lines of sorghum for floury endosperm. Of the 62 lines selected, two exhibited, like opaque-2 corn, a 60 percent increase in seed lysine content (17). A barley line, Hiproly, with a 20 to 30 percent increase in lysine content, was selected from 1000 entries in the world barley collection by a specific lysine-binding technique (18). Other barley lines with a mutagen-induced high lysine content have been isolated but, unlike the Hiproly line, these are usually associated with low crop yields (19). The points to be made here are that nutritionally beneficial changes in the quality of seed protein have been found through laborious examination of many lines, that selections based on endosperm morphology or dye-binding (19) may not select for all beneficial mutants or varieties, and that the numbers of organisms examined are three or four orders of magnitude less than those examined in microbial mutant selection.

Increasing the production of specific amino acids is commonplace in microbiology and is generally accomplished by selecting for mutants resistant to analogs of the given amino acid. This approach has been extended to plant cells, and mutants that display increased endogenous levels of free tryptophan, lysine, and methionine have been recovered (20).

There are several problems with this approach, the first one being that one cannot assume that cells that are resistant to the amino acid analog can be regenerated into whole plants. Another problem is that it cannot be assumed a priori that increased amino acid levels will occur in the edible portions (grain, fruit, tuber, leaf) of the regenerated plant. In any case, increasing free amino acids is probably less desirable than increasing tissue-specific levels of the protein-bound amino acid. Free amino acids are more easily leached out in food preparation and represent only a small fraction of the total amino acids in a cell.

A possible approach to effect increases in specific protein-bound amino acids in grains is to elicit production of specific seed proteins in the culture in vitro. If differentiated states of the plant were better understood and could be manipulated, a perpetual differentiated endosperm could be established in culture. At our present level of expertise, however, it may still be possible to select regulatory mutants in the overproduction of a nutritious, naturally occurring seed protein.

77

Recent work has focused on ways to elevate levels of urease in soybean tissue culture (21). This enzyme is particularly rich in methionine, the nutritionally limiting amino acid in soybeans. Although it is a naturally occurring seed protein in a range of leguminous species, the levels of urease in soybean seeds are only 7 to 13 percent of those found in jack bean. Several general selective systems to recover variants have been established in vitro: (i) the utilization of urea in the presence of specific urease inhibitors; (ii) the utilization of urea in the presence of nonmetabolizable repressors of urease production; and (iii) the conversion of large amounts of urea to ammonia to overcome nitrate poisoning in cell cultures. Although no mutants have been positively identified, there appears to be much leeway in elevating the urease levels. Soybean tissue grows quite well with urea as the sole nitrogen source but under such conditions exhibits urease levels only 0.2 percent of those found in seed extracts.

The production of other nonenzymatic seed proteins, for example, zein, gluten, and glutelin, might be manipulated and analyzed in cultured cells if methods were designed for selecting and assaying the variants.

Net Photosynthetic Efficiency

Could a cell line from a crop plant be established in vitro as an efficient *Chlamydomonas*-like photoautotroph? If so, it would be suited for dissecting and genetically modifying the molecular events of photosynthesis and respiration. Modification of these two processes might be important for improving intrinsic plant productivity (22). Zelitch (22a) has reviewed several reported photoautotrophic cell culture systems. In none of these has the growth rate reached the maximum of cultures supplied with an exogenous carbon source. Selection for photosynthetically efficient cell lines in vitro could be masked by many trivial mutants. Such mutants could be tissue-culture-specific variants, including those with more disperse cell growth allowing better carbon dioxide diffusion, or those with altered optimum hormone requirements for autotrophic growth. Existing photosynthesizing cell lines will have to be altered genetically so that they achieve a higher baseline efficiency before they can be useful in obtaining mutants specific in the carbon-fixing and electron transport processes of photosynthesis.

Net photosynthesis is likely to increase when respiration (essentially the "unfixing" of CO_2) is made more efficient. Photorespiration, which occurs when photosynthetic tissues are illuminated, results from the oxidation of photosynthetic products; it is often more rapid than the rate of respiration in darkness. Evidence suggests that such species as corn and sugarcane are photosynthetically more efficient than other crop plants mainly because they have lower rates of photorespiration; in maize, the primary substrate of photorespiration, glycolic acid, is synthesized more slowly than in species with a rapid photorespiration (22). Inhibition of the synthesis or oxidation of glycolic acid results in increased CO_2 fixation in tobacco leaf disks. Because tobacco has a high rate of photorespiration and is amenable to tissue culture, it would appear ideally suited as a system for genetically modifying net photosynthesis through lowering its rates of photorespiration. A potential selection scheme might be one in which recovered cell lines could utilize glycolate as a carbon source. Such lines might exhibit lowered glycolate oxidase levels, or they might acquire the ability to convert glycolate to other metabolites (malate or glycine, for example). Tobacco cell cultures do not utilize glycolate, even under a variety of conditions. Carboxylic acids in general appear to be poor carbon sources for cultured tobacco cells. As with the selection of photosynthetically efficient mutants, the selection of glycolate-utilizing cell lines will probably not initially yield variants specifically altered in photorespiration, but rather in such functions as glycolate uptake. However, once the genetic and physiological ground-work is more completely defined, the genetic manipulation of glycolate metabolism can be expected.

Some of the respiration that occurs in darkness may also be wasteful. An alternate cyanide- (or antimycin A) insensitive oxidase pathway has been described in the fungi and in plant mitochondria (23). The phosphate to oxygen ratio of this pathway is probably only one, as opposed to three for the main terminal oxidases. The total respiration of spinach leaves is inhibited only 50 percent by cyanids. Likewise, the alternate oxidase pathway can account for more than 50 percent of the total respiration in tobacco suspension cultures. The alternate oxidase pathway can sustain cell viability for at least 24 hours in suspension cultures when the main cytochrome chain is blocked by antimycin A (21). Although a positive selection method for the elimination of the alternate oxidase pathway has been reported for the fungus *Ustilago maydis*, only a negative selection procedure has been devised to select for this mutation in higher plant cells (21). It is difficult to imagine how the plant breeder could select for a specific deficiency in the alternate oxidase pathway at the whole-plant level. Nevertheless, this problem illustrates how plant cell genetics may become increasingly applicable to crop production as the mechanisms of plant biochemistry and physiology are more fully characterized.

Once they have been more fully developed, approaches similar to these should enable plant breeders to obtain useful genetic variants that might be incorporated into crop varieties.

The Future of the Technology

The contributions of cell culture techniques to crop improvement can best be considered as part of an interdisciplinary effort incorporating the approaches of cellular and whole-plant biology. Because cell culture is only one of the many technologies that can be employed in a crop improvement program, the results it yields are dependent upon input analysis, and evaluation from a number of other disciplines. In a real sense, the usefulness of cellular techniques is dependent upon advances in many areas of the plant sciences.

The history of plant breeding contains instances where new techniques have fostered hopes of an easy method to produce the "ideal" plant type. These techniques have become, with time, the standard tools of plant breeders. We believe that cell culture techniques will also become routine tools in the difficult task of plant improvement. The true test of any developing agricultural technology is how its application affects plant characters under actual field conditions. Viewed from this perspective, cell culture techniques have not yet had the chance to prove themselves. They have not yet produced any novel genetic variants or combinations of genetic characters of agricultural importance.

In spite of the advances that have been made in the use of cell cultures, it does not appear likely that these techniques will be of major importance in increasing crop yields over the next decade. Most of the progress during this period must come from a further application of current plant breeding practices. We hope that advances in plant cell culture in the next decade will expand the relevance of this technology for plant breeders.

References and Notes

1. S. H. Wittwer, *BioScience* **24**, 216 (1974).
2. J. R. Harlan, in *Plant Breeding*, K. J. Frey, Ed. (Iowa State Univ. Press, Ames, 1966), p. 55.
3. R. S. Chaleff and P. S. Carlson, *Annu. Rev. Genet.* **8**, 267 (1974).
4. R. F. Heyn, A. Rörsch, R. A. Schilperourt, *Q. Rev. Biophys.* **7**, 35 (1974); Y. P. S. Bajaj, *Euphytica* **23**, 633 (1974).
5. H. E. Street, Ed., *Plant Tissue and Cell Culture* (Univ. of California Press, Berkeley, 1973), vol. 11; *Tissue Culture and Plant Science, 1974, Proceedings of the 3rd International Congress of Plant Tissue and Cell Culture*, Leicester, England (1974).
6. S. Reinert, in *Plant Tissue and Cell Culture*, H. E. Street, Ed. (Univ. of California Press, Berkeley, 1973), vol. 11, p. 338.
7. L. G. Nickell and D. J. Heinz, in *Genes, Enzymes and Populations*, A. M. Srb, Ed. (Plenum, New York, 1973), vol. 2, p. 109; T. Murashige, *Annu. Rev. Plant Physiol.* **25**, 135 (1974).
8. N. Sunderland, in *Plant Tissue and Cell Culture*, H. E. Street, Ed. (Univ. of California Press, Berkeley, 1973), vol. 11, p. 205.
9. G. B. Collins, P. D. Legg, M. J. Kasperhauer, *J. Hered.* **63**, 113 (1972).
10. P. S. Carlson, H. H. Smith, R. D. Dearing, *Proc. Natl. Acad. Sci. U.S.A.* **69**, 2292 (1972).
11. E. C. Cocking and P. K. Evans, in *Plant Tissue and Cell Culture*, H. E. Street, Ed. (Univ. of California Press, Berkeley, 1973), vol. 11, p. 100; E. C. Cocking, *Annu. Rev. Plant Physiol.* **23**, 29 (1972).
12. K. N. Kao, F. Constabel, M. R. Michayluk, O. L. Gamborg, *Planta (Berl.)* **120**, 215 (1974).
13. D. Hess, *Naturwissenschaften* **59**, 348 (1972).
14. D. M. Hegsted, in *Amino Acid Fortification of Protein Foods*, N. S. Scrimshaw and A. M. Altschui, Eds. (MIT Press, Cambridge, Mass., 1971).
15. E. T. Mertz, L. S. Bates, O. E. Nelson, *Science* **145**, 279 (1964).
16. O. E. Nelson, *Adv. Agron.* **21**, 171 (1969).
17. R. Singh and J. D. Axtell, *Crop Sci.* **13**, 535 (1973).
18. L. Munck, K. E. Karlson, A. Magberg, *Proceedings of the Symposium on Barley Genetics* (Washington State Univ. Press, Pullman, 1971), p. 544.

19. J. Ingversen, A. J. Andersen, M. Doll, B. Køie, in *Use of Nuclear Techniques for the Improvement of Seed Protein* (International Atomic Energy Agency, Vienna, 1970).
20. R. S. Chaleff and P. S. Carlson, *John Innes Symposium*, in press; J. M. Widholm, *Biochim. Biophys. Acta* **279**, 48 (1972).
21. J. C. Polacco, unpublished results.
22. I. Zelitch, *Photosynthesis, Photorespiration and Plant Productivity* (Academic Press, New York, 1971).
22a.———, *Science* **188**, 626 (1975).
23. A. M. Lambowitz, C. W. Slayman, C. L. Slayman, W. D. Bonner, Jr., *J. Biol. Chem.* **247**, 1536 (1972); J. T. Bahr and W. D. Bonner, Jr., *ibid.* **248**, 3441 (1972).
24. We thank Drs. T. Rice, P. Day, I. Zelitch, and S. Anagnostakas and Ms. J. Roberts for helpful comments and suggestions.

three

Genetic Aspects of Environmental Hazards

9. The Detection of Environmental Mutagens
James V. Neel and Arthur D. Bloom [1969]

10. A Combined Bacterial and Liver Test System for Detection and Classification of Carcinogens as Mutagens
Bruce N. Ames [1974]

11. Cytogenetic Studies with Cyclamate and Related Compounds
D. R. Stoltz, K. S. Khera, R. Bendall, and S. W. Gunner [1970]

12. Chromosome Studies on Patients (in Vivo) and Cells (in Vitro) Treated with Lysergic Acid Diethylamide
Margaret J. Corey, J. C. Andrews, M. Josephine McLeod, J. Ross MacLean, and W. E. Wilby [1970]

13. Chromosome Breakage in Humans Exposed to Methyl Mercury Through Fish Consumption
Staffan Skerfving, Kerstin Hansson, and Jan Lindsten [1970]

14. Vinyl Chloride Exposure and Human Chromosome Aberrations
Alan Ducatman, Kurt Hirschhorn, and Irving J. Selikoff [1975]

15. Delayed Radiation Effects in Atomic-Bomb Survivors
Robert W. Miller [1969]

The environment is known to be made genetically harmful (mutagenic) as a result of the action of chemicals and radiation. Both of these types of mutagens can cause physiological as well as genetic malfunctions. Although individuals may suffer and die from mercury poisoning, low levels of mercury that are not toxic may still cause mutation. This is shown by the report of Skerfving and coworkers in this section. (In 1974 Skerfving et al. reported on a larger study which substantiated their earlier results.)

The discovery in 1927 by H. J. Muller, and independently by L. J. Stadler, that X-irradiation causes mutations was the first indication that environmental agents under human control can increase the burden of genetic defects in human populations. Muller immediately pointed out that the indiscriminate use of radiation (in medicine, industry, etc.) was a potential health hazard.

Geneticists are now speaking up forcefully about the potential genetic hazards of a variety of environmental "additives." They increasingly announce their concern, undertake systematic studies, and advise reductions of these hazards. The readings in this section start off with a paper by Neel and Bloom that clearly outlines the major concerns and problems about environmentally induced mutagens. This paper is a valuable guide to those that follow, which give examples of various categories of known mutagenic hazards. The paper by Ames shows, by means of a new, elegant technique, that many cancer-causing chemicals are potent mutagens. More recent work by his group has shown that cigarette smoke and hair dyes also contain significant quantities of mutagens (Keir et al., 1974; Ames et al., 1975). The paper by Stoltz and coworkers shows that cyclamates are mutagenic; this artificial sweetener was banned in many countries in 1970, but pressure is still on to permit its use (Holden, 1974). The sweetener saccharin is also under fire; mutagen testing has given conflicting results (Kramers, 1975). Other papers in this section are on the famous psychoactive drug, LSD (Corey et al.), on vinyl chloride (Ducatman et al.), and on mercury (Skerfving et al.). Vinyl chloride is a gas used in producing plastics found in phonograph records, floor tiles, and many other consumer products. Vinyl chloride is mutagenic in Ames's test system (McCann et al., 1975). Mercury is an industrial raw material (paper mills, transistor batteries, dental fillings, house paints) and an industrial waste product, as in the burning of coal. The last paper in this section is on radiation effects in atomic-bomb survivors (Miller).

Within this century the mutagenic content of our environment has been multiplied by "progress" countless times. Through the rapid growth of our industrial society we may now be exposed to 500,000 alien substances, most of them unanalyzed with respect to physiological and genetic effects (Williamson, 1969). The acceptance of radiation and chemicals, including defoliants, as weapons of war only enlarges the problem (see, for example, Bennett, 1970, and Neidlands et al., 1972). Mutagenicity is thus a potential health hazard on a global scale.

Here we can consider the action on human beings of only a few environmental mutagens. No one knows which are the most detrimental nor, indeed, the mechanisms by which many cause genetic damage. For example, in the literature opinion is divided on whether or not LSD causes chromosome breaks in humans, but studies thus far have by no means established this substance as a source of genetic damage (Long, 1972). Another drug of obvious public interest, marijuana, is also in an ambivalent position, some studies suggesting mutagenicity (Maugh, 1974; Stenchever et al., 1974) and others emphatically ruling it out (Nichols et al., 1974; Rubin and Comitas, 1975). Furthermore, many chemicals that seem to be harmless to the genetic material are in fact mutagenic when they are metabolized within the body into other compounds. In addition, the fact that certain human viruses cause chromosome breaks (German, 1970) has not yet been tied into the general problem of environmental mutagens. Clearly, we are still ignorant about most of the genetic (and physiological) effects of our polluted world. Thus, the smattering of environmental hazards chosen for discussion here necessarily represents an arbitrary selection.

An important point is that new substances, or substances now in much wider use than formerly, should be thoroughly studied with respect to their possible genetic effects on man, bearing in mind that some of these effects may only appear in later generations. Old substances in wide use, such as caffeine, which have not been assayed systematically for biological ill effects, also merit immediate attention. It is much more important to discover low mutagenicity in a substance widely used than high mutagenicity in a substance used only in rare circumstances. Far more people can be affected by the former. Caffeine is a case in point: It may very well be a weak mutagen, despite conflicting results

(Challis and Bartlett, 1975; Kihlman, 1974). But it is a concern because it is consumed in such large quantities in coffee, tea, and colas. The same argument and concern applies to a variety of fumigants, medicines (Hulbert et al., 1974), pesticides, paint additives, plastic hardeners, and whatnot. "Progress" is not likely to be without its price.

Addendum on Statistical Tests

In some of the following papers the writers are trying to find out what effect some chemical (or other) treatment has on a group of individuals. To do this, they compare their experimental (treated) group with a control (untreated) group. Suppose that the treated group shows a higher level of mutations, or whatever is being measured, than the control group. The scientist then must decide in his own mind if the difference between the two groups is simply the result of chance events or is significant, that is, the result of the effects of the treatment. (Similarly, a blackjack card player has to decide if the dealer's "luck" in dealing himself five blackjacks in a row is caused by chance events or a "treatment.") The scientist makes up his mind by doing a statistical test rather than making an offhand guess. Time-honored statistical testing techniques are taught to all scientists so that there is a common standard for deciding if research results are significant. In these tests one calculates the odds, or probability (P), that, say, the average scores for two groups could be as different as they are because of chance alone. If the P value so calculated is 5 percent or less (only 1 chance in 20), then scientists conclude that the differences are indeed significant.

Bibliography

CAFFEINE

Brogger, A. Caffeine-induced enhancement of chromosome damage in human lymphocytes treated with methyl-methane-sulphonate, mitomycin C, and X-rays. *Mutation Res.*, **23**:353–360, 1974.

Challis, B. C., and C. D. Bartlett. Possible cocarcinogenic effects of coffee constituents. *Nature*, **254**:532–533, 1975.

Kihlman, B. A. Effects of caffeine on the genetic material. *Mutation Res.*, **26**:53–71, 1974. (A review.)

Weinstein, D., I. Mauer, M. L. Katz, and S. Kazmer. The effect of caffeine on chromosomes of human lymphocytes:

Non-random distribution of damage. *Mutation Res.*, **20**: 441–443, 1973.

CYCLAMATE AND SACCHARIN

Holden, C. FDA turns back bid to reinstate cyclamates. *Science*, **186**:422, 1974.

Kramers, P. G. The mutagenicity of saccharin. *Mutation Res.*, **32**:81–92, 1975.

LSD

Long, S. Y. Does LSD induce chromosomal damage and malformations? *Teratology*, **6**: 74–90, 1972.

Maugh, T. H. LSD and the drug culture: New evidence of hazard. *Science*, **179**:1221–1222, 1973.

Sram, R. J., and P. Goetz. The mutagenic effect of lysergic acid diethylamide. III. Evaluation of the genetic risk of LSD in man. *Mutation Res.*, **26**:523–528, 1974.

MARIJUANA

Goode, E. Effects of cannabis in another culture. *Science*, **189**:41–43, 1975. A review of *Ganja in Jamaica* by V. Rubin and L. Comitas.

Maugh, T. H. Marijuana: New support for immune and reproductive hazards. *Science*, **190**:865–867, 1975.

Nichols, W. W., R. C. Miller, W. Heneen, C. Bradt, L. Hollister, and S. Kanter. Cytogenetic studies on human subjects receiving marijuana. *Mutation Res.*, **26**:413–417, 1974.

Rubin, V., and L. Comitas. *Ganja in Jamaica*. Scotch Plains, N.J.: Mouton/MacFarland Publications, 1975.

Stenchever, M. A., T. J. Kunysz, and M. A. Allen. Chromosome breakage in users of marijuna. *Amer. J. Obstetrics Gynecology*, **118**:106–113, 1974.

MERCURY

Abelson, P. Methyl mercury. *Science*, **169**:237, 1970.

Dorozynski, A. Mediterranean poison fish forecast. *Nature*, **254**: 549–551, 1975. A mercury disaster is brewing in the sea.

Edwards, T., and B. C. McBride. Biosynthesis and degradation of methyl mercury in human feces. *Nature*, **253**:462–464, 1975.

Grant, N. Mercury in man. *Environment*, **13**:2–15, 1971.

Jagiello, G., and J. S. Lin. An assessment of the effects of mercury on the meiosis of mouse ova. *Mutation Res.*, **17**:93–99, 1973.

Ramel, C. The mercury problem—A trigger for environmental pollution control. *Mutation Res.*, **26**:341–348, 1974.

Skerfving, S., K. Hansson, C. Mangs, J. Lindsten, and N. Ryman. Methyl-mercury-induced chromosome damage in man. *Environmental Res.*, **7**:83–98, 1974.

NITRITE

Valencia, P., S. Abrahamson, P. Wagoner, and L. Mansfield. Testing for food additive-induced mutations in *Drosophila melanogaster*. *Mutation Res.*, **21**:240–241, 1973.

Wolff, I. A., and A. E. Wasserman. Nitrates, nitrites, and nitrosamines. *Science*, **177**:15–19, 1972. These compounds pose hazards for cancer and mutation. See also report of " Committee 17 " listed under " General."

PESTICIDES, HERBICIDES, AND CBW

Bennett, I. L. The significance of chemical and biological warfare for the people. *Proc. Nat. Acad. Sci.*, **65**:271–279, 1970.

Bridges, B. A. The mutagenicity of captan and related fungicides. *Mutation Res.*, **32**:3–34, 1975. Concludes that captan might be hazardous to man, but further study is needed.

Epstein, S. S., and M. S. Legator. The mutagenicity of pesticides: Concepts and evaluation. *Heredity*, **29**:111, 1972.

Heden, C. G. Defenses against biological warfare. *Annual Rev. Microbiol.*, **21**:639–676, 1967.

Kelly-Garvert, F., and M. S. Legator. Cytogenetic and mutagenic effects of DDT and DDE in a Chinese hamster cell line. *Mutation Res.*, **17**:223–229, 1973.

Lederberg, J. Biological warfare: A global threat. *Amer. Scientist*, **59**: 195–197, 1971.

Niedlands, J. B., G. H. Orians, E. W. Pfeiffer, A. Vennema, and A. H. Westing. *Harvest of Death: Chemical Warfare in Vietnam and Cambodia*. New York: Free Press, 1972.

Siebert, D., and E. Lemperle. Genetic effects of herbicides: Induction of mitotic gene conversion in *Saccharomyces cerevisiae*. *Mutation Res.*, **22**:111–120, 1974.

VINYL CHLORIDE

Edsall, J. T. Scientific freedom and responsibility. *Science*, **188**:687–693, 1975. Cites suppression of data on carcinogenicity of vinyl chloride.

Kramer, B. Vinyl chloride scare points up dangers of other chemicals. *Wall St. Journal*, Oct. 7, 1974.

McCann, J., V. Simmon, D. Streitwiser, and B. N. Ames. Mutagenicity of chloroacetaldehyde, chloroethanol, vinyl, chloride, and cyclophosphamide. *Proc. Nat. Acad. Sci.*, **72**:3190–3193, 1975.

MISCELLANEOUS CHEMICALS

Ames, B. N., H. O. Kammen, and E. Yamasaki. Hair dyes are mutagenic: Identification of a variety of mutagenic ingredients. *Proc. Nat. Acad. Sci.*, **72**:2423–2427, 1975.

Badr, F. M., and R. S. Badr. Induction of dominant lethal mutation in male mice by ethyl alcohol. *Nature*, **253**: 134–136, 1975.

Clive, D. Mutagenicity of thioxanthenes (hycanthone, lucanthone, and four indazole derivatives) at the TK locus in cultured mammalian cells. *Mutation Res.*, **26**:307–318, 1974.

deSerres, F. J. AF-2: Food preservative or genetic hazard? *Mutation Res.*, **26**:1–2, 1974. AF-2, widely used in Japanese foods until recently, is strongly mutagenic *in vitro*.

Green, S., F. M. Sauro, and M. S. Legator. Cytogenetic effects of hycanthone in the rat. *Mutation Res.*, **17**:239–244, 1973.

Hulbert, P. B., E. Bueding, and P. E. Hartman. Hycanthone analogs: Dissociation of mutagenic from antischistosomal effects. *Science*, **186**:647–648, 1974. This drug is mutagenic and carcinogenic, but related drugs are not.

Keir, L. D., E. Yamasaki, and B. N. Ames. Detection of mutagenic activity in cigarette smoke condensates. *Proc. Nat. Acad. Sci.*, **71**: 4159–4163, 1974.

Kermode, G. Food additives. *Scientific American*, **226**: 15–22, 1972.

Legator, M. S., T. H. Connor, and M. Stoeckel. Detection of mutagenic activity of metronidazole and niridazole in body fluids of humans and mice. *Science*, **188**:1118–1119, 1975. These two medical drugs are in wide use.

McCann, J., N. E. Spingarn, J. Kobori, and B. N. Ames. Detection of carcinogens as mutagens: Bacterial tester strains with R factor plasmids. *Proc. Nat. Acad. Sci.*, **72**:979–983, 1975. See also letters, *Science*, **191**:241–244, 1976.

Mauer, I., D. Weinstein, and H. M. Solomon. Acetylsalicylic acid: No chromosome damage in human leucocytes. *Science*, **169**:198–201, 1970. Aspirin caused no damage in this test.

Riehm, H., and E. Schleiermacher. Genetic counseling of children with acute leukemia cured by intensive chemotherapy. *Mutation Res.*, **21**:179, 1973. Leukemia-cured children have had long-term exposure to mutagenic chemicals; the genetic hazard of this bears watching.

Shapley, D. Nitrosamines: Scientists on trail of prime suspect in urban cancer. *Science*, **191**:268–270, 1976.

Sincock, A., and M. Seabright. Induction of chromosome

changes in Chinese hamster cells by exposure to asbestos fibers. *Nature*, **257**:56–58, 1975.

Tazima, Y. Naturally occurring mutagens of biological origin: A review. *Mutation Res.*, **26**:225–234, 1974. A variety of plants and molds produce mutagenic substances.

RADIATION

Advisory Committee on the Biological Effects of Ionizing Radiations [BEIR report]. *The Effects on Populations of Exposure to Low Levels of Ionizing Radiation.* Washington, D.C.: National Academy of Sciences–National Research Council, 1972.

Barcinski, M., M. Abreu, J. de Almeida, J. Naya, L. Fonseca, and L. Castro, Cytogenetic investigation in a Brazilian population living in an area of high natural radioactivity. *Am. J. Hum. Genet.*, **27**:802–806, 1975.

Blot, W. J., and R. W. Miller. Mental retardation following in utero exposure to the atomic bombs of Hiroshima and Nagasaki. *Radiology*, **106**:617–619, 1973.

Boffey, P. M. Hiroshima/Nagasaki. *Science*, **168**:679–683, 1970.

Boffey, P. M. Nuclear war: Federation disputes Academy on how bad effects would be. *Science*, **190**:248–250 and 640, 1975.

Brewen, J. G., and R. J. Preston. Cytogenetic effects of environmental mutagens in mammalian cells and the extrapolation to man. *Mutation Res.*, **26**:297–305, 1974.

Brewen, J. G., R. J. Preston, and N. Gengozian. Analysis of X-ray induced chromosomal translocations in human and marmoset spermatogonial stem cells. *Nature*, **253**:468–470, 1975.

Edsall, J. T. Scientific freedom and responsibility. *Science*, **188**:687–693, 1975. Supports the Gofman-Tamplin argument (see Gofman citation) that prior to the BEIR report levels of radiation exposure permitted in the United States were unreasonably high.

Gofman, J. W., and A. R. Tamplin. Radiation: The invisible casualties. *Environment*, **12**:12–19 and 49, Apr. 1970.

Holden, C. Radiation standards: The last word or at least a definitive one. *Science*, **178**:966–967, 1972. Report on the BEIR report.

Lapp, R. E. Radiation risks. *New Republic*, Feb. 27, 1971, pp. 17–21. An analysis of Gofman–Tamplin.

Miller, R. W. Late radiation effects: Status and needs of epidemiologic research. *Environ. Res.*, **8**:221–233, 1974.

Morgan, K. Z. Never do harm. *Environment*, **13**:28–38, Jan.–Feb. 1971. Reports that the abundant exposure of humans to X-rays for medical reasons is dangerous and exposure should be greatly reduced.

Nader, C. The dispute over safe uses of X-rays in medical practice. *Health Physics*, **29**:181–206, 1975.

Rose, D. J. Nuclear eclectic power. *Science*, **184**:351–359, 1974. Lists estimates of genetic damage from nuclear power.

Rugh, R. Radiology and the human embryo and fetus. In *Medical Radiation Biology*, D. Whymple et al., eds. Philadelphia: Saunders, 1973, pp. 83–96.

Sagan, L. A. Human costs of nuclear power. *Science*, **177**: 487–493, 1972.

GENERAL

Committee 17. Environmental mutagenic hazards. *Science*, **187**: 503–514, 1975. (A review.)

Drake, J. W. *The Molecular Basis of Mutation*. San Francisco: Holden-Day, 1970.

Ehrlich, P. R. Looking backward from 2000 A.D. *The Progressive*, Apr. 1970, pp. 23–25. Scenario of mutagenic and other catastrophes in the near future.

Epstein, S. S., et al., eds. *Drugs of Abuse: Their Genetic and Other Chronic Non-Psychiatric Hazards*. Cambridge, Mass.: M.I.T. Press, 1971.

Fishbein, L., W. G. Flamm, and H. L. Falk. *Chemical Mutagens*. New York: Academic Press, 1970.

German, J. Studying human chromosomes today. *Amer. Scientist*, **58**:182–201, 1970.

Hollaender, A., ed. *Chemical Mutagens: Principles and Methods for Their Detection*, 3 vols. New York: Plenum Press, 1971 and 1973.

Neel, J. V. Developments in monitoring human populations for mutation rates. *Mutation Res.*, **26**:319–328, 1974.

Prakash, L., F. Sherman, M. W. Miller, C. W. Lawrence, and H. W. Taber. *Molecular and Environmental Aspects of Mutagenesis*. Springfield, Ill.: Thomas, 1975.

Sutton, E., and M. Harris, eds. *Mutagenic Effects of Environmental Contaminants*. New York: Academic Press, 1972.

Vogel, F., and G. Rohrborn, eds. *Chemical Mutagenesis in Mammals and Man*. New York: Springer-Verlag, 1970.

Williamson, F. Population pollution. *Bioscience*, **19**:979–983, 1969.

9   The Detection of Environmental Mutagens

James V. Neel and Arthur D. Bloom

It is a cornerstone of genetic theory that the spontaneous mutation rate of any species has through selection reached a near optimum, serving to introduce mutations at a rate sufficient to permit ongoing evolution but not so rapidly as to exceed the ability of selective factors to eliminate deleterious mutations. This permits the species to maintain its adaptive position. Accordingly, the possibility of any increase in human mutation rates is viewed with concern by the geneticist. Since World War II, stimulated by the advent of controlled nuclear fission, much of the concern has been with radiation. Recently, however, with a lesser environmental contamination from fall-out than earlier feared, and with the development of x-ray techniques resulting in lower exposures per film, attention has increasingly shifted to chemical mutagenesis. That certain chemicals could produce mutations in experimental organisms has been known since Auerbach and Robson[1] reported the mutagenicity of a naturally occurring compound, mustard oil, in 1944, and the implications of this fact for human populations were noted some years ago by Lederberg[35] and Barthelmess.[4] A Macy Conference in 1960 considered this subject at some length.[46] Today such exposures are a subject of increasing scrutiny, as witnessed, for example, by a recent report of the Genetics Study Section of the National Institutes of Health[22] and the series of papers by Vogel and colleagues.[56]

The chemicals to which man is exposed which have been shown to be mutagenic in one or more experimental organisms include such agents as the fungicide captan, the plant-growth inhibitor maleic hydrazide, the artificial sweetner cyclamate, the food preservatives sodium nitrite and sodium nitrate, certain of the streptomyces-derived antibiotics, various insect chemosterilants, such as triethylene phosphoramide and triethylene melamine, the benzapyrene found in smog, and a variety of alkylating agents. These latter, such as ethylenimine, dimethyl and diethyl sulfate, methyl and ethyl methanesulfonate, and others, are widely used in industry in methylating or ethylating phenols, amines, and other compounds. These chemicals are reaching our respiratory and alimentary tracts in minute amounts by way of air and water pollution, food preservatives, soft drinks, residual pesticides, and medications; the extent to which they eventually reach our germ cells is unknown.

Although in the context of this volume our interest is primarily in the mutagenic effects of drugs, this specific issue must be seen as one aspect of the larger issue of chemical mutagenesis. We shall, in this brief essay, concentrate on the problems inherent in the study of chemical mutagenesis in mammals, especially man. This is not intended as an exhaustive review of the literature. It will be argued that critical, direct evidence will be extremely difficult to obtain, so much so that unless a

From *Medical Clinics of North America*, **53**, no. 6 (Nov. 1969), 1243–1256. Copyright 1969 by the W. B. Saunders Company. Reprinted with permission of the authors and the publisher.

truly massive study is launched, the findings will almost surely be inconclusive. The kinds of indirect evidence pertinent to the problem will be described and the present state of this indirect evidence reviewed briefly.

Direct Approaches to the Problem of Chemical Mutagenesis in Man

It is conventional and operationally convenient to distinguish between two types of mutations—point and chromosomal. Point mutations are characterized by minute alterations in the genetic codons and cannot be visualized with the light microscope. Chromosomal mutations are characterized by grosser changes in chromosomes, usually gains or losses of chromosomes or parts thereof, which can be visualized with the light microscope. However, especially in man, it is not always readily possible to distinguish between the two. At the present time, there are three general *direct* approaches to detecting an increased mutation rate in man. These we shall discuss in the order of the rigorousness with which the data can be interpreted. The first two approaches will detect a mixture of point and chromosomal mutations; the third divides into two subapproaches—one primarily detecting the results of point mutations, the other detecting the results of chromosomal mutations.

The *first and least precise approach* consists in comparing the children being born in two populations, one exposed to a particular noxious agent, the other not, with respect to a number of readily recorded traits which have a genetic component. These traits include abortion and stillbirth rates, the frequency of congenital malformations, birth weight, sex ratio, neonatal and childhood death rates, and physical growth and development. Although there is undoubtedly a genetic component in each of these, its precise nature is unknown, and the trait is also influenced by multiple exogenous factors. A change in one of these indicators could be a consequence of newly induced mutations, but it could also be due to a change in the environment. It is thus a matter of great importance either that the control population be similar in every possible respect or that extensive demographic and socioeconomic data be collected to permit an analysis that takes differences between the populations into consideration.

This is the approach that was utilized in the study of the potential genetic effects of the use of atomic bombs on Japan. At the time the study was organized, some 20 years ago, it was clear that the number of irradiated survivors did not permit the next approach to be described, and the technology for a really precise approach was not yet available. The various publications on those studies detail the kinds of problem that can arise.[32, 39, 40, 47] Although, other factors being equal, an unequivocal increase in congenital defects or infantile and childhood death rates or an alteration of the sex ratio would be presumptive evidence of genetic damage, it is clear that the confounding factors are so numerous that, wherever possible, other approaches must be pursued. An example of a current effort to utilize this approach is the study of Rohrborn[43] on the sex ratio in relation to caffeine consumption in man; preliminary data show no effect.

The *second approach, of intermediate precision,* is characterized by efforts to measure in a defined population the rate of occurrence of certain "sentinel pheno-types." These are defined as attributes that have a high probability of being the result of a dominant germinal mutation when displayed by isolated individuals in a family. These phenotypes include such entities as achondroplasia, aniridia, retino-blastoma, partial albinism with deafness, microphthalmos-anophthalmos without mental defect, Marfan's syndrome, and multiple neurofibromatosis. Collectively they occur with a frequency of about 50 per 100,000 births.[38] Although several of these traits may at times result from events other than dominant mutation—e.g., phenocopies or recessive inheritance—the bulk may safely be attributed to a mutational event. Let us say that we wish to detect a 20 per cent increase in the frequency of the mutational event responsible for isolated cases of these defects. This requires intensive surveillance of a rather large population. For instance, at a crude birth rate of 20 per 1000, a population of 10,000,000 will have 200,000 births per year. In this population, on the order of 100 children would "normally" develop the above enumerated defects. With this numerical base, an increase from 100 to 120 would approach statistical significance (5 per cent level). It would in theory require a population of this size to detect that increase, but because of variations in such aspects as ascertainment of affected children and diagnostic standards (i.e., noise in the system), one would be reluctant to accept as valid an increase of less than 40 per cent. Many of these phenotypes will not develop or cannot be accurately diagnosed for some years after birth; thus, there will be a considerable time lag in the system. Finally, the organizational problems inherent in such a large-scale study are obvious. This requirement for large numbers means that the approach is impractical where the exposed population is relatively small, as, for example, was the case in Hiroshima and Nagasaki.

As mentioned earlier, *the more precise approaches are of two types*—one based on cytogenetics, the other on biochemistry. With respect to the *cytogenetic approach*, basically there are two possible effects which mutagenic agents may have on the chromosomes of the germ cells. First, they may act on the meiotic apparatus and cause, in progeny, a consistent numerical abnormality of the chromosomes in all cells. Examples of this in man include the trisomy states ($2n + 1$, where n is the human haploid number of 23 chromosomes), such as Down's syndrome or trisomy 21, the trisomy D (13–15) syndrome, and the trisomy E (16–18) syndrome. There is, thus far, only one reported case of an autosomal monosomy in man ($2n - 1$), and that is for number 21. Secondly, a structural rearrangement may be induced in the meiotic chromosome itself, and if this rearrangement is stable, as a translocation or inversion, it may be transmitted to all of the cells of the progeny. The classic description of the numerical aberrations is that of Bridges in 1916,[11] who first noted that in *Drosophila* both of the homologous chromosomes of the germ cells occasion-ally passed into the same nucleus, failing to separate normally. Thus, the frequency of detectable nondisjunctional events may be used as a measure of the effect of a given agent on chromosome behavior in meiosis. The resulting abnormalities of chromosome number may be readily determined in man by the routine use of chromosome analysis of the peripheral blood cells.

A variety of other factors have been suggested in man as potentially capable of producing the biologically and often clinically drastic nondisjunctional event. These include maternal and/or paternal irradiation,[48, 55] delayed fertilization resulting from diminished frequency of coitus,[27] possible genic control, and the general effect of aging.[27] Further, thyroid autoantibodies have been found by Fialkow et al.[25] to be increased in the cytogenetically normal mothers of patients with Down's syndrome, but what effect these antibodies have on meiotic cell division is not known. However, even the effect of these factors is not unequivocal. For instance, in the Hiroshima and Nagasaki populations, where the estimated maternal radiation doses often exceeded 100 or 200 rads, there was no increased incidence of Down's syndrome[49] or of other congenital malformations.[44] If we exclude the increase in frequency of virtually all trisomics seen with increased maternal age, it is clear that no other factor has as yet been definitely established as having a nondisjunctional effect in man.

Court Brown[20] has estimated that at least 0.5 per cent of persons in the general population are so-called structural heterozygotes; i.e., they have in their somatic cells one member of a pair of homologous chromosomes that is structurally different from its normal mate. This heterozygosity may be the result of the transmission of an abnormal chromosome from a carrier parent, or alternatively, it may be the result of a chromosomal rearrangement occurring de novo in a meiotic cell of one parent. For survival through the meiotic division, and through the subsequent mitotic divisions as well, the rearrangement must not significantly alter the normal movement of the chromosomes, especially during anaphase. Thus far in man, no agent, physical or chemical, has been shown to increase this type of abnormality. Further, the indirect detection of this kind of damage to the meiotic chromosomes might well require many thousands of offspring, even if the level of meiotic chromosome sensitivity to the offending agent were still appreciable. Again, in Hiroshima and Nagasaki, this approach is currently being used to assess possible A-bomb effects on the meiotic chromosomes.[3] To date, no evidence has been found that an aberration was produced and transmitted. We must recognize, however, that in such an approach, the chances of detecting an induced chromosomal mutation are lessened because of (1) possible death of the affected germ cell, (2) possible selection against the affected germ cell, and (3) undetected loss of nonviable zygotes.

Today approximately 1 per cent of children who reach term have a gross chromosomal abnormality. They are the residuum of a much larger number lost during pregnancy. Suppose we wished the same ability to detect an increase as discussed with reference to sentinel phenotypes. If one screened 10,000 births a year, one could detect at the 5 per cent level of statistical significance an increase from the "normal" 100 with gross chromosomal defect to 120, but again, the noise in the system being what it is, a rate of 140 would be much more convincing. With current techniques, such a screening program would require about 20 technicians. Under exceptional circumstances, a much smaller series directed towards the study of spontaneous abortions in which the frequency of chromosomal abnormalities is relatively high, might be adequate to flag a developing problem. Thus, Carr[14] has reported in a small series a striking increase in the frequency of polyploidy (triploidy and tetraploidy) in the spontaneously aborted fetuses of women who became pregnant within 6 months of discontinuing the use of oral contraceptives.

The *biochemical approach*, the last and most precise of the direct approaches, involves a search, with all the laboratory automation possible, for mutations characterized by changes in protein structure or function. For instance, one could screen for changes in specific polypeptides, unselected except on the basis that they are polypeptides that can be identified by electrophoretic and similar procedures. These changes presumably result from alterations in the codons of genes determining protein structure. Here, in contrast to a possibility that exists for the sentinel phenotypes, there should be no bias for traits with high mutability. A possible practical approach is to screen cord blood for some 10 different proteins, such as transferrin, albumin, haptoglobin, hemoglobin, carbonic anhydrase, acid phosphatase, glucose-6-phosphate dehydrogenase, phosphoglucomutase 1 and 2, adenylate kinase, lactate dehydrogenase, and malate dehydrogenase. An unusual variant could be followed up by appropriate family studies to determine whether it was inherited from one or the other parent or arose by mutation. Appropriate studies regarding illegitimacy are, of course, necessary in any case of possible mutation, with a very serious question of invasion of privacy. Collecting a sample of the mother's blood at the time of delivery could greatly reduce the necessary fieldwork. The attractiveness of this approach is that one is much closer to the mutational event than where he is dealing with a complex syndrome. Lack of experience with this approach makes an estimate of the magnitude of the required effort rather difficult. Plainly multitrait screening systems of relatively high resolution must be developed.

It is quite clear that, no matter which of these approaches is employed, once a probable increase in the general mutation rate is detected, one is confronted with a baffling piece of detective work. How does one track down, out of all the possibilities, the one or several offending agents? Establishing suitable paired study populations—one on the pill, the other not; one on a popular tranquilizer, the other not; one employing streptomyces-derived antibiotics, the other exposed to the penicillin antibiotics—on a large scale, demands a type of organization of biological investigation never before achieved. The effort to detect an increase in mutation rates is not beyond the resources of a society that can put men on the moon, but the effort to pinpint the cause of that same increase in mutation rates might prove to be unmanageable. Incidentally, although the emphasis currently is on detecting increased mutation rates, these same approaches would of course also detect a decrease in mutation rates. If, for instance, some air pollutants are mutagenic and current efforts to diminish air pollution are successful, this could be reflected in the results of one of these detection programs.

Indirect Approaches to Estimating Increases in Mutation Rates

In addition to the foregoing direct approaches to the problem, there are a number of indirect approaches, so called either because they deal with organisms other than man or because, employing human material, they involve phenomena related to the mutational event, but not germinal mutations per se. We shall consider three, namely, chemical mutagenesis in experimental mammals, the production of mutants in mammalian cell culture material, and observations on the production of chromosomal damage in cells in culture or in intact mammals, including man. All these methods are being actively pursued. With respect to *chemical mutagenesis in experimental mammals*, despite the really extensive literature on chemical mutagenesis in microorganisms and *Drosophila*,[2, 26, 33] there is a paucity of data on mammalian material. Most of the studies that have appeared concern the induction of "dominant lethals" in the mouse or rat by chemical treatments. A dominant lethal is defined as a dominant genetic change which is incompatible with the survival of the conceptus. In the untreated mouse, from 1 to 10 per cent of zygotes that implant fail to complete development, the precise number varying with the strain. These "spontaneous" losses are thought to be primarily due to some kind of trauma to the zygote or a variety of genetic causes, including spontaneous dominant lethals. The frequency of induced dominant lethals is usually estimated from the increase over the control value in the frequency of such losses among the offspring from treated males and untreated females.

The early mammalian work in this field centered on the action of alkylating agents. Cattanach,[15] in 1958, employed the dominant lethal method to demonstrate the mutagenicity of the antileukemia agent triethylene melamine (TEM) in the mouse; Steinberger et al.[50] and Bateman[6] demonstrated this to be the case for the rat as well. In these experiments, male animals were given doses of the magnitude of 0.1 mg. per kg., intraperitoneally. There is no mention of increased mortality in treated animals. A bit later, Partington and Jackson[41] and Partington and Bateman[42] demonstrated the production of dominant lethals in rats and mice by another group of alkylating agents, including such alkanesulfonic esters as methyl ethanesulfonate, ethyl methanesulfonate, isopropyl methanesulfonate, and tetramethylene-1,4-dimethanesulfonate (busulfan). Doses were of the order of 50 to 100 mg. per kg., with no mention of increased mortality in the treated animals. "Clinical doses" of trenimone and cytoxan also have been shown to increase the dominant lethal rate in mice,[43] but urethane, a well-known mutagen in plants and *Drosophila*, did not produce dominant lethals in the mouse.[6] Cattanach[16] showed that in mice TEM also produced paternal sex chromosome loss, aberrations of the paternal X, and gene mutations at specific autosomal loci. These latter were detected in the special mouse strains developed by W. L. Russell to demonstrate the mutagenic effects of x-rays—as these results are extended, a comparison of the mutational spectrum induced by chemicals with that induced by x-rays will become possible.

More recently, also employing the dominant lethal tests, Epstein and Schafner[24] have confirmed and extended the above findings, demonstrating that aflatoxin, benzo(α)-pyrene, certain of the ethylene-imine alkylating agents, and methyl methanesulfonate induce significant numbers of dominant lethals in the offspring of male mice tested with LD_5 doses intraperitoneally. On the other hand, under similar circumstances, butylated hydroxytoluene; three xanthines (including caffeine); the pesticides captan, maleic hydrazide, and DDT; a variety of other carcinogens (including methane); and a number of pharmaceuticals, such as chloramphenicol, chlorpromazine, and griseofulvin, were not mutagenic. This was a screening program in which the number of mice treated with any one chemical was small; in general, increases less than four- or fivefold would not emerge as statistically significant. Ehling, Cumming, and Malling[23] also found methyl methanesulfonate (MMS), as well as ethyl methanesulfonate (EMS), to be highly effective in the production of dominant lethals in male mice when injected intraperitoneally in doses of 50 and 100 mg. per kg. for MMS and 200 mg. per kg. for EMS—doses which produce no mortality. On the other hand, in comparable doses, neither N-methyl-N'-nitro-N''-nitrosoguanidine (MNNG) nor an acridine mustard known as ICR-170 produced mutations.

While the dominant lethal approach is convenient and inexpensive, there are potential ambiguities in its interpretation. Most geneticists will be more satisfied with data on gains and losses of specific chromosomes or specific locus mutations, data which will undoubtedly be shortly forthcoming. But the principle that in mammals a variety of chemicals administered in doses which increase mortality little or not at all have genetic effects seems established. The doses are, for the most part, high by the standards of possible human exposures to these agents; but experiments with lower doses will also be forthcoming. Much of the work cited has been concerned with the stage in spermatogenesis of maximal sensitivity and the comparability of the results with x-ray effects. These subjects have not been considered in this survey.

With respect to *mammalian cell culture material*, in theory it should be possible to develop methods for handling mammalian cells in culture-like suspensions of bacterial cells and for recovering mutants induced in such cells by treatment with various mutagens. Although care is necessary in extrapolating from such results to the effects of these same mutagens at similar doses in intact organisms, the method should be useful for screening purposes. The pioneering work of Szybalski,[51] Szybalski and Smith,[52] and Lieberman and Ove[36, 37] established the feasibility of testing for the spontaneous rate of appearance in cell cultures of cells resistant to such compounds as puromycin and 8-azaguanine, but since long-established human and mouse cell lines whose chromosomal constitution varied from cell to cell were employed, it is not clear how comparable these rates are to mutation rates in the usual sense. The early efforts in this field have been reviewed by Szybalski and Szybalska.[53]

Recently, methods have been reported for studying the rate of occurrence of "mutant" cells in cell culture material treated with mutagens. Although Chinese hamster cells were utilized, there seems to be no reason why the method cannot be extended to human cell lines. Kao and Puck,[31] employing aneuploid Chinese hamster cell lines with 20 and 21 chromosomes (the diploid number is 22), treated cell cultures with such alkylating chemical mutagens as EMS and MNNG and then by an ingenious technique isolated mutant cells deficient in the ability to grow without such compounds as glycine, hypoxanthine, or inositol in the media. In a spontaneously occurring proline-dependent line, "reverse mutants" (to proline independence) could also be induced by treatment with these two compounds. In diploid tissue, one would expect to detect only dominant nutritional mutants, if indeed such occur, unless the animal from which the tissue was derived was already heterozygous for an appropriate mutation. The frequency of mutants for glycine dependence was so much higher than the frequency of mutants for other nutritional requirements that it was suggested that the genetic locus in question might be in a chromosome represented only once in the cell line (thus permitting recessive mutants as well to find expression). Studies with euploid cell lines would of course help test that suggestion. The 6-day lapse between treatment and testing for mutants renders precise quantitative statements concerning absolute or relative mutation rates difficult.

Chu and Malling[17,18,19] have independently developed a selective system for recovering certain induced mutants in a Chinese hamster cell line, also aneuploid (23 chromosomes). The mutant types involve auxotrophy for L-glutamine and resistance to 8-azaguanine, and the system permits quantitation of mutation rates. Following treatment with EMS, MMS, and MNNG, the frequency of mutant colonies increased more than fiftyfold. For example, whereas at a concentration of 5 micrograms per ml. control cultures showed 2.6 mutants per 10^5 surviving cells, cultures treated with 10^{-5} molar MNNG exhibited 181.4 mutant cells per 10^5 surviving cells. The frequency of reverse mutation from azaguanine resistance to azaguanine sensitivity could also be increased by these agents. When mutant glutamine-requiring and azaguanine-resistant cells were plated together in an experimental medium in which neither parental cell type could survive alone, a cell hybrid obtained through fusion established an essentially tetraploid clone. Both the glutamine requirement and the azaguanine resistance were absent in this tetraploid; i.e., the traits behaved as " recessive."

The third indirect method of the evaluation of the mutagenicity of chemicals in man involves the demonstration of the induction of *chromosomal damage in somatic tissue*. The effects of drugs on somatic cell chromosomes are now being extensively examined. The Food and Drug Administration is now conducting studies on the in vitro effects of new pharmaceuticals on human chromosomes. It must, in fact, be recognized that somatic cell chromosome abnormalities represent in many instances the only effects known for a variety of agents on the genetic material of human cells. Because of the sensitivity of, in particular, human lymphocyte chromosomes to exogenous agents, this system appears to be the most practicable one for the systematic screening of new drugs and environmental contaminants.

A few cautionary notes must, however, be sounded. First, different laboratories employ different culture methods, as well as different culture times.[8] Variations in methods may well affect the "spontaneous" aberration frequencies, and these must be taken into account, in the form of control data for the population of individuals or cells, when evaluating the possible mutagenic effect of a given agent. Secondly, large scale chromosome aberration studies in human populations are still reasonably subjective, and until such time as automated chromosome analysis is feasible, this subjectivity must be minimized by coding of slides and "blind" readings. Lastly, there is at present no standard approach to the reporting of data, with some individuals using primarily the breaks-per-cell approach while others give the raw data.[7, 20] This latter approach was recommended at the International Symposium on Human Radiation Cytogenetics, convened in Edinburgh in 1966, and is clearly the approach of choice. But because of the lack of uniformity in these three areas, and even in the interpretation of the data itself, there is often confusion as to whether a particular drug is or is not capable of producing chromosome damage in human cells. Further, this entire field is complicated by the fact that many kinds of simple chemical and physical treatments cause nonspecific chromosomal alterations that are not likely to be associated with point mutations in germ cells. While, in general, in experimental organisms, agents that produce visible damage in somatic cell chromosomes also produce genic mutation in germ cells, this correlation has not as yet been definitively or even suggestively proven for man.

In attempting to understand how various drugs might produce chromosome aberrations, we must remember that the vast majority of somatic cells are non-dividing. As such, a drug may act on the chromosome (1) when it is single-stranded, i.e., in the G_1 stage of interphase, (2) when it is actively synthesizing DNA, in the S period of interphase, or (3) when it is already duplicated, or in the G_2 period of interphase. Practically speaking, the aberrations formed in S or G_2 may be called chromatid in type, affecting usually one chromatid of the chromosome, while aberrations formed in G_1 may be called chromosome in type, affecting both chromatids.

Space permits our considering only a few examples of chemical agents that are capable of inducing chromosome breakage. We have selected for discussion those in which the mechanism of induction is, if not firmly established, at least biologically reasonable on theoretical grounds. In fact, a considerable number of compounds that might be expected to act as both chromosomal and specific locus mutagens have now been studied. Among these are several base analogues of the nucleic acids. For example, 5-bromodeoxyuridine (BUdR) and 5-fluorodeoxyuridine (FUdR) have been described as capable of inducing chromosome aberrations.[10] BUdR apparently acts by becoming incorporated into the DNA of the chromosome without blocking DNA synthesis, while FUdR has been shown to act by inhibiting thymidylate synthetase, an enzyme which converts deoxyuridylate to thymidylate. Some investigators have suggested that this blocked conversion results in an inhibition of DNA synthesis which is responsible, in some way, for the chromosome damage, while others have suggested that FUdR acts after DNA synthesis, i.e., in the G_2 period of interphase. This latter idea correlates well with the types of observed aberrations, i.e., chromatid, after FUdR treatment.

Experiments on the antimetabolite cytosine arabinoside (CA)[10] have confirmed the induction of chromosomal aberrations by this agent in cultured peripheral blood leukocytes. One action of this drug appears be as an inhibitor in the conversion of cytidine diphosphate to deoxycytidine diphosphate. CA can induce chromosomal breaks at all stages of the cell cycle, and it has been concluded, therefore, that the inhibition of DNA synthesis is not the cause of the chromosomal abnormality. One possibility is that CA causes an increase in a cellular nuclease which in turn is responsible for the chromosomal aberrations.[10]

Other antimetabolites have also been studied. Methotrexate is commonly used in dermatology for the treatment of severe psoriasis. The drug is usually given either as 25 mg. once a week or as 0.625 mg. four times daily for 5 of 7 days. Voorhees et al.[57] have found that in vitro, at levels exceeding the concentrations obtained in the blood at these dosages, chromosome breaks can be produced in 4 to 18 per cent of cells when normal lymphocytes are drug-exposed for 6 to 72 hours. The types of breaks seen were infrequently encountered in nonexposed cells. In vivo, however, methotrexate-treated psoriatic patients showed no significant increase in chromosome breaks, *at these doses*. Higher doses, which are sometimes used in practice, might be effective in breaking chromosomes, as were the higher levels tried in vitro. Methotrexate is known to act on thymidine biosynthesis by competing with dihydrofolic acid for the enzyme dihydrofolic acid reductase. The action of this enzyme is essential for the synthesis of a cofactor in the TMP synthetase reaction. The resulting deficiency of thymidylate may effectively block the cell cycle at the S phase. Alternatively, the cell may proceed, through interphase and mitosis, with gross aberrations present, which, in some as yet unknown fashion, may be the result of the thymidylate deficiency.

Many other antimetabolites, radiomimetic drugs (such as the nitrogen mustards), and antibiotics (such as puromycin) have been tested for their effects on human somatic cell chromosomes. Many act either on protein or nucleic acid synthesis, though the way in which this kind of metabolic effect causes chromosomal abnormalities is not known.

The first reports of the effect of lysergic acid diethylamide (LSD 25)[21,30] on human chromosomes prompted interest in the possible mutagenic effect of psychoactive drugs, ranging from the hallucinogens, e.g., LSD and marijuana, to the tranquilizers, e.g., chlorpromazine. Results of the cytogenetic studies on LSD have been conflicting, with some investigators reporting positive effects[29] and others showing no effects,[54] both in vivo and in vitro. Similarly, in *Drosophila*, the published data have not been consistent. Grace, Carlson, and Goodman[28] found no mutations or chromosome breaks in premeiotic, meiotic, or postmeiotic sperm. Browning,[12] on the other hand, found an increase in recessive lethal mutations after massive doses of LSD were given to *Drosophila* males.

A recent cytogenetic survey of heroin addicts, amphetamine and barbiturate users, and marijuana users by Bloom et al.[9] suggested that some drug users do have more chromosomal aberrations than one might expect, but it was unclear whether these aberrations were drug induced or whether they were the result of other factors in the way of life of these people. It is likely that many agents in man's environment, even more socially acceptable ones that involve ingestion of caffeine, acetylsalicylic acid, or similar agents,[44] are capable of inducing structural chromosomal abnormalities in somatic cells in vitro.

It cannot at present be argued that agents that induce somatic cell chromosome aberrations will also induce point mutations in germ cells or that they are harmful in any other way. It is nonetheless prudent to assume that chromosomal aberrations, at best, do no good. Therefore, while we cannot now be precise in our reasons for recommending that individuals avoid known chromosomal mutagens, we may reasonably urge that such exposures be minimized.

Conclusions

The increasing exposure of man to chemicals, including certain drugs, known to be mutagenic in lower organisms suggests the need for some type of monitoring system. Some of the problems inherent in such systems have been discussed. Detecting an increase in human germinal mutation rates will require an extensive surveillance network, and even when an increase is detected, the problem of pinpointing the responsible agent remains formidable. The most nearly comparable previous experience has been the effort to demonstrate the mutagenic effects of x-rays in man. In this case, although there can be no doubt that radiation is mutagenic for man, and although exposed populations are available, there is still no clear demonstration of the genetic effects of radiation in man.

Indirect approaches to evaluating the potential mutagenicity of certain chemicals for man have been discussed. Significant guidelines concerning suspect chemicals can be obtained from genetic experiments with such mammals as the mouse and from the study of induced mutations or chromosomal damage in tissue culture material derived from man or other mammals. While the ultimate test of mutagenicity of a chemical in man should be its effect in " physiological" doses in the intact organism, it is probable that we will have to be guided in the immediate future in large measure by these experimental systems. Of these, somatic cell chromosome damage is the main tool presently available for study of agents which may be mutagenic in man. It must be added, however, that the demonstration of chromosomal damage does not automatically lead to inferences concerning mutation effects in germ cells.

The demonstration of even a small genetic effect of a drug in the first post-exposure generation implies a larger effect to be spread over subsequent generations; if we introduce mutagens into the environment, it is not so much ourselves as subsequent generations who pay the price. On the other hand, some of these potential mutagens are so important to pest control and industry that, even if mutagenicity were demonstrated, as in the case of radiation, one might be forced to attempt to weigh probable gains to society against probable genetic loss.

References

1. Auerbach, C., and Robson, J. M. Production of mutations by allyl isothiocyanate. Nature, *154*: 81, 1944.
2. Auerbach, D. Mutagenesis, with particular reference to chemical factors. *In* Schull, W. J. (ed.): Mutations: Second Macy Conference on Genetics. Ann Arbor, The University of Michigan Press, 1962.
3. Awa, A. A., Bloom, A. D., Yoshida, M. C., Neriishi, S., and Archer, P. G. Cytogenetic study of the offspring of atom-bomb survivors. Nature, *218*: 367–368, 1968.
4. Barthelmess, A. Mutagene Arzneimittel. Arzneimittel forschung, *6*: 157–168, 1956.
5. Bateman, A. J. The induction of dominant lethal mutations in rats and mice with triethylenemelamine (TEM). Genet. Res., *1*: 381–392, 1960.

6. Bateman, A. J. A failure to detect any mutagenic action of urethane in the mouse. Mutat. Res., *4/5*:710–712, 1967.

7. Bloom, A. D., Neriishi, S., Awa, A. A., et al. Chromosome aberrations in leukocytes of older survivors of the atomic bombings of Hiroshima and Nagasaki. Lancet, *2*:802–805, 1967.

8. Bloom, A. D., and Iida, S. Two-day leukocyte culture for human chromosome studies. Jap. J. Human Genet., *12*:38–42, 1967.

9. Bloom, A. D., Gilmour, D., Lele, K., et al. Drug usage and chromosome aberrations. In preparation.

10. Brewen, J. G., and Christie, N. T. Studies on the induction of chromosomal aberrations in human leukocytes by cytosine arabinoside. Exp. Cell Res., *46*:276–291, 1967.

11. Bridges, C. B. Non-disjunction as a proof of the chromosome theory of heredity. Genetics, *1*: 1–52, 107–163, 1916.

12. Browning, L. S. Lysergic acid diethylamide: Mutagenic effects in Drosophila. Science, *161*:1022–1023, 1968.

13. Carr, D. H. Cytogenetics and the pathology of hydatidiform degeneration. Obstet. Gynec., *33*:333–342, 1969.

14. Carr, D. H. Chromosomes after oral contraceptives. Lancet, *2*:830–831, 1967.

15. Cattanach, B. M. The sensitivity of the mouse testes to the mutagenic action of triethylenemelamine. Z. Vererb., *90*:1–6, 1959.

16. Cattanach, B. M. Induction of paternal sex-chromosome losses and deletions and of autosomal gene mutations by the treatment of mouse post-meiotic germ cells with triethylene-melamine. Mut. Res., *4*:73–82, 1967.

17. Chu, E. H. Y., and Malling, H. V. Mammalian cell genetics. II. Chemical induction of specific locus mutations in Chinese hamster cells in vitro. Proc. Nat. Acad. Sci. Wash., *61*:1306–1312, 1968.

18. Chu, E. H. Y., and Malling, H. V. Chemical mutagenesis in Chinese hamster cells in vitro. Proc. XII Intern. Congr. Genetics, *1*:102, 1968.

19. Chu, E. H. Y., Bremer, P., Jacobson, K. B., and Merriam, E. V. Mammalian cell genetics. I. Selection and characterization of mutations auxotrophic for 1-glutamine or resistant to 8-azaguanine in Chinese hamster cells in vitro. Genetics, *62*:359–377, 1969.

20. Court Brown, W. M. Human Population Cytogenetics. Amsterdam, North-Holland Publishing Company, 1967.

21. Cohen, M. M., Hirschhorn, K., and Frosch, W. A. In vivo and in vitro chromosomal damage induced by LSD-25. New Eng. J. Med., *277*:1043–1049, 1967.

22. Crow, J. F. Chemical risk to future generations. Scientist and Citizen, June–July, pp. 113–117, 1968.

23. Ehling, U. H., Cumming, R. B., and Malling, H. V. Induction of dominant lethal mutations by alkylating agents in male mice. Mut. Res., *5*:417–428, 1968.

24. Epstein, S. S., and Shafner, H. Chemical mutagens in the human environment. Nature, *219*: 385–387, 1968.

25. Fialkow, P. J., Uchida, I. A., et al. Increased frequency of thyroid autoantibodies in mothers of patients with Down's syndrome. Lancet, *2*:868–870, 1965.

26. Freese, E. Molecular mechanisms of mutation. *In* Taylor, J. H. (ed.): Molecular Genetics. New York, Academic Press Inc., 1963.

27. German, J. Mongolism, delayed fertilization and human sexual behavior. Nature, *217*: 516–518, 1968.

28. Grace, D., Carlson, E. A., and Goodman, P. Drosophila melanogaster treated with LSD: Absence of mutation and chromosome breakage. Science, *161*: 694–696, 1968.

29. Hungerford, D. A., Taylor, K. M., Shagass, C., et al. Cytogenetic effects of LSD–25 therapy in man. J.A.M.A., *206*: 2287–2291, 1968.

30. Irwin, S., and Egozcue, J. Chromosomal abnormalities in leukocytes from LSD–25 users. Science, *157*: 313–314, 1967.

31. Kao, F.-T., and Puck, T. T. Genetics of somatic mammalian cells. VIII. Induction and isolation of nutritional mutants in Chinese hamster cells. Proc. Nat. Acad. Sci. Wash., *60*: 1275–1281, 1968.

32. Kato, H., Schull, W. J., and Neel, J. V. A cohort-type study of survival in the children of parents exposed to atomic bombings. Amer. J. Hum. Genet., *18*: 339–373, 1966.

33. Kreig, D. R. Specificity of chemical mutagenesis. *In* Davidson, J. N., and Cohn, W. E. (eds.): Progress in Nucleic Acid Research, New York, Academic Press Inc., 1963.

34. Krooth, R. S. Genetics of cultured somatic cells. Med. Clin. N. Amer., *53*: 795–811, 1969.

35. Lederberg, J. Letter to the Editor. Bull. At. Scientists, December, 1955.

36. Lieberman, I., and Ove, P. Isolation and study of mutants from mammalian cells in culture. Proc. Nat. Acad. Sci. Wash., *45*: 867–872, 1959.

37. Lieberman, I., and Ove, P. Estimation of mutation rates with mammalian cells in culture. Proc. Nat. Acad. Sci. Wash., *45*: 872–877, 1959.

38. Neel, J. V. Mutations in the human population. *In* Burdette, W. J. (ed.) Methodology in Human Genetics. San Francisco, Holden-Day, Inc., 1962.

39. Neel, J. V. Changing Perspectives on the Genetic Effects of Radiation. Springfield, Illinois, Charles C. Thomas, 1963, pp. viii and 97.

40. Neel, J. V., and Schull, W. J. The Effect of Exposure to the Atomic Bombs on Pregnancy Termination in Hiroshima and Nagasaki. Washington, D.C., National Academy of Sciences—National Research Council Publ. 461, 1956, pp. xvi and 241.

41. Partington, M., and Jackson, H. The induction of dominant lethal mutations in rats by alkane sulphonic esters. Genet. Res., *4*: 333–345, 1963.

42. Partington, M., and Bateman, A. J. Dominant lethal mutations induced in male mice by methyl methane sulphonate. Heredity, *19*: 191–200, 1964.

43. Rohrborn, G. Chemical mutagenesis. I. Outline of a research program and examinations in human populations. II. Induction of dominant lethals in mice. Abstracts of Contributed Papers. Third Intern. Congr. Human Genetics, Chicago, 1967, p. 85.

44. Sax, K., and Sax, H. J. Radiomimetic beverages, drugs, and mutagens. Proc. Nat. Acad. Sci., U.S.A., *55*: 1431–1435, 1966.

45. Schull, W. J., and Neel, J. V. Atomic bomb exposure and the pregnancies of biologically related parents. Amer. J. Pub. Health, *49*: 1621, 1959.

46. Schull, W. J. (ed.). Mutations: Second Macy Conference on Genetics. Ann Arbor, The University of Michigan Press, 1962, pp. x and 248.

47. Schull, W. J., Neel, J. V., and Hashizume, A. Some further observations on the sex ratio among infants born to survivors of the atomic bombings of Hiroshima and Nagasaki. Amer. J. Hum. Genet., *18*: 328–338, 1966.
48. Sigler, A. T., Lilienfeld, A. M., Cohen, B. H., et al. Radiation exposure in parents of children with mongolism. Bull. Johns Hopkins Hosp., *117*: 374–399, 1965.
49. Slavin, R. E., Kamada, N., and Hamilton, H. B. A cytogenetic study of Down's syndrome in Hiroshima and Nagasaki. Jap. J. Human Genet., *12*: 17–28, 1967.
50. Steinberger, E., Nelson, W. O., Boccabella, A., and Dixon, W. J. A radio-mimetic effect of TEM in reproduction in the male rat. Endocrinology, *65*: 40–50, 1959.
51. Szybalski, W. Genetics of human cell lines. II. Method for determination of mutation rates to drug resistance. Exp. Cell Res., *18*: 588–591, 1959.
52. Szybalski, W., and Smith, M. J. Genetics of human cell lines. I. 8-azaguanine resistance, a selective "single-step" marker. Proc. Soc. Exp. Biol. Med., *101*: 662–666, 1959.
53. Szybalski, W., and Szybalska, E. H. Drug sensitivity as a genetic marker for human cell lines. *In* Merchant, D. J., and Neel, J. V. (eds.) Approaches to the Genetic Analysis of Mammalian Cells. Ann Arbor, University of Michigan Press, 1962.
54. Tjio, J. H., Pahnke, W. N., and Kurland, A. A. Pre- and post-LSD chromo-somal aberrations: A comparative study. Presented to the American College of Neuropsychopharmacology, San Juan, Puerto Rico, December 20, 1968.
55. Uchida, I., and Curtiss, E. A possible association between maternal radiation and mongolism. Lancet, *2*: 848–851, 1961.
56. Vogel, F., Krüger, J., Röhrborn, G., Schleiermacher, E., and Schroeder, T. M. Mutationen durch chemische Einwirkung bei Sänger und Mensch. Statistische Untersuchungen beim Menschen und zusammenfassende Betrachtungen. Dtsch. Med. Wschr., *92*: 2382–2388, 1967.
57. Voorhees, J. J., Janzen, M. K., et al. Cytogenic evaluation of methotrexate-treated psoriatic patients. Archiv. Dermat., *100*: 269–274, 1969.

10   A Combined Bacterial and Liver Test System for Detection and Classification of Carcinogens as Mutagens

Bruce N. Ames

We have described our test system for mutagen and carcinogen detection in a series of recent papers and I will only give a summary of some of these results, omitting references to other work which is credited in our original papers. Mutagens

From *Genetics*, **78** (Sept. 1974), 91–95. Reprinted with permission of the author and the publisher.

are detected using a special set of bacterial tester strains. A simple procedure for combining human (or rat) liver for carcinogen activation and the bacteria for detection and classification is used. Using this combined system a wide variety of carcinogens—aflatoxin B_1, benzo(a)pyrene, 2-acetylaminofluorine, etc.—are detected as active frameshift mutagens. The structural features of these caricinogens have been discussed on the basis of the theory of frameshift mutagenesis. Other carcinogens have been detected as mutagens causing base pair substitutions. We postulate that carcinogens cause cancer by somatic mutation and suggest that the combined bacteria/liver system be used as a simple procedure for carcinogen detection. A simple test for detecting mutagens in urine has also been developed.

Construction of Strains for Testing Mutagens

During the last seven years we have developed a set of tester strains of *Salmonella typhimurium* for detecting mutagens (Ames, Lee and Durston 1973; Ames 1971, 1972; Ames and Whitfield 1966). This system is super-sensitive and simple to use. The principle of the tester strains is to use mutants caused by a known type of DNA damage (base pair substitutions, and the various kinds of frameshift mutations) for detecting mutagens by the highly sensitive and convenient back mutation test (reversion to prototrophy). A deletion of one of the genes of the excision repair system has been introduced in all of our tester strains and this has made them hundreds of times more sensitive to most mutagens. A great improvement in the sensitivity of the four tester strains has been made by introducing into each strain a mutation which eliminates the lipopolysaccharide which coats the surface of these bacteria and acts as a partial barrier to the penetration of compounds to the bacterial membrane. This is especially important for large compounds such as the aflatoxins.

A bacterial test system has many practical and theoretical advantages in the direction of mutagens, among which are the small genome (about 4×10^6 base pairs), the large number of organisms exposed (about 10^9/plate), and the positive selection of the mutated organisms. The set of Salmonella tester strains has three additional advantages: the lack of excision repair, the loss of the lipopolysaccharide barrier, and the scoring of mutations in "hot spots" for frameshift mutagenesis (e.g., the CGCGCGCG sequence in tester strain TA1538). The scoring of reversion in an easily mutated "hot spot," combined with the smallness of the genome, aid the test by maximizing reversion relative to killing.

New methodology (Ames, Lee and Durston 1973) for testing mutagens by forward mutagenesis has also been developed using strains lacking the excision repair system and the lipopolysaccharide as an adjunct to our set of tester strains. We also compare (Ames, Lee and Durston 1973) the zones of inhibition on two strains, both of which lack the lipopolysaccharide and one of which lacks repair.

Use of These Strains for Detecting and Classifying Carcinogens as Mutagens

It has long been known that many simple alkylating agents, as well as radiation (UV, x-rays, etc.) are both carcinogenic and mutagenic, and we have shown that a variety of these agents can be detected with our tester strain that detects base pair substitutions (Ames, Lee and Durston 1973; Ames 1971, 1972). We have found that a large variety of chemical carcinogens can also be detected as mutagens with our frameshift mutation tester strains. We have identified these carcinogens as members of a special class of mutagens: reactive frameshift mutagens. An addition or deletion of a base pair from the DNA (a frameshift mutation) can occur when there is a shifted pairing in a string of nucleotides $\left(\text{e.g., } \begin{array}{c} \text{-G-G-G-G-} \\ \text{-C-C-C-C-} \end{array}\right)$ during DNA replication. This process is enormously increased by the addition of an acridine compound that intercalates in the DNA base pair stack and stabilizes the shifted pairing. A number of years ago we discovered that when an intercalating agent (an acridine) also has a side chain that can react with DNA (a nitrogen half mustard) it is a more potent mutagen for causing frameshifts by one or two orders of magnitude (Ames and Whitfield 1966, Creech et al. 1972).

Polycyclic hydrocarbons intercalate in DNA and it is thought that polycyclic hydrocarbons are metabolized in mammals to the epoxides which are believed to be the primary carcinogens. We have shown that epoxides of the carcinogenic polycyclic hydrocarbons are extremely potent frameshift mutagens (Ames, Sims and Grover 1972 and manuscript in preparation) of the reactive type.

In addition to these carcinogens we have shown that metabolites of a variety of aromatic amine carcinogens are frameshift mutagens in our system (Ames et al. 1972). 2-Nitrosofluorene, a known metabolite of 2-aminofluorene in the rat (and a more effective carcinogen), is thousands of times more effective as a frameshift mutagen than the parent compound. The nitroso group is a reactive group and we believe, by analogy with the acridine half mustards, that nitrosofluorene is both intercalating and reacting with DNA. Because of this we have tested the nitroso metabolites of a variety of aromatic amine carcinogens and have shown that they are very potent frameshift mutagens (Ames et al. 1972). Among these compounds are nitrosonaphthalene, nitrosobiphenyl, nitroso-trans-stilbene, nitrosoazobenzene, and nitrosophenanthrene. We have also repeated the work of Hartman and shown that 4-nitroquinoline-N-oxide, a well-known carcinogen, is another member of this class (Ames, Lee and Durston 1973).

The Coupling of a Carcinogen Activation System from Liver to the Bacterial Test System

The true or primary carcinogen or mutagen is often a metabolic product of the secondary carcinogen or mutagen that was originally ingested. It is known that mammalian liver microsomal hydroxylase systems are responsible for activating many classes of carcinogens and mutagens; among these are aflatoxin, aromatic amines, and polycyclic hydrocarbons.

We have now coupled a rat (and human liver) microsomal hydroxylase system to our tester strains in a simple manner (Ames et al. 1973). A mitochondria-free supernatant preparation from rat (or human) liver is used and we have found that the preparation is stable for months in the freezer. The conditions for spreading this

107

microsomal system on the petri plates along with our tester bacteria have been worked out so that the variety of metabolites that are made by the microsomal system from a putative carcinogen or mutagen can be tested. With this coupled system a large number of carcinogens that are not directly mutagenic can now be detected as mutagens. Among the carcinogens that are metabolized by liver to active frameshift mutagens under these conditions are: aminobiphenyl, amino and dimethylamino *trans*-stilbene, 2-amino and 2-acetylamino fluorene, aminoanthracene, benzidine, aminopyrene, aminochrysene, dimethyl-benzanthracene, benzyprene and aflatoxin B_1. H. Malling has shown that dimethyl-nitrosamine is activated to a product that causes base pair substitutions in one of our tester strains. We have been able to detect as a mutagen almost every carcinogen we have tested.

The Salmonella test system can detect carcinogens with great sensitivity. In any system for detecting mutagens one only scores mutations in a small part of the genome. Thus the revertant colonies we see represent only a tiny fraction (10^{-2} to 10^{-4}) of the bacteria mutated. Nevertheless, because of the sensitivity of the tester strains and the potency as mutagens of the activated carcinogens one can detect nanograms of carcinogen: e.g., 0.5 μg of 2-aminoanthracene gives 11,000 revertant colonies per petri plate, compared to a control of about 30 colonies (Ames *et al.* 1973). A simple bacterial test can thus be used to see if a compound is likely to be a carcinogen for humans (mammalian testing is of course expensive and takes years).

Carcinogens Are Somatic Mutagens

We believe that the simplest interpretation of our own and others' work on carcinogens as mutagens is that carcinogens cause cancer because of their action as mutagens. The principle of the mutagenesis for a large group of the carcinogens can be explained by the theory of frameshift mutagenesis, and hopefully will enable the prediction of what will be carcinogenic to be put on a more rational basis.

The Bearing of the Test Results on Evaluating Human Hazards

Are all compounds that are mutagens for bacteria going to be mutagens for mammals? We would be astonished if this were so. In animal cells, many additional factors will be involved, such as unique metabolic processes, which will be likely to interfere with the mutagenic activity of some compounds. Compounds which give a positive response in the test, however, should be considered potentially hazardous for man and should be scrutinized as to benefit, risk, and need for extensive testing.

It seems unlikely that any single test will uncover every carcinogen. We believe, from our present results, that a large proportion of the carcinogens that have been reported can be detected using this test, but in certain cases there may be other factors involved such as complex activation processes or the existence of carcinogens which work through mechanisms other than somatic mutation. We also know that the test as currently used is not complete and we are working to improve it still further.

The ultimate test of any *in vitro* screening procedure such as this will be not that it is 100% correlated with mutagenicity or carcinogenicity in man (this seems unlikely for any one test), but that it has a predictive value. I think this has already been demonstrated for this test, and will become more apparent by its use. Clearly a variety of tests should be used. This one is cheap, technically simple, and quite effective when calibrated against the known carcinogens and mutagens. Another reason for turning to microorganisms is the importance of the potential hazard for humans and the technical difficulties, expense, and time in determining mutagenicity and carcinogenicity in higher organisms.

Specificity and Characterization of Tester Strains

We have made good progress in characterizing our frameshift tester strains (Ames, Lee and Durston 1973). One of the frameshift esters has a repetitive C sequence at the site of the mutation and another has been shown by Isono and Yourno to have a repetitive CGCGCGCG sequence. The different carcinogens and mutagens causing frameshift mutations have a great deal of specificity as to the tester strains they revert and thus for the repetitive DNA sequence in each tester strain. Our set of three tester strains for detecting frameshift mutagens is not yet complete. We have developed a methodology and are attempting to complete the theoretically expected set of six frameshift tester strains containing long DNA sequences repetitive for one or two bases. We are also lengthening our present known repetitive sequence. A program has been started to characterize the DNA sequences at the site of mutations by sequencing the various mutant proteins.

Urine and Mammalian Metabolism

Metabolites of drugs and of food additives and other ingested compounds are likely to appear in the urine. We have found that we can couple urine (rat or human) and liver extract and our tester strains in a sensitive assay, and this is now feasible as a mass screen of the human population (Durston and Ames 1974).

This work was supported by U.S. AEC Grant AT(04–3)34, P.A. 156. I am indebted to numerous collaborators (mentioned in LITERATURE CITED) who have contributed enormously to this work and to P. E. Hartman and J. R. Roth for many strains and for stimulating discussions.

Literature Cited

Ames, B. N., 1971   The Detection of Chemical Mutagens with Enteric Bacteria. pp. 267–282. In *Chemical Mutagens: Principles and Methods for Their Detection*. Vol. 1. Edited by A. Hollaender. Plenum Press, New York, N.Y. ————, 1972 A Bacterial System for Detecting Mutagens and Carcinogens. pp. 57–66. In *Mutagenic Effects of Environmental Contaminants*. Edited by E. Sutton and M. Harris. Academic Press, New York, N.Y.

Ames, B. N. and H. J. Whitfield, Jr., 1966   Frameshift Mutagenesis in Salmonella. Cold Spring Harbor Symp. Quant. Biol. **31**: 221–225.

Ames, B. N., F. D. Lee and W. E. Durston, 1973   An Improved Bacterial Test System for the Detection and Classification of Mutagens and Carcinogens. Proc. Natl. Acad. Sci. U.S. **70**: 782–786.

Ames, B. N., P. Sims and P. L. Grover, 1972   Epoxides of Polycyclic Hydrocarbons Are Frame Shift Mutagens. Science **176**: 47–49.

Ames, B. N., W. E. Durston, E. Yamasaki and F. D. Lee, 1973   Carcinogens Are Mutagens. A Simple Test System Combining Liver Homogenates for Activation and Bacteria for Detection. Proc. Natl. Acad. Sci. U.S. **70**: 2281–2285.

Ames, B. N., E. G. Gurney, J. A. Miller and H. Bartsch, 1972   Carcinogens as Frameshift Mutagens: Metabolites and Derivatives of Acetylaminofluorene and Other Aromatic Amine Carcinogens. Proc. Natl. Acad. Sci. U.S. **69**: 3128–3132.

Creech, H. J., R. K. Preston, R. M. Peck, A. P. O'Connell and B. N. Ames, 1972   Antitumor and Mutagenic Properties of a Variety of Heterocyclic Nitrogen and Sulfur Mustards. J. Med. Chem. **15**: 739–746.

Durston, W. E. and B. N. Ames, 1974   A Simple Method for the Detection of Mutagens in Urine Studies with the Carcinogen 2-Acetylaminofluorene. Proc. Natl. Acad. Sci. U.S. **71**: 737–741.

11 Cytogenetic Studies with Cyclamate and Related Compounds

D. R. Stoltz, K. S. Khera, R. Bendall, and S. W. Gunner

ABSTRACT. *Cyclamate, cyclohexylamine, N-hydroxycyclohexylamine, and dicyclohexylamine can induce chromosomal damage in human leukocyte cultures.*

Chemically induced chromosomal aberrations are being used as a rapid indicator of potential carcinogenic, mutagenic, and teratogenic activity. Although complete justification for scoring chromosomal aberrations as a toxicological parameter is not available, cytologists' premonitions are sometimes confirmed by oncologists' discoveries. Such may be the case for the artificial sweetener, cyclamate (sodium and calcium salts of cyclohexanesulfamic acid). Three reports of cytogenetic damage (*1*) induced by cyclamate and by its degradation product, cyclohexylamine, preceded the announcement of cyclamate's possible carcinogenic capabilities and the subsequent restrictions on dietary consumption. We report here an analysis of cytogenetic effects after exposure of human lymphocyte cultures to sodium cyclamate, cyclohexylamine sulfate, *N*-hydroxycyclohexylamine hydrochloride (a metabolite of cyclohexylamine), and dicyclohexylamine sulfate (an occasional contaminant of cyclamate).

Chromosome preparations were made from human whole blood cultures. All

From *Science*, **167** (Mar. 13, 1970), 1501–1502. Copyright 1970 by the American Association for the Advancement of Science. Reprinted with permission of the authors and the publisher.

cultures were prepared with the same ingredients by mixing together medium, serum, phytohemagglutinin M (2), and blood, and by distributing aliquots by automatic syringe into culture bottles. Cultures were exposed to the four compounds at three concentrations (10^{-3}, 10^{-4}, and 10^{-5} M) during two periods of growth, the final 5 and 25 hours of culture. Duplicate cultures of cells were treated with colchicine (1 μg/ml) for 4 hours to give a total incubation time of exactly 72 hours. Fifty cells were analyzed from each culture.

All chromosomal abnormalities were scored. The majority of aberrations were gaps and breaks (3); exchange figures and unusual chromosomes of unknown derivation were observed infrequently (4) and only in slides from treated cultures. The occurrence of the more unusual aberration types was otherwise quite random, being associated neither with a particular compound nor with treatment concentration nor duration. There was essentially no difference between the results of 5- and

Figure 11–1. Total chromosomal abnormalities as a function of dose. Each point represents the analysis of 100 cells. The variation between cultures, if any, is shown. (A) Cyclamate. (B) Cyclohexylamine. (C) N-Hydroxycyclohexylamine. (D) Dicyclohexylamine. Solid line, 25-hour treatment. Broken line, 5-hour treatment. Dotted line, control.

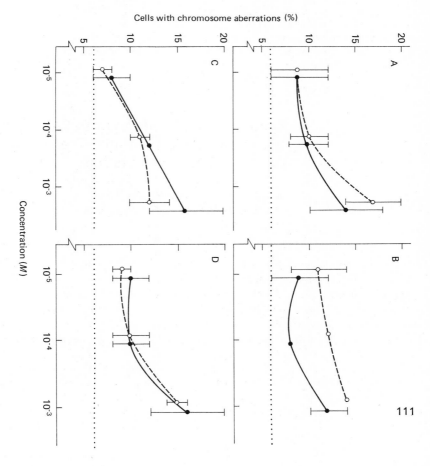

Cells with chromosome aberrations (%)

Concentration (M)

111

25-hour exposures. There appeared to be little if any difference among the effects of these compounds (Figure 11-1). These results confirm reports that cyclamate and cyclohexylamine at high concentrations can induce chromosomal aberrations. In addition, N-hydroxycyclohexylamine and dicyclohexylamine have similar effects.

Drug-induced cytogenetic damage may be associated with either mutagenesis or cytotoxicity. Although these two possible actions appear indistinguishable on the basis of evidence obtained from short-term leukocyte cultures, the technique apparently has value in the selection of potential carcinogens for long-term testing programs.

References and Notes

1. K. Sax and H. J. Sax, *Jap. J. Genet.* **43**, 89 (1968); D. Stone, E. Lamson, Y. S. Chang, K. W. Pickering, *Science* **164**, 568 (1969); M. S. Legator, K. A. Palmer, S. Green, K. W. Petersen, *ibid.* **165**, 1139 (1969).
2. Medium 199 was supplemented with 20 percent calf serum (Grand Island Biological Co.) and 2.5 percent (by volume) phytohemagglutinin M (Difco Laboratories). Blood was obtained by venipuncture from a 25-year-old male with no recent viral infection, medication, or diagnostic irradiation.
3. Gaps and breaks are defined as achromatic lesions without and with displacement of the fragment, respectively.
4. Mode 1 percent; range 0 to 3 percent per culture.
5. We thank Drs. A. B. Morrison, W. P. McKinley, and H. C. Grice for advice and encouragement.

12   Chromosome Studies on Patients (in Vivo) and Cells (in Vitro) Treated with Lysergic Acid Diethylamide

Margaret J. Corey, J. C. Andrews,
M. Josephine McLeod, J. Ross MacLean, and
W. E. Wilby

ABSTRACT. *In a prospective study of 10 patients given d-lysergic acid diethylamide 25, there was no difference in frequency of chromosome breakage between samples obtained immediately before and 24 hours after treatment. In 11 patients, treated over periods ranging from 24 hours to eight years before sampling, the frequency of chromosomal breaks did not differ from that found in untreated controls. In an in vitro study the frequency of chromosomal breaks was increased in replicate cultures from each of*

From *New England Journal of Medicine*, **282** (Apr. 23, 1970), 939–943. Copyright 1970 by the Massachusetts Medical Society. Reprinted with permission of the authors and the publishers.

10 subjects when 1 μg per milliliter of d-lysergic acid diethylamide 25 was added during the last 24 hours of culture. There is no cytogenetic evidence that d-lysergic acid diethylamide 25 given therapeutically produces chromosomal damage. The chromosomal aberrations found after illicit use of the drug remain unexplained.

The possibility that d-lysergic acid diethylamide 25 (d-LSD 25) ingestion induces chromosomal abnormalities was raised in 1967 by Cohen and his associates[1] after in vitro studies. Since that time numerous reports of the effects of LSD on chromosomes of circulating lymphocytes have appeared. Some have indicated an increased frequency of chromosome breakage,[2-5] others have found no increase,[6-8] and one report has suggested a slight transient increase.[9]

In addition to the lack of agreement of reports from different laboratories, there are some unusual features within the results of studies that have reported an increase in chromosomal damage. Within each study, increased aberrations have not been found in all subjects exposed, and the frequency of aberrations appears to have no relation to the dose or to the time elapsed between exposure and testing.[2, 5] Although this suggests a variation in response to the drug the relation of the variation to the nature of the drug is difficult to assess because these studies have used experimental groups in which information concerning the source, composition, frequency of use and dosage of the drug could not be accurately determined.

In an attempt to clarify some of these unexplained effects, a series of studies were carried out using as experimental subjects patients who were treated under clinical conditions with known amounts of d-LSD 25. This report deals with the results from three separate studies of the frequency and types of chromosomal aberrations, including in vivo prospective and retrospective studies and an in vitro study.

Materials and Methods

EXPERIMENTAL SUBJECTS. All subjects included in the study were interviewed to obtain information concerning exposure to known chromosome-breaking agents. None had received extensive therapy with or exposure to radiation or antitumor agents. Exposure to routine x-rays was noted, but those who had received routine chest or dental x-rays were not excluded from the study. All subjects had normal karyotypes.

All treated patients were drawn from a private psychiatric-hospital population. Details of selection for each experiment are presented with the results.

DRUG ADMINISTRATION. The LSD used in all experiments was d-lysergic acid diethylamide 25 (Sandoz). The "mescaline" was mescaline sulfate (Light).

The d-LSD 25 was dissolved in distilled water and administered orally. The mescaline sulfate was administered in capsule form. No other drugs were administered to modify or terminate the effects of the psychoactive drug.

113

CYTOGENETIC TECHNICS. All chromosome analyses were carried out on short-term leukocyte cultures in a commercially prepared medium (GIBCO chromosome medium 1A). In all experiments cultures designated as controls were incubated in the same medium and at the same time.

Ten slides from each treated and each control culture were coded to ensure that the examiner could not identify the source of a slide, and, where possible, 10 cells from each slide were analyzed. Cells for analysis were selected with a low-power objective, and once a cell had been selected, it was included in the final analysis. Cells were analyzed under the miscrocope and photographed, and a second analysis carried out on a photographic print. Analysis of each cell included a chromosome count as well as examination for structural aberrations. For the purpose of this study, a break was defined as a lesion in which the proximal end of the acentric fragment had been displaced from a position of alignment with the distal end of the centric fragment. Achromatic lesions that did not meet this criterion were scored as gaps and were not included as breakage events in the analysis. Chromatid and isochromatid breaks, acentric fragments and deletions were scored as single-break events, whereas exchange configurations, dicentric chromosomes, ring chromosomes and translocations were scored as two-break events. Slides were not decoded until scoring had been completed.

Although specific types of aberrations were scored individually, the values were too low for statistical analysis. However, an analysis was carried out on total single-break aberrations and total two-break aberrations as well as total breaks.

Results

IN VIVO PROSPECTIVE STUDY—FREQUENCY AND TYPE OF CHROMOSOMAL ABERRATIONS BEFORE AND AFTER TREATMENT WITH LSD. Ten subjects with no prior use of d-LSD 25, other "psychedelic" agents or narcotics were drawn in consecutive order of treatment with d-LSD 25.

Blood samples for chromosome analysis were obtained immediately before LSD therapy and a second sample was obtained 24 hours later. In this experiment each sample obtained before the initiation of treatment was used as a control for the sample obtained from the same subject 24 hours after treatment, and the difference between the two samples was used as a measure of the effect of treatment. Drug doses ranged from 200 to 600 μg.

Before treatment the frequencies of total chromosome breakage events ranged from 1 to 9 breaks per 100 cells, with a mean of 5.7. After treatment the frequencies ranged from 2 to 8 breaks per 100 cells, with a mean of 4.9.

The differences between total breaks per 100 cells before and after LSD therapy ranged from −4.0 to +4.0, with a mean of −0.80. Single-break aberration differences ranged from −4.0 to 5.0, with a mean value of 0.20 breaks, and two-break aberration differences from 0 to −4.0, with a mean of −1.0 per 100 cells scored.

The results of paired t-test analysis are presented in Table 1. There was no significant change in the frequency of single-break aberrations, total breaks or gaps.

114

Table 1. Prospective in Vivo Study—Paired t-Test Analysis of Chromosomal Aberrations Before and After Treatment with d-LSD 25.

| | Difference Between Before and After Samples* | | | |
	Gaps	1-Break aberrations	2-Break aberrations	Total breaks
Mean	−0.30	0.20	−1.00	−0.80
t-value	−0.131	0.200	−2.236	−0.873
Probability	0.867	0.825	0.050†	0.409

* 100 cells analyzed from each before and each after samples
from 10 patients.
† Significant.

The analysis did indicate a significant reduction in the frequency of aberrations attributed to two-break events. The frequency of these events in both samples was low, with 1.8 per 100 cells before and 0.8 per 100 cells after treatment. Two-break events were never observed after treatment in persons who had not displayed similar events before treatment, and in four out of five, the frequency was decreased in the sample obtained 24 hours after treatment.

IN VIVO RETROSPECTIVE STUDY—FREQUENCY AND TYPE OF CHROMOSOMAL ABERRATIONS IN PATIENTS TREATED WITH LSD, MESCALINE OR LSD AND MESCALINE AND A CONTROL GROUP. The treatment center from which the patient population was being drawn had available persons who had been treated with varying LSD dosages over a number of years, and a pilot project was carried out to determine whether or not some indication of dose-time response might be elicited. Since those who had received the largest dosage of LSD had also received mescaline, a group who had been treated solely with mescaline was also included.

This study included four groups: five patients treated with d-LSD 25 only; five treated with mescaline sulfate only; six treated with both d-LSD 25 and mescaline sulfate; and 13 control subjects exposed to neither drug.

Illness and treatment with other drugs were uncontrolled and largely undocumented although four patients in the treated group are known to have used prescribed or illicit drugs.

Patients included in this study had LSD dosages varying from 200 to 4350 μg over periods ranging from 24 hours to eight years. The total number of breaks per 100 cells ranged from six to 12 for LSD, two to 13 for mescaline, two to 11 for LSD plus mescaline and two to 17 in the control group.

With the exception that the maximum number of single breaks in an LSD-treated patient was 12, and the maximum in the control population was 11 breaks per 100 cells, the range of aberration frequencies in each of the categories for all treated groups was within the range observed in the control group. The results of an analysis of variance (Table 2) indicated no significant differences between treated

115

Table 2. Retrospective in Vivo Study—Analysis of Variance.

Group	No. of Patients	Total Cells	1-Break Aberrations	2-Break Aberrations	Total Breaks	Gaps
			Mean per 100 cells			
LSD	5	500	6.200	1.600	7.800	20.000
Mescaline	5	500	3.600	2.000	5.600	7.400
LSD + mescaline	6	594	4.708	1.600	6.400	15.418
Control	13	1300	5.923	1.077	7.000	16.923
F value			1.06	0.38	0.31	2.35
p value			0.386	0.773	0.819	0.095

and control groups for single-break, two-break or total-break events. Within the LSD-treated population there was no relation between the frequency of aberrations and either LSD dosage or the time elapsed between treatment and sampling.

IN VITRO STUDY—CHROMOSOMAL ABERRATION FREQUENCIES IN REPLICATE SAMPLES CULTURED WITH AND WITHOUT THE ADDITION OF d-LSD 25. The experimental group for this study consisted of 10 university student and faculty members who had never ingested LSD. Blood samples obtained from each were used to initiate replicate cultures. d-LSD 25 in aqueous solution was added to one replicate of each culture in a final concentration of 1 μg per milliliter 24 hours before the culture was terminated. An equal volume of sterile distilled water was added to the control replicates.

Table 3. In Vitro Study—Paired t-Test Analysis of Chromosomal Aberrations in Replicate Samples Cultured With and Without d-LSD 25.*

	Difference Between Treated and Untreated Replicates			
	Gaps	1-Break aberrations	2-Break aberrations	Total breaks
	per 100 cells			
Mean	7.04	3.96	0.69	4.65
t-value	3.481	9.764	1.428	9.598
Probability	0.007†	0.000†	0.185	0.000†

* Approximately 100 cells analyzed from each treated and each untreated replicate from 10 subjects (1010 treated cells, 1094 untreated cells).
† Highly significant.

The number of breaks per 100 cells in the control cultures ranged from 0 to 15.12, with a mean of 4.72. In the treated replicates the number of breaks per 100 cells ranged from 4.0 to 18.70, with a mean of 9.37.

An increase in both single-break aberrations and total breaks was observed in each of the 10 treated replicates. The differences between treated and control replicates for single breaks ranged from 2.0 to 6.0, with a mean of 3.96, and the differences in total number of breaks per 100 cells scored ranged from 3.0 to 7.9, with a mean of 4.65. Two-break aberrations were increased in four treated replicates and decreased in one, and no difference was observed in five samples. Differences ranged from −2.0 to 3.3, with a mean of 0.69 per 100 cells scored.

A paired *t*-test analysis (Table 3) indicated a significant increase in single breaks and total breaks but not in two-break aberrations.

Discussion

The in vitro study reported here partially duplicates the experiments reported by Cohen, Marinello and Back[1] and Cohen, Hirschhorn and Frosch.[2] Using culture, treatment and scoring technics similar to those described by the above authors, we obtained very similar results. Cohen, Hirschhorn and Frosch[2] reported a range of 3.7 to 17.5 breaks per 100 cells in the treated cultures as compared to an average of 3.8 breaks per 100 cells in control cultures. In the study reported here the frequency of breaks in treated cultures ranged from 4.0 to 18.7, with a mean control value of 4.7 breaks per 100 cells.

An analysis of differences between untreated and treated cultures indicated a consistent and highly significant increase of chromosomal breakage in the treated replicate cultures. Although the in vitro concentration was much greater than any known ingested dosage the mean increase of 4.65 breaks per 100 cells is small as compared to the range of breakage frequencies (0.0 to 15.2) observed in the untreated cultures. The deviation range of less than 5 breaks per 100 cells indicated no variation in response from subject to subject.

Using the same culture and scoring technics, we found no indication of an increased frequency of chromosomal breakage in subjects who had ingested LSD. Neither the direct comparison of samples obtained immediately before and again 24 hours after treatment nor the study of persons exposed to varying doses over a given time indicated any increased frequency in chromosome breaks. The latter group is similar to those reported by Cohen, Hirschhorn and Frosch[2] and Egozcue, Irwin and Maruffo[5] with one exception: these previous reports involved subjects who had ingested self-administered illicit "LSD," whereas the present study was carried out on patients treated with pharmaceutical d-LSD 25 under clinical conditions.

Since the results of the in vitro studies indicate that the scoring systems are sufficiently alike to produce similar results, some difference between the experimental groups appears to be the most likely source of lack of concurrence in the results of in vivo studies.

There is further evidence that differences in the experimental groups influence the results. To date all indications that LSD ingestion is associated with significant increase in the frequency of chromosomal breakage are based on the study of subjects who have ingested "LSD" under nonclinical conditions.[2-5] On the other hand, studies on patients treated with LSD under clinical conditions have not indicated similar increases.[7, 8]

One serial study on four clinically treated subjects[9] indicated a slight transient increased breakage frequency that returned to normal on follow-up observation. In that report LSD was injected, and treatment was in some cases combined with chlorpromazine. In the present prospective study LSD was administered by mouth, with no modifying drugs. In the larger population involved in this study the pretreatment sample shows a greater variation in aberration frequencies than is indicated in the study reported by Hungerford and his associates.[9] However, the deviation between samples before and after treatment had a mean value of −0.80 breaks per 100 cells and was not significant.

Although both Hungerford et al.,[9] in a prospective in vivo study, and Cohen, Hirschhorn and Frosch,[2] in an in vitro study, have reported the occurrence of two-break aberrations (that is, dicentric, exchange figures and so forth) only in samples from treated patients or cultures, a low frequency of similar aberrations was observed in samples from untreated subjects in both the prospective and the in vitro studies reported here. This discrepancy may be due to the selection of subjects. In the present study the effect of LSD was measured by the difference between treated and untreated cultures from the same person, and study of possible variation in response was one of the objectives. Thus, the use of rigid selection criteria was both unnecessary and undesirable. However, the control values are not necessarily representative of a "normal" or a "drug-free" population.

In the prospective study the frequency of two-break events was significantly reduced in samples obtained after treatment, and two-break aberrations were observed after treatment only in patients who had similar aberrations before treatment. In the in vitro study the frequency of two-break aberrations was not significantly increased.

The reduction in two-break aberrations in the prospective study is difficult to explain. However, the 24-hour interval between samplings is remarkable in that the patients were exposed to virtually nothing but LSD. Assuming that the LSD had no influence on the chromosomes, the reduction might be due to the absence of other common environmental agents that contribute to the breakage frequencies observed in control groups.

In conclusion, there appears to be no cytogenetic evidence that d-LSD 25 given therapeutically produces chromosomal damage. On the other hand, there is increasing evidence that populations ingesting "illicit" drugs should be further investigated to identify the source of the increased frequency of chromosome aberrations.

We are indebted to Mr. J. Bjerring, of the University of British Columbia computer center, for assistance with the statistical analysis of the data.

References

1. Cohen, M. M., Marinello, M. J., Back, N. Chromosomal damage in human leukocytes induced by lysergic acid diethylamide. Science 155:1417–1419, 1967.
2. Cohen, M. M., Hirschhorn, K., Frosch, W. A. In vivo and in vitro chromosomal damage induced by LSD-25. New Eng J Med 277:1043–1049, 1967.
3. Cohen, M. M., Hirschhorn, K., Verbo, S., et al. The effect of LSD-25 on the chromosomes of children exposed in utero. Pediat Res 2:486–492, 1968.
4. Irwin, S., Egozcue, J. Chromosomal abnormalities in leukocytes from LSD-25 users. Science 157:313–314, 1967.
5. Egozcue, J., Irwin, S., Maruffo, C. A. Chromosomal damage in LSD users. JAMA 204:214–218, 1968.
6. Loughman, W. D., Sargent, T. W., Israelstam, D. M. Leukocytes of humans exposed to lysergic acid diethylamide: lack of chromosomal damage. Science 158:508–510, 1967.
7. Bender, L., Sankar, D. V. S. Chromosome damage not found in leukocytes of children treated with LSD-25. Science 159:749, 1968.
8. Sparkes, R. S., Melnyk, J., Bozetti, L. P. Chromosomal effect in vivo of exposure to lysergic acid diethylamide. Science 160:1343–1345, 1968.
9. Hungerford, D. A., Taylor, K. M., Shagass, C., et al. Cytogenetic effects of LSD 25 therapy in man. JAMA 206:2287–2291, 1968.

13   Chromosome Breakage in Humans Exposed to Methyl Mercury Through Fish Consumption

Staffan Skerfving, Kerstin Hansson, and Jan Lindsten

ABSTRACT. *Chromosome analysis was performed on cells from lymphocyte cultures from nine subjects with increased levels of mercury in their red blood cells and in four healthy controls. The elevated mercury levels were likely to have originated from dietary fish with high levels of methyl mercury. A statistically significant rank correlation was found between the frequency of cells with chromosome breaks and mercury concentration. The biological significance of these findings is at present unknown.*

A large number of chemicals are known to induce chromosome breakage when added to cell cultures in sufficiently high concentrations for a sufficient length of time.[1] Cells grown in vitro from subjects exposed to chemicals, mainly cytostatic drugs, ionizing radiation, and certain viruses, have also shown an increassd frequency of broken chromosomes. Azathioprine (Imuran) and methotrexate (Amethopterin [Britain]) have been observed to cause chromosome breakage in vivo, as

From *Arch. Environ. Health*, **21** (Aug. 1970), 133–139. Copyright © 1970 by the American Medical Association. Reprinted with permission of the authors and the publisher.

studied in bone marrow cells.[2] It is not yet known if substances to which large human populations are exposed may have such an effect, although cyclamate and cyclohexamine induced chromosome breakage in rats exposed in vivo[3] and in human cells exposed in vitro.[4, 5]

Fish and shellfish contaminated with methyl mercury in polluted waters have caused two epidemics of neurological intoxication in Japan.[6, 7] In contrast to methyl mercury, phenyl, methoxyethyl, and inorganic mercury compounds are considered less toxic.[8] The latter compounds are widely used in certain industrial processes and are discharged into seas, lakes, etc. However, recent studies have shown that they are metabolized into methyl mercury by microorganisms in bottom sediments.[9] Furthermore, considerably increased levels of methyl mercury have been demonstrated during the last five years in fish caught in large areas of Swedish,[10, 11] Finnish,[12] Norwegian,[13] and Danish[14] lakes and coastal waters contaminated by industrial waste of phenyl and inorganic mercury. Human subjects consuming such fish show elevated blood concentrations of mercury.[15, 17] Consequently, selling of fish containing more than 1 mg of mercury per kilogram (1 ppm) of fish for human consumption has been inhibited in Sweden since 1967. The Swedish health authorities have also officially advised the public not to eat fish from other, less polluted areas more than once a week.

Cytologic studies on *Allium* roots have shown that methyl, ethyl, butyl, phenyl, and methoxyethyl mercury compounds induce C-mitosis, polyploidy, and aneuploidy.[18, 19] Methyl and phenyl mercury compounds also induced chromosome fragmentation.[18] Furthermore, nondisjunction of the X chromosome in *Drosophila* eggs has been observed after treatment with these compounds.[20] A C-mitotic effect has also been shown in human leukocytes exposed in vitro to methyl and methoxyethyl mercury chloride.[21] Thus, several alkyl mercury compounds, phenyl mercury, and methoxyethyl mercury have been shown to induce chromosome abnormalities in different organisms. We here report on a preliminary study on the occurrence of an increased frequency of chromosome breaks in cells from lymphocyte cultures from nine subjects known to have increased blood levels of mercury.

Materials and Methods

Nine subjects who had consumed fish contaminated with methyl mercury at least three times a week for more than five years were studied. The fish consumption habits have been studied in detail through dietary histories. The mercury levels of the fish consumed were in the range of 1 to 7 ppm.[15, 22-26] Three of the subjects had stopped consuming contaminated fish 5 to 18 months before the chromosome analysis, while the other six were still exposed (Table 1). Four somewhat older healthy subjects were studied simultaneously as controls. None of them had a history indicating consumption of contaminated fish. They had had sea fish (mercury level about 0.05 ppm or less) once a week or even less frequently.

Ten of the subjects, including the controls, had apparently been healthy and a physical examination revealed no signs of disease. None of them had an occupational history of exposure to mercury or other chemicals, and none reported a recent history of viral infection, drug consumption or diagnostic x-ray examination. None had been subjected to radiological treatment. Subject 1 was on a regular medication of isoproterenol and ethacrynic acid (Edecrin) because of bronchial asthma and hypertension. Subject 5 occasionally used pentobarbital as a sleeping pill. Subject 7 was on a regular digoxin (Lanacrist [Sweden]) treatment on account of slight heart decompensation.

Mercury concentrations in whole blood, red blood cells (RBC) and plasma were determined by activation analysis.[2] Chromosome analysis was performed on lym-

Figure 13–1. Mitotic lymphocytes from blood cultures. Top left, Subject 12 with balanced D/D translocation. Four D chromosomes only and three 3 chromosomes, one of which is involved in formation of dicentric chromosome (arrow) together with B chromosome. Top right, Subject 5 with chromatid breakage on one chromosome, 3, and an acentric fragment (arrows). Bottom left, Subject 9 with two dicentric chromosomes and four acentric double fragments (arrows). Bottom right, subject 1 with one dicentric chromosome (arrow) probably involving chromosomes 1 and 3.

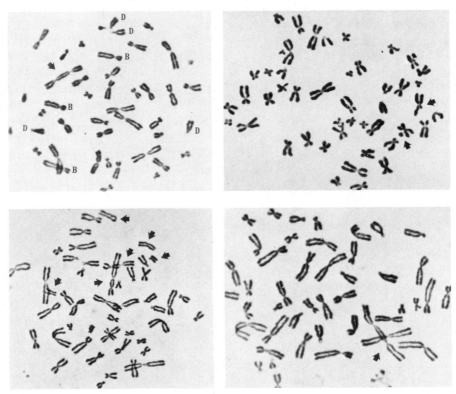

Table 1. Blood Levels of Mercury and Results of the Cytogenetic Analysis.

Subject*	Age at Examination (yr)	Mercury Levels (mμg/gm)				Lymphocyte Culture		No. of Analyzed Cells						
		Date	Whole blood	RBC	Plasma	Date	Time of incubation (hr)	Total	Hypo-diploid†	Hyper-diploid†	Chromatid breaks only	Chromosome breaks and/or structural re-arrangements	Total no. of cells with breaks	%
1 F	47	3/28/68	124	190	22	10/19/68	96	200	16	2	2	2	4	2.0
		10/19/68	—	330	33		120	200	14	4	7	7	14	7.0
2 M	53	3/28/68	43	86	8	10/19/68	96	200	4	0	5	10	15	7.5
		10/19/68	—	200	16		120	200	9	2	2	12	14	7.0
3 M	55	12/3/66	650	—	—	4/24/68	48	200	5	0	4	7	11	5.5
		1/4/68	108	178	23		96	200	9	1‡	0	12	12	6.0
4 M	64	12/15/67	66	126	14	3/12/68	72	200	6	4	6	19	25	12.5
5 F	66	3/28/68	113	242	19	10/19/68	96§	200	6	1	3	5	8	4.0
		10/19/68	—	370	27		120	200	10	2	3	16	19	9.5
6 F	66	12/15/67	84	160	17	3/18/68	72	200	7	2	6	9	15	7.5
7 M	68	1/24/68	98	181	17	6/10/68	72	200	8∥	0	5	1	6	3.0
		6/10/68	114	186	19									
		11/18/68	—	80	—		96	156	9	0	0	2	2	1.3
8 M	72	1/26/68	42	82	7	4/22/68	48	65	0	0	0	3	3	4.6
		6/10/68	—	48	5	9/23/68	72	200	8∥	0	0	1	1	0.5
		11/26/68	—	21	—		96	200	8∥	0	0	2	2	1.0
							120	200	7∥	1	0	2	2	1.0

Subject	Sex	Age	Date				Date	Hrs	Cells						Index
9	M	78	1/26/68	27	43	8									
			6/7/68	50	100	10									
			11/26/68	—	89	—									
							6/7/68	48	159	5	0	2	2	4	2.5
								72	162	11	0	0	1	1	0.6
								96	200	4	0	1	0	1	0.5
								120	200	14	0	0	4	4	2.0
10	F	56	11/14/68	—	12	3									
							11/11/68	96§	200	9	3	1	3	4	2.0
								120	200#	3	2	0	4	4	2.0
11	M	80	6/24/68	7	11	2									
			10/2/68	—	6	—									
							6/24/68	72	71	3	0	0	0	0	0
								96	36	7	0	0	0	0	0
								120	39	2	0	1	1	2	5.1
12	M	81	6/24/68	11	17	2									
			11/14/68	—	13	—									
							9/30/68	72§	200	4	2	1	1	2	1.0
							11/11/68	96	200	3	0	1	5	6	3.0
							12/2/68	48	60	1	0	0	1	1	1.7
								72	200	5	3	2	9	11	5.5
								96	200	7	1‡	2	6	8	4.0
								120	200	6	0	1	4	5	2.5
13	M	83	6/24/68	4	5	1									
							10/1/68	48	200	5	1‡	1	6	7	3.5
								72	200	5	1‡	1	5	6	3.0
								96	200	15	3‡	2	5	7	3.5

* Subjects 10 to 13 were controls; 12 M turned out to have a balanced D/D translocation; subjects 3, 4, and 6 finished consumption of contaminated fish in November 1966, August 1967, and October 1967, respectively.
† Calculated as number of centromeres.
‡ One cell had an XYY constitution.
§ Mercury analysis on the cell culture medium with and without cells was performed in subjects 5, 10, and 12; 9 and 5 mμg per gram, respectively, was found for subject 5, 0.4 and 0.3 for subject 10, and 0.7 and 0.5 for subject 12.
|| Six cells lacked the Y chromosome.
¶ Three cells lacked the Y chromosome.
# Six cells were endoreduplications.

phocytes grown in vitro for different periods of time (Table 1). All slides from exposed, as well as unexposed, subjects were coded and analyzed blindly. The chromosomes in 200 consecutive cells were counted if possible and analyzed for breakage and for structural rearrangements. Only clearly dislocated chromosome sections were scored as fragments and breaks (Figure 13–1), whereas gaps and doubtful abnormalities were disregarded.

Results

Blood levels of mercury and the frequency of chromosome abnormalites are shown in Table 1 and Figure 13–2. Mercury levels analyzed prior to or after the chromosome analysis have also been included to show variations in exposure with time.

The mercury concentrations in RBC are considered to be the most reliable index available of exposure to methyl mercury.[8] The control subjects had levels in agreement with earlier studies on larger materials from the general population with the same fish-consumption habits.[28] The exposed subjects had mercury levels in RBC between 21 and 370 mμg per gram, i.e., about 2 to 40 times higher than unexposed subjects.

The frequency of cells with chromosome breakage varied considerably, both within and between individuals. However, there was no systematic difference between cell cultures incubated for different periods of time. Therefore, all cells from each individual were pooled in the statistical analysis. There was no statistic-

Figure 13–2. Mercury levels in RBC and mean frequency of cells with breaks for each of (see number) exposed (open circles) and control (filled circles) subjects. The range of variation is given when more than one analysis was made.

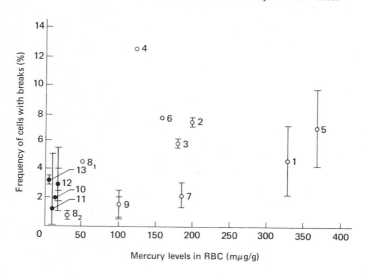

Table 2. Distribution of Breaks on Different Chromosomes in Comparison to the Relative Length of the Chromosomes.*

Chromosome†	1	2 p	2 q	2 cen	3	3 p	B q	B cen	B p	C q	C cen	D q	D p	E q	E cen	F	G q	Y q	Total
Mercury exposed subjects (%)	12 (8.6)	2 (1.4)	16 (11.5)	3 (2.6)	30 (21.6)	1 (0.7)	8 (5.8)	2 (1.4)	5 (3.6)	29 (20.9)	5 (3.6)	14 (10.1)	0	6 (4.3)	0	0	5 (3.6)	1 (0.7)	139
Controls (%)	4 (8.3)	0	3 (6.3)	1 (2.1)	12 (25.0)	0	2 (4.2)	1 (2.1)	3 (6.3)	12 (25.0)	2 (4.2)	2 (4.2)	1 (2.1)	1 (2.1)	1 (2.1)	1 (2.1)	0	2 (4.2)	48
Chromosome length in % of the total complement	9.08	3.3	5.15	—	7.06	3.49	9.19	—	14.19	26.29	—	10.55	3.02	6.12	—	4.85	3.51	1.96	—

* Nomenclature and mean values of arm and chromosome lengths from the Chicago Conference (1966) were used.
† p. short arm; q. long arm; cen. centromere.

ally significant difference in the frequency of chromosome breaks between the control and the exposed group. The difference in RBC mercury levels between the control and exposed groups approaches statistical significance at the $P = 0.05$ level. The control group is small and the standard deviation for RBC mercury levels in the exposed group is large; the lowest values do not differ much from the controls. So, the two groups were pooled in the further statistical analysis. The correlation ($r = 0.6$; $P < 0.05$) between the frequency of cells with chromosome breaks and mercury concentration in RBC, using Spearman's rank correlation test, was statistically significant. The frequency of chromosome breakage in cultured control cells agrees with the reports published by other authors (Court Brown[29]). When the chromosome breaks were divided into chromatid breaks on the one hand, and chromosome breaks and structural rearrangements on the other, only the latter type of breaks proved to be significantly correlated ($r = 0.6$; $P < 0.05$) with mercury concentration. No significant correlation could be demonstrated between mercury concentration and aneuploid or polyploid cells or endoreduplications.

The distribution of breaks in different chromosomes is shown in Table 2. Cells with fragments of unknown origin have been excluded and those with breaks in the centromere region of a chromosome have been listed separately. There was no significant difference in the distribution of breaks between the exposed and the control groups. Furthermore, the number of breaks corresponded well with the relative length of the chromosomes or group of chromosomes with one exception, chromosome 3. This chromosome, which is only about 7% in length of the total complement, contributed, respectively, 22% (exposed group) and 25% (control group) of the cells with breaks. These results are in agreement with earlier findings.[26]

Comment

The frequencies of cells with chromosome breaks showed great variation at the different mercury levels; this makes the interpretation of the results difficult. One possible explanation for this finding could be the variation in mercury levels with time. These levels depend on the consumption of contaminated fish and the biological half-life (50 to 80 days, as determined by whole body counting[31, 32] and blood counting[32] after intake of methyl mercuric nitrate Hg 203 or methyl mercuric proteinate Hg 203, and as calculated from repeated blood mercury analyses in intoxicated persons[33] for several months after the end of exposure). Some subjects had stopped consumption while others continued and even showed increasing mercury levels. If methyl mercury induces chromosome aberrations which persist in the body for a long period of time like those induced by ionizing radiation,[34] after the end of exposure we might get high frequencies in spite of low RBC mercury concentrations. Thus, the values obtained for subjects 3, 4, and 6, who had stopped consumption of contaminated fish 5 to 18 months before the chromosome analysis, might reflect earlier, higher mercury levels. The stability of the chromosome abnormalities observed may be studied in such subjects.

The biological significance of the present preliminary observation of a correlation between the frequency of cells with chromosome breaks and mercury concentration in RBC is as yet unknown. Studies on bone-marrow cells should show whether the same correlation occurs also in vivo. A long-term clinical follow-up study of exposed subjects with a high frequency of cells with chromosome breaks might indicate whether any diseases can be ascribed to the finding of cells with broken chromosomes.

Sulfhydryl groups, which are present also in the spindle apparatus of the dividing cell, bind mercury compounds. This has been regarded as the cause of the increased frequency of polyploidy and aneuploidy in *Allium* and *Drosophila*.[18, 20] In the present study we were unable to confirm these observations. Whether this is due to in vitro selection of such cells can also be invesitgated by direct analysis of bone-marrow cells.

In contrast to inorganic mercury and phenyl mercury, methyl mercury easily passes the placenta barrier at least in some species.[35, 36] The occurrence of prenatal intoxication in children shows that this is the case also in man.[6, 37] It remains to be investigated whether tissues from exposed fetuses have an increased frequency of cells with broken chromosomes. It is not yet known whether women, with high consumption of contaminated fish before or during pregnancy, have an increased frequency of spontaneous abortions; we have not been able to perform chromosome analysis on children of exposed mothers. Theoretically, methyl mercury-induced chromosome damage in germline cells could give rise to an abnormal offspring.

This investigation was supported by a grant from the Expressen's Prenatal Research Fund and grant 7–29/68 from the Research Board of the Swedish National Nature Conservancy Office.

Technical assistance was provided by Anita Tillberg.

Nonproprietary and Trade Names of Drugs

Isoproterenol—*Norisodrine.*
Pentobarbital—*Nembutal.*

References

1. Kihlman, B. A. *Action of Chemicals on Dividing Cells.* Englewood Cliffs, NJ, Prentice-Hall Inc. 1966.
2. Krogh, Jensen M. Chromosome studies in patients treated with azathioprine and amethopterin. *Acta Med Scand* 182: 445–455, 1967.
3. Legator, M. S., Palmer, K. A., Green, S., et al. Cytogenetic studies in rats on cyclohexylamine, a metabolite of cyclamate. *Science* 165: 1139–1140, 1969.
4. Stone, D., Lamson, F., Chang, I. S., et al. Cytogenetic effects of cyalamate on human cells in vitro. *Science* 164: 568–569, 1969.
5. Stoltz, D. R., Khera, K. S., Bendall, R., et al. Cytogenetic studies with cyclamate and related compounds. *Science* 167: 1501–1502, 1970.
6. *Minamata Disease.* Study Group of Minamata Disease, Kumamoto University, Japan, 1968.
7. *Special Report on Cases of Mercury Intoxication in Niigata Prefecture.* Tokyo, Ministry of Health and Welfare, 1967.

8. Maximum allowable concentrations of mercury compounds. *Arch Environmental Health* **19**: 891–905, 1969.

9. Jensen, S., Jernelöv, A. Biosynthesis of mono- and dimethyl-mercury. *Nature* **223**: 1453–1454, 1969.

10. Westöö, G. Methyl mercury compounds in animal foods, in Miller, W., Berg, G. B. (eds.). *Chemical Fallout.* Springfield, Ill., Charles C Thomas Publisher, 1969, pp. 75–89.

11. Johnels, A. G., Westermark, T. Mercury contamination of the environment in Sweden, in Miller, W., Berg, G. B. (eds.). *Chemical Fallout.* Springfield, Ill., Charles C Thomas Publisher, 1969, pp. 221–239.

12. Sjöblom, V., Häsänen, E. The mercury level in fish in Finland. *Nord Hyg T* **50**: 37–53, 1969.

13. Underdal, B. Investigation of mercury in food. *Nord Hyg T* **50**: 60–63, 1969.

14. Dalgaard-Mikkelsen, S. Mercury in the environment in Denmark. *Nord Hyg T* **50**: 34–36, 1969.

15. Birke, G., Johnels, A. G., Plantin, C.-O., et al. Mercury intoxication through consumption of fish? *Lakartidningen* **64**: 3628–3637, 1967.

16. Tejning, S. Mercury Levels in Red Blood Cells, Plasma and Hair in Heavy Consumers of Fish From Different Areas of Lake Vänern and Correlation Between Mercury Levels in These Organ Elements and Mercury Level in Fish and a Proposal About the Use for Fish and Fish Products From International Food Hygiene Limit Value, report 670831. Clinic of Occupational Medicine, University Hospital, Lund, Sweden, 1967.

17. Sumari, P., Backman, A.-L., Karli, P., et al. Investigations of health among fish consumers in Finland. *Nord Hyg T* **50**: 97–102, 1969.

18. Ramel, C. Genetic effects of organic mercury compounds: I. Cytological investigations on Allium roots. *Hereditas* **61**: 208–230, 1969.

19. Fiskesjö, G. Some results from Allium tests with organic mercury halogenides. *Hereditas* **62**: 314–322, 1969.

20. Ramel, C., Magnusson, J. Genetic effects of organic mercury compounds: II. Chromosome segregation in Drosophila melanogaster. *Hereditas* **61**: 231–254, 1969.

21. Fiskesjö, G. The effect of two organic mercury compounds on human leukocytes in vitro. *Hereditas* **64**: 142–146, 1970.

22. Westöö, G. Mercury in fish. *Vår Föda* **19**: 1–7, 1967.

23. Norén, K., Westöö, G. Methyl mercury in fish. *Vår Föda* **19**: 13–17, 1967.

24. Johnels, A., Olsson, M., Westermark, T. Mercury in fish. *Vår Föda* **19**: 65–103, 1967.

25. Westöö, G., Norén, K. Mercury and methyl mercury in fish. *Vår Föda* **19**: 137–178, 1967.

26. Westöö, G., Rydälv, M. Mercury and methyl mercury in fish and crayfish. *Vår Föda* **21**: 19–111, 1969.

27. Sjöstrand, B. Simultaneous determination of mercury and arsenic in biological and organic materials by activation analysis. *Anal Chem* **36**: 814–819, 1964.

28. Tejning, S. Mercury Concentrations in Red Blood Cells and Plasma in Grown Up Normal Persons, report 670210. Clinic of Occupational Medicine. University Hospital, Lund, Sweden.

29. Court Brown, W. M. *Human Population Cytogenetics.* Amsterdam, North-Holland Publishing Co., 1967.

30. Lubs, M. A., Samuelson, J. Chromosome abnormalities in lymphocytes from normal human subjects: A study of 3,720 cells. *Cytogenetics* **6**:402–411, 1967.

31. Åberg, B., Ekman, L., Falk, R., et al. Metabolism of methyl mercury (^{203}Hg) compounds in man. *Arch Environ Health* **19**: 478–484, 1969.

32. Miettinen, J. K., Rahola, T., Hattula, T., et al. Retention and excretion of ^{203}Hg-labelled methyl mercury in man after oral administration of CH$_3$ ^{203}Hg biologically incorporated into fish muscle protein: Preliminary results. Fifth Radioactivity in Scandinavia Symposium, Helsinki, 1969.

33. Tsubaki, cited by Berlin, M., Ramel, C., Swenson, Å. Intoxication Through Consumption of Fish Containing Methyl Mercury Compound: Report from a Study Trip to Japan in September 1968. mimeograph. Sweden, National Institute of Public Health, 1969.

34. Buckton, K. E., Smith, P. G., Court Brown, W. M. The estimation of lymphocyte lifespan from studies on males treated with x-rays for ankylosing spondylitis, in Evans, H. J., Court Brown, W. M. (eds): *Human Radiation Cytogenetics*. Amsterdam, North-Holland Publishing Co., 1967.

35. Berlin, M., Ullberg, S. Accumulation and retention of mercury in the mouse: III. An autoradiographic comparison of methylmercuric dicyandiamide with inorganic mercury. *Arch Environ Health* **6**: 610–616, 1963.

36. Moriyama, H. A study on congenital Minamata disease. *Kumamoto Igk Z* **41**: 506–532, 1968.

37. Engleson, G., Herner, T. Alkyl mercury poisoning. *Acta Paediat* **41**: 289–294, 1952.

14

Vinyl Chloride Exposure and Human Chromosome Aberrations

Alan Ducatman, Kurt Hirschhorn and Irving J. Selikoff

Summary

Examination of lymphocyte cultures from 11 vinyl chloride polymerization workers and 10 controls revealed a significantly higher incidence of aberrations in the exposed population. Most of the excess damage was of the "unstable" variety and involved the grossest kinds of changes such as fragments or rearrangements. When these complex changes were regarded as the product of two breaks, the incidence of all breaking events was also significantly increased in the workers.

* Supported by USPHS Center Grants ES 00928 and GM 19443, and Research Grant HD 02552.

From *Mutation Research*, **31** (1975), 163–168. © Elsevier Scientific Publishing Company Amsterdam, The Netherlands.

The results indicate the presence of chromosome damage in vinyl chloride exposed workers.

Introduction

Since the pioneering radiation studies of the early 1960's it has been clear that chromosome morphology examination could uncover evidence of genetic damage in man[1-3]. The earliest of these studies discovered the now well-known link between carcinogenesis and chromosome changes (for recent review, see ref. 9). These were soon followed by the discovery that a number of chemicals are clastogenic (chromosome breaking), notably alkylating agents and cytostatic drugs, and DNA base analogues.[6] In recent years, occupational exposure to various chemicals has been recognized as another potential source of genetic change in man. Chromosome examination following human exposures had revealed the probable clastogenicity of some pesticides and herbicides.[10,17] Industrial exposure to benzene has also been implicated by a number of investigators in both leukemogenesis and clastogenesis (refs. 7, 8, 16).

Within the last year, a new industrial carcinogen has been identified.[4] Vinyl chloride, the monomer used for production of the common polyvinyl chloride, has been held responsible for at least 22 angiosarcomas of the liver in exposed workers.[10] There is additional suggestion from both animal and human data of an increased risk for cancers of the central nervous system, the respiratory system, and of the blood forming tissues.[10,13,15] With an association between clastogenicity and carcinogenesis established for other agents, it appeared that this new carcinogen might also cause chromosome damage. We therefore undertook a blind study of lymphocyte chromosomes from vinyl chloride-exposed workers and nonexposed controls.

Exposure to Vinyl Chloride

The 11 subjects studied were male workers who had received repeated exposure to vinyl chloride in an upstate New York polyvinyl chloride polymerization plant. Major exposures came from leaks of unreacted vinyl chloride gas, fumes from polyvinyl chloride slurry, and from polymerization reactor cleaning operations. Reactor cleaning involved skin contact to polyvinyl chloride and inhalation of vinyl chloride gas residues; this operation accounted for the most intense exposures. Duration of recurrent occupational exposure in the 11 men ranged from 4–28 years with an average of 15 years (Table 1). There is no record of ambient gas levels in the factory, but it is assumed that these must have exceeded 500 ppm at times, based on reports of odor detection, dizziness, and headaches.

Of the 10 healthy male controls, 4 were from within the same factory population, without known vinyl chloride exposure. Nevertheless, as long-term employment implied the possibility of some vinyl chloride exposure, albeit at low levels, we also selected 6 older controls from outside the factory environment. The average age of the controls was 27, of the subjects 40.

Table 1. Age and Degree of Vinyl Chloride Exposure of Selected Cases and Controls.

Case No.	Age	Years Exposed	Cells Examined	Control No.	Age	Years Exposed	Cells Examined
1	61	23	50	1	43	0	50
2	50	28	50	2	30	0	50
3	47	19	50	3	29	0	50
4	44	11	50	4	29	0	50
5	40	13	50	5	28	0	50
6	39	17	50	6	28	0	50
7	36	17	50	7	27	0	50
8	36	12	50	8	20	0	50
9	33	11	50	9	19	0	50
10	32	10	50	10	18	0	50
11	25	4	50				
Total	443	165	550		271	0	500
Average	40	15	50		27	0	50

Methods

The chromosome studies were performed on cultures of peripheral blood lymphocytes incubated at $37°$ for 65–68 h with phytohemagglutinin (Wellcome). Harvests were performed according to micromethod modifications of procedures first described by Moorhead et al.[14] Media used in this study came from a single batch (Gibco) in order to insure uniform pH and other culture conditions.

In each individual studied, 50 metaphases were counted directly under the microscope at $1600\times$. The A, B, D, E, F, G, and Y chromosomes were evaluated separately so far as possible. Suspected aberrations and some normal metaphases were photographed; karyotypes were performed where helpful. From the total of 1050 cells examined there were 281 photographs taken and 140 karyotypes made from the photographs.

Confirmed aberrations were classified according to two systems in order to discern if a pattern of breakage might exist. The system of Buckton et al.[3] and Court Brown[5] has three categories of aberrations: B, cells with simple aberrations such as breaks and gaps; C_u, cells with "unstable" chromosome changes such as fragments, dicentrics, and rings; C_s, cells with "stable" chromosome changes such as monosomies, trisomies, deletions, and exchanges. The distinction between cells with stable and unstable aberrations is related to their tendency to either remain in or disappear from the circulation.

The system of Hirschhorn and Cohen[11] differs in two respects. It considers all breakage, including the total number of breaks in cells with more than one aberration, and it gives weighted consideration to those aberrations which are the apparent result of two "hits." Therefore, complex breakage (C) such as rings, di-

131

centrics, and exchanges is counted twice in the total, whereas simple breaks and deletions (S) are counted once.

Results

Gaps and breaks were the predominant aberrations among the 1050 cells examined, as seen in the B column of Table 2. Subjects have nonsignificant increases of such aberrations when compared to controls by a t-test for comparison of the means $(0.1 > P > 0.05)$, and marginally significant increases by an F-ratio for comparison of the variances $(0.05 > P > 0.01)$. The difference in cells with stable aberrations, or those that Buckton *et al.* and Court Brown found to persist in

Table 2. Vinyl Chloride Exposure and Chromosome Aberrations (Classification System of Court Brown[5] and Buckton *et al.*[3])

Case No.	B^a	C_u^a	C_s^a
1	7	2	3
2	8	2	3
3	6	0	3
4	3	1	3
5	5	1	3
6	3	0	0
7	5	2	2
8	8	2	6
9	4	0	4
10	8	4	5
11	5	3	5
Total	62.0	17.0	37.0
Mean($\pm$S.D.)	5.64($\pm$1.91)	1.55($\pm$1.29)	3.36($\pm$1.63)
Comparison	$t = 1.84$	$t = 2.863$	$t = 0.56$
	$0.1 > p > 0.05$	$p = 0.01$	$0.6 > p > 0.5$
Control No.			
1	5	1	3
2	4	0	7
3	4	0	1
4	5	0	3
5	2	1	6
6	5	0	3
7	5	0	3
8	5	1	2
9	4	0	0
10	5	0	1
Total	44.0	3.0	29.0
Mean($\pm$S.D.)	4.40($\pm$0.97)	0.30($\pm$0.48)	2.90($\pm$2.18)

[a] B, Breaks and gaps; C_u, "unstable" changes (fragments, dicentrics, rings); C_s, "stable" changes (monosomy, trisomy, deletions, exchanges).

Table 3. Chromosome Break Events (in 50 Cells/Individual) and Vinyl Chloride Exposure (System of Hirschhorn and Cohen).

Case No.	S^a	C^a	Total Break Events $(S + 2C)$
1	4	3	10
2	2	3	8
3	2	0	2
4	1	2	5
5	0	1	2
6	0	0	0
7	3	2	7
8	1	2	5
9	4	0	4
10	1	4	9
11	3	5	13
Total	21	22	65
Mean($\pm$S.D.)	1.91($\pm$1.44)	2.00($\pm$1.67)	5.91($\pm$3.91)
Comparison	$t = 1.12$	$t = 2.80$	$t = 2.75$
	$0.3 > p > 0.2$	$0.02 > p > 0.01$	$0.02 > p > 0.01$
Control No.			
1	3	1	5
2	1	0	1
3	1	0	1
4	0	0	0
5	2	2	6
6	2	0	2
7	0	0	0
8	1	1	3
9	2	0	2
10	1	0	1
Total	13	4	21
Mean($\pm$S.D.)	1.30($\pm$0.95)	0.40($\pm$0.70)	2.10($\pm$2.02)

[a] S, Single hit events (breaks, deletions) ; C, complex events (rings, dicentrics, exchanges).

circulating lymphocytes[3, 5], is also nonsignificant. In our study, stable aberrations were usually cells with random chromosome loss. However, cells with unstable aberrations were observed significantly more frequently in the cultures from exposed workers: t, $P = 0.01$; F, $0.01 > P > 0.001$.

Table 3 focuses on breaking events only. The total of simple breaks, including those from multiaberrant cells, is increased but not significantly in the subjects. This is shown in the S column. Complex breaking events in C are significantly more frequent in those exposed, with: t, $0.02 > P > 0.01$; and F, $0.01 > P > 0.001$. The total of breaking events $(S + 2C)$ shows a similarly significant increase in the subjects (t, $0.02 > P > 0.01$; F, $0.01 > P > 0.001$). Not included in the charts is a combined total of all breaks, gaps, and deletions. There was a marginally significant difference in these with t, $0.05 > P > 0.02$; F was not significant.

Exaggerated secondary constrictions of the No. 9 chromosome were easily noted because of their high degree of visibility. The average subject was observed to have 3.91 ($\pm$2.47 S.D.) and the average control had 2.00 ($\pm$1.63) in the 50 cells examined per individual. Significance was marginal at most, $P < 0.05$ by a t-test; F not significant.

Discussion

There are obvious perils in drawing strong conclusions from small samples, and it would clearly be preferable to have an age-matched control group despite experimental evidence that age is generally unrelated to chromosome changes other than chromosome loss.[5] Also, the relatively high degree of breaks and gaps in controls as well as subjects is somewhat disconcerting, although subjects do have more. Within these limitations, it is clearly indicated that chronic high level exposure to vinyl chloride is clastogenic. Much of the increased damage was of the unstable variety as defined by Buckton et al.[3] The difference between cases and controls is most evident for unstable aberrations in general and fragments in particular. Hirschhorn and Cohen's system, which is a better overall index of chromosome damage, shows that long-term exposure is associated with an evidently significant increase in all breaking events.

In this small sample it has not been possible to correlate the degree of damage with either the degree of exposure or with vinyl chloride disease symptoms. The former may never be possible, as the best estimate of total exposure for any individual is only a crude guess. In general, the majority of subjects selected from this factory exhibited increased breakage rates, and those with the shortest duration of exposure showed damage similar to those with the longest duration.

Stating that chronic vinyl chloride exposure almost certainly damages chromosomes leaves us with two questions. First, can chromosome damage studies be at all predictive of environmentally induced carcinogenesis? Ionizing radiation and now vinyl chloride exposures have been studied for genetic properties *after* association with induced neoplasms. If we can reverse the order of events, perhaps in animal studies, we will know with more certainty if chromosome examination has an important role to play in predicting environmental carcinogenesis.

The second question is: what kind of genetic studies should be done with vinyl chloride? We feel strongly that chromosome study of individual concerned fathers is unwarranted. The degree of damage discovered here is unlikely to yield meaningful findings in any single individual. Women are not employed in polymerization work. However, they may work in polyvinyl chloride processing industries, in which some exposure to unreacted vinyl chloride may occur. This could be important in light of experimental findings of transplacental carcinogenesis.[13]

Carefully controlled examination of larger groups with vinyl chloride and polyvinyl chloride exposure are obviously needed, first to provide the larger data base necessary to confirm clastogenicity for vinyl chloride exposure, and also to evaluate dose-response relationships. Mutagenicity is being assessed in other test systems, including bacterial studies, insect and animal studies, along with *in vitro* chromosome studies.

The case for studying other genetically suspect chemicals is now stronger than ever.

References

1. Bender, M. A., and P. C. Gooch, Persistent chromosome aberrations in irradiated human subjects, *Radiat. Res.*, 16 (1962) 44–53.
2. Bloom, A. D., and J. H. Tho, *In vivo* effects of diagnostic X-irradiation on human chromosomes, *New Engl. J. Med.*, 270 (1964) 1341–1344.
3. Buckton, K. E., P. A. Jacobs, W. M. Court Brown and R. Doll, A study of the chromosome damage persisting after X-ray therapy for ankylosing spondylitis, *Lancet*, ii (1962) 676–682.
4. Creech, J. L., and M. N. Johnson, Angiosarcoma of liver in the manufacture of vinyl chloride, *J. Occup. Med.*, 16 (1974) 150–151.
5. Court Brown, W. M., Human population cytogenetics, in A. Neuberger and E. L. Tatum (Eds.), *Frontiers of Biology*, Vol. V, North-Holland, Amsterdam, 1967, pp. 1–31.
6. Evans, H. J., Population cytogenetics and environmental factors, in Patricia A. Jacobs, W. H. Price and Pamela Law (Eds.), *Human Population Cytogenetics*, Williams and Wilkins, Baltimore, 1970, pp. 191–216.
7. Forni, A. M., A. Coppellini, E. Pacifico and E. C. Vigliani, Chromosome changes and their evolution in subjects with past exposure to benzene, *Arch. Environ. Health*, 23 (1971) 385–391.
8. Forni, A. M., E. Pacifico and A. Limonta, Chromosome studies in workers exposed to benzene or toluene or both, *Arch. Environ. Health*, 22 (1971) 373–378.
9. German, J. (Ed.), *Chromosomes and Cancer*, Wiley, New York, 1974.
10. Heath, C., and J. Wagoner, *Report of a Working Group on Vinyl Chloride, Lyon, 24–25 June, 1974*, IARC No. 74/005. World Health Organization, pp. 18–19.
11. Hirschhorn, K., and M. M. Cohen, Drug-induced chromosomal aberrations, *Ann. N.Y. Acad. Sci.*, 151 (1968) 955–987.
12. Hoopingamer, R., and A. W. Bloomer, Lymphocyte chromosome analysis of pesticide exposed individuals, in *7th Int. Congr. Plant Protection, Paris, 1970*, p. 772.
13. Maltoni, C., and G. Lefemine, Carcinogenicity bioassays of vinyl chloride, I. Research plan and early results, *Environ. Res.*, 7 (1974) 387–405.
14. Moorhead, P. S., P. C. Nowell, W. J. Mellman, D. M. Battips and D. A. Hungerford, Chromosome preparations of leukocytes cultured from human peripheral blood, *Exp. Cell Res.*, 20 (1960) 613–616.
15. Nicholson, W. J., E. C. Hammond, H. Seidman and I. J. Selikoff, Mortality experience of a cohort of vinyl chloride-polyvinyl chloride workers, *Ann. N.Y. Acad. Sci.*, 255 (1975) 225–230.
16. Tough, I. M., and W. M. Court Brown, Chromosome aberrations and exposure to ambient benzene, *Lancet*, i (1965) 684.
17. Yoder, J., M. Watson and W. W. Benson, Lymphocyte chromosome analysis of agricultural workers during extensive occupational exposure, *Mutation Res.*, 21 (1973) 335–340.

15   Delayed Radiation Effects in Atomic-Bomb Survivors

Robert W. Miller

The pace at which radiation effects in the Japanese survivors of the atomic bombs are being reported has recently quickened. In this article I seek to put into perspective the major findings of the Atomic Bomb Casualty Commission (ABCC).

Immediately after World War II, a joint commission of the U.S. Army and Navy made observations concerning the immediate effects of exposure to the atomic bombs in Hiroshima and Nagasaki. Upon completion of its work, the joint commission recommended that the National Academy of Sciences—National Research Council conduct a study of the long-range biomedical effects of the exposures. The Council convened an advisory group, whose study of the situation in Japan led to a Presidential directive authorizing the National Research Council (NRC) to establish an organization to evaluate the delayed effects of exposure to the bombs. Thus the Atomic Bomb Casualty Commission came into existence (*1*). Its large-scale study, begun in 1948, is a cooperative venture between the NRC, representing the United States, and the Japanese National Institute of Health. The Commission's present staff of 725 Japanese and 36 foreign nationals, including 18 U.S. professionals, is collecting and analyzing data from periodic comprehensive medical examinations, from postmortem findings, and from a review of vital certificates as they are generated.

Genetic Effects

It is commonly thought that congenital anomalies are the only measure of genetic effects among children conceived after one or both of the parents have been exposed to ionizing radiation. The studies conducted at the ABCC, however, concerned six indicators of genetic damage in the F_1 generation.

Pregnancies were ingeniously ascertained (*1*). In postwar Japan, when food was in short supply, pregnant women were allowed an extra ration of rice, beginning in the fifth month of pregnancy. When such women registered for this supplement in Hiroshima or Nagasaki, they were entered in the ABCC genetic study. From 1948 to 1953, 71,280 pregnancies were ascertained in this way, 93 percent of all that went to term in that interval. Midwives notified ABCC of each delivery they attended, and the newborns were examined in their homes by ABCC staff physicians. About 40 percent of the children were reexamined at the clinics when they were 8 to 10 months old. The results were distributed according to five levels of radiation exposure for each parent. No influence of radiation was demonstrable in this study, which, statistical tests have shown, was likely to detect a 2-fold increase in rates of malformation or a 1.8-fold increase in rates of stillbirths and deaths of newborns.

Table 1. Frequencies, by Age Group, of Complex Cytogenetic Abnormalities Among Japanese Survivors of the Atomic Bombs.

Age Group at Time of the Bomb	Dose (rad)	Exposed		Control		Reference
		Examined (no.)	Affected (%)	Examined (no.)	Affected (%)	
≤30 years	200+	94	34	94	1	(5)
Over 30 years	200+	77	61	80	16	(6)
In utero	100+ *	38	39	48	4	(7)
Not yet conceived	150+ *	103	0			(8)
Not yet conceived	100+ †	25	0			(8)

* Maternal dose.
† Dose received by at least one parent.

Moreover, there was no effect on birth weight or on anthropometric values at 8 to 10 months attributable to radiation exposure. The sex ratio (the proportion of males to females) for children conceived after exposure of one parent to radiation will, in theory, be diminished if the mother was irradiated, and increased if the father was (2). In a study of about 120,000 births, such shifts in the sex ratios were found to occur in the first 10 years following detonation of the atomic bombs but not thereafter (3). No effect has been found on the mortality of children conceived after exposure of their parents to the bombs in Hiroshima or Nagasaki (4). Thus, though laboratory experimentation leaves no doubt that irradiation is mutagenic, the effect could not be demonstrated in the F_1 generation studied by ABCC.

Cytogenetic Abnormalities

In Table 1 are summarized the results obtained in a series of studies of chromosomal abnormalities, by age group, of individuals exposed to the atomic bombs in Japan. It should be noted that, among those 30 years of age or younger at the time of the bomb who received a dose of at least 200 rad, 34 percent had complex cytogenetic abnormalities 20 years later, as compared with 1 percent of the controls (5). The percentage of individuals so affected who were over 30 years old at the time of the bomb was almost double the percentage for the younger group, and 4 of the 77 individuals in the older group had clones of cytogenetically abnormal cells (6). The frequency of complex cytogenetic abnormalities apparently increases naturally with age, from 1 percent in the younger group of the controls to 16 percent in the older group (Table 1). The complex chromosomal aberrations which occurred with increasing frequency among persons who had been over 30 years old at the time of the bomb consisted of translocations, pericentric inversions, deletions, chromatid exchanges, and centromere breaks.

Among persons who were *in utero* at the time of the bomb and whose mothers received a dose of a least 100 rad, 39 percent displayed complex chromosomal abnormalities as compared with 4 percent of the controls (7). Finding these abnor-

Table 2. Effects of Intrauterine Exposure to the Hiroshima Atomic Bomb [From Wood *et al.* (*13*)].

Gestational Age (week)	Total Number Exposed*	Total Number Examined	Number with Small Head Circumference†	
			Mental retardation	Normal intelligence
≤15	57‡	56‡	9	14
>15	109	105	2§	4‖

* Exposure within 1800 meters of the hypocenter.
† Circumference 2 or more standard deviations below average.
‡ Excludes two with preexistent Down's syndrome.
§ Exposed at 21 and 24 to 25 weeks, respectively.
‖ Exposed at 16, 32, 32 and 36 to 40 weeks, respectively.

malities even among persons exposed during the first trimester *in utero* indicates that radiation can induce long-persisting changes in the lymphocyte precursors. In contrast to the cytogenetic defects observed following intrauterine (postconception) exposure, no such defects were found following preconception irradiation (exposure of either parent before conception of the child) (*8*).

These observations revealed that, in man, long-persisting chromosomal damage was induced even though an effect on the F_1 generation was not demonstrable.

Effects on the Embryo

Not long after the discovery of x-rays, case reports began to appear in the medical literature describing mentally retarded children with heads of small circumference born of mothers who had received pelvic radiotherapy during early pregnancy. Fourteen of these reports were described in a publication by Murphy in 1928 (*9*). Goldstein and Murphy (*10*) identified 16 additional cases from replies to a questionnaire sent to, and completed by, a substantial number of obstetricians. In view of these findings it is not surprising to learn that the same abnormalities were observed among children born of women who were exposed to the atomic bomb while pregnant (*11—13*). The effect was primarily among children of women who were exposed within 15 weeks of their last menstrual period (Table 2). Of the individuals examined, 56 in this category were born of mothers who had been within 1800 meters of the hypocenter. A head circumference 2 or more standard deviations below the mean for age and sex was expected (on the basis of a normal distribution) in 1.4 persons (2.5 percent) of this group but observed in 23, of whom 9 were mentally retarded. Among the 105 individuals exposed *in utero* more than 15 weeks after the mother's last menstrual period, small head circumference was expected in 2.6 but observed in 6, of whom 2 were mentally retarded. The usual frequency of comparable mental retardation among the nonexposed of the same age in Hiroshima and Nagasaki was about 1 percent (*14*). Table 3, derived from the most recent data on

Table 3. Radiation Effect on Head Circumference and Intelligence Following Intrauterine Exposure to the Hiroshima Atomic Bomb Within 15 Weeks of the Mother's Last Menstrual Period. S.D., standard deviation. [From Wood *et al.* (*13*).]

| Distance from Hypocenter (meter) | Retarded (no.) | | Normal (no.) | | |
	Head circumference > 3 S.D. below mean	Head circumference −2 to −3 S.D.	Head circumference > 3 S.D. below mean	Head circumference −2 to −3 S.D.	Total Exposed*
≤ 1200	6	2	1	1	11
1201–1500	0†	0	2	6	23†
1501–1800	0	1	0	5	23‡
1801–2200	0	0	0	0	21‡

* Some children in the study were normal with respect to both intelligence and head circumference, thus the numbers in columns 2 to 5 do not add up to the totals in column 6.
† Excludes two with preexistent Down's syndrome.
‡ One not examined.

the group exposed in the early weeks of pregnancy (*13*), reveals a dose-response relationship; that is, the effect diminishes in frequency and severity as the distance from the hypocenter increases.

These findings are in accord with the results of animal experimentation and with the clinical observations by Goldstein and Murphy cited above (*10*). The malformation occurred excessively only in association with high radiation dosage and not in association with the more extensive areas of heavy destruction and economic loss, which extended far beyond the high-dosage area. Estimates of the teratogenic dose range in man have not as yet been published, but about half of the mothers of affected children reported that they had signs of severe acute radiation sickness (*12*). No other anomalies occurred excessively among the survivors (*12, 13*), although many others have been observed in animals exposed experimentally to x-irradiation (*15*).

Fetal and infant mortality following exposure to the atomic bomb was not evaluated until 6 years after the event. Among women who were pregnant when exposed within 2000 meters of the Nagasaki hypocenter and who said they had had major signs of acute radiation sickness, 43 percent reported such mortality as compared with 9 percent of pregnant women in the same distance category who had not had acute radiation sickness (*16*). The excess is highly significant statistically ($P < .001$).

Growth

In 1951, as part of a comprehensive medical examination, 12 anthropometric determinations were obtained on about 2400 Hiroshima children 6 to 19 years old who had been exposed to the bomb 6 years earlier, and comparison was made with an equal number who had not been exposed (*17*). About 78 percent in each group were reexamined in 1952, and 53 percent were reexamined in 1953.

Multivariate analysis revealed that as radiation exposure increased, there were small but statistically significant decreases in body measurements at all age levels, and in growth rate at postpubertal age levels (*18*). To some extent these differences may be due to variables other than radiation exposure—for example, to economic loss due to the blast and fires.

Nagasaki adolescents who were exposed *in utero* to radiation from the atomic bomb have been studied with respect to the mean values for several anthropometric variables (*19*). The sample of heavily exposed subjects was small. Only 16 boys and 15 girls were estimated to have received doses of 50 rad or more, and only 9 of these were exposed in the first trimester of pregnancy. Some significant differences were found which were consistent with a radiation effect.

Eye Findings

In 1963 and 1964, ophthalmologic examinations were made on 1627 residents of Hiroshima and 841 residents of Nagasaki, of whom an estimated 40 percent had received doses in excess of 200 rad (*20*). In the high-dose group there were significantly more axial opacities seen by ophthalmoscope and confirmed by slit-lamp examination than in groups more distantly exposed. Most of these lesions were small, and only one was regarded as a mature radiation cataract. Another finding, more in the nature of a measure for biologic dosimetry than anything else, was a polychromatic sheen, sometimes granular, in the posterior subcapsular area of the lens as visualized by slit-lamp biomicroscopy. A dose-response relationship was demonstrated. These abnormalities did not affect visual acuity. In previous ophthalmologic examinations by ABCC, less than a dozen survivors were classified as having severe radiation cataracts, and in none was visual acuity worse than 20/70 (*21*).

In the most recent survey (*20*), there was a suggestion of a dose-related impairment of visual acuity among children who were *in utero* at the time of exposure to the bomb, but the sample size was too small for the test to be of statistical significance. In brief, though some ophthalmic effects of irradiation have been noted among atomic-bomb survivors, impairment of vision has been rare, and relatively mild.

Leukemogenesis

The leukemogenic effect of radiation in man was suspected at about the same time that teratogenic effects were—again, from case reports (*22*). Lymphoma was experimentally induced in mice in 1930 (*23*). Then, by simply reviewing the death notices published weekly in the *Journal of the American Medical Association*, Henshaw and Hawkins (*24*) found that leukemia was reported 1.7 times more often as a cause of death among U.S. physicians, a group occupationally exposed to x-rays, than among the general population of adult white males. Using the same source, Ulrich and March independently found that U.S. radiologists died significantly more often of leukemia than other physicians did (*25*). In consequence, a leukemogenic effect of exposure to atomic radiation was expected among the survivors in Hiroshima and Nagasaki, and it was found (26). A dose-response relationship was observed which can be attributed to no variable except radiation. A peak in occurrence was reached in 1951, more marked for the acute leukemias than for chronic granulocytic leukemia. Chronic lymphocytic leukemia, rare among the Japanese (27), did not increase in frequency. In children, leukemia was generally acute, as it usually is in children, the lymphocytic form being as common as the granulocytic. In all age groups acute leukemia continued to occur at higher than usual rates through 1966, whereas chronic granulocytic leukemia had fallen to near-normal rates. The ABCC study leaves no doubt that whole-body exposure to ionizing radiation at sufficiently high doses can induce leukemia in man.

Human leukemia may also be induced by partial-body irradiation, as indicated by the dose-response effect observed in British men given radiotherapy for ankylosing spondylitis (28). Again, the peak was reached about 5 years after the first exposure (the first course of therapy). The predominant form in these adults was granulocytic; no increase occurred in the frequency of chronic lymphocytic leukemia. In all, 52 cases of leukemia were observed, as compared with the 5.48 expected on the basis of national mortality rates for England and Wales, and 15 persons developed aplastic anemia (perhaps subclinical leukemia) as compared with the 0.51 expected. This study, in conjunction with the ABCC study, revealed that ionizing radiation can induce more than one form of leukemia in man, but not all forms, the notable exception being chronic lymphocytic leukemia.

Irradiation is but one of several circumstances that carry exceptionally high risk of leukemia (29). At the highest risk yet known is the child whose identical twin develops leukemia before the age of 6 years. The probability is 1 in 5 that the second twin will develop the disease within weeks or months after the first child falls ill. In about the same category are the persons with polycythemia vera treated with x-ray or phosphorus-32 (or both) and persons with Bloom's syndrome or Fanconi's anemia. The probability of developing leukemia was substantially less for heavily exposed Hiroshima survivors—about 1 in 60 individuals were so affected within 12 years of exposure. At still lower risk of leukemia are children with Down's syndrome (1 in 95 for children under 10 years old) and radiation-treated patients with ankylosing spondylitis (1 in 270 were so affected within 15 years after radiotherapy). These groups are alike in that each has a distinctive genetic feature, but these features are not of a single type. Identical twins have identical genes; Bloom's and Fanconi's syndromes are heritable disorders characterized by chromosomal fragility; in radiation-treated polycythemia vera, aneuploidy has been described in a substantial proportion of cases before radiation, and chromosomal breaks are regularly found following radiotherapy; atomic-bomb survivors (and persons exposed to ionizing radiation from other sources) exhibit long-lasting chromosomal breaks; and in Down's syndrome there is congenital aneuploidy.

Leukemia in patients with polycythemia vera has been attributed to radiotherapy (30). It should be noted, however, that the probability of occurrence of the neoplasm was 10 times as high in these patients as it was among heavily exposed survivors of the Hiroshima bomb and 45 times as high as it was in radiation-treated ankylosing spondylitis. One must conclude either that radiation exposure or damage is greater in polycythemia vera than in the other two instances or that polycythemia vera predisposes to leukemia in the absence of radiation exposure.

Several studies, considered individually, suggest that very small exposures to radiation before conception or during pregnancy may increase by 50 percent the child's risk of leukemia. When these studies are considered collectively, however, there is reason to suspect that some fault in the methods, difficult or impossible to escape, may be implicating radiation spuriously.

The individual results are as follows. In 1958, Stewart and her associates (31) described a study of 677 leukemic children in England and Wales in which the proportion of mothers who reported abdominal exposure to *diagnostic* radiation during the relevant pregnancy was twice the proportion for mothers of normal children living in the same area. A similar difference was reported with respect to 739 children with neoplasms of other kinds. It is possible that mothers of children with cancer reported their radiologic exposures more fully than the mothers of healthy children did. MacMahon (32) avoided this potential bias in histories obtained through interviews by studying obstetric records for irradiation during pregnancy among mothers of 304 leukemic children and 252 children with other cancer as compared with records for a 1-percent sample of all other births in the area (New England). He found a 40-percent excess of (i) leukemia and (ii) all other neoplasia among children whose mothers' records showed diagnostic radiation of the abdomen during pregnancy. Similar results were obtained by Graham *et al.* (33) with respect to leukemia, and by Stewart and Kneale (34) for each of six categories of childhood cancer. A causal relationship would be indicated if a dose-response effect could be demonstrated, if the results were consistent with those from animal experimentation, and if concomitant variables could be excluded. The exposures involved were too small to permit evaluation of a dose-response effect, there are no data from animal studies which support the observations in man, and the condition being treated by the radiologic procedure, rather than the x-ray exposure itself, could, in theory at least, be the oncogenic factor.

Recently Graham and his associates (33) described an excess of leukemia among children whose mothers *or* fathers gave histories of diagnostic radiation exposure up to a decade before the children were conceived. Again, there are no animal studies to support this observation. Moreover, in a prospective study (35) of 22,400 children conceived after their parents had been heavily exposed to radiation from the atomic bombs in Hiroshima or Nagasaki, no excess of leukemia was found.

Table 4 summarizes the results following very small doses of x-ray, and indicates that such irradiation was equally oncogenic whether exposure occurred before conception or during pregnancy, whether the neoplasm studied was leukemia or any other major cancer of childhood and whether the study was based on interviews, which may be biased, or on hospital records. Taken in the aggregate, the similarity of results in the absence of a dose-response effect or of supporting data from animal experimentation raises a question about the biologic plausibility of a causal relationship. In particular one must ask, in the absence of demonstrable mutagenic or cytogenetic abnormalities in the F_1 generation, if irradiation of the parent prior to conception is likely to induce leukemia in the child.

Table 4. Relative Risk of Various Childhood Cancers Following Intrauterine or Preconception Exposures to Diagnostic Radiation.

Neoplasm	Relative Risk [*]
Intrauterine exposure	
Stewart and Kneale (*33*)	
Leukemia	1.5
Lymphosarcoma	1.5
Cerebral tumors	1.5
Neuroblastoma	1.5
Wilms' tumor	1.6
Other cancer	1.5
MacMahon (*31*)	
Leukemia	1.5
Central nervous system tumors	1.6
Other cancer	1.4
Graham *et al.* (*32*)	
Leukemia	1.4
Preconception exposure	
Graham *et al.* (*32*)	
Leukemia	
Mother exposed	1.6
Father exposed	1.3

[*] Relative risk in controls = 1.0.

Other Cancer

Wanebo et al. (*36*) have recently reported that "accumulated information . . . strongly suggests that exposure to ionizing radiation has increased the risk of lung cancer among atomic bomb survivors." These investigators observed 17 such cases, as compared with 9 expected (dose, 90 rad or more). A weakness in the report was the finding that the lung cancers induced were nonspecific as to histologic type, rather than of the undifferentiated or small-cell type, as in U.S. uranium miners and in workers heavily exposed to mustard gas, a radiomimetic chemical (*37*).

Wanebo et al.(*38*) have reported that "information on breast cancer among survivors of the atomic bombings of Hiroshima and Nagasaki has now accumulated to the point where a fairly definite carcinogenic effect seems established." Six cases were observed among women who were exposed to 90 rad or more, as compared with 1.53 cases expected—an excess of only 4.5 cases. There was no specificity as to histologic type.

It may be difficult or impossible to avoid certain biases that could produce such a small excess—for example, unequal detection of cases with respect to exposure category, or dissimilar cancer risks in relation to some variable other than radiation which distinguished the heavily exposed from others in the study. Wanebo et al. considered the possibility of biases and believed that none were present. The absence of a dose-responsive relationship makes interpretation of the results difficult, as does the small or uncertain effect observed in studies of other exposed persons.

Wood *et al.* (*39*) have recently described an excess of thyroid cancer among Japanese survivors of the atomic bomb. The increase was greater in women than in men, the effect being proportionate to the radiation dose, but no specificity as to cell type was found. These observations are in accord with the results of animal experimentation and with the increase in frequency of thyroid cancer following therapeutic radiation early in childhood (*40*).

One may conclude that, among the Japanese survivors of the atomic bomb, only leukemia and thyroid cancer have been shown to be radiation-induced. The evidence pertaining to cancer of the breast or lung is still very much in doubt.

Mortality

Animal experimentation has shown that ionizing radiation can induce a shortening of life span which is attributable to no specific disease but to an accelerated occurrence of disease in general (*41*). The ABCC has conducted a study of life span among the survivors of the atomic bombs. The most recent published analysis concerns deaths in Hiroshima and Nagasaki, in the decade 1950 to 1960, in a sample of 99,393 persons—survivors of all ages—and a similar group of individuals not exposed to the bomb (*42*). When leukemia was excluded as a cause of death, the mortality ratios for exposed persons who had been within 1200 meters of the hypocenter in Hiroshima and Nagasaki were elevated by about 15 percent, an increase that was statistically significant when the data for both sexes and both cities were evaluated through a combined test. The increase was greater for women than for men and faded with time, reaching near-normal rates in about 1955. In another analysis of mortality, now in progress, the data through 1966 are being evaluated to determine if, after an extended period of latency, mortality may again be increased.

Summary

Since 1948 the ABCC has been evaluating the health of survivors of the atomic bombs in Hiroshima and Nagasaki. In a study of about 70,000 children conceived after the explosion, six indicators of genetic damage failed to reveal an unequivocal effect of radiation. Furthermore, this group displayed no evidence of cytogenetic abnormality, in contrast to the increased frequency of complex chromosomal aberrations found among those exposed *in utero* or at any time during the entire life span. The effect was most pronounced among persons whose exposures occurred when they were 30 years of age or older.

Although a wide variety of congenital malformations have been produced in experimental animals by irradiation of the pregnant mother, the only anomaly observed among the Japanese survivors to date has been small head circumference associated with mental retardation, the effect being proportionate to the radiation dose.

The ABCC study leaves no doubt that whole-body irradiation in sufficient dose is leukemogenic in man. A similar effect following partial-body irradiation has been observed among British men given radiotherapy for ankylosing spondylitis. In both studies the effect was proportionate to the dose, the peak occurred about 6 years after first exposures, and the increase was in acute leukemias and chronic granulo-cytic leukemia, not in the chronic lymphocytic form of the disease.

In the past few years, a high risk of leukemia has been associated with several human attributes and with radiation exposure. These circumstances have in common an unusual genetic feature, though not of a single type.

In several studies conducted in the United States or Great Britain, very small doses of x-ray were reported to be equally oncogenic whether exposure occurred before conception or during intrauterine life; whether the neoplasm studied was leukemia or any other major cancer of childhood; and whether the study was based on interviews, which are subjective, or on hospital records, which are not. Among the features that argue against a causal relationship are the similarity of results despite the dissimilarity of subject matter and, with regard to radiation before the child's conception, the failure, in a prospective study by ABCC, to find an excess of leukemia in 22,400 children conceived after their parents had been heavily exposed from the atomic bomb.

Increases in cancers other than leukemia have recently been reported among the Japanese survivors. Twice the normal frequency of lung cancer was found among persons exposed to doses of 90 rad or more, in a study handicapped by failure to demonstrate specificity with regard to histologic type, as in U.S. uranium miners. A report of an excess of breast cancer was based on 6 cases observed as compared with 1.53 expected among women who were exposed to doses of 90 rad or more. Certain biases, difficult or impossible to avoid, could produce this small excess. Thyroid cancer, on the other hand, does appear to have been induced by radiation, since a dose-response relationship was apparent and the results are consistent with those observed following therapeutic irradiation.

Other effects attributable to radiation but relatively small in magnitude were an increase in general mortality, exclusive of death from leukemia, during the first 10 years after exposure; a statistically significant but biologically small retardation in growth and development; infrequent radiation cataracts, none of which greatly diminished visual acuity; and a polychromatic sheen on the posterior subcapsule of the lens of the eye, which caused no disability but was related to radiation dose.

References and Notes

1. J. V. Neel and W. J. Schull. " The Effect of Exposure to the Atomic Bombs on Pregnancy Termination in Hiroshima and Nagasaki." *Nat. Acad. Sci. Nat. Res. Counc. Publ. No. 461* (1956).
2. J. V. Neel. *Changing Perspectives on the Genetic Effects of Radiation* (Thomas, Springfield, Ill., 1963).
3. W. J. Schull, J. V. Neel, A. Hashizume. *Amer. J. Hum. Genet.* **18**, 328 (1966).
4. H. Kato, W. J. Schull, J. V. Neel. *Ibid.* p. 339.
5. A. D. Bloom, S. Neriishi, N. Kamada, T. Iseki, R. J. Keehn. *Lancet* **1966–II**, 672 (1966).

6. A. D. Bloom, S. Neriishi, A. A. Awa, T. Honda, P. G. Archer. *Ibid.* **1967–II**, 802 (1967).

7. A. D. Bloom, S. Neriishi, P. G. Archer. *Ibid.* **1968–II**, 10 (1968).

8. A. A. Awa, A. D. Bloom, M. C. Yoshida, S. Neriishi, P. G. Archer. *Nature* **218**, 367 (1968).

9. D. P. Murphy. *Surg. Gynecol. Obstet. Int. Abstr. Surg.* **47**, 201 (1928).

10. L. Goldstein and D. P. Murphy. *Amer. J. Roentgenol. Radium Ther. Nucl. Med.* **22**, 322 (1929).

11. G. Plummer. *Pediatrics* **10**, 687 (1952).

12. R. W. Miller. *Ibid.* **18**, 1 (1956).

13. J. W. Wood, K. G. Johnson, Y. Omori. *Ibid.* **39**, 385 (1967).

14. J. W. Wood, K. G. Johnson, Y. Omori, S. Kawamoto, R. J. Keehn. *Amer. J. Public Health Nat. Health* **57**, 1381 (1967).

15. R. Rugh. *Ann. Rev. Nucl. Sci.* **9**, 493 (1959).

16. J. N. Yamazaki, S. W. Wright, P. M. Wright. *Amer. J. Dis. Child.* **87**, 448 (1954).

17. E. L. Reynolds. *Atomic Bomb Casualty Comm. Tech. Rep.* (1954), pp. 20–59.

18. J. V. Nehemias. *Health Phys.* **8**, 165 (1962).

19. G. N. Burrow, H. B. Hamilton, Z. Hrubec. *J. Amer. Med. Ass.* **192**, 97 (1965).

20. M. D. Nefzger, R. J. Miller, T. Fujino. *Amer. J. Epidemiol.* **89**, 129 (1968).

21. D. G. Cogan, S. F. Martin, H. Ikui. *Trans. Amer. Ophthalmol. Soc.* **48**, 62 (1950); R. J. Miller, T. Fujino, M. D. Nefzger. *Arch. Ophthalmol.* **78**, 697 (1967).

22. C. E. Dunlap. *Arch. Pathol.* **34**, 562 (1942).

23. C. Krebs, H. C. Rask-Nielsen, A. Wagner. *Acta Radiol. Suppl.* **10**, 1 (1930).

24. P. S. Henshaw and J. W. Hawkins. *J. Nat. Cancer Inst.* **4**, 339 (1944).

25. R. C. March. *Radiology* **43**, 276 (1944); H. Ulrich. *N. Engl. J. Med.* **234**, 45 (1946).

26. J. H. Folley, W. Borges, T. Yamawaki. *Amer. J. Med.* **13**, 311 (1952); A. B. Brill, M. Tomonaga, R. M. Heyssel. *Ann. Intern. Med.* **56**, 590 (1962); O. J. Bizzozero, Jr., K. G. Johnson, A. Ciocco. *N. Engl. J. Med.* **274**, 1095 (1966).

27. S. C. Finch, T. Hoshino, T. Itoga, M. Ichimaru, R. H. Ingram, Jr. *Blood* **33**, 79 (1969).

28. W. M. Court Brown and R. Doll. *Leukaemia and Aplastic Anaemia in Patients Irradiated for Ankylosing Spondylitis* (Her Majesty's Stationery Office, London, 1957); *Brit. Med. J.* **1965–II**, 1327 (1965).

29. R. W. Miller. *Cancer Res.* **27**, 2420 (1967).

30. B. Modan and A. M. Lilienfeld. *Medicine* **44**, 305 (1965).

31. A. Stewart, J. Webb, D. Hewitt. *Brit. Med. J.* **1958–I**, 1495 (1958).

32. B. MacMahon. *J. Nat. Cancer Inst.* **28**, 1173 (1962).

33. S. Graham, M. L. Levin, A. M. Lilienfeld, L. M. Schuman, R. Gibson, J. E. Dowd, L. Hempelmann. *Nat. Cancer Inst. Monogr.* **19**, 347 (1966). Subsequent reanalysis of the data for children under 4 years old suggested that cofactors are involved—for example, maternal history of fetal mortality, or virus infection in the child more than 12 months before the diagnosis of leukemia [R. W. Gibson, I. D. J. Bross, S. Graham, A. M. Lilienfeld, L. M. Schuman, M. L. Levin, J. E. Dowd. *N. Engl. J. Med.* **279**, 906 (1968)].

34. A. Stewart and G. W. Kneale. *Lancet* **1968–I**, 104 (1968).

35. T. Hoshino, H. Kato, S. C. Finch, Z. Hrubec. *Blood* **30**, 719 (1967).

36. C. K. Wanebo, K. G. Johnson, K. Sato, T. W. Thorslund. *Amer. Rev. Resp. Dis.* **98**, 778 (1968).
37. J. K. Wagoner, V. E. Archer, F. E. Lundin, Jr., D. A. Holaday, J. W. Lloyd. *N. Engl. J. Med.* **273**, 181 (1965); S. Wada, Y. Nishimoto, M. Miyanishi, S. Kambe, R. W. Miller. *Lancet* **1968–I**, 1161 (1968).
38. C. K. Wanebo, K. G. Johnson, K. Sato, T. W. Thorslund. *N. Engl. J. Med.* **279**, 667 (1968).
39. J. W. Wood, H. Tamagaki, S. Neriishi, T. Sato, W. F. Sheldon, P. G. Archer, H. B. Hamilton, K. G. Johnson. *Amer. J. Epidemiol.* **89**, 4 (1969).
40. S. Lindsey and I. L. Cheikoff. *Cancer Res.* **24**, 1099 (1964); L. H. Hempelmann. *Science* **160**, 159 (1968).
41. J. B. Storer. *Radiation Res.* **25**, 435 (1965).
42. S. Jablon, M. Ishida, M. Yamasaki. *Ibid.* p. 25.

four Genes and Behavior

149

Behavior stands at the pinnacle of human adaptation. It includes all of culture. But behavior is an intractable subject because it involves a maximum integration of morphological and functional components. Whereas the nature of DNA, enzymes, or muscles can be described without reference to a specific environment, behavior can only be described as an organism-environment relationship. This is not to deny that imprinting and other gene-determined, unlearned actions exist. On the contrary, it emphasizes their evolutionary adaptation. And although behavioral patterns are quite variable, the successful behaviors of man and other animals lead to quite specific ends—to adjustment to the immediate environment (getting food, keeping warm) or, in terms of sex, to evolutionary success through reproduction.

The gamut of human behavior is discussed from a genetic vantage point in a spate of recent books (Glass, 1968; Rosenthal, 1970; McClearn and DeFries, 1973). The readings in this section introduce two approaches to the relation between genetics and behavior: the analytical approach (Erlenmeyer-Kimling; Heston) and the evolutionary approach (Gilula and Daniels). One facet of behavior reserved for detailed discussion later (Section Eight) is intelligence.

The analytical approach of behavioral genetics is straightforward. Its adherents strive to determine the genetic bases of the complex activities we call behavior. They investigate both normal and pathological behavior, and they recognize that the two forms intergrade. Moreover, what is normal in one culture may be considered pathological in another. Attempts to study abnormal behavior (psychoses and neuroses) by genetic analysis have seen only modest success, despite the sincere pronouncements of some psychologists. A notable exception is the genetic analysis of schizophrenia in a paper by Heston (1970), not included in this section. Schizophrenia, the commonest psychotic disorder in many populations, affects about 1 per cent of the people in the United States at some time during their lives. Heston in his 1970 paper provides at least suggestive evidence that schizophrenia and the related "schizoid personality" are the result of a single dominant gene. But the metabolic consequences of this gene's malfunction have not been substantiated (Shaskan and Becker, 1975; Wyatt et al., 1975). The genetic status of another well-known psychosis, manic-depression, is, I think, less well established, although twin and family studies support a genetic input. Mendlewicz and Rainer (1974) have gone so far as to opt for a dominant X-linked major gene for manic-depression (but see Akiskal and McKinney, 1973). In addition, scrutiny of the neuroses is now underway (Miner, 1975; Inouye, 1972). Personality differences also receive some attention from time to time. For example, persons with Down's syndrome differ in personality from persons with Huntington's chorea (see paper by Stern in Section Eight).

Another abnormal behavior of widespread concern is alcoholism. Wolff (1972, 1973) has shown that Japanese, Taiwanese, Koreans, and American Indians become flushed and have symptoms of intoxication after drinking beer, much more so than do whites. Fenna and colleagues (1971) have shown that the level of alcohol in the blood after alcohol consumption drops more slowly in Eskimos and Canadian Indians than in whites. Does this indicate a genetic basis for alcoholism? Omenn (1975) has reviewed pertinent biochemical and genetic studies and outlined the metabolic and nervous-system factors that may contribute to alcoholism (see also Hasumura et al., 1975; Bennion and Li, 1976; Stamatoyannopoulos et al., 1975) but no one has yet uncovered a successful line of attack on its underlying causation. Nevertheless, alcoholism in biological relatives of adopted alcoholic individuals is several times higher than in members of the adopting family (Schuckit et al., 1972).

Behavioral differences among ethnic groups, other than those related to alcohol consumption, are only sparsely studied. Freedman and Freedman (1969) have shown differences in temperament between Chinese and white infants with the former being more imperturbable than the latter. Other information in this area is largely at the anecdotal level.

The evolutionary approach to studying behavior is popular and often simplistic (see reviews by Berkowitz, 1969, and Scott, 1967). The subject that receives the most attention from the point of view of adaptation is aggression. The most disquieting views are held about aggression: that it is, through the agency of war, the paramount organizing and adhesive force of society (Lewin, 1967; Carneiro, 1970); that it is indivisible from human love (Lorenz, 1966); that it is the most powerful obstacle to culture (Sigmund Freud), and so forth. The paper by Gilula and Daniels presented here examines such theories in the general context of evolutionary adaptive actions. In addition to the points these authors make, it is important to recognize that evidence of aggression as an evolutionary adaptation requires that an aggressor be highly fit for reproduction. This problem is considered, somewhat obliquely, in the paper by Hardin in Section Six.

Are there any behavioral correlates of the human chromosome abnormalities? These abnormalities are reported more often to have behavioral problems than are point-mutational traits, although the latter also include examples of secondary behavioral effects (PKU producing mental retardation, for example). The behavior of persons with XO, XYY, XXY, and Down's syndrome has been reported on (e.g., Bekker, 1974; Hook, 1973, 1974; Money et al., 1974). The XYY condition has received the most attention (e.g., Culliton, 1974, 1975). Current opinion is divided among geneticists as to whether males with an extra Y run significantly greater risk of behavioral problems than do those with XXY or XY constitutions. Fortunately, the press has now lost interest in XYY's as "congenital criminals," an untruth that disparages the numerous XYY's who have no behavioral problems. Studies of all chromosome anomalies may thus proceed dispassionately, I would hope.

Bibliography

Akiskal, H. S., and W. T. McKinney. Depressive disorders: Toward a unified hypothesis. *Science*, **182**: 20–29, 1973.

Bekker, F. J. Personality development in XO-Turner's syndrome. In *The Genetics of Behavior*, J. van Abeelen, ed. New York: American Elsevier, 1974, pp. 273–290.

Bennion, L. J., and T. Li. Alcohol metabolism in American Indians and Whites. *New Eng. J. Med.*, **294**: 9–13, 1976.

Benzer, S. Genetic dissection of behavior. *Sci. Amer.*, **229**: 24–37, 1973.

Berkowitz, L. Simple views of aggression. *Amer. Scientist*, **57**: 372–383, 1969. Puts K. Lorenz and other popularizers of ethology at some distance from careful sociological and behavioral-genetics studies.

Carneiro, R. L. A theory of the origin of the state. *Science*, **169**: 733–738, 1970.

Culliton, B. J. Patients' rights: Harvard is site of bitter battle over X and Y chromosomes. *Science*, **186**: 715–717, 1974.

Culliton, B. J. XYY: Harvard researcher under fire stops newborn screening. *Science*, **188**: 1284–1285, 1975.

Daniels, D. N., M. F. Gilula, and F. M. Ochberg, eds. *Violence and the Struggle for Existence*. Boston: Little, Brown, 1970.

Elston, R. C. Methodologies in human behavior genetics. *Soc. Biol.*, **20**: 276–279, 1973.

Ewing, J. A., B. A. Rouse, and E. D. Pellizzari. Alcohol sensitivity and ethnic background. *Amer. J. Psychiatry*, **131**:206–210, 1974.

Fenna, D., L. Mix, O. Schaefer, and J. A. L. Gilbert. Ethanol metabolism in various racial groups. *Canad. Med. Assoc. J.*, **105**:472–475, 1971.

Freedman, D. G., and N. C. Freedman. Behavioral differences between Chinese-American and European-American newborns. *Nature*, **224**:1227, 1969.

Fuller, J. L. Suggestions from animal studies for human behavior genetics. In *Methods and Goals in Human Behavior Genetics*, S. G. Vandenberg, ed. New York: Academic Press, 1965, pp. 245–253.

Glass, D. C., ed. *Genetics*. Biology and Behavior Series. New York: Rockefeller University Press and Russell Sage Foundation, 1968.

Gottesman, I. I., and L. L. Heston. Human behavioral adaptations—Speculations on their genesis. In *Genetics, Environment, and Behavior: Implications for Educational Policy*, L. Ehrman and G. S. Omenn, eds. New York: Academic, 1972.

Hasumura, Y., R. Teschke, and C. S. Lieber. Acetaldehyde oxidation by hepatic mitochondria: Decrease after chronic alcohol consumption. *Science*, **189**: 727–729, 1975.

Heston, L. L. The genetics of schizophrenic and schizoid disease. *Science*, **167**:249–256, 1970.

Hirsch, J., ed. *Behavior-Genetic Analysis*. New York: McGraw-Hill, 1967.

Hook, E. B. Behavioral implications of the XYY genotype. *Science*, **179**:139–150, 1973.

Hook, E. B. Racial differentials in the prevalence rates of males with sex chromosome abnormalities (XXY, XYY) in security settings in the United States. *Amer. J. Hum. Genet.*, **26**:504–511, 1974.

Hunt, M. Man and beast. *Playboy*, July 1970.

Inouye, E. Genetic aspects of neurosis. *Internat. J. Mental Health*, **1**:176–189, 1972.

Inouye, E. Some considerations in the methodology of human behavior genetics. *Soc. Biol.*, **20**:241–245, 1973.

Karlsson, J. L. Inheritance of schizophrenia. *Acta Psychiatrica Scandinavica Suppl.*, **247**, 1974.

Kessler, S., and R. H. Moos. The XYY karyotype and criminality: A review. *J. Psychiatric Research*, **7**:153–170, 1971.

Kety, S. S., D. Rosenthal, P. H. Wender, F. Schulsinger, and B. Jacobsen. Mental illness in the biological and adoptive families of adopted individuals who have become schizophrenic. In *Genetic Research in Psychiatry*, R. Fieve, D.

Rosenthal, and H. Brill, eds. Baltimore: Johns Hopkins University Press, 1975.

Lewin, L. C. *Report from Iron Mountain on the Possibility and Desirability of Peace*. New York: Dial, 1967. See especially Section 5, "The Functions of War."

Lorenz, K. *On Aggression*. New York: Harcourt, 1966.

McClearn, G. E., and J. C. DeFries. *Introduction to Behavioral Genetics*. San Francisco: Freeman, 1973.

Manosevitz, M., G. Lindzey, and D. D. Thiessen, eds. *Behavioral Genetics*. New York: Appleton, 1969.

Mendlewicz, J., R. R. Fieve, and F. Stallone. Relationship between the effectiveness of lithium therapy and family history. *Amer. J. Psychiat.*, **130**:1011–1013, 1973.

Mendlewicz, J., and J. D. Rainer. Morbidity risk and genetic transmission in manic-depressive illness. *Amer. J. Hum. Genet.*, **26**:692–701, 1974.

Miner, G. D. The evidence for genetic components in the neuroses: A review. *Archives Gen. Psychiatry*, **29**:111–118, 1975.

Money, J., C. Annecillo, B. Van Orman, and D. S. Borgaonkar. Cytogenetics, hormones and behavior disability: Comparison of XYY and XXY syndromes. *Clin. Genet.*, **6**:370–382, 1974.

Morrison, J. R., and M. A. Stewart. Evidence for polygenetic inheritance in the hyperactive child syndrome. *Amer. J. Psychiatry*, **130**:791–792, 1973.

Motulsky, A. G., and G. S. Omenn. Human behavior, genes, and society. In *Heredity and Society*, I. H. Porter and R. G. Skalko, eds. New York: Academic, 1973, pp. 93–113.

Omenn, G. S. Alcoholism: A pharmacogenetic disorder. *Mod. Problems Pharmacopsych.*, **10**:12–22, 1975.

Rieder, R. O. The offspring of schizophrenic parents: A review. *J. Nervous Mental Disease*, **157**:179–190, 1973. Psychic disorders were found in 25 to 45 per cent of these offspring.

Rosenthal, D. *Genetic Theory and Abnormal Behavior*. New York: McGraw-Hill, 1970.

Schuckit, M. A., D. A. Goodwin, and G. Winokur. A study of alcoholism in half-siblings. *Amer. J. Psychiat.*, **128**:1132–1136, 1972.

Schwartz, M., and J. Schwartz. No evidence for heritability of social attitudes. *Nature*, **225**:429, 1975.

Scott, J. P. That old-time aggression. *The Nation*, Jan. 9, 1967, pp. 53–54. Shows how K. Lorenz is "pre-Mendelian."

Shaskan, E. G., and R. E. Becker. Platelet monoamine oxidase in schizophrenia. *Nature*, **253**:659–660, 1975.

Spuhler, J. N., ed. *Genetic Diversity and Human Behavior*. Chicago: Aldine, 1967.

Stamatoyannopoulos, G., S. Chen, and M. Fukui. Liver alcohol dehydrogenase in Japanese: High population frequency of a typical form and its possible role in alcohol sensitivity. *Amer. J. Hum. Genet.*, **27**:789-796, 1975.

Thiessen, D. D. The biology of aggression: Evolution and physiology. In *Challenging Biological Problems*, J. A. Behnke, ed. New York: Oxford University Press, 1972, pp. 168–192.

Thoday, J., and A. Parkes, eds. *Genetic and Environmental Influences on Behavior*. New York: Plenum, 1968.

Wolff, P. H. Ethnic differences in alcohol sensitivity. *Science*, **175**:449–450, 1972.

Wolff, P. H. Vasomoter sensitivity to alcohol in diverse mongoloid populations. *Amer. J. Hum. Genet.*, **25**:193–199, 1973.

Wyatt, R. J., M. A. Schwartz, E. Erdelyi, and J. D. Barchas. Dopamine beta-hydrolase activity in brains of chronic schizophrenic patients. *Science*, **187**:368–370, 1975.

Zerbin-Rudin, E. The genetics of schizophrenia: An international survey. *Psychiat. Quarterly*, **46**:371–383, 1972.

16 Genetics, Interaction, and Mental Illness: Setting the Problem

L. Erlenmeyer-Kimling

Probably no other approach to the understanding of behavior has met more enduring opposition than that which includes the possibility of hereditary contributions underlying man's intellectual or emotional development, "normal" or "abnormal." Nearly everyone has found it easy to acknowledge the coexistence of inherited and environmental factors in the instinctive behavioral patterns of insects, birds, and other animals, and not too difficult, either, to concede a possible role played by heredity in some higher-order behaviors of animals other than man. In a few specific disorders of human behavior, such as Huntington's chorea or phenylketonuria, also, a genetic basis has long been recognized. But, on the whole, the idea that genes could have much to do with the complex activities of human mental life has gained only slow acceptance in most of the behavioral sciences and mental health professions. To many workers in these fields, the idea is still remote, despite a growing tendency in the literature to give at least a nod to the concept of gene-environment interaction.

The Problem and the History

This long-standing outcast status of genetics is, obviously, explained in part by history (see Dunn, 1962; Hirsch, 1963, 1967; Rosenthal, 1970). From their beginnings, psychology, psychiatry, and the other social sciences have leaned heavily on philosophical foundations in which the biological properties of man were de-emphasized. Carrying freight such as the body-mind dualism theory, or the common misrepresentation (see Dobzhansky, 1967) of Lockes' *tabula rasa* concept of the human mind at birth, or, later, for example, behaviorism's notion of the individual as an "empty organism" or "black box" describable entirely in terms of its reinforcement history, the behavioral sciences have had scant preparation to admit into their mainstream theories about genotypic diversity and human behavioral variations.

The credibility of such theories, moreover, was weakened when some of their early proponents tended to oversell their wares and to push the claims of "nature" over "nurture" far beyond reasonable bounds. Yet overzealousness is not unusual in young sciences; and, in this, equal match was given by some of the proponents of "nurture" as the sole determinant of behavior. Given the undeveloped state of knowledge and methodology in both human genetics and the behavioral sciences in the early years of this century, the excessively eager claims of both sides are perhaps understandable. Given time to develop their respective scientific crafts in an unemotional climate, the two sides might soon have discovered the fallacy of the "heredity *or* environment" argument. An unemotional climate, however, was not to be had. From outside the field of genetics itself, extreme views about the over-riding importance of heredity's influences upon intelligence, character, and mental health were seized upon and made pernicious by racists and the advocates of dubious eugenics movements—and, later, with sweeping and awful consequences, by Hitler & Co.

Acquiring guilt by association, *any* interest in the heritable aspects of psychological traits came to be considered " beyond the pale " of respectability. It has taken some time for sound, scientific work in human behavioral genetics to win free of the stigma. It would be particularly unfortunate if new voices, now borrowing the name but not the knowledge of modern genetics in speculative claims about inherent differences among races and other population groups, were again to bring legitimate research and theory under the pall of suspicion.

Misunderstandings and Their Implications

History aside, much of the unpopularity of genetic hypotheses within the disciplines traditionally concerned with man's behavior grows out of basic misunderstandings about the meaning of the hypotheses themselves. The confusion centers mainly around the distinction between genotype and phenotype—or, rather, around the failure to make such a distinction. Thus many people, in the behavioral sciences as well as in the public at large, continue to hold to the notion that characteristics with a hereditary basis are inherited full-blown and ready-made to follow a fixed, unalterable course.

Viewed in this way, of course, genetic theories seem threatening at a personal, as well as a professional, level. As Gottesman (1965, p. 72) has observed: " There is a natural reluctance to accept the supposed determinism that is associated with views that human behavior is genetically influenced. The former is especially true when one thinks of oneself. The ego defenses aroused are in part due to the values placed on free will and equality which are part and parcel of our democratic way of life "— and, he might have added, in part due to the pessimistic belief that heredity is incompatible with possibilities for modification or amelioration. The heart of the problem is exposed in the words of a colleague to whom I was recently introduced: " Genetics is such a discouraging subject. After all, if something is determined by heredity, there's nothing you can do to change it." Discouraging, indeed, if this were an accurate picture of the impact of genes upon phenotypes, behavioral or otherwise! Fortunately, it is not.

What is not understood, then, is the interaction of genotypes and environments. It is true, as mentioned earlier, that increasingly frequent reference is made to the concept in the recent literature on mental health and mental disorder. In general, however, the implications of interaction have not been integrated meaningfully into the major theoretical contexts of the behavioral sciences. For geneticists, by contrast, it is a key concept. Heredity is not *the* determinant of phenotypic characters, but *a* determinant factor thereof, along with aspects of the environment. No more lucid explanation can be found than the following, offered by Dobzhansky (1962, p.137): "To say that health, or intelligence, or musical ability are inherited does not mean that those who inherit them will be healthy even in unhealthy environments, intelligent without opportunity to develop intelligence, or will necessarily be musicians."

To acknowledge the possible influence of genes upon behavioral pathologies such as those discussed in this journal issue, therefore, by no means constitutes a surrender of hopes for their prevention or cure at the phenotypic level. By no means does the documentation of genetic evidence devalue attempts to find solutions through nongenetic means. Examples of environmental management of conditions with large hereditary components are becoming more and more numerous. For instance, the adverse extremes of phenylketonuria and galactosemia may be prevented by dietary regulation, chemical or physical therapies offer correction or control of diabetes, hemophilia, certain forms of cancer and heart diseases, adrenogenital syndrome, severe myopia, a variety of congenital anomalies, etc. Treatment methods for sickle-cell anemia, a disorder of known genetic origin, are beginning to be extensively researched. Epilepsies and the affective psychoses discussed in the present collection of papers are coming under the relatively successful management of medications; and even for schizophrenia, it is clear that existing drug therapies often afford relief.

But what of intervention and remedial methods that are not based on physical or chemical manipulations? As Meehl (pp. 10–27 this issue)* explains, psychological intervention—whether it takes the form of psychotherapy, special education programs, rehabilitation, or other approaches to psychosocial adjustment—is not contraindicated, not made "pragmatically useless," by the demonstration of genetic influences in behavioral disturbances. The point has been aptly made by MacMahon (1968, p. 394) in a discussion of the interaction of heredity and environment in schizophrenia: ". . . even if the concordance of monozygous twins were 100% and that of dizygous twins close to zero, we should not give up the search for environmental determinants that have preventive significance."

If genetic evidence is not a justification for alarm on the part of theoreticians and clinicians who deal with environmental causation and control, it nevertheless cannot be safely ignored. Interactions are often highly specific to a given genotype and a particular feature of what is broadly called "the environment." Some persons, for example, react with a severe hemolytic crisis following exposure to a particular drug or food or other agent— and *only* following such exposure—whereas persons with other genotypes are not adversely affected by the same item. Deprivations, too, affect different people differently; and so do methods of treatment, whether for the common cold or for mental illness. "The crux of the matter," says Dobzhansky (1962, p. 137), "is that an environment optimal for one genotype may be mediocre for another and adverse for a third." Workers engaged in the problems of healing and preventing disruptions of mental life cannot afford *not* to know something about genotypes and their environmental ranges.

Internat. J. Mental Health, vol. 1, 1972.

About This Issue

A number of internationally known authorities in psychiatric genetics have contributed to this issue* on the role of hereditary factors in the etiology of mental disorders. Some of the papers offer original research data, others provide timely reviews of findings as they now stand, and others focus on topics of theoretical interest. There has been no attempt to cover all of psychiatric genetics.

It will be seen that the collection places more emphasis on schizophrenia than on any other category of disorder. This is fitting, perhaps, because schizophrenia is the prototypical mental illness (Rosenthal, 1970, p. xi), the disorder—or group of disorders—that exacts the highest toll in personal and social costs, that generates the greatest amount of research and dialectics, and that is most likely to confront the majority of workers in the mental health professions.

Perhaps, as students of human behavior, we all have something to learn from the strategies of modern ecology. Ecology has come to grips with the fact that the systems it works with are complexes of numerous interacting components; accordingly, they must be studied and managed as such. Attention limited to one or two components will not do. To know how to cure an ailing pond or defend the stability of a threatened one, to be able to optimize conditions so that the system may thrive, or simply to describe its functioning at a point in time, ecologists must learn about the members of the pond's community as well as about inputs of matter and energy flowing in from streamlets, the air, and other sources. It is the web of interactions and feedback phenomena among these components that determines how the life of the pond has come to be as it is and how it is likely to progress in the future. Ponds and other natural systems become understandable and potentially manipulable as these synergistic patterns are gradually traced out.

Human behavior, too, is the product of many converging factors. As in ecosystem management, sound solutions to the prevention and treatment of behavioral pathologies, or to the enhancement of psychological assets, depend upon our knowing first about the workings of these root factors and their ways of interacting. For clinical practitioner, social counselor, educator, theoretician, and basic research worker alike, the need is to understand the components from which behavior is shaped. The ecological model needs to be taken to heart by everyone.

References

Dobzhansky, T. (1962) Genetics and equality. *Science,* **137,** 112.

Dobzhansky, T. (1967) On types, genotypes, and the genetic diversity in populations. In J. N. Spuhler (Ed.), *Genetic diversity and human behavior.* Chicago: Aldine.

Dunn, L. C. (1962) Cross currents in the history of human genetics. *Amer. J. Hum. Genet.* **14,** 1.

Gottesman, I. I. (1965) Personality and natural selection. In S. G. Vandenberg (Ed.), *Methods and goals in human behavior genetics.* New York: Academic Press.

*Ibid.

Hirsch, J. (1963) Behavior genetics and individuality understood. *Science*, **142**, 1436.

Hirsch, J. (1967) Intellectual functioning and the dimensions of human variation. In J. N. Spuhler (Ed.), Op. cit.

MacMahon, B. (1968) Gene-environment interaction in human disease. In D. Rosenthal & S. S. Kety (Eds.), *The transmission of schizophrenia.* Oxford: Pergamon, P. 393.

Rosenthal, D. (1970) *Genetic theory and abnormal behavior.* New York: McGraw-Hill.

17 Psychiatric Disorders in Foster Home Reared Children of Schizophrenic Mothers

Leonard L. Heston

Introduction

The place of genetic factors in the aetiology of schizophrenia remains disputed. Several surveys have demonstrated a significantly higher incidence of the disorder in relatives of schizophrenic persons as compared to the general population. Furthermore, the closer the relationship, the higher the incidence of schizophrenia. The studies of Kallmann (1938) and Slater (1953) are especially significant and the research in this area has been thoroughly reviewed by Alanen (1958).

Although the evidence for a primarily genetic aetiology of schizophrenia is impressive, an alternative explanation—that schizophrenia is produced by a distorted family environment—has not been excluded. A close relative who is schizophrenic can be presumed to produce a distorted interpersonal environment and the closer the relationship the greater the distortion.

This study tests the genetic contribution to schizophrenia by separating the effects of an environment made "schizophrenogenic" by the ambivalence and thinking disorder of a schizophrenic parent from the effects of genes from such a parent. This is done by comparing a group of adults born to schizophrenic mothers where mother and child were permanently separated after the first two postpartum weeks with a group of control subjects.

Selection of Subjects

The Experimental subjects were born between 1915 and 1945 to schizophrenic mothers confined to an Oregon State psychiatric hospital. Most of the subjects were

This research was supported by the Medical Research Foundation of Oregon. From the *British Journal of Psychiatry*, **112**: 819–825, 1966. Reprinted with permission of the author and the publisher.

born in the psychiatric hospital; however, hospital authorities encouraged confinement in a neighbouring general hospital whenever possible, in which case the children were delivered during brief furloughs. All apparently normal children born of such mothers during the above time span were included in the study if the mother's hospital record (1) specified a diagnosis of schizophrenia, dementia praecox, or psychosis; (2) contained sufficient descriptions of a thinking disorder or bizarre regressed behaviour to substantiate the diagnosis; (3) recorded a negative serologic test for syphilis and contained no evidence of coincident disease with known psychiatric manifestations; and (4) contained presumptive evidence that mother and child had been separated from birth. Such evidence typically consisted of a statement that the mother had yielded the child for adoption, a note that the father was divorcing the mother, the continued hospitalization of the mother for several years, or the death of the mother. In practice these requirements meant that the mothers as a group were biased in the direction of severe, chronic disease. No attempt was made to assess the psychiatric status of the father; however, none were known to be hospital patients. The 74 children ascertained as above were retained in the study if subsequent record searches or interviews confirmed that the child had had no contact with its natural mother and never lived with maternal relatives. (The latter restriction was intended to preclude significant exposure to the environment which might have produced the mother's schizophrenia.)

All of the children were discharged from the State hospital within three days of birth (in accordance with a strictly applied hospital policy) to the care of family members or to foundling homes. The records of the child care institutions made it possible to follow many subjects through their early life, including, for some, adoption. The early life of those subjects discharged to relatives was less completely known, although considerable information was developed by methods to be described.

Sixteen subjects were dropped because of information found in foundling home records; 6 children, 4 males and 2 females, died in early infancy. Ten others were discarded, 8 because of contact with their natural mother or maternal relatives, one because of multiple gastrointestinal anomalies, and one because no control subject whose history matched the bizarre series of events that complicated the Experimental subject's early life could be found. The remaining 58 subjects comprise the final Experimental group.

A like number of control subjects, apparently normal at birth, were selected from the records of the same foundling homes that received some of the Experimental subjects. The control subjects were matched for sex, type of eventual placement (adoptive, foster family, or institutional), and for length of time in child care institutions to within ± 10 per cent up to 5 years. (Oregon State law prohibited keeping a child in an institution more than five years. Subjects in institutions up to this maximum were counted as "institutionalized" regardless of final placement.) Control subjects for the Experimental children who went to foundling homes were selected as follows: When the record of an Experimental subject was located, the admission next preceding in time was checked, then the next subsequent, then the second preceding and so on, until a child admitted to the home within a few days of

Table 1

	Experimental		Control	
	Male	Female	Male	Female
Number	33	25	33	25
Died, infancy or child-hood	3	2		5
Lost to follow-up		6		3
Final groups	30	17	33	17

birth and meeting the above criteria was found. Those Experimental subjects who were never in child care institutions were matched with children who had spent less than three months in a foundling home. The above method of selection was used with the record search beginning with an Experimental child's year of birth. The above restrictions regarding maternal contacts were applied to the control group. Oregon State psychiatric hospital records were searched for the names of the natural parents (where known) of the control subjects. In two cases a psychiatric hospital record was located and the children of these persons were replaced by others. All of the children went to families in which both parental figures were present.

Exact matching was complicated by the subsequent admission of several subjects to other child care institutions and by changes of foster or even adoptive homes. However, these disruptions occurred with equal frequency and intensity in the two groups and are considered random.

Table 1 gives the sex distribution of the subjects and the causes of further losses. Fifteen of the 74 Experimental subjects died before achieving school age. This rate is higher than that experienced by the general population for the ages and years involved, but not significantly so.

Follow-Up Method

Starting in 1964, it proved possible to locate or account for all of the original subjects except five persons, all females. During this phase of the research, considerable background information of psychiatric import was developed. The records of all subjects known to police agencies and to the Veterans' Administration were examined. Retail credit reports were obtained for most subjects. School records, civil and criminal court actions, and newspaper files were reviewed. The records of all public psychiatric hospitals in the three West Coast States were screened for the names of the subjects and the records located were reviewed. Enquiries were directed to psychiatric facilities serving other areas where subjects were living, and to probation departments, private physicians, and various social service agencies with which the subjects were involved. Finally, relatives, friends, and employers of most subjects were contacted.

In addition to information obtained from the above sources, for most subjects the psychiatric assessment included a personal interview, a Minnesota Multiphasic Personality Inventory (MMPI), an I.Q. test score, the social class of the subject's first home, and the subject's current social class. As the subjects were located, they were contacted by letter and asked to participate in a personal interview. The interview was standardized, although all promising leads were followed, and was structured as a general medical and environmental questionnaire which explored all important psychosocial dimensions in considerable depth. Nearly all of the interviews were conducted in the homes of the subjects, which added to the range of possible observations. The short form of the MMPI was given after the interview. The results of an I.Q. test were available from school or other records for nearly all subjects. If a test score was not available, the Information, Similarities, and Vocabulary subtests of the Wechsler Adult Intelligence Scale (WAIS) was administered and the I.Q. derived from the results. Two social class values were assigned according to the occupational classification system of Hollingshead (1958). One value was based on the occupation of the father or surrogate father of the subject's first family at the time of placement, and a second on the subject's present occupational status or, for married females, the occupation of the husband. The social class values move from 1 to 7 with decreasing social status.

All of the investigations and interviews were conducted by the author in 14 States and in Canada.

Evaluation of Subjects

The dossier compiled on each subject, excluding genetic and institutional information, was evaluated blindly and independently by two psychiatrists. A third evaluation was made by the author. Two evaluative measures were used. A numerical score moving from 100 to 0 with increasing psychosocial disability was assigned for each subject. The scoring was based on the landmarks of the Menninger Mental Health-Sickness Rating Scale (MHSRS) (Luborsky, 1962). Where indicated, the raters also assigned a psychiatric diagnosis from the American Psychiatric Association nomenclature.

Evaluations of 97 persons were done. Seventy-two subjects were interviewed. Of the remaining 25 persons, six refused the interview (7.6 per cent of those asked to participate), eight were deceased, seven are inaccessible (active in Armed Forces, abroad, etc.), and four were not approached because of risk of exposure of the subject's adoption. It did not seem reasonable to drop all of these 25 persons from the study, since considerable information was available for most of them. For instance, one man was killed in prison after intermittently spending most of his life there. His behavioural and social record was available in prison records plus the results of recent psychological evaluations. A man who refused the interview was a known, overt, practising homosexual who had a recent felony conviction for selling narcotics. All persons in the Armed Forces were known through letters from their Commanding Officers or medical officers to have been serving honourably without psychiatric or serious behavioural problems. One 21-year-old man, the least known of any of the subjects, had been in Europe for the preceding 18 months in an uncertain capacity. He is known to have graduated from high school and to have no adverse behavioural record. In a conference the raters agreed that it would be misleading to discard any cases, and that all subjects should be rated by forced choice.

The MHSRS proved highly reliable as a measure of degree of incapacity. The Intraclass Correlation Coefficient between the scores assigned by the respective raters was 0.94, indicating a high degree of accuracy. As expected, several differences arose in the assignment of specific diagnoses. In disputed cases a fourth psychiatrist was asked for an opinion and differences were discussed in conference. The only differences not easily resolved involved distinctions such as obsessive-compulsive neurosis versus compulsive personality or mixed neurosis versus emotionally unstable personality. All differences were within three diagnostic categories: psychoneurotic disorders, personality trait, or personality pattern disturbances. The raters decided to merge these categories into one: "neurotic personality disorder." This category included all persons with MHSRS scores less than 75—the point on the scale where psychiatric symptoms become troublesome—who received various combinations of the above three diagnoses. In this way, complete agreement on four diagnoses was achieved: schizophrenia, mental deficiency, sociopathic personality, and neurotic personality disorder. One mental defective was also diagnosed schizophrenic and another sociopathic. Only one diagnosis was made for all other subjects.

Results

Psychiatric disability was heavily concentrated in the Experimental group. Table 2 summarizes the results.

The MHSRS scores assess the cumulative psychosocial disability in the two groups. The difference is highly significant with the Experimental group, the more disabled by this measure. However, the difference is attributable to the low scores achieved by about one-half (26/47) of the Experimental subjects rather than a general lowering of all scores.

The diagnosis of schizophrenia was based on generally accepted standards. In addition to the unanimous opinion of the three raters, all subjects were similarly diagnosed in psychiatric hospitals. One female and four males comprised the schizophrenic group. Three were chronic deteriorated patients who had been hospitalized for several years. The other two had been hospitalized and were taking antipsychotic drugs. One of the latter persons was also mentally deficient: a brief history of this person follows.

A farm labourer, now 36 years old, was in an institution for mentally retarded children from age 6-16. Several I.Q. tests averaged 62. He was discharged to a family farm, where he worked for the next 16 years. Before his hospitalization

Table 2

	Control	Experimental	Exact Probability
Number	50	47	
Male	33	30	
Age, mean	36.3	35.8	
Adopted	19	22	
MHSRS, mean (total group mean = 72.8, S.D. = 18.4)	80.1	65.2	0.0006
Schizophrenia (morbid risk = 16.6%)	0	5	0.024
Mental deficiency (I.Q. < 70)	0	4	0.052
Sociopathic personality	2	9	0.017
Neurotic personality disorder	7	13	0.052
Persons spending > 1 year in penal or psychiatric institution	2	11	0.006
Total years institutionalized	15	112	
Felons	2	7	0.054
Armed forces, number serving	17	21	
Armed forces, number discharges, psychiatric or behavioural	1	8	0.021
Social group, first home, mean	4.2	4.5	
Social group, present, mean	4.7	5.4	
I.Q., mean	103.7	94.0	
Years school, mean	12.4	11.6	
Children, total	84	71	
Divorces, total	7	6	
Never married, > 30 years age	4	9	

One mental defective was also schizophrenic.
Another was sociopathic.
Considerable duplication occurs in the entries below Neurotic Personality Disorder.
* Fisher Exact Probability Test.

at age 32 he was described as a peculiar but harmless person who was interested only in his bank account: he saved $5,500 out of a salary averaging $900 per year. Following a windstorm that did major damage to the farm where he worked he appeared increasingly agitated. Two days later he threatened his employer with a knife and accused him of trying to poison him. A court committed him to a psychiatric hospital. When admitted, he talked to imaginary persons and assumed a posture of prayer for long periods. His responses to questions were incoherent or irrelevant. The hospital diagnosis was schizophrenic reaction. He was treated with phenothiazine drugs, became increasingly rational, and was discharged within a month. After discharge he returned to the same farm, but was less efficient in his work and spent long periods sitting and staring blankly. He has been followed as an out-patient since discharge, has taken phenothiazine drugs continuously, and anti-depressants occasionally. This man exhibited almost no facial expression. His responses to questions, though relevant, were given after a long and variable latency.

The age-corrected rate for schizophrenia is 16.6 per cent, a finding consistent with Kallmann's 16.4 per cent (Weinberg's short method, age of risk 15-45 years). Hoffman (1921) and Oppler (1932) reported rates of from 7 to 10.8 per cent of schizophrenia in children of schizophrenics. No relationship between the severity and sub-type of the disease in the mother-child pairs was evident.

Mental deficiency was diagnosed when a subject's I.Q. was consistently less than 70. All of these persons were in homes for mental defectives at some time during their life and one was continuously institutionalized. His I.Q. was 35. The other mentally deficient subjects had I.Q.s between 50 and 65. No history of CNS disease or trauma of possible causal importance was obtained for any of these subjects. The mothers of the mentally defective subjects were not different from the other mothers and none were mentally defective.

Three behavioural traits were found almost exclusively within the Experimental group. These were: (1) significant musical ability, 7 persons; (2) expression of unusually strong religious feelings, 6 persons; and (3) problem drinking, 8 persons.

The results with respect to the effects of institutional care, social group, and type of placement will be discussed in a later paper. None of these factors had measurable effects on the outcome.

Discussion

The results of this study support a genetic aetiology of schizophrenia. Schizophrenia was found only in the offspring of schizophrenic mothers. The probability of this segregation being effected by chance is less than 0.025. Furthermore, about one-half of the Experimental group exhibited major psychosocial disability. The bulk of these persons had disorders other than schizophrenia which were nearly as malignant in effect as schizophrenia itself. An illustration is provided by the 8 of 21 Experimental males who received psychiatric or behavioural discharges from the armed services. If three subjects who were rejected for service for the same reasons are added, the ratio becomes 11:24, or essentially 1:2. Only three of these 11 subjects were schizophrenic and one schizophrenic served honourably. Kallmann's (1938) rate for first degree relatives and Slater's (1953) for dizygotic twins of schizophrenic persons who developed significant psychosocial disability not limited to schizophrenia are slightly lower, though in the same range, as those found in the present study.

The association of mental deficiency with schizophrenia has been reported by Hallgren and Sjögren (1959) who noted an incidence of low-grade mental deficiency (I.Q. < 50–55) in schizophrenic subjects of about 10.5 per cent. Kallmann (1938) found from 5–10 per cent mental defectives among his descendants of schizophrenic persons, but did not consider the finding significant. The association of mental deficiency with schizophrenia—if such an association exists—remains uncertain.

Two sub-groups of persons within the impaired one-half of Experimental subjects exhibited roughly delineable symptom-behaviour complexes other than schizophrenia or mental deficiency. The personalities of the persons composing these groups are described in aggregate below.

The first group is composed of subjects who fit the older diagnostic category, "schizoid psychopath." This term was used by Kallmann (1938) to describe a significant sub-group of his relatives of schizophrenic persons. Eight males from the present study fall into this group, all of whom received a diagnosis of sociopathic personality. These persons are distinguished by anti-social behaviour of an impulsive, illogical nature. Multiple arrests for assault, battery, poorly planned impulsive thefts dot their police records. Two were homosexual, four alcoholic, and one person, also homosexual, was a narcotics addict. These subjects tended to live alone—only one was married—in deteriorated hotels and rooming houses in large cities, and locating them would have been impossible without the co-operation of the police. They worked at irregular casual jobs such as dishwasher, race-track tout, parking attendants. When interviewed they did not acknowledge or exhibit evidence of anxiety. Usually secretive about their own life and circumstances, they expressed very definite though general opinions regarding social and political ills. In spite of their suggestive life histories, no evidence of schizophrenia was elicited in interviews. No similar personalities were found among the control subjects.

A second sub-group was characterized by emotional liability and may correspond to the neurotic sibs of schizophrenics described by Alanen (1963). Six females and two males from the Experimental group as opposed to two control subjects were in this category. These persons complained of anxiety or panic attacks, hyper-irritability, and depression. The most frequent complaint was panic when in groups of people as in church or at parties, which was so profoundly uncomfortable that the subject was forced to remove himself abruptly. Most subjects described their problems as occurring episodically; a situation that they might tolerate with ease on one occasion was intolerable on another. The women reported life-long difficulty with menses, especially hyper-irritability or crying spells, and depressions coincident with pregnancy. These subjects described themselves as "moody," stating that they usually could not relate their mood swings to temporal events. Four such subjects referred to their strong religious beliefs much more frequently than other respondents. Psychophysiological gastrointestinal symptoms were prominent in five subjects. The most frequent diagnoses advanced by the raters were emotionally unstable personality and cyclothymic personality, with neurosis a strong third.

Of the 9 persons in the control group who were seriously disabled, 2 were professional criminals, careful and methodical in their work, 2 were very similar to the emotionally labile group described above, one was a compulsive phobia-ridden neurotic, and 4 were inadequate or passive-aggressive personalities.

The 21 Experimental subjects who exhibited no significant psychosocial impairment were not only successful adults but in comparison to the control group were more spontaneous when interviewed and had more colourful life histories. They held the more creative jobs: musician, teacher, home-designer; and followed the more imaginative hobbies: oil painting, music, antique aircraft. Within the Experimental group there was much more variability of personality and behaviour in all social dimensions.

Summary

This report compares the psychosocial adjustment of 47 adults born to schizophrenic mothers with 50 control adults, where all subjects had been separated from their natural mothers from the first few days of life. The comparison is based on a review of school, police, veterans, and hospital, among several other records, plus a personal interview and MMPI which were administered to 72 subjects. An I.Q. and social class determination were also available. Three psychiatrists independently rated the subjects.

The results were:

(1) Schizophrenic and sociopathic personality disorders were found in those persons born to schizophrenic persons in an excess exceeding chance expectation at the 0.05 level of probability. Five of 47 persons born to schizophrenic mothers were schizophrenic. No cases of schizophrenia were found in 50 control subjects.

(2) Several other comparisons, such as persons given other psychiatric diagnoses, felons, and persons discharged from the Armed Forces for psychiatric or behavioural reasons, demonstrated a significant excess of psychosocial disability in about one-half of the persons born to schizophrenic mothers.

(3) The remaining one-half of the persons born to schizophrenic mothers were notably successful adults. They possessed artistic talents and demonstrated imaginative adaptations to life which were uncommon in the control group.

References

Alanen, Y. O. (1958). "The mothers of schizophrenic patients." *Acta pyschiat. neurol. scand.*, Suppl. 1227.

———, Rekola, J., Staven, A., Tuovinen, M., Takala, K., and Rutanen, E. (1963). "Mental disorders in the siblings of schizophrenic patients." *Acta psychiat. scand.*, Suppl. 169, 39, 167.

Hallgren, B., and Sjögren, T. (1959). "A clinical and genetico-statistical study of schizophrenia and low grade mental deficiency in a large Swedish rural population." *Acta psychiat. neurol. scand.*, Suppl. 140. Vol. 35.

Hoffman, H. (1921). "Studien über Vererbung und Entstehung geistiger Störungen. II. Die Nachkommenschaft bei endogenen Psychosen." Berlin: Springer.

Hollingshead, A. B., and Redlich, F. C. (1958). *Social Class and Mental Illness: A Community Study*. New York: J. Wiley.

Kallmann, F. J. (1938). *The Genetics of Schizophrenia*. New York: J. J. Augustin.

Luborsky, L. (1962). "Clinicians' judgements of mental health: a proposed scale." *Arch. gen. Psychiat. (Chic.)*, 7, 407.

Oppler, W. (1932). "Zum Problem der Erbprognosebestimmung." *Z. Neurol.*, 141, 549–616.

Slater, E., with Shields, J. (1953). *Psychotic and neurotic illnesses in twins*. Medical Research Council Special Report Series No. 278. London: H.M. Stationery Office.

18   Violence and Man's Struggle to Adapt

Marshall F. Gilula and David N. Daniels

The need is not really for more brains, the need is now for a gentler, a more tolerant people than those who won for us against the ice, the tiger, and the bear (*1*).

Violence waits in the dusty sunlight of a tenement yard and in the shadows of a distraught mind. Violence draws nearer in the shouts of a protest march and in ghetto rumblings. Violence erupts from Mace-sprinkled billy clubs and a homemade Molotov cocktail. Violence of war explodes the peace it promises to bring. Hourly reports of violence bring numbness, shock, confusion, sorrow. We live in a violent world (*2*).

Violence surrounds us, and we must try to understand it in the hopes of finding alternatives that will meet today's demand for change. Do we benefit from violence? Or is violence losing whatever adaptive value it may once have had? We present two theses. (i) Violence can best be understood in the context of adaptation. Violence is part of a struggle to resolve stressful and threatening events—a struggle to adapt. (ii) Adaptive alternatives to violence are needed in this technological era because the survival value of violent aggression is diminishing rapidly.

The shock of Robert F. Kennedy's death prompted the formation of a committee on violence (*3*) in the Department of Psychiatry, Stanford University School of Medicine. We committee members reviewed the literature on violence and then interpreted this literature from the point of view of psychiatrists and psychologists. We discussed our readings in seminars and sought answers to questions about violence. This article presents a synthesis of our group's findings and observations and reflects our view of adaptation theory as a unifying principle in human behavior.

We define pertinent terms and describe the adaptation process before we examine violence as it relates to individual coping behavior and collective survival. We then describe three theories of aggression and relate them to adaptation. Next, we discuss relevant examples of violence as attempted coping behavior and factors that foster violence and illustrate the urgent need for other ways of expressing aggression. Finally, we consider the changing nature of adaptation and suggest ways of coping with violence.

Definition of Terms

Two groups of terms require definition: (i) aggression and violence; and (ii) adaptation, adjustment, and coping. We found that these terms have quite different meanings for different disciplines.

From *Science*, **164** (Apr. 25, 1969), 396–405. Copyright 1969 by the American Association for the Advancement of Science. Reprinted with permission of the authors and the publisher.

We here define aggression (*4, 5*) as the entire spectrum of assertive, intrusive, and attacking behaviors. Aggression thus includes both overt and covert attacks, such defamatory acts as sarcasm, self-directed attacks, and dominance behavior. We extend aggression to include such assertive behaviors as forceful and determined attempts to master a task or accomplish an act. We choose a broad definition of aggression rather than a restrictive one because relations between the underlying physiological mechanisms and the social correlates of dominant, assertive, and violent behavior are still poorly understood. Hence, our definition encompasses but is broader than the definition of aggression in animals that is used in experimental biology (*6, 7*), which says that an animal acts aggressively when he inflicts, attempts to inflict, or threatens to inflict damage upon another animal. Violence (*4*) is destructive aggression and involves inflicting physical damage on persons or property (since property is so often symbolically equated with the self). Violent inflicting of damage is often intense, uncontrolled, excessive, furious, sudden, or seemingly purposeless. Furthermore, violence may be collective or individual, intentional or unintentional, apparently just or unjust.

By adaptation we mean the behavioral and biological fit between the species and the environment resulting from the process of natural selection (*8, 9*). In man, adaptation increasingly involves modifying the environment as well. Here we want to stress that behavior, especially group-living behavior in higher social species like man, is a crucial element in natural selection (*10*). Adaptive behaviors are those that enhance species survival and, in most instances, individual survival. In contrast, we define adjustment as behavior of a group or individual that temporarily enhances the way we fit with the immediate situation. By definition, adjustment is often a passive rather than active process and does not result in an enduring alteration of behavior structure or patterns (*4, 11*). In fact, adjustment may have biologically maladaptive consequences in the long run. In addition, rapid environmental change or extraordinary environmental circumstances may render formerly adaptive behaviors largely maladaptive (*10*), that is, behaviors appropriate to past environmental conditions can work against survival in "new" or unusual environments.

We define coping as the continuing and usually successful struggle to accomplish tasks and goals with adaptive consequences. Put another way: "Behavior may be considered to serve coping functions when it increases the likelihood (from a specified vantage point with respect to a specified time unit) that a task will be accomplished according to standards that are tolerable both to the individual and to the group in which he lives" (*12*). Whereas each specific sequence of task-oriented behaviors may or may not have adaptive value, coping taken as a whole is an adaptive rather than adjustive human process.

Definition of Human Adaptation

Every culture prescribes the range of coping behaviors available to its people, but within this range individual adaptive behavior is forged and tested in times of stress. Stressful or new situations paradoxically offer us both the danger of failure and the opportunity for learning. Stress can be dangerous when it overwhelms the individual or group. Either the situation itself or unpleasant feelings about the situation (including massive anxiety) may block our usual resources and prevent problem solving, and aggressive reactions that are both indiscriminate and protective may occur. We may show primitive forms of behavior: passive adjustment, withdrawal, falsely blaming others, indiscriminate rage, violence, or confusion.

Alternately, stressful events provide a constructive challenge and expanded opportunity for learning. In a stressful situation that is not overwhelming, we seek information helpful in dealing with the situation and try to apply this information (13). From information seeking and subsequent exploratory behavior come not only greater use of information and eventual mastery of new situations but also a sense of heightened self-awareness, enhanced coping skills, and personal growth.

A number of commonly occurring stressful life situations that may challenge and develop our coping skills have been recognised (13). These are associated with the transitions in life and include adolescence, separation from parents, and marriage. Other challenging transitions involve cultural stresses, such as war and the threat of war; rapid technological change; and physical events, such as drought, earthquakes, and famine. These transition points in life are important because they provide opportunity for learning and developing more sophisticated ways of coping with problems.

We have marvelous adaptive abilities for coping with varying, even extreme, situations. These abilities result from cultural evolution interacting with our biological evolution. Culturally we survive through complex communal living. Through our living groups we obtain satisfaction, develop identity, and find meaning to life. Basic social values are of special cultural importance, for they determine the limits of acceptable behavior, especially during times of stress. Biologically we are uniquely endowed for complex communal living. Such biological characteristics as aggression, the upright posture, prehension, speech, prolonged infancy and maturation, and profound development of the brain—all favor and allow for rich, dynamic, and complex living. Development of the cerebral hemispheres has played an especially important role in adaptation, for the cerebrum constitutes the biological basis of higher intelligence, self-awareness, complex language, and flexibility (8).

Thus through the interaction of biological evolution and cultural evolution, we have the equipment for adapting to and molding diverse environments. But this ability to adapt by manipulating the environment is now our cause for greatest concern, for in changing the environment, man changes the conditions necessary for his survival. We now are seeing an unprecedented acceleration of various man-made changes which call for accompanying changes in man, changes which we are having difficulty in making. While biological change is extremely slow, cultural change theoretically occurs at least every generation, although some aspects of culture (such as technology) change faster than others (for example, beliefs and customs). The term "generation gap" not only describes how we today view the battle of the generations but also alludes to the speed of cultural change and how people have trouble keeping pace. Living in the electronic age, we watch televised accounts of preagricultural-age violence and feel our industrial-age mentality straining to cope with the environment.

Since survival results from the long-range adaptiveness of our behavior, knowledge of adaptive mechanisms is important for understanding the role of violence in human behavior and survival. In the section that follows we shall relate three theories of aggression to adaptation.

Adaptation and Theories of Aggression

Aggression has helped man survive. Aggression in man—including behaviors that are assertive, intrusive, and dominant as well as violent—is fundamental and adaptive. Violence is not a result of aggression but simply a form of aggression. Nor is all violence necessarily motivated by destructive aggression. For instance, in the sadistic behavior of sexual assaults, violence is evoked in part by sexual motives. In other instances, violence can occur accidentally or without conscious intent, as in many auto accidents. Currently there are three main views of aggression—all involving adaptation—but each suggests a different solution to the problem of violent behavior. Broadly labeled, these theories are (i) the biological-instinctual theory, (ii) the frustration theory, and (iii) the social-learning theory.

1) *The biological-instinctual theory* (*14–16*) holds that aggressive behavior, including violence, is an intrinsic component of man resulting from natural selection: Man is naturally aggressive. It is hard to imagine the survival of man without aggressiveness, namely because aggression is an element of all purposeful behavior and, in many cases, provides the drive for a particular action. This theory says that aggression includes a wide variety of behaviors, many of which are constructive and essential to an active existence. Stimulus-seeking behavior (for example, curiosity or the need to have something happen) is certainly at least as important a facet of human behavior as avoidance behavior and need-satisfaction. Seeking the novel and unexpected provides much of life's color and excitement. Aggression can supply much of the force and power for man's creative potential.

Psychiatric and psychoanalytic case studies are one source of evidence supporting this theory (*14–17*). Examples range from individuals with destructive antisocial behavior who express violent aggression directly and often impulsively, to cases of depression and suicide in which violent aggression is turned against the self, and to seriously inhibited persons for whom the expression of aggression, even in the form of assertion, is blocked almost entirely. Psychiatrists and other mental-health professionals describe many disordered behaviors as stemming from ramifications and distortions of the aggressive drive (*14*).

Animal studies (*6, 15, 18*) (including primate field studies), studies of brain-damaged humans, and male-female comparisons provide behavioral, anatomical, and hormonal data illustrating the human predisposition to aggression. Among nonhuman mammals, intraspecies violence occurs less frequently than with humans (*7*). When violent aggressive behaviors do occur among members of the same species, they serve the valuable functions of spacing the population over the available land and maintaining a dominance order among the group members. Uncontrolled aggression in animals generally occurs only under conditions of overcrowding. Aggression in humans, even in the form of violence, has had similar adaptive value historically.

The biological-instinctual theory suggests that since aggression is inevitable, effective controls upon its expression are necessary, and reduction of violence depends upon providing constructive channels for expressing aggression.

2) *The frustration theory* (*19*) states that aggressive behavior comes from interfering with ongoing purposeful activity. A person feels frustrated when a violation of his hopes or expectations occurs, and he then tries to solve the problem by behaving aggressively. Frustrations can take various forms: threats to life, thwarting of basic needs, and personal insults. This theory often equates aggression with destructive or damaging violent behavior. Major factors influencing aggressive responses to frustration are the nature of the frustration, previous experience, available alternatives for reaction (aggression is by no means the only response to frustration), the person's maturity, and the preceding events or feelings. Even boredom may provoke an aggressive response. As a response to frustration, aggression is often viewed as a learned rather than an innate behavior. According to this theory, frustration-evoked aggression aims at removing obstacles to our goals; hence the frustration theory also ties in with adaptation. The aggressive response to frustration often is a form of coping behavior that may have not only adjustive but also long range consequences.

The frustration theory suggests that control or reduction of violence requires reducing existing frustrations as well as encouraging constructive redirection of aggressive responses to frustration. This reduction includes removing or improving frustrating environmental factors that stand between personal needs and environmental demands. Such factors include violation of human rights, economic deprivation, and various social stresses.

3) *The social-learning theory* (*20*) states that aggressive behavior results from child-rearing practices and other forms of socialization. Documentation comes from sociological and anthropological studies and from observing social learning in children. Aggressive behavior can be acquired merely by watching and learning—often by imitation—and does not require frustration. Aggressive behaviors rewarded by a particular culture or subculture usually reflect the basic values and adaptive behaviors of the group. In American culture, where achievement, self-reliance, and individual self-interest are valued highly, we also find a relatively high emphasis on military glory, a relatively high incidence of personal crime, and a society characterized by a relatively high degree of bellicosity. Similar patterns occur in other cultures. From this theory we infer that as long as a nation values and accepts violence as an effective coping strategy, violent behavior will continue.

The social-learning theory of aggression suggests that control and reduction of violence require changes in cultural traditions, child-rearing practices, and parental examples. Parents who violently punish children for violent acts are teaching their children how and in what circumstances violence can be performed with impunity.

Other changes in cultural traditions would emphasize prevention rather than punishment of violent acts and, equally important, would emphasize human rights and group effort rather than excessive and isolated self-reliance. The first step toward making the changes that will reduce violence is to examine our values. We must decide which values foster violence and then begin the difficult job of altering basic values.

In reality, the three theories of aggression are interrelated. Proclivities for social learning and for frustration often have a biological determinant. For example, the biology of sex influences the learning of courting behavior. Regarding violence, from these theories of aggression we see that the many expressions of violence include man's inherent aggression, aggressive responses to thwarted goals, and behavior patterns imitatively learned within the cultural setting. All three theories of aggression and violence fit into the adaptation-coping explanation. Violence is an attempt to cope with stressful situations and to resolve intolerable conflicts. Violence may have short-run adjustive value, even when the long-run adaptive consequences may in fact be adverse. It is the sometimes conflicting natures of adjustment and adaptation that are confusing and insufficiently appreciated. In some instances violence emerges when other more constructive coping strategies have failed. In other instances violence is used to enhance survival. Our species apparently has over absorbed violence into our cultures as a survival technique. Children and adolescents have learned well the accepted violent behaviors of their elders.

All three theories help us understand violent behavior and hence suggest potential ways of reducing violence. In the following sections we consider current examples of violence from the perspective of those factors in our society that foster violence and from the standpoint of how these examples reflect the changing nature of adaptation.

Phenomenon of Presidential Assassination

Assassination is not an isolated historical quirk, eluding comprehension or analysis. The event is usually overdetermined by multiple but equally important factors: personal qualities of the assassin, a fatalistic posture assumed by the victim, and such factors in the social environment as political stereotypes, murder sanctions, and the symbolic nature of high offices.

Although assassination can strike down anyone, we have restricted our examination to assassination of presidents in America (21) by studying the personal qualities of " successful " assassins and of others who almost succeeded. Of the eight assassination attempts on American presidents, four have been successful. The following facts emerge. (i) All the assassination attempts were made with guns, all but one with pistols. (ii) All the assassins were shorter and weighed less than average men of the period. (iii) All the assassins were young adult Caucasian males. (iv) All the assassination attempts but one were made by individuals who were seriously disturbed or even paranoid schizophrenics (22). The exception was the final attempt of two Puerto Rican nationalists to kill President Harry S. Truman. The successful assassins, for the most part, were mentally unbalanced and had persecutory and grandiose delusions.

Assassination provides a method for instantly satisfying a need for personal importance. The delusional assassin very probably had a fantasy that once the act was committed, an outcry of favorable opinion and acclaim would vindicate what he had done. In most of the instances of attempted or successful assassination, escape plans were inadequate or nonexistent.

The life pattern of most of the assassins included extreme resentment toward others—a resentment aggravated by a long history of isolation and loneliness. Often the isolation stemmed from poor and inconsistent relations with parents and others early in life, which resulted in most of the assassins having resentment and mistrust of parental figures. Their resentment toward parental figures might have included the President (political symbol of parenthood) as the head of the federal government. In response to imagined unfair treatment from others and a distortion of his own inadequacies, the assassin turned his anger on the chief of state.

Typically the assassin had struggled for importance, success, and manliness, but had failed. At the time of the attempted presidential assassination, the assassin was on a downward life course. Haunted by resentment and failure and plagued with disordered thinking and distortions of reality, the assassin took action. Shooting the President was thus an attempt to resolve conflicts with which he apparently could not otherwise cope. Providing an alternate outlet for his violent dissatisfaction would be one way of preventing the potential assassin from killing. Perhaps the ombudsman (public complaint receiver) system would allow the would-be assassin to voice his grievances against his intended victim, thereby lessening his pent-up frustrations and reducing the likelihood that he would kill.

Our discussion of another important determinant of assassination—the victim's fatalistic attitude—is not restricted to presidential assassinations. The fatalistic thinking and actions of several assassination victims are reflected in their strong disinclination toward taking precautionary measures despite recognizing the existence of violent impulses in others toward presidents and presidential candidates. Robert Kennedy stated a view that he shared with Abraham Lincoln, Martin Luther King, Jr., and John F. Kennedy: "There's no sense in worrying about these things. If they want you, they can get you" (23). This attitude often leads to dangerous negligence that is an exaggerated form of denying that one is actually afraid of physical harm. Lincoln has been described as "downright reckless" (24) about personal safety. Robert Kennedy was quoted as saying, "I'll tell you one thing: If I'm President, you won't find me riding around in any of those awful [bullet-proof] cars" (23). The fatalistic attitude illustrated by statements like this is encouraged by our tradition of expecting physical courage in our leaders. Men who repeatedly and publicly proclaim their vulnerability may be unwittingly encouraging assassination by offering an invitation to the delusional, grandiose, and isolated person who dreams of accomplishing at least one important and publicly recognized act in his life. "Mixing with the people" is firmly embedded in the American political tradition, but it is also an accomplice to assassination. One way to cope with this problem would be legislation to restrict the contact and exposure of a President with crowds when his presence has been announced in advance.

Mass Media and Violence

Television could be one of our most powerful tools for dealing with today's violence. It could provide education and encourage, if not induce, desired culture modification. Unfortunately, it does little of either today, perhaps because the harmful effects of televised violence have been glossed over. However, all the mass media do little to discourage and much to encourage violence in America. The Ugly American as a national stereotype is rapidly being displaced in the eyes of the world by the Violent American, his brother of late. This stereotype is fostered by the media but is sustained by the violent acts of some of our citizens. Armed with shotgun, ignorance, frustrations, or hunger, this Violent American can be seen today throughout our society. We are not all violent Americans, but mass media are giving us the violence we seem to want.

What effect do the mass media have (25)? All of us are probably affected by the media to some degree, but most research has focused on children, since an immature and developing mind is usually less capable of discrimination when responding to a given stimulus. One comprehensive review (26) described short-term effects that include the child's emotional reactions to what he views, reads, and hears. Long-term effects, what the child actually learns as a result of his exposure, may include vocabulary, factual information, belief systems, and such altered personality characteristics as increased aggressiveness. No one selects all the media materials available, nor does anyone absorb or retain the selected materials consistently or completely. Prior information, differing needs, and quality of life adjustment also help to filter the child's processing of the offered materials. Mass media effects also depend somewhat on the applicability of the learned material to the child's own life situation.

Similarly, as shown by another researcher (27), frustration, the anger evoked by it, the overall situation, the apparent severity and justification of the violence viewed in a film—all relate to whether or not children use these aggressive responses.

A large study in Great Britain (28) showed that certain portrayals of violence are more disturbing to children than others. Unusual motives, settings, and weapons are more disturbing then stereotyped violence. For example, knives or daggers are more upsetting than guns or fist fights. Similarly, seeing violence or disasters in newsreels bothered children more than dramatized violence.

Another study (29) found that the average American child from 3 through 16 years old spends more of his waking hours watching television than attending school. First-graders spend 40 percent and sixth-graders spend 80 percent of their viewing time watching "adult" programs, with Westerns and situation comedies being most popular. By the eighth grade, children favor crime programs.

Can we justifiably say that the media teach violence? Television teaches more than vocabulary and factual information to the impressionable young viewer, who learns by identification and social imitation. Learning theorists have shown that children readily mimic the aggressive behavior of adults and that the degree of imitation is comparable whether the behavior is live or televised. In another study (*30*) nursery school children watched a film of adults aggressively hitting an inflatable plastic figure, a Bobo doll. Later these and other children were first mildly frustrated and then led individually into a room in which they found the Bobo doll and other materials not shown in the film. Those who had seen the film imitated precisely the film's physical and verbal aggression and made more aggressive use of other toys, such as guns, that had not been in the film. Film-watchers showed twice as much aggressiveness as those who had not seen the film.

These children were all from a "normal" nursery school population, and all showed some effect. This finding seriously questions the claim that such violence is learned only by deviant individuals. The findings apply equally to real, fictional, and fantasy violence. The impact on children observing aggressive behavior has been further corroborated in experiments in which live models, cartoons, and play materials were used. The idea that watching television satisfactorily releases pent-up aggressions (the catharsis theory) loses credibility in the face of these data from social-learning experiments. Watching dramatized violence may actually lead to subsequent aggressive behavior.

A tendency toward repeating certain behaviors viewed in the media clearly exists. The mass media teach the alphabet of violence, but whether or not the actual performance of violent behaviors occurs depends on personality, subcultural values, and other factors. The research to date indicates that the learning of violence must be distinguished from the performance of it. One fear we have is that restraints and taboos against violent behavior may diminish as the result of observing prohibited behavior being condoned and rewarded on the screen. Violence depicts a way of life; it is disguised by a cloak of history or locale and becomes acceptable. We are never taught "in this School for Violence that violence in itself is something reprehensible" (*31*).

Even with the portrayed violence, the screen environment may be more desirable than the viewer's actual environment. In the culturally deprived American household the underfed, underoccupied, undereducated person may be an apt pupil of the school for violence. Such pupils more readily accept as real a violent world made of movies, newsprint, comic books, and video. The blurred line between fiction and reality grows fainter when there is nothing for dinner. Ghetto violence is one way of at least temporarily adjusting to intolerable personal frustrations and an unbearable environment.

Given the effectiveness of the mass media in achieving culture modification, we should determine whether the content of the media produces desirable or undesirable modification. How frequently is violent content offered in our media? According to a 1951 New Zealand study (*32*), 70 American films had roughly twice as much violence per film as did 30 films from other countries. A 1954 study of network television programs (*33*) found an actual doubling from one year to the next in the number of acts or threats of violence, with much of the increase occurring during children's viewing hours. These studies were all conducted before the documentary and news depiction of violence became common, and thus these studies dealt essentially with fictional violence. More recent studies reflect the same trends, however. A *New York Times* headline from July 1968, reads "85 Killings Shown in $85\frac{1}{2}$ TV Hours on the 3 Networks" (*34*).

Thus the media's repetitive, staccato beat of violence and the evidence of its impact upon the most impressionable members of our society show that violence is valued, wanted, enjoyed. In teaching that violence is a good quick way to get things done, television and other media teach that violence is adaptive behavior.

Part of the tragedy is that the mass media could effectively promote adaptive behaviors like nonviolent protest and other alternatives to violence. The communications personnel and we consumers alike share the responsibility for seeing that our mass media develop their own constructive educational potential. At the very least, violence in the media must be reduced. The statement is hackneyed, the conclusion is not.

Mental Illness, Violence, and Homicide

What is the relationship between mental illness and violence (*35*)? Generally the stereotype of the mentally ill person as a potentially dangerous criminal is not valid. The act of homicide often raises the question of psychosis, but only a relatively few psychotic individuals are potential murderers. The stereotype is kept alive, however, by the sensationalist news coverage of the few homicides committed by psychotics.

Mental illness does not usually predispose one to commit violent acts toward others. The patient with severe mental illness (psychosis) is frequently so preoccupied with himself and so disorganized that he is more likely to commit suicide than homicide. A main exception is the fairly well-organized paranoid patient with persecutory delusions concerning one or more particular individuals, intense hostility and mistrust for others, and a pervasive tendency to blame his troubles on the world. However, this type of mentally disordered person constitutes a small minority and does not greatly increase the low incidence of violent acts committed by those identified as mentally ill. In fact, several comparative studies indicate that patients discharged from mental hospitals have an arrest rate considerably lower than that of the general population. In a Connecticut state mental hospital (*36*) the felony arrest rate was 4.2 per 1000 patients, whereas among the general population it was 27 per 1000. Compared to an arrest rate of 491 per 100,000 among the general population, New York state mental hospitals (*37*) reported a figure of 122 per 100,000 for male patients discharged during 1947. Ten thousand patients were studied. One state-wide survey of Maryland psychiatric hospitals (*38*) showed that the mentally ill are involved in criminal behavior about as often as the general population.

Since mental illness of itself is not predictive of violence or homicide, we must look for other predisposing conditions. Predicting specifically who will murder is difficult because over 90 percent of the murders committed are not premeditated and 80 percent involve an acquaintance or family member (*39*). One often demonstrated factor related to homicide is the excessive use of alcohol (40). Overindulgence in alcohol has been cited as one feature of the "pre-assaultive state" (40). Persons who are preassaultive usually show some combination of the following five factors: (i) difficulty enjoying leisure time often associated with the heavy use of alcohol; (ii) frequent clashes with close friends, spouse, and others; (iii) history of many fistfights and evidence of past violence (such as scars) reflecting difficulty with impulse control; (iv) fondness for guns and knives; and (v) being relatively young, usually under 45 years old. Comparing homicide rates for males and females universally indicates that a potential murderer is more often male than female. This difference reflects more frequent use of guns and knives ("male" weapons) for murdering as well as sex differences in expressing aggression.

Case histories of homicide reveal repeatedly that a person uses murder as a means of conflict resolution in an unbearable situation for which he can find no other solution. Predisposing factors for homicide include alcoholism, subcultural norms accepting violence as a means of settling conflict, a setting in which the individual experiences intolerable frustration or attack, helplessness resulting from the unavailability of or the inability to perceive alternative actions, intense emotions, and distortion of reality (perhaps even to the point where reality disappears because of personality disintegration). In the instance of blind rage, a person sometimes murders without realizing what he is doing.

The acts of homicide may be viewed as attempted coping behavior. Homicide eliminates the immediate problem at a time when there seems to be no future or when the future seems unimportant, and the long-range consequences of the act are not considered. Put another way, homicide has adjustive rather than adaptive value.

Firearms Control and Violence

Violence by firearms has recently caused great concern (41, 42). The question of whether there is a gun problem is complicated by regional variations in both the actual incidence and the reporting of crime and multiple psychosocial variables, such as individual "choice" of homicide, population density, age, race, socioeconomic status, religion, and law-enforcement effectiveness.

Even so, the following statistics (39, 43) estimating the involvement of guns in various forms of violence in America indicate that a problem does exist. In 1967 firearms caused approximately 21,500 deaths—approximately 7,700 murders, 11,000 suicides, and 2,800 accidental deaths. In addition, there were also about 55,000 cases of aggravated assault by gun and 71,000 cases of armed robbery by gun. Between 1960 and 1967, firearms were used in 96 percent (that is, 394) of 411 murders of police officers. More than 100,000 nonfatal injuries were caused by firearms during 1966. A study in Chicago (44) in which assaults with guns were compared to those with knives shows many more equally serious assaults with knives than with guns; but more of the gun assaults were fatal. Another study (27) convincingly shows that the mere presence of a gun serves as a stimulus to aggression, that is, "The finger pulls the trigger, but the trigger may also be pulling the finger."

The number of guns owned by citizens is unknown, but estimates run from 50 to 200 million (*39*). In 1967 approximately 4,585,000 firearms were sold in the United States, of which 1,208,000 were imports (*43*). Lately, data from a 1963 World Health Organization survey of 16 developed countries (*39*) give America an overwhelming lead in death rates for both homicide and suicide by firearms.

These data speak for themselves. What they do not show are the steady increases in all categories for gun-related mortality cited during the past few years. Firearms sales increased by 132 percent between 1963 and 1967.

Responsibility for legal restrictions on guns has generally been left to the states. Consequently, regulations on the sale of guns vary greatly. The lack of uniform laws and the ability (until recently) to buy guns in one state and transport them to another state have made it difficult to compare accurately the gun laws of different states. Even the so-called strict gun laws may not possess sufficient strength to reduce gun killings significantly.

Until 1968 there were only two federal laws of note (*45*). The National Firearms Act of 1934 imposes a tax on the transfer of certain fully automatic weapons and sawed-off shotguns. The Federal Firearms Act of 1938 requires a license for interstate sale of firearms and prohibits interstate shipment of guns to convicted felons, fugitives, and certain other persons. Two bills passed in 1968 go somewhat further but do not include firearm registration (*41*). The Omnibus Crime Control and Safe Streets Act restricts interstate and foreign commerce in hand guns. The Gun Control Act also adds mailorder sale of rifles and shotguns to this restriction and prohibits over-the-counter sales to out-of-state residents, juveniles, convicted felons, drug users, mental defectives, and patients committed to mental hospitals.

Although the data do not provide an ironclad indictment against weak, inconsistent legislation, we believe that they make a convincing argument. What is more, more than two-thirds of the American people continue to favor stronger gun-control legislation (*42*). Even the frightening regularity of assassination has not resulted in strong legislation (that is, legislation requiring registration of guns and owners). How then can we account for the successful opposition to strong gun legislation?

Diverse groups comprise the one-third or less Americans who do not favor stricter gun control laws. The most visible opposition group is the large (about 1 million members), well-organized National Rifle Association (NRA). With an immense operating budget (approximately $5.7 million in 1967), the NRA is an especially effective "gun lobby" (*46*). Another group, the Black Panthers, sees arms as necessary for survival. Eldridge Cleaver, Defense Minister of the Black Panthers, wrote, "We are going to keep our guns to protect ourselves from the pigs [police]" (*47*). Protection is also the issue in Dearborn, Michigan, where housewives are arming against the potential rioter and looter who might "invade" Dearborn from Detroit. Tragic escalation continues around the interplay of urban and suburban action and reaction.

Arguments opposing gun legislation can be divided into five overlapping categories.

1) Gun control would cause the loss of rights and possessions. This argument

takes various forms: Restrictive legislation is an effort to disarm American sports-men and law-abiding citizens; legislation would result in the loss of the so-called basic American freedom, "the right of the people to keep and bear arms"; and maintaining an armed citizenry ensures the protection of American liberties, especially against tyrannies from the political right or left. A common fear is that gun laws could lead from registration to discrimination and finally to confiscation of all firearms.

Our traditional frontier and rural ways of life are disappearing, and with this change has come a decrease in our traditional freedom and individualism. For many opposing gun legislation, the actual and potential loss of a way of life and its prized symbol—the gun—make gun legislation a concern basic to the adaptiveness of our society. These opponents assume that restrictions on the "right to bear arms" endanger our way of life.

2) Guns represent protection from dangers. The gun is seen as providing personal protection from and means of coping with life-threatening dangers and de-structive evil forces, be they criminals, drug addicts, rapists, communists, other sub-versives, mental patients, rioters, police, or racists. The NRA promotes this coping strategy in its official publication, *The American Rifleman* (48). A monthly NRA column, "The Armed Citizen," states that "law-enforcement officers cannot at all times be where they are needed to protect life or property in danger of serious violation. In many such instances, the citizen has no choice but to defend himself with a gun" (48). The power of this argument depends upon a person's feelings of helplessness and mistrust in the face of danger.

Many people in urban areas or changing neighborhoods fear the rising crime rate and the breakdown of law and order. However, there is no documentation that an armed citizenry provides greater individual or group protection than an unarmed citizenry. On the contrary, the potential danger of such individual armed protection in our congested urban society includes harm to innocent bystanders, accidental shootings and the increased likelihood of impulsive violence, which already accounts for over 90 percent of homicides in America.

3) Crime is reduced by punishment and not by gun control. Several forms of this argument state that gun-control legislation simply is not an effective way of reducing crime and violence: (i) Guns don't kill people, people kill people; (ii) when guns are outlawed, only outlaws will have guns (because they steal them anyway); (iii) crime is not associated with guns but with such social factors as population density, population composition, economic status, and strength of police; and (iv) effective enforcement of present laws has not been tried.

Using stronger and even cruel punishment to cope with gun-using criminals has to date not been proven as an effective deterrent, and its use, we believe, is morally indefensible. The "crime and punishment" thesis ignores data showing that more than three out of four homicides and two out of three criminal assaults occur among family and friends, that is, most murders are committed by "law-abiding citizens." In addition, criminals can and do purchase weapons from legal sources.

4) A gun represents strength and manliness. Gun literature usually implies this argument. Acts of heroism and bravery are associated with gun usage. Members of

the NRA receive distinguished fighting medals. Pictures and advertisements reflect manliness, and imply that gun usage means "standing up for your rights."

Guns may serve as a source of power, pride, and independence (the "equalizer"—for feelings of inferiority or inadequacy) and as the symbol of manliness and potency. Guns can and do represent these qualities in our culture, even to a pathological degree in some of us.

5) Guns provide recreation and support the economy. Arguments here portray citizens as being restricted from and deprived of healthy outdoor life, the hobby of gun collecting, family recreation, and the fellowship associated with hunting and target shooting. For example, an article in *The American Rifleman* entitled "Happiness Is a Warm Gun" (*49*) depicts a close father-son relationship based on shooting. Additionally, gun sales and fees are held to be important economic factors supporting hunting states and conservation programs.

These arguments indicate that the issue of gun legislation is pragmatic, ideological, psychological, and economic, and is not based upon sound empirical data. The fervor of the arguments accurately reflects the deep emotional attachments at stake. Indeed, the specific content of proposed gun laws often seems irrelevant. Tragically, the arguments confuse ideology with issues of violence that must be solved. If strictly pragmatic issues of protection were involved, better police protection and increased communication with the feared group or groups should diminish the fear.

Finally, we have found that the "statistics game" is often played by both sides of this particular controversy. By presenting selected statistics and invalid inferences, both sides have obscured the more important goals of reducing gun killings and violence.

Yet, on balance, data document the need for strong and more uniform firearms legislation. We know of no single issue concerning violence that reflects more clearly the changing nature of adaptation. Challenges of the complex urban society in which we live cannot be met with old frontier means of survival—every man protecting himself with his own gun. Yet, gun legislation is no panacea. While reflecting America's desire for action, focusing or relying on legislation alone tends to obscure basic issues of violence and how we persist in using both individual and collective violence as a means of resolving conflict.

Collective and Sanctioned Violence

An additional dilemma is that killing is neither legally nor socially defined as an unequivocally criminal act. The existence of capital punishment and war gives qualified sanction to violence as a means of resolving conflict. Both the general public and their leaders always seem to be able to justify any violence perpetrated on their fellow man. Thus in practice the legitimacy of violence is arbitrary and depends more on the will of powerful men than on moral, ethical, or humane considerations. In a sense, all sanctioned violence is collective, since it has group social approval. Certainly the existence of sanctioned violence abrades the concept of law and order.

We desperately need research on the psychological processes that permit an individual or group to view some violence as good (and presumably adaptive) and other forms of violence as bad (and presumably maladaptive). Although the history of violence in man is polymorphous, there likely are psychological mechanisms common to all cultures and times. For instance, the psychology of sanctioned violence everywhere depends on attributing evil motives to the "outsiders." Then because "they" are violent (evil), "we" *have* to be violent, or (twisted even further) because "they" are violent, it is *good* for "us" to be violent.

Thus people who have seen sanctioned violence being committed in the name of law, order, justice, moral obligation, and duty come to use violence themselves as a "just" means of solving their own problems. The people are acting as their government's representatives have acted—if the cause is just, the grievance real, then unlimited power and force can be used.

Nowhere do we better find this thinking reflected than in the actions of rioters (*50*). Study of the 1967 Detroit uprising (*51*) showed that the rioters (young, better educated men who had experienced frustration of their rising expectations) viewed violence against the "system" as justified. Not surprisingly, their views of what justifies violence differed greatly from those of the law enforcers and of the middle-aged black citizens. To the rioters violence was a means of accomplishing goals seemingly not attainable by nonviolent means. Their belief in the power of violence is understandable. Civil disorders are serving in part as a catalyst for change and an instrument of achievement. Some uprising participants reported that violence provided a sense of manliness and strength. But do these supposed gains outweigh the damage of escalations of counterviolence and potential suppression? At least the hypothesis that violence purifies, enhances manliness, and strengthens identity is subject to empirical study.

The results of social-psychiatric field investigations like those in Detroit and at Brandeis University's Lemberg Center for the Study of Violence are useful steps toward understanding the psychological processes and conditions evoking collective violence. For instance, a Lemberg report (*52*) cited four socio-psychological antecedents to ghetto uprisings: (i) a severe conflict of values between dominant and minority groups; (ii) a "hostile belief system" held by the aggrieved group, based considerably on reality; (iii) a failure of communication between the aggrieved and dominant groups; and (iv) a failure in social control resulting from either over-control or undercontrol. In short, these studies show that psychiatrists and psychologists can and must help to resolve the crisis of violence through field studies, facilitating communication between opposing groups, and making recommendations for social change.

But what of war? Behavioral scientists have grasped at all sorts of explanations for this species' warring behavior. Perhaps even this attempt to explain war is a cause of war; our ability to justify any form of violence is part of man's magnificent cerebral endowment. Many causes of war have been suggested: contiguity, habituation, social learning, predation, psychological defenses (for example, rationalization, blaming, denial, counterphobic tendencies among others), the host of fears associated with the human condition, territoriality and power, intolerable frustration, biologically rooted aggressive instincts, and sadism (53–55). One wonders whether the mere distance and speed with which we kill are factors rendering meaningless the signals of submission that other animals use to halt violent encounters (54). Often we literally no longer have to touch the results of our violence. The impersonal factor shows up in another way. Since war is an activity between organized nation states rather than angry individuals, decisions producing war often are made in a calculated manner by those who do not participate directly in any personal acts of violence.

The evidence of history is that war proves everything and nothing. An adequate analysis of the Vietnam war and of the myriad of other wars dotting history is far too great a task for this discussion, despite the relevance of war to the current crisis of violence (55).

Although preventive measures are difficult to administer in the face of the contradicting sanctioned and unsanctioned violence, there are remedies to violence, and we have discussed some of them. More effort could be expended trying to understand the all-important relation between the excessive use of alcohol and homicide. Disseminating currently available information on how to identify a potential murderer will help. Despite Americans' conflicting feelings about guns, there is a gun-death problem today, and more effective and uniform gun legislation can keep guns out of the hands of those who are likely to act impulsively. The mass media can play an increasingly responsible and educational role, while reducing the amount of violence for violence's sake. Many positive potentials of the media have not been tapped. Citizen complaint agencies can be established, of which one possibility might be homicide prevention centers along the lines of the suicide prevention centers. Frustrated minority groups will become less frustrated when they are not blocked from responsible participation and self-determination. Peaceful resolution of conflict (56) such as nonviolent protest and negotiation, reducing the amount of sanctioned violence, encouraging a shared sense of humanity, and moving toward rehabilitation rather than retribution in dealing with crime—all these are promising directions. Violence must be studied scientifically so that human behavior can be sustained by knowledge.

Changing Nature of Human Adaptation: Some Speculations

Violence is unique to no particular region, nation, or time (55). Centuries ago man survived primarily as a nomadic hunter relying on violent aggression for both food and protection. Even when becoming agricultural and sedentary, man struggled against nature, and survival still required violent aggression, especially for maintaining territory when food was scarce.

Then in a moment of evolution man's energies suddenly produced the age of technology. Instead of adapting mainly by way of biological evolution, we are now increasingly subject to the effects and demands of cultural evolution. Instead of having to adapt to our environment, we now can adapt our environment to our needs. Despite this potential emancipation from biological evolution, we retain the adaptive mechanisms derived from a long history of mammalian and primate

184

evolution, including our primitive forms of aggression, our violence, bellicosity, and inclination to fight in a time of emergency. Where these mechanisms once responded more to physical stress, they now must respond more to social, cultural, and psychological stresses, and the response does not always produce adaptive results. Where violent aggressive behavior once served to maintain the human species in time of danger, it now threatens our continued existence.

In this new era, culture changes so rapidly that even time has assumed another dimension—the dimension of acceleration. Looking to the past becomes less relevant for discerning the future.

In the current rapidly expanding technological era, many once useful modes of adaptation are transformed into threats to survival. Territorial exclusivity is becoming obsolete in an economy of abundance. Vast weapons, communication, and transportation networks shrink the world to living-room size and expand our own backyard to encompass a "global village." Yet war and exclusivity continue. Our exploitation of natural resources becomes maladaptive. Unlimited reproduction, once adaptive for advancing the survival of the species, now produces the over-crowded conditions similar to those that lead to destructive and violent behavior in laboratory experiments with other species.

The rate at which we change our environment now apparently exceeds our capacity for adapting to the changes we make. Technological advances alter our physical and social environments, which in turn demand different adaptive strategies and a reshaping of culture. The accelerated civilization of technology is crowded, complex, ambiguous, uncertain. To cope with it we must become capable of restructuring knowledge of our current situation and then applying new information adaptively. Several factors give us reason to hope that we can succeed.

1) Our social organization and intellectual abilities give us vast potential for coping. Knowledge and technology can be harnessed to serve goals determined by man. Automation makes possible the economics of abundance, but only our cultural values can make abundance a reality for all people. Medicine permits us to control life, but we have not yet seen fit to use this power to determine the limits of population. The technologies of communication and travel shrink the world, but man has not yet expanded the horizon of exclusion. We can learn to unite in goals that transcend exclusivity and direct cultural evolution in accordance with adaptive values and wisdom. The past need not be master of our future.

2) Violence can be understood and controlled. The crisis is one of violence, not of aggression, and it is violence that we must replace. Aggression in the service of adaptation can build and create rather than destroy. The several theories of aggression and current issues of violence suggest many complementary ways of controlling and redirecting aggression. We have suggested some in this article. Furthermore, our brief review of theory and issues points to many possibilities for multidimensional research—an approach that we believe is needed rather than "one note" studies or presentations.

3) Greater attention can be focused on both social change and adaptation processes. Cultural lag in the technological era produces not stability but a repetitious game of "catch up" characterized by one major social crisis after another and by

behaviors that are too often only adjustive in that they bring relief of immediate problems while doing little to provide long-range solutions. Expanding our knowledge of the processes of social change and understanding resistance to change are of highest priority. Unforeseen change produces intolerable stress, anxiety, and increased resistance to rational change. These reactions inhibit solution-seeking behavior; evoke feelings of mistrust, loss, and helplessness; and lead to attacks on the apparent agents of change. We must develop the ability to foresee crises and actively meet them. We must dwell more on our strengths, assets, and potential as the really challenging frontier.

Conclusion

The current examples of violence and the factors encouraging it reflect our vacillation between the anachronistic culture of violence and the perplexing culture of constant change. We feel alienated and experience social disruption. Current demands for change are potentially dangerous because change activates a tendency to return to older, formerly effective, coping behaviors. Social disruption caused by change tends to increase violence as a means of coping at a time when violence is becoming a great danger to our survival.

America's current crises of violence make it difficult for us to cope with our changing world. Today's challenge, the crisis of violence, is really the crisis of man. This crisis is especially difficult because violence, a once useful but now increasingly maladaptive coping strategy, seems to be firmly rooted in human behavior patterns. We conquer the elements and yet end up facing our own image. Adaptation to a changing world rests on how effectively we can understand, channel, and redirect our aggressive energies. Then man can close his era of violence.

Summary

We are uniquely endowed both biologically and culturally to adapt to our environment. Although we are potentially capable of consciously determining the nature of our environment, our outmoded adaptive behavior—our violent aggression —keeps us from doing so.

Aggression is viewed as multidetermined. It is inherent, caused by frustration, or learned by imitation. Violent aggression is a form of attempted coping behavior that we in America, as others elsewhere, use despite its potentially maladaptive and destructive results. Current examples of violence and the factors fostering it include assassination, the mass media, mental illness and homicide, firearms and resistance to restrictive gun legislation, and collective and sanctioned violence. These examples are considered from the perspectives of the changing nature of adaptation and the opportunities they offer for research. Among recommendations for resolving or reducing violence, the need for thoughtful research by behavioral scientists is stressed. But the major obstacle to removing violence from our society is our slowness to recognize that our anachronistic, violent style of coping with problems will destroy us in this technological era.

References and Notes

1. L. Eiseley, *The Immense Journey* (Random House, New York, 1946).
2. D. N. Daniels, M. F. Gilula, F. M. Ochberg Eds., *Violence and the Struggle for Existence* (Little, Brown, Boston, 1970).
3. Dr. T. Bittker, C. Boelkins, Dr. P. Bourne, Dr. D. N. Daniels (co-chairman), Dr. J. C. Gillin, Dr. M. F. Gilula, Dr. G. D. Gulevich, Dr. B. Hamburg, Dr. J. Heiser, Dr. F. Ifeld, Dr. M. Jackman, Dr. P. H. Leiderman, Dr. F. T. Melges, Dr. R. Metzner, Dr. F. M. Ochberg (co-chairman), Dr. J. Rosenthal, Dr. W. T. Roth, Dr. A. Siegel, Dr. G. F. Solomon, Dr. R. Stillman, Dr. R. Taylor, Dr. J. Tinklenberg, Dr. Edison Trickett, and Dr. A. Weisz.
4. *Webster's Third New International Dictionary* (Merriam, Springfield, Mass., 1966).
5. J. Gould and W. L. Kolb, *A Dictionary of the Social Sciences* (Free Press, New York, 1964); L. E. Hinsie and R. J. Campbell, *Psychiatric Dictionary* (Oxford Univ. Press, ed. 3, New York, 1960).
6. R. C. Boelkins and J. Heiser, "Biological aspects of aggression," in *Violence and the Struggle for Existence*, D. N. Daniels, M. F. Gilula, F. M. Ochberg, Eds. (Little, Brown, Boston, 1970).
7. *The Natural History of Aggression*, J. D. Carthy and F. J. Ebling, Eds. (Academic Press, New York, 1964).
8. Th. Dobzhansky, *Mankind Evolving* (Yale Univ. Press, New Haven, 1962).
9. G. G. Simpson, "The study of evolution: Methods and present states of theory," in *Behavior and Evolution*, A. Roe and G. G. Simpson, Eds. (Yale Univ. Press, New Haven, 1958).
10. D. A. Hamburg, "Emotions in the perspective of human evolution," in *Expression of the Emotions in Man*, P. D. Knapp, Ed. (International Universities Press, New York, 1963).
11. C. Kluckhohn, "The limitations of adaptation and adjustment as concepts for understanding cultural behavior," in *Adaptation*, J. Romano, Ed. (Cornell Univ. Press, Ithaca, New York, 1949).
12. E. Silber, D. A. Hamburg, G. V. Coelho, E. B. Murphey, M. Rosenberg, L. I. Pearlin, *Arch. Gen. Psychiat.* 5, 354 (1961).
13. D. A. Hamburg and J. E. Adams, *ibid.* 17, 277 (1967).
14. O. Fenichel, *The Psychoanalytic Theory of Neurosis* (Norton, New York, 1945).
15. K. Lorenz, *On Aggression* (Harcourt, Brace and World, New York, 1966).
16. A. Storr, *Human Aggression* (Atheneum, New York, 1968).
17. G. F. Solomon, "Case studies in violence," in *Violence and the Struggle for Existence*, D. N. Daniels, M. F. Gilula, F. M. Ochberg, Eds. (Little, Brown, Boston, 1970).
18. J. P. Scott, *Aggression* (Univ. of Chicago Press, Chicago, 1958).
19. L. Berkowitz, *Aggression: A Social-Psychological Analysis* (McGraw-Hill, New York, 1962); J. Dollard, L. W. Doob, N. E. Miller, O. H. Mowrer, R. R. Sears, *Frustration and Aggression* (Yale Univ. Press, New Haven, 1939).
20. A. Bandura and R. H. Walters, *Social Learning and Personality Development* (Holt, Rinehart and Winston, New York, 1963);, F. Ilfeld "Environmental theories of aggression," in *Violence and the Struggle for Existence*, D. N. Daniels, M. F. Gilula, F. M. Ochberg, Eds. (Little, Brown, Boston, 1970);

M. E. Wolfgang and F. Ferracuti, *The Sub-Culture of Violence* (Barnes and Noble, New York, 1967).

21. R. Taylor and A. Weisz, "The phenomenon of assassination," in *Violence and the Struggle for Existence*, D. N. Daniels, M. F. Gilula, F. M. Ochberg, Eds. (Little, Brown, Boston, 1970).

22. L. Z. Freedman, *Postgrad. Med.* **37**, 650 (1965); D. W. Hastings, *J. Lancet* **85**, 93 (1965); *ibid.*, p. 157; *ibid.*, p. 189; *ibid.*, p. 294.

23. "It's Russian roulette every day, said Bobby," San Francisco *Examiner* (6 June 1968).

24. J. Cottrel, *Anatomy of an Assassination* (Muller, London, 1966).

25. A. E. Siegel, "Mass media and violence," in *Violence and the Struggle for Existence*, D. N. Daniels, M. F. Gilula, F. M. Ochberg, Eds. (Little, Brown, Boston, 1970); O. N. Larsen, Ed., *Violence and the Mass Media* (Harper and Row, New York, 1968).

26. E. A. Maccoby, "Effects of the mass media," in *Review of Child Development Research*, L. W. Hoffman and M. L. Hoffman, Eds. (Russell Sage Foundation, New York, 1964).

27. L. Berkowitz, *Psychol. Today* 2 (No. 4), 18 (1968).

28. H. T. Himmelweit, A. N. Oppenheim, P. Vince, *Television and the Child* (Oxford Univ. Press, New York, 1958).

29. W. Schramm, J. Lyle, E. B. Parker, *Television in the Lives of Our Children* (Stanford Univ. Press, Stanford, Calif., 1961).

30. A. Bandura, D. Ross, S. Ross, *J. Abnorm. Soc. Psychol.* **63**, 575 (1961); *ibid.* **66**, 3 (1963).

31. F. Wertham, *A Sign for Cain* (Macmillan, New York, 1966).

32. G. Mirams, *Quart. Film Radio Television* **6**, 1 (1951).

33. Purdue Opinion Panel, *Four Years of New York Television* (National Association of Educational Broadcasters, Urbana, Ill., 1954).

34. Associated Press report of 25 July 1968; *Christian Science Monitor* article; New York *Times* (29 July 1968).

35. G. D. Gulevich and P. Bourne, "Mental illness and violence," in *Violence and the Struggle for Existence*, D. N. Daniels, M. F. Gilula, F. M. Ochberg, Eds. (Little, Brown, Boston, 1970).

36. L. H. Cohen and H. Freeman, *Conn. State Med. J.* **9**, 697 (1945).

37. H. Brill and B. Malzberg, *Mental Hospital Service (APA) Suppl. No. 153* (August 1962).

38. J. R. Rappaport and G. Lassen, *Amer. J. Psychiat.* **121**, 776 (1964).

39. C. Bakal, *No [sic] Right to Bear Arms* (Paperback Library, New York, 1968).

40. C. A. deLeon, "Threatened homicide—A medical emergency," *J. Nat. Med. Assoc.* **53**, 467 (1961).

41. J. C. Gillin and F. M. Ochberg, "Firearms control and violence," in *Violence and the Struggle for Existence*, D. N. Daniels, M. F. Gilula, F. M. Ochberg, Eds. (Little, Brown, Boston, 1970).

42. D. N. Daniels, E. J. Trickett, J. R. Tinklenberg, J. M. Jackman, "The gun law controversy: Issues, arguments, and speculations concerning gun legislation," in *Violence and the Struggle for Existence*, D. N. Daniels, M. F. Gilula, F. M. Ochberg, Eds. (Little, Brown, Boston, 1970).

43. Criminal Division, U.S. Department of Justice, *Firearms Facts* (16 June 1968);

based in large part on the Federal Bureau of Investigation, *Uniform Crime Reports* (1967) (U.S. Government Printing Office, Washington, D.C., 1968).

44. F. Zimring, "*Is Gun Control Likely to Reduce Violent Killings?*" (Center for Studies in Criminal Justice, Univ. of Chicago Law School, Chicago, 1968).

45. *Congressional Quarterly*, "King's murder, riots spark demands for gun controls" (12 April 1968), pp. 805–815.

46. R. Harris, *The New Yorker* **44** (20 April 1968), p. 56.

47. E. Cleaver, *Ramparts* **7** (15 June 1968), p. 17.

48. *Amer. Rifleman* **116** (Nos. 2–5) (1968), various writings.

49. W. W. Herlihy, *ibid.* **116** (No. 5), 21 (1968).

50. T. E. Bittker, "The choice of collective violence in intergroup conflict," in *Violence and the Struggle for Existence*, D. N. Daniels, M. F. Gilula, F. M. Ochberg, Eds. (Little, Brown, Boston, 1970).

51. P. Lowinger, E. D. Luby, R. Mendelsohn, C. Darrow, "*Case study of the Detroit uprising: The troops and leaders*" (Department of Psychiatry, Wayne State Univ. School of Medicine, and the Lafayette Clinic, Detroit, 1968); C. Darrow and P. Lowinger, "The Detroit uprising: A psychosocial study," in *Science and Psychoanalysis, Dissent*, J. H. Masserman, Ed. (Grune and Stratton, New York, 1968), vol. 13.

52. J. Spiegel, *Psychiat. Opinion* **5** (No. 3), 6 (1968).

53. J. D. Frank, *Sanity and Survival: Psychological Aspects of War and Peace* (Random House, New York, 1967); I. Ziferstein, *Amer. J. Orthopsychiat.* **37**, 457 (1967).

54. D. Freeman, "Human aggression in anthropological perspective," in *The Natural History of Aggression*, J. D. Carthy and F. J. Ebling, Eds. (Academic Press, New York, 1964).

55. L. F. Richardson, *Statistics of Deadly Quarrels* (Boxwood Press, Pittsburgh, 1960).

56. F. Ilfeld and R. Metzner, "Alternatives to violence: Strategies for coping with social conflict," in *Violence and the Struggle for Existence*, D. N. Daniels, M. F. Gilula, F. M. Ochberg, Eds. (Little, Brown, Boston, 1970).

57. We thank Dr. D. A. Hamburg and Dr. A. Siegel for their review and critique of this paper and M. Shapiro, C. DiMaria, and R. Franklin for their contributions in preparing this manuscript.

five # Heredity and Aging

Some people act as if they think, "Since I have never died, I never will." But age and death pursue us all. In some ways genetics has contributed strikingly to our understanding of human development—from egg to old age. So far the most important practical consequences of developmental genetics has been the treatment of a few metabolic abnormalities. But the essence of development—the pattern of differentiation and physiological maintenance and decline—remains somewhat elusive. There is an abundance of developmental theory; it is perhaps most intriguing when applied to the process of aging. In this connection evidence is slowly accumulating on aging mechanisms, although no one yet is predicting any magical cure. As Burch (1969) put it, "In real life there are no Peter Pans."

The paper by Comfort in this section reviews the major ideas on aging. Following his paper there is a potpourri of aging reports from contrary points of view. These papers underline the concern expressed by Hayflick (1974) when he said, "There is probably no other area of scientific inquiry that abounds with as many untested or untestable theories as does the biology of aging." The papers by Chetsanga et al., Dykhuizen, Baird et al., and Holliday in this section are each in their own way plausible and at least somewhat persuasive. But they have fundamentally different premises. One might conclude from these papers that there is a need for less theory and for crisper data of all sorts on the enigma of aging.

Several of the papers here mention accumulated somatic mutations underlying normal aging. If this is true the mutagenic debris of civilization, discussed earlier, deserves even more somber scrutiny than it is given now—not to legitimize dreams of immortality, but rather to maximize for a healthy old age. Indeed the "genetic bomb" has repercussions in many spheres.

Radiation and other nonspecific treatments that shorten the life span seem to be only tangentially informative about the process of aging. If—and this has not yet been accomplished for humans—the life span could be significantly extended by selection for a specific gene or by some experimental technique (Comfort, this section), research on aging might be more of a bandwagon than heretofore has been the case.

One aspect of the problem of aging and death not considered in depth in the papers here is the shape versus the end point of the survival curve. The survival curves of various human groups differ greatly for childhood and adulthood—early death being far commoner in India than, for instance, among the U.S. middle class. But the various curves all end at about the same point (85–90 years), showing that the maximum life span is very similar in different populations (but see Kyucharyants, 1974; Leaf, 1973; Hayflick, 1974). This implies that longevity is genetically programmed and further improvements in sanitation and disease immunization and treatment will only cut down on early deaths but will not extend the maximum life span. It is now clear that as cancer and other well-known "killers" are overcome, the population will instead die increasingly of respiratory infections and accidents (Kohn, 1971). Disquietingly, since some forms of cancer have a partially genetic basis, there is an evolutionary danger—for those reproductively active—in perpetuating the causative genes by sanctioning medical treatment for cancer (see Section Seven). In addition, the unhappy plight of the semi-well aged in our society accentuates the fact that medicine's attempts to extend life are a mixed blessing.

One other consideration: Some investigators suspect that the maximum human life span is based on maximum heterozygosity and, therefore, it is a genetic trait that natural selection cannot increase (Hollingsworth, 1967).

Bibliography

Baird, M. B., and H. R. Massie. A further note on the Orgel error hypothesis and senescence. *Gerontologia*, **21**: 240–243, 1975.

Burch, P. R. J. *An Inquiry Concerning Growth, Disease, and Aging*. Toronto: University of Toronto Press, 1969.

Burnet, F. M. A genetic interpretation of aging. *Lancet*, Sept. 1, 1973, pp. 480–483.

Burnet, F. M. *Intrinsic Mutagenesis. A Genetic Approach to Aging*. New York: Wiley, 1974.

Comfort, A. Aged Equadoreans. *Nature*, **258**:41, 1975.

Comfort, A. The position of aging studies. *Mechanisms Age. Devel.*, **3**:1–32, 1974.

Cotzias, G. C., S. T. Miller, A. R. Nicholson, W. H. Matson, and L. C. Tang. Prolongation of the lifespan in mice adapted to large amounts of L-Dopa. *Proc. Nat. Acad. Sci.*, **71**:2466–2469, 1974.

Curtis, H. J. The role of somatic mutations in aging. In *Topics in the Biology of Aging*, P. L. Krohn, ed. New York: Wiley, 1966, pp. 63–74.

Danes, B. S. Progeria: Reduced growth of human progeric-mouse hybrids. *Exp. Gerontol.*, **9**: 169–172, 1974. Progeria is a genetic condition showing accelerated aging.

Fialkow, P. J., G. M. Martin, and C. A. Sprague. Replicative lifespan of cultured skin fibroblasts from young mothers of subjects with Down's syndrome: Failure to detect accelerated aging. *Amer. J. Hum. Genet.*, **25**: 317–321, 1973.

Greenberg, L. J., and E. J. Yunis. Immunologic control of aging: A possible primary event. *Gerontologia*, **18**: 247–266, 1972. A consideration of the central role of the thymus gland in aging.

Guthrie, R. D. Senescence as an adaptive trait. *Persp. Biol. Med.*, **12**: 313–324, 1969.

Hart, R. W., and R. B. Setlow. Correlation between deoxyribonucleic acid excision repair and lifespan in a number of mammalian species. *Proc. Nat. Acad. Sci.*, **71**: 2169–2173, 1974.

Hayflick, L. Cell biology of aging. *BioScience*, **25**: 629–637, 1975.

Hayflick, L. Cytogerontology. In *Theoretical Aspects of Aging*, M. Rockstein, ed. New York: Academic, 1974, pp. 83–103.

Hollingsworth, M. J. Genetic studies of longevity. In *Social and Genetic Influences on Life and Death*, R. Platt and A. S. Parkes, eds. New York: Plenum Press, 1967, pp. 194–214.

Kohn, R. R. Aging and cell division (letter). *Science*, **188**: 203–204, 1975. Disputes Hayflick model.

Kohn, R. R. *Principles of Mammalian Aging.* Englewood Cliffs, N.J.: Prentice-Hall, 1971.

Kormendy, C. G., and A. D. Bender. Chemical interference with aging. *Gerontologia*, **17**: 52–64, 1971. A technical review of possible elixirs of youth (see also Cotzias et al., cited above).

Kyucharyants, V. Will the human lifespan reach one hundred? *Gerontologist*, **14**: 377–380, 1974. Discusses oldsters in the U.S.S.R. (but see Hayflick, op. cit.).

Leaf, A. Getting old. *Sci. Amer.*, **229**: 45–52, 1973. Discusses oldsters around the world.

Marx, J. L. Aging research: I, Cellular theories of senescence; II, Pacemakers for aging? *Science*, **186**: 1105–1107 and 1196–1197, 1974.

Orgel, L. Aging of clones in mammalian cells. *Nature*, **243**: 441–445, 1973.

Rockstein, M., ed. *Theoretical Aspects of Aging.* New York: Academic, 1974.

Ross, M. H., and G. Bras. Food preference and length of life. *Science*, **190**: 165–167, 1975.

Spiegel, P. M. Theories of aging. In *Developmental Physiology and Aging*, P. S. Timiras, ed. New York: Macmillan, 1972, pp. 564–580.

Strehler, B. L. The understanding and control of the aging process. In *Challenging Biological Problems*. J. A. Behnke, ed. New York: Oxford University Press, 1972, pp. 133–147.

Wilson, D. L. The programmed theory of aging. In *Theoretical Aspects of Aging*, M. Rockstein, ed. New York: Academic, 1974, pp. 11–21.

Yielding, K. L. A model for aging based on differential repair of somatic mutational damage. *Persp. Biol. Med.*, **17**: 201–208, 1974.

19   Basic Research in Gerontology

A. Comfort

The assignment of experimental gerontology is to explain the processes which cause the human body to deteriorate with time, and to see whether, and how, the rate and character of this deterioration can be interfered with. The deterioration is multiform, but its rate, as measured by the force of mortality, is highly stable. This suggests the possibility of identifying an accessible " clock " or clocks, by tampering with which we might modify the process of aging and postpone some or all of the vigour loss which goes with it. Gerontology is the applied study of these matters, with the specific objective of modifying aging in Man.

The conviction that such a project is worthy of serious operational attack has grown steadily over the last twenty years. In several countries investment decisions are now being taken, or are about to be taken, about the scale of effort which should be devoted to it. Its credibility as a " growth stock " in medicine depends on the observed effects of scientific and social advance in conventional therapeutics on the survival curve. Because of the rapid and various failure of homoeostasis with age we are already in sight of the practical limit of public health in prolonging active life. On the other hand, if it were to prove feasible to tamper with the " clock " controlling information loss, whatever that may be, the possible outcome of such intervention would be to move the time of onset of the multiple vigour loss, and with it of almost all major deteriorations and malignancies which reflect it, to higher ages. Mammalian experiments involving relatively simple manipulations such as calorie restriction [review, see Comfort, 1960] or the administration of hormones [Bellamy, 1968; Friedman and Friedman, 1964] indicate that this type of intervention is at least practicable in certain circumstances, and that the mammalian lifespan is in principle modifiable. This is probably so in Man, but if the leading process is, for example, one of exposure of uneliminated late-acting genes or otherwise an evolutionary or statistical entity only, it might prove inaccessible by reason of diversity [Medawar, 1945; Strehler, 1962] and research money would be better devoted to the piecemeal alleviation of senile disability through prosthetics and geriatrics. It now seems highly likely that the choice between these hypotheses could be made within a reasonably short time, given adequate research effort. At the moment the balance of evidence suggests that the first alternative—integral modifiability—is likely but by no means certain.

It seems rational to approach fundamental age research in the spirit of a stock-market analyst, starting with the main theoretical approaches, and thereafter selecting a portfolio. The projects which look most rewarding at the moment, from the viewpoint of understanding and modifying age processes, have in common the assumption that aging represents an information loss, and several involve a second

From *Gerontologia*, **16**:48–64, 1970. Reprinted with permission of the author and S. Karger, Basel, Switzerland.

assumption, that this loss occurs at a cellular level. If this second assumption is correct, it becomes necessary next to decide (a) whether the information loss is predominantly in fixed cells such as neurons, or in clonally dividing cells; and (b) whether the leading process in either case is one of noise accumulation in homoeostatic and copying processes, or whether it is secondary to differentiation, depending on the irreversible switching-off of synthetic capacities with morphogenesis. Clearly, if we are dealing with a gramophone record which becomes scratched with repeated use to the point of unintelligibility, we might be able to lubricate the needle, or otherwise reduce the rate of noise-injection. If we are dealing with a record which, once played, cannot be restarted, we require to devise a means of running it more slowly, though not so much more slowly as to spoil the music. Neither model is exclusive nor exhaustive—thus if switching-off is involved (as at some point in ontogeny it must be) we are still thereafter dealing with information loss depending on spoilage of irreplaceable structures. The site of such spoilage offers further choices, and may include molecular, chromosomal, organelle or membrane changes, or all of these. Any model of information loss, whether cellular, molecular or intercellular, seems moreover mathematically and logically to require some kind of self-aggravating or feedback process if it is to explain the exponential character of aging. Apart from verifying any model as a whole, one important experimental problem is accordingly to determine the size and site of the feedback loop. Another is to determine if there is one such loop or several. We need accordingly to look for evidence of mis-specified or nonsense cells or materials accumulating with age, and, in particular, for sites at which mis-specification would be self-propagating: for evidence of a nonsense material which either produces more of itself (an aging pseudovirus, in other words) or one which produces increasing disorder of specification. The feedback loop could also lie in the community of cells, even if the original error is molecular.

"Primary Error" Hypotheses

The simplest hypothesis is still that information is being lost from DNA—either through mutation, through macromolecular damage, or through epigenetic masking. The mutational theories arise from the consideration of the supposed aging effects of radiation, though the identity of these with natural aging remains debatable. Simple point mutation, until recently a popular hypothesis, is probably ruled out, both by the mathematics of the process [Maynard Smith, 1959] and by the failure of chemical mutagens to induce radiation-type life-shortening unless they are capable of producing changes more extensive than point mutation—crosslinking, for example [Alexander, 1967; Alexander and Connell, 1960]. Macromolecular damage by cross links, irreparable strand breakage and the like is more acceptable, and would square with the postulated effects of radiation, mutagens and free radicals. Crosslinking in longterm molecules, much favoured by Bjorksten [1958] has drawn some popularity as a source of DNA damage from the large body of work on natural aging of collagen, begun by Verzár and lately reviewed in detail by Hall [1968] and by Deyl [1968]. Attempts to demonstrate it directly in "old" DNA have been made [von Hahn and Verzár, 1963; von Hahn and Fritz, 1966], but there are experimental difficulties in the direct comparison of function in "old" and "young" DNA. "Old" DNA has been said to differ from "young" in protein binding power [Samis and Wulff, 1969] and possibly in unwinding capacity [Devi *et al.*, 1966]. Its template activity has been variously reported [Devi *et al.*, 1966; Pyhtilä and Sherman, 1968; Samis and Wulff, 1969] but seems to depend far more on RNA polymerase concentration than on donor age. Direct studies of nuclear DNA, conducted with gerontology in mind, have so far been made by less sophisticated techniques than those employed by the genetic chemists. There is difficulty in avoiding mixed nuclear populations of old fixed, younger fixed, and clonally competent cells even in a tissue such as myocardium.

There is a wide range of subsidiary theory about possible causes and manifestations of DNA error, involving e.g. immunity [Burch, 1968], or escape of lysosomal DNAses [Allison and Paton, 1965; Allison, 1966]. At the same time, the direct demonstration that young DNA differs or does not differ from old seems likely to come from the further general advance of genetic chemistry rather than from *ad hoc* research, the role of gerontologists being to raise the possibility. There are *a priori* objections to a "primary" DNA lesion, moreover—including the need to explain why some clones, such as those of William pears, are stable, and why the rate of damage accumulation measured by mortality is 50 times as great in the mouse as in Man [Sacher, 1968]. Masking is another theoretical possibility: epigenetic switching-on and switching-off must, it is held, occur at the nuclear level during differentiation. It could be timed by outside intercellular signals, and could even include an evolved lifespan fixing, "destructor" mechanism [Medvedev, 1967]. Some processes once switched off might be irrecoverable within the differentiated state, and the life of cells in that state could accordingly be fixed by the lifespan of irreplaceable secondary copies, enzymes and the like.

Non-DNA Error Theories

At the present moment it seems clear (1) that error accumulation in primary DNA has been neither demonstrated nor excluded as a cause of aging and (2) that all the arguments and mechanisms adduced for it could apply with equal force to error accumulation at a post-DNA level. This could be chromosomal (involving nuclear structures rather than molecular DNA) or it could involve later transcription steps (RNA, synthetases and proteins). Some argument has been based on the rate of visible nuclear error-accumulation in long- and short-lived mouse strains [Curtis, 1967]. By far the most coherent location of the likely primary error site, however, puts it at the RNA-synthetase step. This theory was developed by Orgel [1963] and is now having a profound influence on the study of limitation in clones. According to Orgel, stochastic error, either in RNA specification or in the production of synthetases from RNA templates, both cytoplasmic in site, would lead to precisely the required accummulation of further error through the machine-tool role of the synthetases (RNA polymerases and the like) in specifying both secondary templates and themselves. This would produce an eventual error crisis, bad copies driving out good—by which time mis-specification would be present in all but the primary copies. If incorrect synthetases produce both further errors and more of themselves, they could exert a virus-like action both within the cell and, if transferred, in other cells. A model of this kind is in agreement with the work which has been done to locate error sites through the search for runaway synthesis of RNA, DNA and protein, as possible compensation for nonsense material production [Samis, Wulff and Falzone, 1964; Wulff, Quastler and Sherman, 1962; Clarke and Maynard Smith, 1966]. It also agrees extremely well with the results obtained in the studies of spanned clones (described below), and would gain still further in plausibility if it could be linked experimentally with one of the forms of chemical-stochastic attack incriminated by the free-radical and crosslinking schools.

The main investigable consequences of an Orgel-type error model would be (1) increasing clonal divergence with age, (2) aggravation by error agents such as unconventional aminoacids which act directly on protein specification and (3) cytoplasmic transmissibility.

It becomes profitable at this point to switch from the consideration of hypotheses to that of empirical research projects. In view of the large diversity of possible error sites inside and outside the cell, it now seems at least marginally more likely that the exact site, the size of the feedback loop, and the choice between primary DNA and later transcription faults will be settled from observing empirical agents acting on the lifespan rather than the other way round. It is now possible to outline a scientifically accountable research " portfolio," which, while neither sufficient nor exhaustive, covers the market on reasonable investment principles. This is based on (1) maximum number of options, having regard to the theories laid down, and (2) the preferential use of life-prolongation as the critical experiment: nonspecific life-shortening is easy to produce, but prolongation of the active life of outbred animals much less so.

Apart from the detailed search for individual error processes, the most important projects now going forward in gerontology fall under 5 main headings:

1. The investigation of substances, and of procedures such as calorie restriction, which are known to be able to prolong mammalian life—with special reference to those which are theoretically accountable already, such as antiradiation agents.

2. The investigation of clonal senescence and the nature of spanned clones—with special reference to the Hayflick phenomenon (the apparent limit of potential further division in somatic cells).

3. The investigation of autoimmune age effects, both as contributors to the pathology of observed aging in Man, and for the light they throw on clone divergence and quality control in body cells.

4. The investigation of developmental agents at the whole-animal level. Here I include both those which retard the program, antimetabolic and antidifferentiant agents (calorie restriction again being one of these) and the normal hormones.

5. Continued work on the nature of the differences between animals which determine relative lifespans, bearing in mind Sacher's warning [1968] against overcommitment to the " molecular paradigm."

These project headings provide a convenient view of practical age research: they

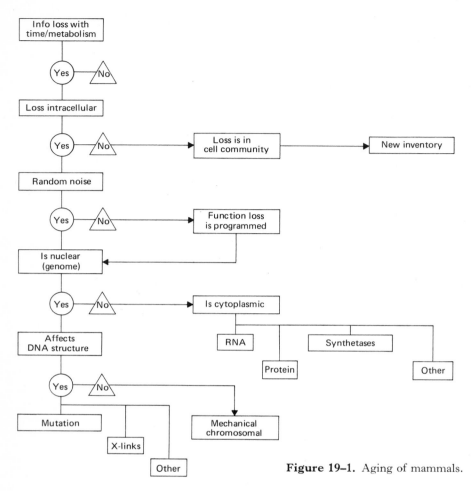

Figure 19–1. Aging of mammals.

supplement fundamental molecular biology but do not require prior choices on the site (DNA, RNA, protein, organelles, membranes) of the "senile lesion," while offering the possibility of making such a choice more evident. They also overlap constructively, autoimmune studies throwing light on clonal phenomena, and both on error theory; protectant, dietary and hormonal studies throwing light on the determinants of molecular processes such as collagen aging. From the practical standpoint, these approaches provide a simple overview of research possibilities, once the underlying theoretical choices are known (Fig. 19–1). An overall pattern is, moreover, beginning to emerge.

Protectant Substances

If information loss from cells is stochastic, and is comparable to that produced by gamma rays, then it ought to be possible to affect its rate chemically without further particularisation of site. The earliest test of the chemical-stochastic hypothesis, other than by looking for life-shortening, was suggested by Harman [1961, 1962]. If radiation accelerates aging, then radioprotectants, and especially antioxidants, might reasonably be expected to delay it in the intact animal. No studies appear to have been conducted to determine the sensitivity of longterm radiation effects to these agents, or to see if they are oxygen sensitive, and the experiment has practical difficulties. The critical experiment, however, would be to define an antiradiation drug which delays aging in the unirradiated animal. In recent experiments Harman [1968] has administered nontoxic antioxidant drugs (2-mercaptoethylamine, tert-butyl hydroxytoluene, ethoxyquin) to mice, with lifespan increases, measured as mean expectation, of up to 40% over controls. The nature of this increase is not yet clear. Inspection of curves suggests that it may represent the suppression of one predominant cause of senile death, possibly associated with lipid intoxication, which is known to be prevented by antioxidants in broiler chicks and in mink, rather than a full postponement of age changes. The effect is small compared with that of calorie restriction. In mice fed on large doses of ethoxyquin [Comfort, unpublished] weight gain is materially reduced: it is accordingly possible that we are spoiling the appetite of the mice or limiting fat uptake, and pair-feeding studies are needed. The results are nevertheless interesting, and might still provide evidence for the role of free radical damage as a contributor to the aging process, which has been persuasively argued by Tappel [1968]. Any such demonstration would lead not only to a re-examination of natural molecular protection by such agents as tocopherols and—SH groups, but to a fresh look at the stability of "living" molecules against the kind of attack already familiar in food and materials chemistry and in the deterioration of synthetics. Within the same-project-range we may also include the extensive work done on lipofuscin pigment (as a probable product of intracellular lipid oxidation) and upon the lysosomes themselves, which appear to be concerned with it [Bjorkerud, 1967; Strehler, 1964; Barber and Bernheim, 1967].

Other theoretical considerations suggest other agents which might affect the lifespan by modifying crosslinking [LaBella, 1966, 1968; Kohn and Leash, 1967], lysosome stability [Allison, 1966] or protein synthesis. Some of these approaches have already been reviewed [Comfort, 1966]—a repertoire of substances undoubtedly exists by means of which a high proportion of the subsidiary theories of noise-accumulation with age could be tested, and the search for protectants is clearly a valid approach to the fundamental questions which require answers (Table 1). If, in toxicity testing of new substances—particularly preservatives, anticancer, immunosuppressant, antimetabolic and radioprotectant drugs—investigators can be persuaded to run the experiment until natural death in both control and treated groups, we might get a great deal of valuable information. The large effects of single hormones in some mouse strains [Bellamy, 1968] open another field of investigation. The agent used by Bellamy (prednisolone) has been shown to prolong the postmitotic lifespan of human cultured ammion cells [Yuan and Chang, 1969]. Concentration on the details of error theory, which may itself prove erroneous, should not exclude the possibility that aging can be delayed and lifespan modified at the somatic level: the nature of the delay in calorie-restricted rodents, long known

Table 1. Substances Which Might Modify the Rate of Aging.

Nature	Theoretical Basis	Findings
Antioxidants	Scavenge free radicals, prevent attack on DNA or some other system	"Prolong life," may or may not shift specific age (Harman)
Radioprotectants	Assume aging similar in nature to radiation damage	Equivocal—not fully tested lifelong
Protein synthesis inhibitors	Break "vicious circle" if faulty synthesis involved	—
Lysosome stabilisers	Prevent escape of enzymes (including lysosome DNAse)	—
Immunosuppressants	Abolish any aging effects due to immune divergence	Imuran prolongs mouse life slightly (Walford) Predict ALS will limit some age changes but increase clonal divergence by stopping clone deletion
Anti-crosslinking agents	Aging due to cross links in long-term molecules	Under test (LaBella, 1966, 1968) BAPN, penicillamine (Kohn and Leash, 1967)
Hormonal agents	Modify chemical allometry, retard senescent program	
Anabolics	Prevent decline protein storage and muscular strength	Effective clinically
Somatotrophin	Maintain "young" pattern of protein synthesis	Carcinogenic—fails to prolong rat life (Everitt)
Prednisolone	Program slowing, antiautoimmune?	Doubles lifespan in short-lived strain (Bellamy)
17-Ketosteroids	Decline most closely parallels human senescence	—
Antimetabolic drugs	Delay program, simulate calorie restriction, induce "active diapause"	Proposed (Strehler) but not tested

[McCay and Crowell, 1934], has still not been elucidated, but it appears to involve dietary hypophysectomy. More sophisticated metabolic interference by lowering the overall metabolic rate is possible—if such measures fail to delay aging, that result in itself would be theoretically important for our understanding of the "clock."

Limitation in Clones

Carrel thought that somatic fibroblast cultures were immortal—that they were, in other words, stable clones. The facts now appear to be that stable clones exist, particularly for plant cells, but that somatic diploid cells in tissue culture under conventional nutritional regimes are not, or are not always, truly stable: that some if not all of such cultures appear intrinsically limited to approximately the number of doubling divisions which they would undergo in the host animal, and that, judging from Krohn's transplantation experiments [1962], this number differs from cell to cell. The chief authority for the instability of somatic diploid clones is Hayflick [1965]. Argument continues over the nature of the instability observed by him, and over its inherence in somatic diploids. Neither Hayflick nor Hay and Strehler [1967] were able to detect any reduction in the lifespan of clones from old as against young postembryonic donors. In consideration of Hayflick's findings it is still not possible to exclude nutritional and other causes of depression in cultured diploids [Hay *et al.*, 1968]. The latent period of growth initiation, an old index of aging, might prove more manageable in resolving the nature of clonal age changes.

It seems empirically clear that mammals do not die through running out of further mitosis, at least in high-turnover tissues. At the same time, clonal finitude is a highly important concept for aging studies even if some somatic cells only are limited in this way. It is still more significant as evidence in support of the various error theories. In this context the most interesting work is that on cytoplasmic agents controlling clone death. These studies date from the work of Rizet [1957] and Marcou [1957] on fungal clones, and have been continued at Buffalo by Muggleton and Danielli and at the Mill Hill National Institute of Medical Research by Holliday.

Muggleton and Danielli [1968] took stable clones of Amoeba and placed them under conditions which inhibited protein synthesis for long periods. Changes were induced in the treated clones which rendered them unstable in one of two ways— either they displayed stem-cell behaviour, one product of each division dying without further doubling, or the entire clone died synchronously after a fixed number of divisions. By some elegant microtransplanation experiments, they showed that stem-cell behaviour results from the implantation of a "spanned" nucleus in normal cytoplasm, but Hayflick-type limitation was transmissible to stable clones cytoplasmically, by very small inocula. This conforms closely to Rizet's [1957] observation, now confirmed by Holliday [1969], that in *Podospora* hyphal senescence is preceded by the appearance in the cytoplasm of a material which can transmit premature senescence to young hyphae.

Holliday [Harrison and Holliday, 1967] originally attempted to confirm Orgel's hypothesis of synthetase error by the administration of unconventional aminoacids to *Drosophila* larvae, with resulting life-shortening in the adult. He has now [Holliday, 1969] investigated the clonal aging of *Podospora* and *Neurospora* from this viewpoint. He found both an apparent increase in protein variation for "spanned" clones, judged by the appearance of variants in auxotrophes, and the presence of a cytoplasmically transmissible aging factor. It becomes highly important to identify this material, in view of Orgel's suggestion that the stochastic error in aging is cytoplasmic.

There are technical difficulties in extending this observation to mammalian diploid clones, but the attempt is clearly necessary, as a major check on the "faulty machine tool" hypothesis. Clonal death could result primarily from DNA error or masking, even though lower stages of the cycle of mis-specification are self-maintaining and cytoplasmically transmissible. Gofman and co-workers [1968] have claimed that Hayflick-type clones show a loss of E16 and an increase in E17 chromatin. From label incorporation studies Wulff, Quastler and Sherman [1962] had already suggested that faulty synthetase production might originate at the RNA–protein link, through " wear " affecting the sites of m-RNA synthesis. This process they put down to mutation, but it seems equally likely that error in the specification of polymerases could itself be mutagenic. Such mutagenic polymerases have been described [Speyer, 1965] and nucleotide incorporation in the nucleus by their action has been found even in fixed postmitotics [von Borstel, Prescott and Bollum, 1966]. In plant clones Oota [1964] found a separation of ribosomal RNA from protein prior to cell aging. For a validation of Orgel's hypothesis, any stochastic error either in RNA or in synthetase specification will serve, however. Other types of self-propagating error not primarily involving transcription but affecting it later could produce like effects—the self-propagating foci of polysaccharide crosslinking suggested by Field [1967—for the scrapie agent], for example. More recently, it has been shown that irradiated RNA synthetase can mis-specify [Goddard et al., 1969].

Clonal aging studies accordingly become critical in working out possible feedback loops in cellular error theory. A character of Orgel's model is that it is in theory at least reversible, the fault being initially confined to high-turnover protein, leaving the DNA master copies intact until a late stage. Attempts to reverse the "spanning" of Amoeba have so far failed [Muggleton and Danielli, 1968], and Marcou's [1957] reversal of aging in fungi appears to depend on strain selection, but the idea is interesting. It becomes urgent both to determine which somatic cells accord with Hayflick's finding, having due regard to nutritional and other criticisms levelled at it [Hay, 1967; Hay et al., 1968], and to pursue transmissible cytoplasmic factors in clones generally.

Autoimmune Phenomena

If aging reflects cellular divergence, as in fungi, whatever its site, it would be reasonable to look for immunological factors in its development arising from the presence of cells which are " not self." A parallel between aging and autoimmunity has been suggested by a number of investigators [Burch, 1963, 1968; Walford, 1962 (review)]. Autoimmunity is a pleiotropic process of exactly the kind which might give rise to the widely diverse appearances seen in aging. The syndrome of chronic incompatibility ("runt disease") bears, to the eye of faith, certain resemblances both to senescence and to progeria [Tyler, 1960]. There is a suggestive three-cornered relationship between age-processes, neoplastic processes, and various autoimmune or supposedly autoimmune phenomena such as scleroderma, lupus erythematosus and amyloid. It can also be argued that immunology is at present a "growth stock" and should be examined in connection with every unsolved biological problem. All these arguments have been intelligently developed by Walford [1962, 1964]. Direct evidence of immunological diversity with aging is so far not very strong—chiefly because it has not yet been sought by the sophisticated methods now available for histocompatibility analysis. Mariani [Mariani et al., 1960] grafting from male to female Strong "A" strain mice of different ages found signs of age-linked immunological changes in skin: of the old-old grafts, none took. Much of the original interest in autoimmune age changes derives from the idea of somatic mutation, leading to mis-specified cells. This concept has been the basis of elaborate theories [Burch, 1963, 1968], but it can be extended to other forms of clonal divergence or molecular error. Burnet [1959, 1967] has suggested that mutation in lymphocyte clones might cause them to lose their inhibition against the production of antibodies for correctly-specified body tissues—one of the few simple mutative models which satisfies the observed mathematical findings. Another possibility is that the efficacy of the filter mechanism against mis-specification itself declines with age. Teller and co-workers [Aoki and Teller, 1966; Teller *et al.*, 1964] found that the immunological response of old mice to tumours and heterografts is markedly impaired. Albright and Makinodan [1966] found that senile mouse spleen cells were themselves toxic to non-senile, lightly-irradiated recipients, and that in old animals there was a decay in the number of immunologically competent " progenitor " cells; this might reflect exhaustion of a stem-cell store or clonal limitation in the immunocyte line—an important instance, if valid, because impairment here could remove the somatic censorship on other mis-specified cells.

Walford [1964] has found that mouse lifespan can be marginally prolonged by immunosuppressants (Imuran) though far less than e.g. by calorie restriction. Studies on complete immunosuppression with antilymphocytic serum [ALS] have not yet been published, though these may well resolve the problem. Burnet's theory would lead us to expect an amelioration of aging with ALS. It seems more likely, however, that the action of immunosuppressants will prove to be more like that of anti-inflammatory agents, i.e. that they will limit some of the reactive pathology seen in aging, but aggravate any consequences of clonal divergence by removing the filter mechanism. Increased tumor incidence is already reported. What we shall need in this case is a way of enhancing rather than stopping the elimination of mis-specified cells and molecules, and autoimmune diseases could be evidence only of the body's attempts to do this. Energetic application of new immunological knowledge is clearly an important field in age research both theoretically and practically.

The remaining project headings, *metabolic program retardation* and *general comparative studies*, require less explanatory elaboration here. Metabolic retardation other than by calorie restriction is feasible in mammals without reducing activity, and has been suggested by Strehler [unpublished]. Whether it would retard aging or not is not *a priori* predictable, in view of the fact that temperature effects on aging are not apparently invariably additive [Clarke and Maynard Smith, 1961] at least in insect imagoes, though full hibernation retards collagen crosslinking [Hruza *et al.*, 1966]. The dietary restriction phenomenon too remains entirely unexplained at the fundamental level. The sensitivity of collagen crosslinking to dietary, hormonal and other metabolic influences does not necessarily indicate a similar sensitivity for intracellular molecules, but in general agents which delay collagen crosslinking appear to affect other age indices as well. *General comparative studies* provide a repertoire of test cases by which theories can be initially screened. They also act as an insurance against overcommitment to molecular error theories. It could well still prove that whatever the eventual rate of cellular mis-specification, human aging as we see it is primarily timed by some mechanical and somatic process such as cellular membrane alteration. The wise investor will leave room in his portfolio for such eventualities, and the collection of a wide range of lifetable and physiological data for the largest possible variety of living systems is clearly a sound investment. From his comparison of body-weights, metabolic rates and lifespans in mammals, Sacher [1959, 1968; Sacher and Trucco, 1962] points out that longevity correlates with two main characters—positively, with the amount of homoeostatic information in the system, including the brain, and negatively with gram/Calorie metabolism—and theories, whether of error or of general homoeostatic instability, must square with these findings to be acceptable.

The picture of fundamental research in aging has been greatly altered over the last decade. Though it is diverse, there is now a convergence in the diversities. Another decade of application might well see the diversities resolved—even if our present attempts to resolve them are incorrect—and its clinical possibilities revealed.

References

Albright, J. F., and Makinodan, T. Growth and senescence of antibody-forming cells. J. cell. Physiol. *67*: suppl. 1, 185–206 (1966).

Alexander, P. The role of DNA lesions in processes leading to aging in mice. Symp. Soc. exp. Biol., vol. 21, pp. 29–50 (1967).

Alexander, P., and Connell, D. I. Shortening of the lifespan of mice by irradiation with X-rays and treatment with radiomimetic chemicals. Radiat. Res. *12*: 510–525 (1960).

Allison, A. C. The role of lysosomes in pathology. Proc. roy. Soc. Med. *59*: 867–868 (1966).

Allison, A. C., and Paton, G. M. Chromosome damage in human diploid cells following activation of lysosomal enzymes. Nature, Lond. *207*: 1170 (1965).

Aoki, T., and Teller, M. N. Aging and cancerigenesis. III. Effect of age on isoantibody formation. Cancer Res. *26*: 1648–1652 (1966).

Barber, A. A., and Bernheim, F. Lipid peroxidation: its measurement, occurrence and significance in animal tissues. Adv. Geront. Res. vol. 2, pp. 355–403 (Academic Press, New York 1967).

Bellamy, D. Longterm action of prednisolone phosphate on a strain of shortlived mice. Exp. Geront. *4*: 327–334 (1968).

Bjorkerud, S. Isolated lipofuscin granules, a survey of a new field. Adv. Geront. Res., vol. 2, pp. 257–288 (Academic Press, New York/London 1967).

Bjorksten, J. A common molecular basis for the aging syndrome. J. amer. geriat. Soc. *6*: 740–748 (1958).

Burch, P. J. R. Autoimmunity: some aetiological aspects. Lancet *i*: 1253–1257 (1963).

Burch, P. J. R. An inquiry concerning growth, disease and aging (Oliver & Boyd, Edinburgh 1968).

Burnet, F. M. The clonal selection theory of acquired immunity. (University Press, Cambridge 1959).

Burnet, F. M. Concepts of autoimmune disease and their implications for therapy. Perspect. Biol. Med. *10*: 141–152 (1967).

Clarke, J. M., and Maynard Smith, J. Independence of temperature of the rate of aging in *Drosophila subobscura*. Nature, Lond. *190*: 1027–1028 (1961).

Clarke, J. M., and Maynard Smith, J. Increase in the rate of protein synthesis with age in *Drosophila subobscura*. Nature, Lond. *209*: 627–629 (1966).

Comfort, A. Nutrition and longevity in animals. Proc. Nutrit. Soc. *19*: 125–129 (1960).

Comfort, A. The prevention of aging in cells. Lancet *ii*: 1325–1329 (1966).

Curtis, H. J. Radiation and aging. Symp. Soc. exp. Biol., vol. 21, pp. 51–64 (University Press, Cambridge 1967).

Devi, A., Lindsay, P., Rainer, P. L., and Sarkar, N. K. Effects of age on some aspects of the synthesis of RNA. Nature, Lond. *212*: 474–475 (1966).

Deyl, Z. Macromolecular aspects of aging. Exp. Geront. *3*: 91–112 (1968).

Everitt, A. V. The effect of pituitary growth hormone on the aging male rat. J. Geront. *14*: 415–424 (1959).

Field, E. J. The significance of astroglial hypertrophy in scrapie, kuru, multiple sclerosis and old age. Dtsch. Z. Nervenheilk. *192*: 265–274 (1967).

Friedman, S. M., and Friedman, C. L. Prolonged treatment with pituitary powder in aged rats. Exp. Geront. *1*: 37–48 (1964).

Goddard, J. P., Weiss, J. J., and Wheeler, C. M. Error frequency during *in vitro* transcription of poly-V with γ-irradiated RNA polymerase. Nature, Lond. *222*: 670–671 (1969).

Gofman, J. W., Minkler, J. L., and Tandy, R. K. A specific common chromosomal pathway for the origin of human malignancy. Resp. Univ. Calif. Radiation Lab., Livermore, Calif. UCRL 50356 (1967).

Hahn, H. P. von, and Verzár, F. Age-dependent thermal denaturation of DNA from bovine thymus. Gerontologia 7: 105–108 (1963).

Hahn, H. P. von, and Fritz, E. Age-related alterations in the structure of DNA. Gerontologia *12*: 237–249 (1966).

Hall, D. A. The aging of connective tissue. Exp. Geront. *3*: 77–90 (1968).

Harman, D. Prolongation of the normal lifespan and inhibition of spontaneous cancer by antioxidants. J. Geront. *16*: 247–254 (1961).

Harman, D. *In* Shock, Biological aspects of aging (Columbia Univ. Press, New York 1962).

Harman, D. Free radical theory of aging: effect of free radical reaction inhibitors on the mortality rate of male LAF_1 mice. J. Geront. *23*: 476–482 (1968).

Harrison, B. J., and Holliday, R. Senescence and the fidelity of protein synthesis in Drosophila. Nature, Lond. *213*: 990–991 (1967).

Hay, R. J. Cell and tissue culture in aging research. Adv. Geront. Res. *2*: 121–158 (Academic Press, New York 1967).

Hay, R. J., and Strehler, B. L. The limited growth span of cell strains isolated from the chick embryo. Exp. Geront. *2*: 123–136 (1967).

Hay, R. J., Menzies, R. A., Morgan, H. P., and Strehler, B. L. The division potential of cells in continuous growth. Exp. Geront. *3*: 35–44 (1968).

Hayflick, L. The limited *in vitro* lifetime of human diploid cell strains. Exp. Cell Res. *37*: 614–636 (1965).

Holliday, R. Errors in protein synthesis and clonal senescence in fungi. Nature, Lond. *221*: 1224–1228 (1969).

Hruza, Z., Vrzalova, Z., Hrabalová, Z., and HlaváČová, V. The effect of cooling on the speed of aging in collagen *in vitro* and in hibernation of the fat dormouse (*Glis glis*). Exp. Geront. *2*: 29–36 (1966).

Kohn, R. R., and Leash, A. M. Longterm lathyrogen administration to rats, with special reference to aging. Exp. molec. Pathol. *1*: 354–361 (1967).

Krohn, P. L. Heterochronic transplantation in the study of aging. Proc. roy. Soc. B *157*: 128–147 (1962).

LaBella, F. S. The effect of chronic dietary lathyrogen in rat survival. Gerontologist *8*: 13 (1968).

LaBella, F. S. Pharmacological retardation of aging. Gerontologist *6*: 46–50 (1966).

McCay, C. M., and Crowell, M. F. Prolonging the lifespan. Sci. Mon. *39*: 405–414 (1934).

Marcou, D. Rajeunissement et arrêt de croissance chez *Podospora anserina*. C. R Acad. Sci. *244*: 661–662 (1957).

Mariani, T., Martinez, C., Smith, J. M., and Good, R. A. Age factor and induction of tolerance to male skin grafts in female mice subsequent to the neonatal period. Ann. N.Y. Acad. Sci. *87*: 93–105 (1960).

Maynard Smith, J. A theory of aging. Nature, Lond. *184*: 956–968 (1969).

Medawar, P. B. Old age and natural death. Modern Quart. *2*: 30–38 (1945).

Medvedev, Zh. A. Protein biosynthesis. (Oliver & Boyd, Edinburgh/London 1966).

Muggleton, A., and Danielli, J. F. Inheritance of the "life-spanning" phenomenon in *Amoeba proteis*. Exp. Cell Res. *49*: 116–120 (1968).

Oota, Y. RNA in developing plant cells. Ann. Rev. Plant Physiol. *15*: 17–36 (1964).

Orgel, L. E. The maintenance of accuracy of protein synthesis and its relevance to aging. Proc. nat. Acad. Sci., Wash. *49*: 517–521 (1963).

Pyhtilä, M. J., and Sherman, F. G. Age-associated studies on thermal stability and template effectiveness of DNA and nucleoprotein from beef thymus. Biochem. biophys. Res. Comm. *31*: 340–344 (1968).

Rizet, G. Les modifications qui conduisent à la sénescence chez *Podospora*: sont-elles de nature cytoplasmique? C. R. Acad. Sci. *244*: 663–666 (1957).

Sacher, G. A. Relation of lifespan to brain weight and body weight in mammals. CIBA Found. Symp. Aging, vol. 5, pp. 115–132 (Churchill, London 1959).

Sacher, G. A. Molecular versus systemic theories on the genesis of aging. Exp. Geront. *3*: 265–272 (1968).

Sacher, G. A., and Trucco, E. The stochastic theory of mortality. Ann. N.Y. Acad. Sci. *96*: 985–1007 (1962).

Samis, H. V., and Wulff, V. J. The template activity of rat liver chromatin. Exp. Geront. *4*: (in press, 1969).

Samis, H. V., Wulff, V., J. and Falzone, J. A. The incorporation of ^{3}H cytidine into RNA of liver nuclei in young and old rats. Biochem. biophys. Acta *91*: 223–230 (1964).

Simonsen, M., Engelbreth-Holm, J., Jensen, E., and Poulsen, H. A study of the graft-versus-host reaction in transplantation to embryos, F_1 hybrids and irradiated animals. Ann. N.Y. Acad. Sci. *73*: 834–839 (1958).

Speyer, J. F. Mutagenic DNA polymerases. Biochem. biophys. Acta *91*: 223–227 (1965).

Strehler, B. L. Time, cells and aging. (Academic Press, New York 1962).

Strehler, B. L. On the histochemistry and ultrastructure of age pigment. Adv. Geront. Res., vol. 1, pp. 343–348 (Academic Press, New York 1964).

Tappel, A. L. Will antioxidant nutrients slow aging process? Geriatrics *23*: 97–105 (1968).

Teller, M. N., Stokes, G., Curlett, W., Kubisek, M. L., and Curtis, D. Aging and cancerigenesis. I. Immunity to tumor and skin grafts. J. nat. Cancer Inst. *33*: 649–656 (1964).

Tyler, A. Clues to the aetiology, pathology and therapy of cancer provided by analogies with transplantation disease. J. Nat. Cancer Inst. *25*: 1197–1229 (1960).

Von Borstel, R. C., Prescott, D. M., and Bollum, F. J. Incorporation of nucleotides into nuclei of fixed cells by DNA polymerase. J. Cell Biol. *29*: 21–28 (1966).

Walford, R. L. Autoimmunity and aging. J. Geront. *17*: 281–285 (1962).

Walford, R. L. Further considerations towards an immunologic theory of aging. Exp. Geront. *1*: 67–76 (1964).

Wulff, V. J., Quastler, H., and Sherman, F. G. A hypothesis concerning RNA metabolism and aging. Proc. nat. Acad. Sci., Wash. *48*: 1373–1375 (1962).

Yuan, G. C., and Chang, R. S. Testing of compounds for capacity to prolong post-mitotic lifespan of cultured human amnion cells. J. Geront. *24*: 82–85 (1969).

20   Single-Stranded Regions in DNA of Old Mice

C. J. Chetsanga, V. Boyd, L. Peterson, and K. Rushlow

Investigations have shown that senescence is accompanied by molecular changes in the structure of the eukaryotic genome. Some of these changes have been reported to result in the loss of some genetic material in the brain cells of ageing beagles[1]. Furthermore, the molecular weight of rat liver DNA decreases as a function of senescence[2]. If such changes occurred on a large scale, the resulting genetic damage could affect severely some of the crucial gene functions of an organism. We are studying the physical state of the DNA of senescing mice, and report here the accumulation of an increasing amount of single-stranded regions in the DNA with age.

DNA was isolated from the livers of 1, 6, 15, 20, 25, and 30-month-old CBF_1 mice (Charles River Breeding Laboratories). Single-strand-specific nuclease S_1 was used to investigate the presence of single-stranded regions in the native DNA of mice of different ages. Figure 20-1 shows the absence of nuclease S_1 sensitivity in the DNA of mice aged 1–15 months. That of mice 20 months of age or older was increasingly sensitive to the enzyme, which digested 14–25% of it. We cannot rule out the possibility that this sensitivity began before the mice were 20 months old, since we did not test the DNA of 15–20-month-old mice. These results are based on three experiments in each of which the DNA from different animals of the same age group were used.

From *Nature*, **253**, no. 5487 (Jan. 10, 1975), 130–131. Reprinted with permission of the author and publisher.

Figure 20–1. Nuclease S_1 digestion of 1-30-month-old mouse liver DNA. The DNA was prepared essentially by the method of Marmur[3] except that a phenol-chloroform (1 : 1) mixture was used for extracting the proteins. Nuclease S_1 was prepared from *Aspergillus orzae* crude α-amylase (Sigma) by the method of Sutton.[4] The reaction mixture consisted of 30 μg of native or heat-denatured DNA, and 10 μg nuclease S_1 all in 1.5 ml of KZS(0.1 M KCl, 0.1 mM $ZnSO_4$, 0.025 M sodium acetate pH 4.5). Incubation was at 37° C for 30 min. After chilling, 25 μg of carrier calf thymus DNA was added, followed by 0.5 ml of 2 N perchloric acid. The control system (1.5 ml) consisted of 30 μg of liver DNA, 10 μg enzyme in KZS to which 25 μg carrier DNA and 0.5 ml of 2 N chilled perchloric acid were added without incubating. The acid-insoluble material was collected by centrifuging at 17,000 r.p.m. for 20 min. The A_{260} of the supernate was measured. The amount of DNA rendered acid-soluble by nuclease S_1 was obtained by subtracting the A_{260} of the control supernate from that of the corresponding experimental sample. This value represented the amount of DNA digested by nuclease S_1 and was used to calculate the percentage of hydrolysed DNA. ●, Native DNA; ○, denatured DNA.

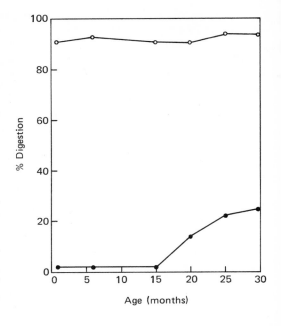

There are already two sets of indirect evidence for single-stranded regions in the DNA of ageing animals. Samis *et al.*[5] reported increased incorporation of ^{3}H-thymidine into the nuclear DNA of senescent rats in the absence of a corresponding increase in mitotic index. This implies that the thymidine, after conversion to dTTP, was used in the repair of nuclear DNA. Calf thymus DNA polymerase has been shown to catalyse a greater incorporation of DNA precursors into the nuclei of old mouse neurones, astrocytes, Kupffer cells and heart muscle fibres[6]. This reaction is known to require single-stranded regions of primer DNA along which the complementary strand can be synthesised[7]. Presumably these single-stranded regions of the DNA from senescent mice were hydrolysed by nuclease S_1 in our experiments.

As we do not know the ultimate fate of the nuclease S_1-sensitive regions in the DNA of old mouse liver, we suggest that the formation of regional single-stranded stretches in the DNA of senescent animals represents a transitional stage in the process leading to the elimination of nucleotide sequences that are to be discarded[1]. Figure 20-2 shows models of the possible structure of the regions in DNA attacked by nuclease S_1. There may be many areas of nuclease S_1 attack on crosslinked DNA molecules of which we present a simpler model in Fig. 2c. The metabolic significance of the existence of single-stranded regions in the DNA of older mice remains unknown. These regions may be the result of increased nuclease activity coinciding in time with a senescence-associated decline in repair enzyme activity in the cells. Loss in enzyme activity during senescence has been observed in other systems[8]. Thermodynamically, this has the advantage of making the energy previously reserved for repair functions available for hydrolytic processes. If the

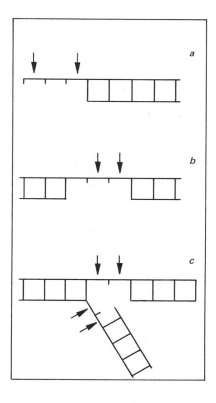

Figure 20–2. Postulated model of the regions of nuclease S_1 attack on the DNA of senescent mice. DNA with single-stranded region (a) terminally located, and (b) internally located; (c) cross-linked DNA with single-stranded regions in the junction region. Arrows indicate points of possible nuclease attack.

defective regions of the DNA occur in the genes for constitutive functions and are not repaired at the appropriate time, the biochemical lesions in the metabolic pathways of the particular cells are likely to result in the death of the organism; the harm can be particularly severe in the cells of non-replenishing tissues such as those of the muscular and nervous systems.

We thank the University of Michigan for grants and the National Institute of Child Health and Human Development for mice.

References

1. Johnson, R., and Strehler, N. L., *Nature*, **240**, 412–414 (1972).
2. Massie, H. R., Baird, M. B., Nicolosi, R. J., and Samis, H. B., *Archs. biochem. Biophys.*, **153**, 736–741 (1972).
3. Marmur, J., *J. molec. Biol.*, **3**, 208–218 (1961).
4. Sutton, W. D., *Biochim. biophys. Acta*, **240**, 522–531 (1971).
5. Samis, H. B., Falzone, J. A., and Wulff, V. J., *Gerontologia, Basel*, **12**, 79–88 (1966).
6. Price, G. B., Modak, S. P., and Makinodan, T., *Science*, **171**, 917–920 (1971).
7. Yoneda, M., and Bollum, F. J., *J. Biol. Chem.*, **240**, 3385–3391 (1965).
8. Wang, K. M., Rose, N. R., Bartholomew, I. A., Balzer, M., Berde, K., and Foldvary, M., *Expl. Cell Res.*, **61**, 357–364 (1970).

21   Evolution of Cell Senescence, Atherosclerosis, and Benign Tumours

Daniel Dykhuizen

It has been proposed that cell senescence is the cause of organismic senescence[1-3] and that senescence is under evolutionary control[4]. Orgel[5] has argued that cell senescence could not be under evolutionary control because it is difficult to imagine how senescence of the organism has a positive evolutionary advantage.

I propose that cell senescence itself, rather than senescence of the organism, is the genetically controlled and programmed event selected by evolution and that any effects on somatic senescence is a pleiotropic effect of this function. The advantage of programmed cell senescence to the organism is that it stops cells which have escaped from normal control from dividing indefinitely. If a cell with potential for further division gets out of place in the body, it is possible for it to settle in a new location where the normal environmental signal inhibiting division is lacking. Were these now uninhibited cells capable of unlimited growth, they would divide to fill the entire space available. A limited number of cell divisions, however, would limit the size of the growth, thus saving the organism.

From *Nature*, **251**, no. 5476 (Oct. 18, 1974), 616–618. Reprinted with permission of the author and publisher.

In vitro normal human fibroblasts are limited in the number of divisions[1,2]. The life of a culture of cells from a tissue explant can be divided into three phases[1]. Phase I ends with the formation of the first monolayer; phase II refers to a period when rapid growth occurs in proper culture conditions, while phase III refers to the period of senescence of the culture. Not only is the percentage of non-dividing cells higher in phase III cells than phase II cells[6], but the growth rate of dividing cells is reduced[7]: so all cells of the culture are affected by senescence. Phase II cells, however, are unaffected by being mixed with phase III cells[2].

Phase II cells can be held for long periods in a non-dividing state without affecting the number of divisions the cells undergo before senescence[8]. Thus, the metabolic age of a culture can be separated from the doubling age. The onset of senescence is shown to be dependent on the doubling age and the doubling age of a culture can be estimated[9].

Growth of a clone which arose from a single cell is restricted to the edge. The cells in the centre are inhibited by contact inhibition. Thus, the cells in the centre are at a younger divisional age than the cells near the edge[10]. As the cells go out from the centre, they become progressively older and growth stops at the outer edge of the colony when the cells there have divided the maximum number of times and senesce.

The evidence also indicates that cells senesce *in vivo*. With increasing age of the human donor there is a reduction in the number of cell doublings before senescence in culture[11]. Serial transplants of normal tissue into young inbred animals die out[12-14] and the death of the tissue is related to number of cell divisions rather than to its metabolic age[15]. In contrast, transformed cells (malignant cells) which do not show contact inhibition and do not senesce in tissue culture, can be transferred *in vivo* indefinitely[12].

Hayflick postulated that organisms age because their cells age[2,3] and this relationship has been supported by studies on Werner's syndrome, a genetic disease characterised by the development in early adulthood of a wide variety of degenerative features similar to those found in old age. Cells from sufferers of this syndrome are already senescent, doubling less than ten times in tissue culture[11], while cells of normal people of similar age double 35 times. Cell senescence is not necessarily the only cause of organismic senescence since humans aged 70–90 yrs are senescent; but fibroblast cultures from them are not senescent, doubling at least twenty times in tissue culture[11], which is over 40% of the doublings found in human embryo cultures.

Atherosclerotic lesions consist of "plaques" of fibroblasts lining the interior of the great arteries. Limitation in the size of these plaques provides a strong selection for cell senescence.

Benditt and Benditt[16] present the following evidence that a single plaque is monoclonal; that is, all of the cells of a plaque are derived from a single dividing precursor cell. They looked at plaques from the three females who were heterozygous for two electrophoretic variants of the X-linked glucose-6-phosphate dehydrogenase. There is random inactivation of one or the other of the X chromosomes at about the 1,000 cell stage of embryo development, consequently each heterozygous

213

female has a mixture of cells with one or the other X chromosome inactivated[17]. When small pieces (0.1 mm^3) were taken from the aorta wall, both enzyme types were present. But when larger pieces from plaques (0.5 mm^3 to 20–30 mm^3) were taken, most of them had only one isozyme present. It is unlikely that these plaques are scars[18] formed where the artery had been injured, since hypertrophic scar tissue is polyclonal[16].

Cells escaping into the blood stream from points where there is injury to the vascular system would be carried away from the wound by blood flow and could settle on the inner wall of arteries. Possibly they have trouble settling or dividing except where there are fatty streaks, thus giving the correlation between fats and atherosclerosis. The fibroblasts, now living in a new place in the body, exposed to a high velocity flow of nutrient rich blood, would no longer be contact inhibited[19] and would start dividing. As the single cell forms a clone, the cells in the centre away from the flow of blood, will cease dividing, while the cells on the edge will continue to divide. If these cells were able to continue dividing they would form quite a large growth and constrict the artery. Since they are limited to a certain number of divisions, however, the size of the plaque is limited in the same way as the size of a colony arising from a single cell in tissue culture[10].

I am postulating that atherosclerotic plaques arise from cells which escape into the arteries, settle and start growing. To keep these plaques at a minimum size, evolution has selected for a " clock " which limits the number of cell divisions. This number is set by the conflicting demands of the large number of divisions required for wound repair and cell replacement and the restriction in the number of cell divisions needed to control atherosclerotic plaques and other abnormal growths.

The theory does make several predictions which can be tested. The first is that plaques will grow quickly and then stop when the cells on the outer surface senesce. Second, with fully grown but relatively young plaques, the cells at the centre of the plaque should be able to divide in tissue culture and the ones at the edge should not. Third, the number of divisions the cells from the centre of a plaque can go through should be correlated with the size of the plaque, given that the plaque started from a single cell; and last, the median final size of a plaque should be correlated with the age of the person at the time the cells were released into the blood stream.

Limitation in the size of some benign tumours could also provide strong selection for cell senescence. It has been shown that leiomyomas are monoclonal[20]. Many leiomyomas seem to originate near arteries[21] and thus may have arisen from cells from arterial walls which have escaped into the lyometrium of the uterus. Similar predictions as those made for atherosclerotic plaques apply to such benign tumours. Other benign tumours particularly the large ones, could originate from other causes, such as by a mutation which destroys the system providing contact inhibition. In this case, all the cells, not just the ones on the edge, would go through the maximum number of divisions.

So far no mechanism has been proposed as to how cell senescence takes place. The theory presented above explains how natural selection can select for cell senescence and implies that cell senescence is under genetic control. Views on cell senescence may be divided into two groups: one favours an evolutionary-genetic basis, as in this paper, and the other believes that senescence is caused by random events which create disturbances within the cell which are not fully correctable and that this accumulation of errors eventually leads to death. Any random error theory, such as the "error catastrophe theory,"[5] must be able to explain why cultures of transformed cells are not severely enough affected by the errors to senesce and why normal cells can not evolve the state of perfection enjoyed by transformed cells. Moreover, any random error theory needs to explain why germ line cells are less affected by the errors than the somatic cells. The simplest explanation to these problems is that senescence is under genetic control and that induction of senescence takes place during differentiation leaving the germ line unaffected.

Any mechanism, such as that proposed by the "error catastrophe theory," could be used by evolution as the means to induce senescence. But a special compound could be produced whose sole function is to induce senescence. Milo[22] has presented evidence that a glycoprotein extracted from the cell wall of senescent phase III cells causes almost immediate senescence when put on phase II cells.

I thank Dr P. Morrow, Dr B. Richardson, Dr J. Campbell and Professor W. Hayes for helpful criticism.

References

1. Hayflick, L., and Moorhead, P. S., *Expl Cell Res.*, **25**, 585–621 (1961).
2. Hayflick, L., *Expl Cell Res.*, **37**, 614–636 (1965).
3. Hayflick, L., *Expl Geront.*, **5**, 291–303 (1970).
4. Comfort, A., *Ageing, The Biology of Senescence* (Routledge and Kegan Paul, London, 1964).
5. Orgel, L. E., *Nature*, **243**, 441–445 (1973).
6. Merz, G. S., and Ross, J. D., *J. Cell Physiol.*, **74**, 219–222 (1969).
7. Macieira-Coelho, A., Pontén, J., and Philipson, L., *Expl Cell Res.*, **42**, 673–684 (1966).
8. Dell'orco, R. T., Martin, J. G., and Kruse, P. F., jun., *Expl Cell Res.*, **77**, 356–360 (1973).
9. Cristofalo, V. J., and Sharf, B. B., *Expl Cell Res.*, **76**, 419–427 (1973).
10. Brunk, U., Esicsson, J. L. E., Pontén, J., and Westermark, B., *Expl Cell Res.*, **79**, 1–14 (1973).
11. Martin, G. M., Sprague, C. A., and Epstein, C. J., *Lab. Invest.*, **23**, 86–92 (1970).
12. Daniel, C. W., DeOme, K. B., Young, J. T., Blair, P. B., and Faulkin, L. J., jun., *Proc. natn. Acad. Sci. U.S.A.*, **61**, 53–60 (1968).
13. Krohn, P. L., *Proc. R. Soc.*, **B157**, 128–147 (1962).
14. Williamson, A. R., and Askonas, B. A., *Nature*, **238**, 337–339 (1972).
15. Daniel, C. W., and Young, L. J. T., *Expl Cell Res.*, **65**, 27–32 (1972).
16. Benditt, E. P., and Benditt, J. M., *Proc. natn. Acad. Sci. U.S.A.*, **70**, 1753–1756 (1973).
17. Ohno, S., *Ann. Rev. Gen.*, **3**, 495–524 (1969).
18. Timiras, P. S., *Developmental Physiology and Ageing*, ch. 25 (Collier-MacMillan Ltd., London, 1972).
19. Stoker, M. G. P., *Nature*, **246**, 200–202 (1973).
20. Linder, D., and Gartler, S. M., *Science*, **150**, 67–69 (1965).

21. Anderson, W. A. D., *Pathology* (2nd ed.), 1089 (C. V. Mos Co., St. Louis, 1953).
22. Milo, G. E., jun., *Expl Cell Res.*, **79,** 143–151 (1973).

22   A Brief Argument in Opposition to the Orgel Hypothesis

M. B. Baird, H. V. Samis, H. R. Massie, and J. A. Zimmerman

ABSTRACT. *The Orgel hypothesis receives considerable attention as a possible explanation for the phenomenon of senescence. Experimental observations which argue in favor of the Orgel hypothesis are discussed, and criticized in part. This is followed by a presentation of experimental data which argue in opposition to the notion. On the basis of the considerable body of data which argue in opposition to the Orgel theory, a call for reappraisal of the applicability of this theory to the phenomenon of senescence is suggested.*

The Orgel [1963] hypothesis receives considerable support as a plausible explanation for the phenomenon of senescence. The hypothesis itself was born at a time when studies directly concerned with cellular information transfer were the epitome of biological research, prior to the current flood of interest in disease states such as cancer and heart disease. The Orgel hypothesis provides a stochastic theory which fills the gap created by the apparent decline in interest in the highly touted somatic mutation theory as a result of experimental evidence which did not offer support for the mutation theory [Curtis and Gebhard, 1960; Clark and Rubin, 1961]. Although error theories in general have been criticized because of the inability of the notions to explain the widely variant life expectancies observed for living organisms [Sacher, 1968], the refutation of any hypothesis ultimately rests in experimental evidence which renders the notion untenable.

The purpose of this brief communication is to present some experimental evidence which tends to support the Orgel notion and some observations which are clearly at odds with the notion and, in fact, argue in opposition to this explanation for senescence.

Orgel proposed that senescence results from the progressive accumulation of altered molecules with the passage of time. That such molecules will be generated during the processes concerned with molecular biogenesis is not seriously doubted, since cellular processes responsible for the generation of molecules would not be expected to operate with absolute fidelity. Orgel further posited that molecular alterations will eventually appear in those molecules which in themselves are concerned with the synthesis of other molecules. This would lead to an "error catastrophe," a rapid progressive increase in the rate of accumulation of aberrant

From *Gerontologia*, **21**:57–63, 1975. Reprinted with permission of the authors and the publisher.

molecules, resulting in the loss of vitality and, ultimately, the death of the organism.

There are at least two major predictions of the Orgel notion. The first is that senescence should be accompanied by an increase in aberrant protein molecules concomitant with the passage of time. Secondly, experimental manipulation of the protein in a manner which will result in an increased frequency of aberrant molecules should mimic the manifestations of aging, and especially result in a decrease in adult lifespan.

There are, at present, several ways in which one might demonstrate the presence of altered molecules in aging organisms. One might observe an age-related decrease in specific activity of an enzyme protein, which could result from a dilution of a specific protein population by altered, nonfunctional or partially functional molecules. Likewise, preparations of enzymic protein may possess components which show alterations in various physicochemical properties, such as in thermostability, substrate-binding properties or altered immunological properties.*

There is in literature a series of papers which offer support for the Orgel hypothesis. Using immunological methods, Gershon and Gershon [1970] showed that the specific activity of the enzyme isocitrate lyase decreased with advancing age in the nematode, *Turbatrix aceti*. This decline in specific enzyme activity was paralleled by an age-related increase in totally inactive isocitrate lyase molecules. Subsequently, Zeelon *et al.* [1973] demonstrated that the specific activity of yet another enzyme in nematodes, fructose-1,6-diphosphate aldolase, decreases with advancing age, and that this decrease in activity is paralleled by an increase in inactive aldolase cross-reacting material (CRM) as detected by immunological methods.

Similar studies were performed in rodents, where it was shown that the specific activity of mouse liver fructose-1,6-diphosphate aldolase decreased with advancing age, concomitant with an increase in inactive liver aldolase CRM [Gershon and Gershon, 1973a]. Likewise, although there was no age-related change in the specific activity of mouse skeletal muscle aldolase, there was a 1.3- to 1.4-fold increase in aldolase protein in senescent organisms as detected by specific immunoprecipitation, the implication being that a large number of the molecules (30–40%) were inactive or partially active in senescent mouse muscle [Gershon and Gershon, 1973b]. (It is important to point out that these studies compared very young animals [e.g. 2 months] with very old animals. It appears plausible to suggest that the observed differences between young and old animals represented differences in developmental state. It would be of interest to see results obtained from young animals that have completed growth [8 months] as well as from "middle-aged" animals [18–20 months] instead of only the comparison between animals 2 and 30 months of age.)

Parallel to these studies were experiments designed to specifically test the validity of the Orgel hypothesis. Harrison and Holliday [1967] showed that when *Drosophila* larvae were fed amino acid analogues, the treatment presumably resulted in an increased frequency of aberrant protein, and resulted in a decrease in the adult life

---

* R. S. Krooth (J. Theor. Biol. *46*:501, 1974) has proposed a mechanism which could permit the differential accumulation of altered molecular forms in aging organisms.

span of the imago. (The conclusions drawn from this observation must be criticized. A decrease in adult life span as a result of an assumed increase in frequency of aberrant molecules does not of necessity demonstrate that altered molecules are the cause of normal aging. Such cause and effect reasoning parallels that which follows: inasmuch as ionizing radiation must produce somatic mutations in exposed organisms, and radiation exposure shortens adult life span, therefore somatic mutations must be the cause of normal aging. Such reasoning is not only unjustified, but unacceptable.)

Holliday and Tarrant [1972] demonstrated that the thermostability of a fraction of the enzymic glucose-6-phosphate dehydrogenase population of MCR-5 male fetal lung fibroblasts decreases progressively with advancing age. This change in thermostability was attributed to the presence of altered enzyme molecules. The presence of such thermolabile molecules was reportedly observed only in cells which were clearly dying. That is, there was no progressive increase in presumably aberrant molecules as a function of increasing cell generations. Inasmuch as one of the salient features of senescence is that the phenomenon is progressive, these results suggest that the accumulation of altered molecules may be a characteristic of dying cells, and not of cells which are simply undergoing senescence. That is, these observations suggest that an accumulation of aberrant molecules may be a signal feature of death and not of senescence. The pertinence of these and other results obtained with cell culture systems to the phenomenon of senescence in the intact organism remains to be demonstrated. It has not been demonstrated unequivocally that there is any relationship between limited cell doubling potential and senescence and longevity in the intact organism [Massie et al., 1974].

The work cited from Holliday's group offers tenuous support for the Orgel hypothesis, and, taken together with the work of the Gershons', offers considerable support for the hypothesis. There is, however, a body of evidence which argues to the contrary.

Data concerning the effect of advancing age upon enzymic activity in rodents have been gleaned from the literature and organized into tables by Finch [1972] and Wilson [1973]. Mere perusal of these tables clearly shows that the activity of *most* enzymes either remains unchanged or rises in senescent organisms. This is highly disturbing, since an accumulation of aberrant molecules should depress the specific activity of a population of enzyme molecules through dilution with inactive or partially active molecular representatives. On the other hand, enzyme activity could remain unchanged by means of increasing the numbers of inactive or partially inactive molecules through compensatory increase in protein synthesis or decrease in rate of degradation, as is apparently the case for mouse skeletal muscle aldolase [Gershon and Gershon, 1973b]. However, if this mechanism is *de rigueur* in senescent organisms, old age should be accompanied by a substantial increase in total protein which does not appear to be borne out by observation in intact organisms [Barrows et al., 1960; Fonda et al., 1973] or in chick fibroblast cells [Hay and Strehler, 1967].

It seems clear that age-related changes in specific activity of a particular enzyme are, by themselves, of questionable gravity. This is especially true in light of the

218

observation that many age-related changes in enzyme activity are reversible with hormonal or pharmacological treatment [Gold and Widnell, 1974; Baird *et al.*, 1974]. It seems likely that many age-related changes observed in enzyme activities, which initially support the notion of dilution of a population of molecules with aberrant molecules, represent in reality a compensatory response to alterations in hormonal or other metabolic factors [Gold and Widnell, 1974].

There is direct evidence that some populations of enzyme molecules do not accumulate altered species in aging organisms. Oliveira and Pfuderer [1973] have presented evidence that the age-related decrease in mouse muscle lactate dehydrogenase activity results from a loss in fully active enzyme molecules, and not from an accumulation of mis-synthesized, partially active or inactive molecules. Grinna and Barber [1972] have demonstrated that although the activity of glucose-6-phosphate dehydrogenase decreases in rat liver with advancing age, the apparent K_m for the enzyme is unaltered, demonstrating that the population for molecules remaining has the same substrate-binding capacity, an observation which does not support the notion that substantial numbers of the molecular representatives are altered. These workers have also presented evidence that physicochemical changes seen in hepatic microsomal NADPH-cytochrome c reductase are the result of changes in the matrix in which the molecules are organized, rather than changes in the enzyme molecules themselves. Zimmerman, Samis, Baird and Massie [manuscript in preparation] have demonstrated that the thermostability of rat liver catalase remains unaltered in senescent animals. Furthermore, immunoprecipitation of catalase activity in young and old homogenates by anti-young catalase produced precipitation curves which are virtually superimposable.

Thus, there is strong evidence that there are at least three populations of molecules which do not accumulate altered representatives in senescent organisms. These observations must be considered in terms of an hypothesis which demands the presence of altered molecules in all protein populations in the senescent organism.

One final observation is perhaps in order. There is a considerable body of literature which shows that old animals can respond to many external stimuli as completely as do their younger counterparts. Thus, hepatic α-glycerol phosphate dehydrogenase activity in female Wistar rats of different ages is induced to similar levels following treatment with thyroid hormone [Bulos *et al.*, 1972]. Although both hepatic zoxazolamine hydroxylase activity and hepatic NADPH-cytochrome c reductase activity are lower in senescent CFN rats, treatment with phenobarbital induced both enzyme functions to the same levels in young and old animals [Baird *et al.*, in press]. It is difficult for us to imagine how animals of different ages can respond to identical levels of stimulation when, according to the hypothesis under consideration, aging must be accompanied by a *loss* in capability for the production of normal molecules.

We have addressed ourselves to a brief consideration of some observations which argue in opposition to the Orgel notion, and have not attempted to comprehensively review the literature pertaining to this notion. It is often relatively easy to obtain experimental support for any particular notion. It behooves the proponents of any hypothesis, however, to consider experimental data which argue in opposition. We

219

conclude that the argument presented in opposition to the Orgel notion as an explanation of senescence indicates that a stark reappraisal of the applicability of this particular hypothesis to the phenomenon of senescence is warranted.

References

Baird, M. B.; Nicolosi, R. J., Massie, H. R., and Samis, H. V.: Microsomal mixed function oxidase activity and senescence. I. Hexobarbital sleep time and induction of components of the hepatic microsomal enzyme system in rats of different ages. Exp. Geront. (in the press).

Baird, M. B.; Zimmerman, J. A., Massie, H. R., and Samis, H. V.: Response of liver and kidney catalase to α-p-chlorophenoxy isobutyrate (clofibrate) in C57BL/6J male mice of different ages. Gerontologia 20: 167–178 (1974).

Barrows, C. H.; Falzone, J. A., and Shock, N. W.: Age differences in the succinoxidase activity of homogenates and mitochondria from the liver and kidneys of rats. J. Geront. 15: 130–133 (1960).

Bulos, B.; Shukla, S., and Sacktor, B.: The rate of induction of the mitochondrial α-glycerolphosphate dehydrogenase by thyroid hormone in adult and senescent rats. Mech. Aging Dev. 1: 227–231 (1972).

Clark, A. M. and Rubin, M. A.: The modification by X-irradiation of the lifespan of haploids and diploids of the wasp, Habrobracon. Radiat. Res. 15: 244–253 (1961).

Curtis, H. J. and Gebhard, K. L.: Aging effect of toxic and radiation stresses; in Strehler, The biology of aging, pp. 162–166 (Amer. Inst. Biol. Sci., Washington 1960).

Finch, C. E.: Enzyme activities, gene function and aging in mammals. Review Exp. Geront. 7: 53–67 (1972).

Fonda, M. L.; Acree, D. W., and Auerbach, S. B.: The relationship of α-amino-butyrate levels and its metabolism to age in brains of mice. Arch. Biochem. Biophys. 159: 622–628 (1973).

Gershon, H. and Gershon, D.: Detection of inactive enzyme molecules in aging organisms. Nature, Lond. 227: 1214–1217 (1970).

Gershon, H. and Gershon, D.: Inactive enzyme molecules in aging mice: liver aldolase. Proc. nat. Acad. Sci., Wash. 70: 909–913 (1973a).

Gershon, H. and Gershon, D.: Altered enzyme molecules in senescent organisms: mouse muscle aldolase. Mech. Aging Dev. 2: 33–41 (1973b).

Gold, G. and Widnell, C. C.: Reversal of age-related changes in microsomal enzyme activities following the administration of triamcinolone, triiodothyronine and phenobarbital. Biochim. biophys. Acta 334: 75–85 (1974).

Grinna, L. S. and Barber, A. A.: Age-related changes in membrane lipid content and enzyme activities. Biochim. biophys. Acta 288: 343–353 (1972).

Harrison, B. J. and Holliday, R.: Senescence and the fidelity of protein synthesis in Drosophila. Nature, Lond. 213: 990–992 (1967).

Hay, R. J. and Strehler, B. L.: The limited growth span of cell strains isolated from the chick embryo. Exp. Geront. 2: 123–135 (1967).

Holliday, R. and Tarrant, G. M.: Altered enzymes in aging human fibroblasts. Nature, Lond. 238: 26–30 (1972).

Massie, H. R.; Baird, M. B., and Samis, H. V.: Prolonged cultivation of primary chick cultures using organic buffers. In Vitro 9: 441–444 (1974).

Oliveira, R. J. and Pfuderer, P.: Test for mis-synthesis of lactate dehydrogenase in aging mice by use of a monospecific antibody. Exp. Geront. *8*: 193–198 (1973).

Orgel, L. E.: The maintenance of the accuracy of protein synthesis and its relevance to aging. Proc. nat. Acad. Sci., Wash. *49*: 517–521 (1963).

Sacher, G. A.: Molecular versus systemic theories on the genesis of aging. Exp. Geront. *3*: 265–271 (1968).

Wilson, P. D.: Enzyme changes in aging mammals. Gerontologia *19*: 79–125 (1973).

Zeelon, P.; Gershon, H., and Gershon, D.: Inactive enzyme molecules in aging organisms. Nematode fructose-1,6-diphosphate aldolase. Biochemistry *12*: 1743–1750 (1973).

23   Testing the Protein Error Theory of Ageing: A Reply to Baird, Samis, Massie, and Zimmerman

R. Holliday

ABSTRACT. *A major prediction of Orgel's theory is that the misincorporation of amino acids into proteins will increase with age. This has not yet been tested experimentally. Indirect methods have been used to search for the presence of altered proteins in ageing cells or organisms, but these would not necessarily detect a low level of mistakes, nor do they distinguish between errors in synthesis and post-synthetic changes. Nevertheless, some experimental results have been obtained from genetic and biochemical studies with fungi and fibroblasts which confirm certain predictions of the protein error theory.*

Orgel's [1963] theory of ageing makes a number of predictions, but these are not necessarily easy to test. A most important one is that the misincorporation of amino acids into proteins should increase during the process of ageing. So far, no one has been able to carry out this test, for reasons that are not hard to see. The only published experiments on the misincorporation of amino acids are by Loftfield[1963] and Loftfield and Vanderjagt [1972], using peptides of ovalbumin or haemoglobin from young animals. They obtained an estimate of error frequency, but if their peptides were not completely pure, the actual frequency could have been very much lower. If there is no accurate measure of the spontaneous error frequency in protein synthesis, how can we know what to expect during ageing? Moreover, it is not known whether a small increase in errors, perhaps twofold, would be lethal to a cell, or whether a hundredfold increase could be tolerated. It must be borne in mind that whereas a high level of errors in a structural protein, or an enzyme present in any

From *Gerontologia*, **21**:64–68, 1975. Reprinted with permission of the author and publisher.

large amount, may be relatively harmless to a cell, a low level in a repressor or some other essential regulatory protein may be extremely damaging.

For these reasons the theory does not predict that the specific activity of an enzyme will show a significant decline with age. There may be a decline which is too small to measure, or there may even be an increase in specific activity if regulatory mechanisms are affected. For example, it is known that a particular amino acid substitution in glucose-6-phosphate dehydrogenase (G6PD) results in a fourfold increase in the quantity of enzyme synthesised [Yoshida, 1970]. In fact, during ageing some enzymes decrease in activity, some increase and others are unchanged. These results, together with all those mentioned in the latter part of the paper by Baird et al. [1975] are in no way inconsistent with Orgel's theory.

Since it is very difficult to measure the actual misincorporation of amino acids during ageing, investigators have used less direct methods for detecting altered molecules. They have either searched for the formation of a heat-labile fraction of an enzyme, or for the appearance of molecules which react with antibody to an enzyme, but which have no enzyme activity (CRM). Unfortunately neither of these procedures provide a critical test of the error theory. The error frequency during ageing may result in the production of less than 5% altered protein, which is probably too low to detect. Alternatively, any altered protein which is seen may be the result of post-synthetic modifications, for instance the loss of amide groups.

With regard to the the experiments carried out in my laboratory [Holliday and Tarrant, 1972], Baird et al. [1975] make no serious attempt to provide a critical evaluation. They state that heat-labile G6PD was detected only in cultures of human fibroblasts which were "clearly dying" rather than senescent. This is a surprising conclusion to draw, since a significant increase of altered enzyme was detected up to 15 passages *before* the cultures finally died out. Moreover, it is stated in our paper that on several occasions "it was possible to predict when a culture will become visibly senescent by the appearance of significantly more than 10% of heat sensitive enzyme." There are three reasons for believing that the altered G6PD might be the result of errors in synthesis, but none are mentioned by Baird et al. [1975].

(1) Cultures were starved by serum deprivation, treated with cycloheximide (an inhibitor of protein synthesis) or held confluent for long periods of time, but in no case was an increase in heat-labile G6PD detected. This suggests that post-synthetic modifications are not a significant factor.

(2) Cells grown in the presence of 5-fluorouracil (5-FU) for many generations become prematurely senescent, and at the same time a significant fraction of heat-labile G6PD appeared. 5-FU is known to induce errors in protein synthesis.

(3) It was shown that a proportion of the heat-labile G6PD has altered substrate specificity. This is more likely to result from amino acid substitutions than post-synthetic modifications. It could be due to mutation, but only if the frequency was extremely high.

Subsequently it has been shown that cells from patients with the premature ageing disease known as Werner's syndrome, which can be passaged only a few times in culture, accumulate as much altered G6PD as senescent cultures of normal cells [Holliday et al., 1974].

A number of other experiments also support the protein error theory, but none are mentioned by Baird *et al.* [1975]. Lewis and Tarrant [1972] detected the appearance of lactic dehydrogenase cross-reacting material and a decrease in enzyme activity during the ageing of fibroblasts in culture. They also presented evidence for a decline in the ability to discriminate against ethionine, an analogue of methionine, which would be expected if the specificity of the amino acyl tRNA synthetase became reduced during ageing. Similar results have been reported by Ogrodnik *et al.* [1974, and personal commun.] during the ageing of mice.

The best evidence that a lethal protein error catastrophe can actually occur comes from experiments with *Neurospora crassa* [Lewis and Holliday, 1970]. The mutant *leu* 5 has an altered leucyl tRNA synthetase, and there are good reasons to believe that this enzyme is responsible for the misincorporation of amino acids other than leucine into proteins, particularly at 35 °C [Printz and Gross, 1967]. The strain can be grown indefinitely at 25 °C, but on transfer to 35 °C it grows at a constant rate for 3–4 days before the whole culture dies. (This process is analogous to the clonal ageing of human fibroblasts; the doubling time for *Neurospora* cells is about 2 h, therefore, 3–4 days growth represents 36–48 doublings.) Examination of the glutamic dehydrogenase (GDH) during this time showed that a high proportion of altered molecules accumulated right at the end of the life span, which would be expected for an exponential build-up of errors. Another ageing mutant, natural death (*nd*), accumulated altered GDH in a similar way. Moreover this mutant was shown to act as a weak missense suppressor during its ageing—precisely what would be predicted from a loss of fidelity in protein synthesis [Holliday, 1969].

Baird *et al.* [1975] mention experiments carried out some years ago in which it was shown that *Drosophila* larvae treated with amino acid analogues produced adults with a reduced life span [Harrison and Holliday, 1967]. They refer to a "conclusion" from this experiment which they say is based on unjustified and unacceptable reasoning. Our conclusion was: "the finding that agents which might be expected to increase the normal frequency of errors in protein synthesis have the effect of shortening the life span does not of course prove that Orgel's hypothesis is correct, but it does confirm one of its main predictions. Other explanations are possibly applicable to our results." And we go on to discuss two of these.

Eight years after publication of the experiments on *Drosophila* (which were, incidentally, the first specifically designed to test the protein error theory), I feel that the conclusion just cited still holds. Many of the predictions of Orgel's theory have been confirmed, but no experiment so far carried out fully establishes its validity. None of the observations cited by Baird *et al.* [1975] provide any evidence against the theory, although some they do not mention may do so—for instance, the lack of discernible effect on viruses grown in senescent cells [Holland *et al.*, 1973].

Finally, I would like to dispel the quite erroneous view, mentioned by Baird *et al.* [1975], that error theories cannot explain the difference in longevity between species. It is known from studies on microorganisms that mutations exist which alter the fidelity of genetic replication or protein synthesis. If the accuracy of these processes is under genetic control, and ageing is due to errors in macromolecules, then different species could evolve quite different longevities.

References

Baird, M. B.; Samis, H. V., Massie, H. R., and Zimmerman, J. A.: A brief argument in opposition to the Orgel hypothesis. Gerontologia *21*: 57–63 (1975).

Harrison, B. J. and Holliday, R.: Senescence and the fidelity of protein synthesis in *Drosophila*. Nature, Lond. *213*: 990–992 (1967).

Holland, J. J.; Kohne, D., and Doyle, M. V.: Analysis of virus replication in ageing human fibroblast cultures. Nature, Lond. *245*: 316 (1973).

Holliday, R.: Errors in protein synthesis and clonal senescence in fungi. Nature, Lond. *221*: 1224–1228 (1969).

Holliday, R.; Porterfield, J. S., and Gibbs, D. D.: Premature ageing and occurrence of altered enzyme in Werner's syndrome fibroblasts. Nature, Lond. *248*: 762–763 (1974).

Holliday, R. and Tarrant, G. M.: Altered enzymes in ageing human fibroblasts. Nature, Lond. *238*: 26–30 (1972).

Lewis, C. M. and Holliday, R.: Mistranslation and ageing in *Neurospora*. Nature, Lond. *228*: 877–880 (1970).

Lewis, C. M. and Tarrant, G. M.: Error theory and ageing in human diploid fibroblasts. Nature, Lond. *239*: 316–318 (1972).

Loftfield, R. B.: The frequency of errors in protein biosynthesis. Biochem. J. *89*: 82–91 (1963).

Loftfield, R. B. and Vanderjagt, D.: The frequency of errors in protein biosynthesis. Biochem. J. *128*: 1353–1356 (1972).

Ogrodnik, J. P.; Wulf, J. H., and Cutler, R. G.: Altered protein hypothesis of mammalian ageing processes. II. Discrimination ratio of methionine versus ethionine in the synthesis of ribosomal protein and RNA of C57BL/6J mouse liver. Exp. Geront., (in press).

Orgel, L. E.: The maintenance of the accuracy of protein synthesis and its relevance to ageing. Proc. nat. Acad. Sci., Wash. *49*: 517–521 (1963).

Printz, D. B. and Gross, S. R.: An apparent relationship between mistranslation and an altered leucyl tRNA synthetase in a conditional lethal mutant of *Neurospora crassa*. Genetics *55*: 451–467 (1967).

Yoshida, A.: Amino acid substitution (histidine to tyrosine) in a glucose-6-phosphate variant (G6PD Hektoen) associated with overproduction. J. molec. Biol. *52*: 483–490 (1970).

six   Population Problems and Genetics

The human population is growing at the rate of 2 per cent per year. This means that the already crowded earth, containing about 3.5 billion people in 1970, will have 7 to 8 billion in the year 2000 (Frejka, 1973; Holden, 1974; but see Ehrlich, 1970). The doubling time for the population is now 35 years, or perhaps even less. Ecological, psychological, and related aspects of this catastrophic growth rate have been widely proclaimed (Ehrlich, 1968; Mesarovic and Pestel, 1974). The genetic consequences have heretofore received less public attention. Unless we succeed in curbing population growth worldwide, the genetic implications of the population problem are unimportant—in the face of impending famine and other possibilities. But if the human species does curb its population growth and survives well into the twenty-first century, the population dynamics of genes become an important consideration.

The current low birthrate in Japan provides some evidence that reproductive sanity might prevail in human society, although the Japanese rate is still too high for population decline to an ecologically balanced size. Japan is overpopulated. Japanese-style small families are genetically beneficial in that, for instance, they reduce the frequency of Down's syndrome by one third, reduce Rh-erythroblastosis by one half, and reduce other miscellaneous genetic defects by about one tenth, according to the estimates of Matsunaga presented in his paper in this section. Similar conclusions are drawn by Graham, Cohen, and Schull for the United States in their report presented here. The decline in Japan of cousin (and other consanguineous) marriages, which results from the decrease in family size and the increase in population mobility, reduces the incidence of homozygous recessive defects, one part of the miscellaneous genetic ills mentioned here.

The trend to small families is eugenic in that it minimizes the level of those mutations which occur increasingly with parental age, including those chromosomal mutations producing Down's syndrome (Matsunaga, 1973). But the trend to small families also decreases the opportunity for natural selection, according to some authors. This is thought to be particularly true in the United States (Kirk, 1968) where mortality is very low (96 per cent of females born in 1960 will live to age 45 as compared with 86 per cent born in 1920) and, for various reasons, sterility is now also low. The homogeneity in time and amount of reproduction in the United States and Japan, along with the decrease in inbreeding, implies a high genetic load—an increase in the level of potentially harmful genes in the population (see Brues, 1969; Vogel, 1972). But there is some controversy on whether the genetic load is favorable or unfavorable in terms of evolutionary fitness (Wills, 1970).

In association with these trends, the voluntary use of induced abortion (or inhibitors of conception) is increasing. This is discussed in a medical-ethical context in Section Eight. The testing of the fetus in utero for certain genetic defects is now possible. This practice, followed by therapeutic abortion of abnormal fetuses, increases the prenatal selection of good phenotypes.

These innovations are beneficial both in minimizing human misery and, for certain defects, in improving the gene pool (Motulsky et al., 1971; Holloway and Smith, 1975; Fraser, 1972). On the other hand, medical treatment of dominant inherited defects, such as the gut obstruction known as pyloric stenosis, could produce a notable increase in the frequency of the defect over several generations (see paper by Crow in Section Eight); this harms the gene pool. We shall consider eugenics further in the remainder of the book.

Another connection between the population crisis and genetics is provided by Hardin in this section (see also Hardin, 1972). Hardin surprised some people by advising a policy of self-interest, for a while yet, on the question of who should reproduce. In this connection it is interesting to note that many years ago J. B. S. Haldane expressed the opinion that should any human gene for altruism arise, it would be relentlessly selected against in this competitive and resource-limited world. Contrary positions have been explored by others (Wilson, 1973; Gadgil, 1975).

Hardin's thoughtful paper aroused much critical interest. He called for a governmentally enforced, mutual compromise on a low level of breeding to solve the population crisis because he saw no technical solution to the crisis (such as a "green revolution," which we discussed earlier). Hardin was answered by Crowe, the author of the following paper in this section, who believes that overbreeding has neither a technical nor a political solution—and indeed may have no solution at all. Crowe thinks this crisis will be ameliorated by foreseeable changes in human values or by enforceable laws. He reasons that the population crisis and other crises of our times may never be resolved because there is no common value system in contemporary American society (as Hardin assumed); rather, the "society" is increasingly a set of camps of warring tribes (students, blacks, WASPs, etc.) whose values are not negotiable, not compromisable.

Moreover, the population crisis must be acknowledged to be worldwide. Hardin's forced compromise on breeding is almost inconceivable on that scale, because ethnocentrism is such a potent force in human affairs. Voluntary restraint on breeding (see also Section Seven) would seem to be the only feasible option. If imperfect voluntary restraint selectively favors genes for selfish overbreeding, as opposed to what Hardin calls conscience, the human species of the future may be different from what we wish it to be—but it may not care.

Other authors who have thought searchingly about optimized individual well-being being destructive of group welfare include Frolich and Oppenheimer (1970), Potter (1974), Schelling (1971), and Miles (1970). Miles argues, in contrast to Hardin, that with today's self-expectations in the United States, children are expensive pleasures; that they are economic liabilities, not assets. He sees "no conflict, therefore, between the economic self-interest of married couples to have small families and the collective need of society to preserve 'the commons.'" Potter argues that cultural moral sanctions imposed a bioethical restraint on use of the commons in England in the past, as well as in the Sahel region of Africa until the introduction of new technology precipitated a drought (in 1974).

What should geneticists do about the population problem? Graham (1971) urges them to turn to relevant research on the regulation of human fertility or to politicking at all levels to keep politicians informed of the magnitude of the problem.

Addendum on Variance*

Variance has been found to be a convenient method to describe the way a group of individuals differs from the average in some measurement. For example, suppose we had two basketball teams, the heights of the men in one being 6, 5, 6, 6, and 7 feet and the heights of men in the other being 7, 7, 5, 6, and 5 feet. The mean is 6 feet in each group but there is more variation in the second group. The variation in each group can be quantified as variance,

* As an aid to understanding Matsunaga's paper.

taking the deviation of each player's height from the mean, squaring it, and summing for the group. This sum divided by the number of observations (N) is the variance.

Team one ($N = 5$)			Team two ($N = 5$)		
Height	d	d^2	Height	d	d^2
6	0	0	7	$+1$	1
5	-1	1	7	$+1$	1
6	0	0	5	-1	1
6	0	0	6	0	0
7	$+1$	1	5	-1	1
30		2	30		4

mean height (feet): $\quad \bar{x} = \dfrac{30}{5} = 6 \qquad \bar{x} = \dfrac{30}{5} = 6$

variance: $\quad \dfrac{\sum d^2}{N} = \dfrac{2}{5} = .4 \qquad \dfrac{\sum d^2}{N} = \dfrac{4}{5} = .8$

We find that the variance of the second group is twice that of the first. This tells us more about the variation within each group than, say, the range does; the range of heights is identical in the two groups (5 to 7 feet).

Bibliography

Bajema, C. J., ed. *Natural Selection in Human Populations.* New York: Wiley, 1971.

Brues, A. Genetic load and its varieties. *Science,* **164** : 1130–1136, 1969.

Cavalli-Sforza, L. L. The genetics of human populations. *Sci. Amer.,* **231** : 81–89, 1974.

Dumond, D. E. The limitation of human population: A natural history. *Science,* **187** : 713–721, 1975.

Ehrlich, P. Looking backward from 2000 A.D. *The Progressive,* pp. 23–25, Apr. 1970.

Ehrlich, P. *The Population Bomb.* New York: Ballantine, 1968.

Emery, A. The prevention of genetic disease in the population. *Intern. J. Environ. Studies,* **3** : 37–41, 1972.

Fraser, G. R. The short-term reduction in birth incidence of recessive diseases as a result of genetic counseling after the birth of an affected child. *Hum. Hered.,* **22** : 1–6, 1972.

Frejka, T. The prospects for a stationary world population. *Sci. Amer.,* **228** : 15–23, 1973.

Frolich, F., and J. A. Oppenheimer. I get by with a little help from my friends: The " free-rider " problem. *World Politics*, **23** : 104–120, 1970.

Gadgil, M. Evolution of social behavior through interpopulation selection. *Proc. Nat. Acad. Sci.*, **72** : 1199–1201, 1975.

Graham, J. B. The relation of genetics to control of human fertility. *Persp. Biol. Med.*, **14** : 615–638, 1971.

Hardin, G. *Exploring New Ethics for Survival. The Voyage of the Spaceship Beagle*. New York: Viking, 1972.

Holden, C. World population: U. N. on the move but grounds for optimism are scant. *Science*, **183** : 833–836, 1974.

Holloway, S. M., and C. Smith. Effects of various medical and social practices on the frequency of genetic disorders. *Amer. J. Hum. Genet.*, **27** : 614–627, 1975.

Kirk, D. Patterns of survival and reproduction in the United States: Implications for selection. *Proc. Nat. Acad. Sci.*, **59** : 662–670, 1968.

Matsunaga, E. Effect of changing parental age patterns on chromosomal aberrations and mutations. *Soc. Biol.*, **20** : 82–88, 1973.

Medawar, P. B. Do advances in medicine lead to genetic deterioration? *Mayo Clinic Proc.*, **40** : 23–33, 1965.

Mesarovic, M., and E. Pestel. *Mankind at the Turning Point*. New York: Dutton, 1974. Scenarios on world doom from the Club of Rome.

Miles, R. Whose baby is the population bomb? *Pop. Bulletin*, **16** : 3–36, 1970.

Motulsky, A. G., G. R. Fraser, and J. Felsenstein. Public health and long-term genetic implications of intrauterine diagnosis and selective abortion. *Symp. on Intrauterine Diagnosis*, D. Bergsma, ed. *Birth Defects*: Original Article Series 7, 22–32, 1971.

Ornstein, L. The population explosion, conservative eugenics, and human evolution. *BioScience*, **17** : 461–464, 1967.

Potter, V. R. The tragedy of the Sahel commons (letter). *Science*, **185** : 813, 1974.

Schelling, T. C. On the ecology of micromotives. *The Public Interest*, **25** : 61–98, 1971.

Vogel, F. Eugenic aspects of genetic engineering. *Adv. in the Biosciences*, **8** : 397–410, 1972.

Westoff, C. F. The decline of unplanned births in the United States. *Science*, **191** : 38–41, 1976.

Wills, C. Genetic load. *Sci. Amer.*, **222** : 98–107, 1970.

Wilson, E. O. Group selection and its significance for ecology. *BioScience*, **23** : 631–638, 1973.

24   Possible Genetic Consequences of Family Planning

Ei Matsunaga

Family planning means to have children in a desired number, each child at a desired time. Although the parental desire in this regard may be different in individual families, a more or less uniform pattern of reproduction would emerge from practicing family planning, since this need is conditioned by life circumstances that are rather common to all members of the society. Thus, if an appreciable fraction of a population practices it, a change in demographic trends, namely, decline in variances would be expected not only in the number of children per family but also in birth order and parental ages in live-birth data. As family planning is usually used to limit family size, a second sign would be decline in the mean number of children and the mean birth order, while the mean of the parental ages may scarcely be affected.

The purpose of this paper is to review the various aspects of demographic data and vital records now available in Japan with special attention to their potential effects upon future generations.

Recent Demographic Transition in Japan

Japan has achieved an unprecedented drop in births during a short period after World War II. It is true that this has been done mostly by induced abortions, legalized by the Eugenic Protection Law in 1948 for economic reasons as well as from physical considerations for maternal health. But less is known about the fact that the trend toward lowered fertility has been, as pointed out by demographers in this country, emerging since around 1920. Table 1 shows the mean and the variance of maternal age for selected years from 1925 to 1960, together with the standardized birth rates, which are from data in the report[1] by the Institute of Population Problems. The birth rate has been steadily declining, from 35.3 in 1925 to 14.6 in 1960, with the exception of 1940, during which the government encouraged population increase, and of the following period of the postwar baby boom. In accord with this trend the variance of maternal age does show a remarkable decline throughout the period; it decreased from 45.5 in 1925 to 19.9 in 1960. It is of some interest to note that the decline in the variance of maternal age was hardly affected by the government's policy in 1940 or by the postwar baby boom. These results clearly indicate that the practice of family planning has been diffusing in Japan since some 40 years ago.

This work, contribution 587 from the National Institute of Genetics, was supported by grant RF 61113 from the Rockefeller Foundation and a grant from the Toyo Rayon Foundation for the Promotion of Science and Technology.

Table 1. Standardized Birth Rates* and Means and Variances of Maternal Age in Live-Birth Data in Japan, 1925–1960.

| Year | Standardized Birth Rate† | Maternal Age, yr | |
		Mean	Variance
1925	35.3	28.4	45.48
1930	32.4	28.6	42.50
1937	29.8	28.9	39.34
1938	26.0	29.0	39.71
1939	25.4	29.3	38.62
1940	27.7	29.3	37.29
1947	30.7	29.1	36.36
1948	30.0	28.7	37.75
1949	29.7	28.5	35.22
1950	25.3	28.2	33.45
1952	20.8	28.1	30.04
1954	17.4	27.8	27.04
1956	15.8	27.6	24.67
1958	15.2	27.3	21.85
1960	14.6	27.1	19.93

* From Institute of Population Problems.[1]
† Rates are given as number of births per 1,000 population.

Although the direct genetic effect of the above change is not immediately obvious, it may come into light if we review what kinds of change have taken place in the distribution of live births by maternal age and birth order. Table 2 illustrates the relevant data for 1947, 1953, and 1960. The number of live births of rank 4 or higher decreased from 36% in 1947 to only 10% in 1960. During the same period mothers

Table 2. Population Trends in Japan, 1947 to 1960.*

Population Trend	1947	1953	1960
Total no. of live births	2,679,000	1,868,000	1,606,000
Total no. of reported cases of induced abortions	—	1,068,000	1,063,000
Mean age of women at first marriage, yr	22.9	23.4	24.2
Percentage of live births of birth order 1 to 3	64.1	75.1	90.3
Percentage of live births by age of mother			
Under 19 yr	2.3	1.7	1.2
Over 35 yr	19.8	12.1	5.8
Infant death rate, per 1,000 live births	76.7	48.9	30.7
Infant death rate from congenital malformations, per 10,000 live births	14.7†	21.1	19.0

* Data are based on the Annual Vital Statistics from the Ministry of Health and Welfare.
† This figure seems to be an underestimate; the corresponding rate was 23.7 in 1950, and thereafter has been declining.

aged 19 or less decreased from 2.3% to 1.2%, and mothers aged 35 or more decreased from 19.8 to 5.8%, while the mean age of women at first marriage was raised from 22.9 to 24.2 years. Briefly, a rapid transition to a family pattern with two to three children, born when the mother's age was 20 to 34 years, has taken place in Japan.

The demographic transition outlined above must have resulted in important changes in the frequencies at birth of those congenital defects which are correlated with parental age and birth order. Since we are concerned here only with the genetic aspects, reference may be made to two categories of the defects; one is of those due to new mutations, on both chromosomal and genic levels, and the other is determined by genetic factors in combination with environmental influence.

Among the first category of defects, the best known is mongolism (Down's syndrome), whose occurrence depends upon maternal age. The increase with maternal age is rather slow until the mother reaches the age of 35 years, and then it rises almost exponentially as the mother approaches the menopause. Similar but less pronounced dependency upon maternal age has been noted for Klinefelter's syndrome (XXY), triple-X female (XXX), and trisomy 18 syndrome.[2, 3] As to the occurrence of gene mutation, there is some evidence of positive correlation with paternal age for certain rare dominant anomalies, notably achondroplasia and acrocephalosyndactyla.[3] However, the data are still insufficient for evaluation of the increased rate of the risk according to paternal age.

The second category includes a variety of congenital defects in addition to Rh-erythroblastosis, for which the risk increases with birth order. Recent studies in Canada[4, 5] have shown increased risks to children of older mothers, apart from birth-order effect, for cerebral palsy and congenital malformations of the circulatory system, and increased risks with advancing birth order, independent of maternal-age effect, for strabismus and other congenital malformations of the nervous system and sense organs. On the other hand, there were apparently no special risks, except for injuries at birth, to firstborns when maternal-age effect was eliminated or to children of very young mothers provided they were firstborns. The risks of diseases of the nervous system and sense organs were increased with higher birth rank to the children of very young mothers, however. The average relative-risk figures for all these defects, in Canadian data, were noted to range from 1.6 to 2.4, as compared with the standard rates either for younger mothers or for children of lower birth rank.

Consequently, the variety of congenital defects mentioned above must have been reduced in Japan as a result of the decreasing frequencies of births of higher rank as well as those of both older and very young mothers. Considering the changes from 1947 to 1960, the reduction value may have been about one third for mongolism, not so much for XXY and XXX types of chromosomal aberrations, more than one half for Rh-erythroblastosis, and perhaps of the order of one tenth for the rest of the defects. It should be noted that the reduction in the defects attributable to new mutations is beyond doubt eugenic for the population, whereas the reduction in other diseases may be dysgenic, since selection intensity against the genes is relaxed. However, so long as the pattern of small family size continues, the manifestation of the diseases would be prevented.

Secular Change in Sibship Size and Frequency of Consanguineous Marriages

While there are a variety of factors which affect the frequencies of consanguineous marriages, one of the most important is the mean size of sibships in a population. It is obvious that with a smaller sibship mean a person has fewer relatives—for example, fewer cousins than was the case when the mean was larger—but how frequencies of consanguineous marriages will be affected by the change in the variance of sibship size is not immediately recognized. Taking account of both mean and variance of sibship size, Nei and Imaizumi[6] were able to formulate the frequencies of various types of consanguineous marriages. We may cite here with a slight modification (in the denominator, they used $N-1$ instead of N), their formula for first-cousin marriages:

$$f = \frac{\bar{x}_2{}^2(V_1 + \bar{x}_1{}^2) - \bar{x}_1(V_2 + \bar{x}_2{}^2)}{\bar{x}_1 \bar{x}_2} \cdot \frac{2}{N},$$

where $\bar{x}_1$, V_1 and $\bar{x}_2$, V_2 stand for the mean and variance of sibship sizes of parental and children's generations, respectively, and N for the so-called isolate size in the sense of Dahlberg[7] of the size of the population within which marriages are contracted at random: the formula (f) represents the expected frequency of first-cousin marriages by chance. It is to be noted that the variability of sibship size in the parental generation increases the frequency of first-cousin marriages, while the variability in the children's generation reduces it. The same argument can be applied to the frequencies of other types of consanguineous marriages. Although the formula is based on certain assumptions (ie, 1:1 sex ratio at maturity, disregarding those who will not marry, no polygamy and no remarriage, and no correlation in sibship size between parents and children), it may be used to give us an idea about the expected change we are concerned with. In a recent survey by the Population Problems Research Council of the Mainichi Newspapers, Tokyo,[8] however, a significantly negative correlation has been found between sibship size of married women and number of their live children.

At present, we have no means to estimate the isolate size, N, without resorting to the observed frequencies of consanguineous marriages in the population studies. It is evident, however, that industrialization and urbanization have been increasingly breaking down the geographical barriers among populations that had been isolated, so that the scope of finding one's mate must have become considerably wider than before. As to the secular change in sibship sizes, two sets of data may be used to deduce the trend; the one is the 1960 census report[9] on the number of children born to all married women with at least one child and the other is the vital statistics of live births by live-birth order from 1950 to 1963 (*Annual Vital Statistics of Japan*, 1950–1963). Table 3 shows that the mean and variance of the number of children had been almost stable for Japanese mothers now older than 55 years, the mean being slightly larger than five and the variance about seven, while for younger mothers the mean has been decreasing gradually and the variance has been decreasing relatively faster. Table 4 shows the secular change in the live-birth order of all newborns during the past 15 years; the mean and variance have decreased

Table 3. Mean and Variance of No. of Children Ever Born to All Married Women with at Least One Child, by Age.*

Age of Mother, yr	No. of Children	
	Mean	Variance
80—	5.159	6.798
75—79	5.132	6.842
70—74	5.261	7.152
65—69	5.209	7.000
60—64	5.191	7.093
55—59	5.130	6.966
50—54	4.846	6.054
45—49	4.249	4·514
40—44	3.563	3.021

*Data are based on the 1960 census report in Japan.

from about three and four in 1950 to less than two and one, respectively, in 1963. Considering the high mortality for children in the earlier times, especially in large families, it may roughly be estimated that the mean and variance of sibship size had been approximately four and six, respectively, until about 1930, and since then both have been steadily declining, at first rather slowly but then very rapidly after World War II. In accord with such secular changes, various values of means and variances of sibship sizes have been substituted in the following formula:

$$e = \frac{\bar{x}_2^2(V_1 + \bar{x}_1^2) - \bar{x}_1(V_2 + \bar{x}_2^2)}{\bar{x}_1 \bar{x}_2}$$

Table 4. Secular Changes in Mean and Variance of Live-Birth Orders of All Live-Born Children, 1950 to 1963.*

Year	Mean	Variance	Variance Square of mean
1950	2.878	3.847	0.465
1954	2.578	2.646	.398
1955	2.492	2.512	.405
1956	2.381	2.338	.412
1957	2.272	2.124	.412
1958	2.148	1.874	.406
1959	2.039	1.664	.400
1960	1.957	1.461	.382
1961	1.890	1.306	.366
1962	1.819	1.106	.334
1963	1.784	0.999	0.314

* Data are based on Annual Vital Statistics from the Ministry of Health and Welfare.

235

Table 5. Changes in Expected No. of Potentially Marriageable First Cousins for Each Person in Hypothetical Population with Trend Toward Smaller Sibship.

	Parental Generation*		Children's Generation†		Expected No. of Potentially Marriageable First Cousins
	$\bar{x}_1$	V_1	$\bar{x}_2$	V_2	
Population with large family	4	6	4	6	16.5
	4	6	3.5	5	14.3
	3.5	5	3	4	10.5
Population in transition to smaller family size	3	4	3	3	9.0
	3	3	2.5	2.5	6.5
	2.5	2.5	2	2	4.0
	2	2	2	1	3.5
Population with small family size	2	1	2	1	2.5

* $\bar{x}_1$ and V_1 = mean and variance of sibship size for parental generation.
† $\bar{x}_2$ and V_2 = mean and variance of sibship size for children's generation.

where e is the expected number of potentially marriageable first cousins for each individual. Table 5 shows that the expected number of such first cousins decreases from 16.5 for an expanding population with the mean of four and the variance of 6 to 2.5 only for a stationary population with the mean of two and the variance of 1; the reduction in the mean noticeably affects the reduction in the number of the first cousins, while the reduction in the variance will have its effect one generation later. It can be shown that the expected rates of reduction are more pronounced for first cousins once removed and for second cousins.

Japan is well-known for the high frequency of consanguineous marriages. Although it varies according to localities investigated, it is usually about 2% to 5% for urban areas and of the order of 10% for rural areas. Table 6 shows as an example the results of our recent survey in the city of Ohdate, Akita Prefecture, in northern Japan.[10] The data cover all couples which were registered in the ward offices and in which the wife's age was in the range from 30 to 40 years. The frequency of the major types of consanguineous marriages was about 2% for couples living in the central part of the city, while it was as high as about 8% for those in the peripheral part. When the data are broken down by age of wives, no secular change is apparent in the frequencies of consanguineous marriages. This is not surprising, because these wives were born during the period from 1922 to 1932, so that their sibship sizes as well as those of their parents must have been still large. There have certainly been other socioeconomic factors, particularly in rural areas, which favored consanguinity. Nevertheless, the rapid reduction in both the mean and the variance of sibship size, on the one hand, and the modern trend toward the breakdown of isolates, on the other hand, should in the near future result in a significant reduction in the frequency of consanguineous marriages. Unless other factors dominate, the reduction rate may presumably be of the order of 4/5 in proportion with the reduction expected in the number of potentially marriageable relatives for each individual, say from 15 to 3.

Table 6. Frequency of Consanguineous Marriages, by Ages of Wives 30 to 40 Years Old, Registered in City of Ohdate (1962).*

Relationship	Age of Wives, yr					
	30–31	32–33	34–35	36–37	38–40	Total
	Center of City (Old City)					
First cousins	3	1	2	0	4	10
First cousins once removed	3 } 2.3%	0 } 0.9%	2 } 2%	1 } 1%	1 } 2.6%	7 } 1.7%
Second cousins	1	2	3	2	3	11
Other	0	0	0	2	2	4
Not related	299	347	350	297	299	1,592
Total	306	350	357	302	309	1,624
	Periphery of City (New City)					
First cousins	6	12	13	13	11	55
First cousins once removed	6 } 5.8%	10 } 9.9%	7 } 8.1%	7 } 8.3%	10 } 7%	40 } 7.8%
Second cousins	9	13	11	9	4	46
Other	9	5	4	5	6	29
Not related	334	313	347	314	324	1,632
Total	364	353	382	348	355	1,802

* In Table 2 presented at the World Population Conference held in Belgrade in 1965,[11] the arrangement of the data was not proper; this has been corrected in this Table.

The genetic consequence would be of advantage for the society, since consanguineous marriages lead to increased risks of illness, premature death, and congenital abnormality among the offspring. The reduction rate in mortality and morbidity for the population as a whole depends upon the present frequencies of consanguineous marriages and their reduction rates, as well as the increased rates in the relative risks to the children from such marriages. The results of recent informative studies[11] show that the increased rates in the relative risks are not very large. In certain Japanese cities child mortality, including stillbirths, was found to be 10% to 11% among the offspring of first cousins, against 8% to 9% among the controls, whereas the risk of death caused by congenital abnormalities was relatively high but less than double. Assuming double average risk for the children from first-cousin marriages and a decrease in their frequency from 5% to 1%, the reduction rate in death rate for the population may be estimated to be about 4%. This may seem of relatively small magnitude but may not be insignificant from the point of view of the community as a whole. On the other hand, the decrease in consanguineous marriages would be dysgenic, as it should result in increase in rare recessive genes carried by heterozygous persons that had otherwise been eliminated in homozygous form.

Possible Dysgenic Effect of Differential Fertility

The practice of family planning usually spreads more rapidly in some social strata than in others, and more rapidly among better-educated than among less-educated portions of the society. This would in all probability result in fertility differences,

Table 7. Secular Changes in Spread of Contraceptive Practice, Not Including Induced Abortions, Among Different Strata in Japan, 1950 to 1965.*

	Percentage of "Current Users"						
	1950	1952	1955	1957	1959	1961	1965
Occupation of the husband							
Farmers and fishermen ⎫		17.0	25.4	30.5	34.9	37.7	47.0
Manual workers ⎪	11.5	23.9	35.8	37.6	40.1	36.7	50.0
Medium and small enterprisers⎬		24.7	37.4	39.0	40.4	39.1	51.0
Salaried workers ⎪	25.9	36.9	39.7	49.1	50.7	50.3	54.4
Others ⎭		35.2	41.0	47.0	—	—	—
Duration of husband's education, yr							
Less than 9	14.2	18.2	28.2	33.4	37.6	37.1	—
10–12	25.4	37.0	37.7	46.5	43.9	45.4	—
More than 13	37.3	47.0	48.8	52.5	54.0	56.2	—
Duration of wife's education, yr							
Less than 9	13.0	20.1	28.2	33.3	35.0	35.6	46.9
10–12	32.4	38.7	46.1	48.4	51.6	49.6	58.1
More than 13	36.0	59.1	47.8	53.2	51.9	60.4	65.2
Total	19.5	26.3	33.6	39.2	42.5	42.5	51.9

* Data are based on surveys by the Population Problems Research Council, Mainichi Newspapers, Tokyo.[8, 12, 13]

and if these were correlated with some genetically determined traits, the genetic composition of the offspring population would be altered.

The first question to be answered is to what extent the family-planning practice in Japan has been unevenly distributed with respect to certain characteristics of the population. In Table 7 some relevant data are reproduced from the reports[7, 12, 13] of seven consecutive surveys conducted by the Population Problems Research Council of the Mainichi Newspapers, showing secular changes in the frequencies of use of various kinds of contraceptive methods, not including induced abortions, among married women aged 16 to 49. For each survey some 3,000 couples were sampled from the whole country by appropriate methods of stratification, so that the results for a specific year are comparable. The term "current users" does not necessarily include those couples who had practiced contraception in the past. From the Table it is clear that the practice of contraception has been rapidly spreading among all social strata; the average frequency of current users has increased from 20% in 1950 to 52% in 1965. There have been, in fact, variations in the frequencies according to husband's occupation and couple's educational background. But perhaps the most remarkable feature of the data is that the increased rate has been particularly high among those portions of the society in which the initial frequencies were the lowest, so that the variation due to social stratification appears to have been considerably lessened. This impression is verified by computing the coefficient of variation for each year of the surveys among categories of husband's occupation and of couple's education (Table 8). There is no doubt that the high literacy of the people, active mass communication of the knowledge about contraception, and

Table 8. Secular Changes in Coefficient of Variation with Respect to Differential Spread of Contraception.

	Coefficient of Variation, %						
Cause of Variation	1950	1952	1955	1957	1959	1961	1965
Husband's occupation	—	30.3	17.2	18.5	15.9	15.4	6.0
Husband's education	45.1	42.9	27.0	22.1	18.3	20.7	—
Wife's education	45.6	49.6	26.7	23.1	20.9	25.6	16.3

mass campaigns by the regional centers for public health services have greatly contributed to the spread of family-planning practice in Japan.

The second question is concerned with the extent of variation in fertility according to the social strata. Since fertility should be measured for women who had completed reproduction, the outcome of the current trends is still to be seen. Here we may refer to the data obtained from the fertility surveys made in 1952 and 1962 by the Institute of Population Problems.[14] Table 9 shows the average number of live births per wife aged over 45 in the two surveys by husband's education and occupation, together with the corresponding coefficients of variation. It is seen that the coefficients of variation in fertility due to the two characteristics of the husbands have both decreased from 16% in 1952 to about 10% in 1962. In this connection, it is to be noted that in Japan the age at marriage tends to be higher with the longer span of education, which in turn correlates positively with the rank in the occupa-

Table 9. Secular Changes in Differential Fertility in Japanese Wives Aged 45 Years, by Husband's Education and Occupation.*

	Average No. of Live Births	
	1952 survey	1962 survey
Total	4.47	3.91
Duration of husband's education, yr		
Less than 9	4.62	4.05
10–12	3.62	3.60
More than 13	3.47	3.21
Coefficient of variation, %	16.0	11.6
Husband's occupation		
Farmers and fishermen	5.06	4.19
Workers on own account in nonprimary industries	4.08	4.02
Manual workers	3.79	3.82
Nonmanual workers	3.57	3.37
Coefficient of variation, %	15.9	9.2

* Data are based on consecutive surveys by The Institute of Population Problems.

tional status. Therefore, it would hardly be possible to evaluate the net effect of family planning upon the differential fertility.

The third question is referred to the possible genetic consequence of the observed differences in fertility. Among a variety of mental traits the best defined and the most important is intelligence. The extent of possible dysgenic effect with respect to intelligence depends, not only upon its heritability and correlation with the length of education or the occupational status, but also upon the extent of variation in fertility differences, both among social strata and within a social stratum. Although present methods of measuring intelligence are imperfect, there seems to be a high correlation between the intelligence quotient and school performance for Japanese children, and those with higher scores are more prone to enter into the higher school courses. We do not know, however, to what extent intelligence is correlated with the length of education or the occupational status. The extent of heritability within the normal range of intelligence still remains to be recognized. Although we do know the extent to which variation in fertility difference among social strata is rapidly decreasing in Japan, we have no information about the differential fertility by intelligence within each stratum. The loss of variation among strata does not necessarily mean the loss of variation within a stratum. In the absence of our knowledge on many important points, it is difficult to evaluate the extent, if any, of dysgenic effect of differential fertility upon the intelligence of the future population.

Possible Relaxation of Selection Intensity

As is well-known, the modern demographic trend toward smaller family size in western Europe as well as in Japan has been accompanied by a reduction in childhood mortality, so that a greater proportion of the children born have survived and reached maturity than had been formerly the case. This must have resulted in a relaxation of natural selection against most, if not all, of the mutant genes affecting viability.

While the above result is mainly attributed to factors other than family planning itself, such a consequence would be brought about if genetic losses of children were compensated as a result of family planning. If every family had exactly the same number of children that survived and reached maturity, natural selection due to genetic difference between families would be removed, the component due only to within-sibship difference remaining; in this model the selection rates against harmful genes affecting fitness are reduced to about one half to two thirds of the normal rate.[15] Though such compensation is likely to occur in some families, the effect would be counterbalanced if contraception were employed in other families where the genetic risk is known to be high to the subsequent child. This aspect will be treated in the next section.

Crow[16] has proposed a method for measuring the total selection intensity by an index, I, representing the ratio of the variance in progeny number to the square of its mean, and has shown that I can be separated into two components, I_m and I_f, due to differential mortality and fertility, respectively. Following his expression,

$$I = I_m + \frac{1}{p_s} I_f,$$

and

$$I_m = \frac{p_d}{p_s}, \quad I_f = \frac{V_f}{\bar{x}_s^2},$$

where p_d (counted at birth) is the proportion of premature deaths, p_s is the proportion of those survived and having varying number of progeny, and $\bar{x}_s$ and V_f are the mean and variance of the number of births per surviving parents. Later, using vital statistics data for different populations, Spuhler[17] found that the total selection intensity remains relatively high in industrialized populations; differential fertility seems to keep it high despite the relatively low preadult mortality. However, it is to be noted that the above formulation is based on the assumption that all variation in mortality and fertility has a genetic basis and fitness is completely heritable; the index provides, therefore, not the net intensity of total selection, but only its upper limit. Further, the genetic evaluation for varying values for I_f may be quite different, depending upon whether the population studied uses birth-control measures or is still under natural conditions.[18] Therefore, caution should be taken in the conclusion of the results.

We have seen that the family-planning practice in Japan has resulted in a rapid reduction in the variance of progeny number relative to its mean. In this situation, it would be pertinent to separate the index I_f further into two components, I_f and I_f', the former being the selection intensity due to infertility and the latter due to the variation in the number of children for fertile parents. Thus, the formula may be written as follows:

$$I_f = I_i + \frac{1}{1 - p_0} I_f',$$

$$I_i = \frac{p_0}{1 - p_0}, \quad I_f' = \frac{V_f'}{\bar{x}_s'^2}$$

where p_0 is the proportion of parents having no children, $\bar{x}_s'$ and V_f' are respectively the mean and variance of the number of births per fertile couple. If family planning were practiced only by potentially fertile parents, this would result in a reduction in the value for I_f' but not for I_i, while if otherwise infertile couples could become fertile by medical-care service, the value for I_i would be affected.

In order to show the secular changes in the respective values for I_i and I_f', Table 10 represents some relevant data that were based on the 1960 census report. It should be mentioned that the census report is limited because of the failure to include those married women who died during the reproductive period. The proportion of married women having no children has been slowly decreasing, from about 10% for the women now aged over 60 to less than 8% for those aged 40 to 49, resulting in a gradual decline in the value for I_i from about 0.12 to 0.08, while

Table 10. Secular Changes in Selection Intensity due to Fertility Differences.*

Age of Married Women	$\bar{x}_s$ †	V_f	p_0	I_i	I'_f	I_f
80–	4.670	8.462	0.096	0.106	0.255	0.388
75–79	4.574	8.661	.109	.122	.260	.414
70–74	4.683	8.992	.108	.121	.258	.410
65–69	4.680	8.717	.100	.111	.258	.398
60–64	4.688	8.791	.098	.108	.263	.400
55–59	4.675	8.458	.088	.096	.265	.387
50–54	4.466	7.320	.080	.087	.258	.367
45–49	3.919	5.455	.076	.082	.250	.353
40–44	3.286	3.682	0.077	0.083	0.238	0.341

* Data are based on the 1960 census report in Japan.
† $\bar{x}_s$ = mean No. of children born to married women; V_f = variance in No. of children born to married women; p_0 = proportion of married women without children; I_i = index of selection intensity due to infertility; I'_f = index of selection intensity due to variation in No. of children for women having at least one child; I_f = index of selection intensity due to differential fertility.

the value for I'_f appears to have been almost constant. These data are concerned with married women who had completed reproduction. On the other hand, it is clear from Table 4, where the expression of variance divided by square of the mean is equivalent to I'_f, that the value of I'_f has been steadily declining for the currently reproducing women from 0.47 in 1950 to 0.31 in 1963.

There is no doubt that the above tendency will become more evident in the near future. It is not clear, however, to what extent the observed reductions in these values are a reflection of the reduction in the net intensity of selection. We do know that some cases of infertility are genetically determined; they are mostly resistant to medical cure, so that the net intensity of selection due to infertility seems unlikely to be subjected to relaxation. Since the loss of variation in the number of children among *fertile* women represents social conformity more than genetic homogeneity, the population may scarcely lose its variability as to fecundity and could resume high fertility if social needs required this.

Further Eugenic Application of Birth-Control Measures

The Eugenic Protection Law primarily aims at two quite different objects, the prevention of hereditary diseases and the protection of maternal health. In Table 11 are summarized all reported cases of induced abortions in Japan during five years from 1960 to 1964, classified by the reasons stated in the Table (*Annual Reports for the Eugenic Protection Statistics*, 1960–1964). The total number of cases amounts to about 4.9 million, of which only 4,004 were reported to be "for prevention of hereditary diseases." Taking account of the negligible frequency, it may appear that the law had scarcely contributed to the initial purpose of preventing hereditary diseases. However, there are some reasons to believe that there may be many more cases of induced abortions for some kind of eugenic reasons than stated in the

Table 11. Number of Induced Abortions Reported, 1960 to 1964, by Stated Reason.*

Year	Stated Reason					Total
	Hereditary disease	Leprosy	Maternal health	Violence	Unknown	
1960	1,109	191	1,059,801	310	1,845	1,063,256
1961	995	225	1,031,910	284	1,915	1,035,329
1962	698	85	982,296	226	2,046	985,351
1963	556	93	952,142	166	2,135	955,092
1964	646	99	875,808	243	1,952	878,748
Total	4,004	693	4,901,957	1,229	9,893	4,917,776
%	0.08	0.01	99.68	0.03	0.20	100

* Data are based on statistics from the Ministry of Health and Welfare.

official report. Because of the traditional family system, a hereditary disease in the family is particularly shameful for the Japanese. Further, the distinction in the reasons for "prevention of hereditary diseases" from "protection of maternal health" seems to play for the physician practically no role. These considerations make it very hard to evaluate an official report of this nature.

In countries like Japan, where family planning is practiced on a large scale, it may be hoped that various measures for birth control could readily be used for eugenic purpose. This should be conditioned by popular education on genetic matters of public-health importance and by genetic counseling as an integral part of medical-care services. If the birth of a child with some hereditary disorder could deter the parents from further reproduction, this would reduce not only the absolute number but also the relative frequency of the affected individuals in the population.[19] The opportunity for such selective limitation would be large enough for a number of defects that are recognizable early in infancy. For many other diseases with relatively late onset, the affected individuals may survive and marry, but they would still have the opportunity for using birth control. A preliminary result of our recent attempt to find out whether the parents having had a child with mental defect would limit further reproduction failed to provide evidence that such selection is occurring to an appreciable extent. But we may hope that the increased opportunities for eugenic application of family-planning practice, if guided by proper genetic counseling, could have an important effect upon future generations.

Summary and Conclusions

The recent demographic transition in Japan must have resulted in some reduction in the frequencies of a variety of congenital defects that are correlated with parental age and birth order. These defects may be classified into two categories, one comprising new mutations and the other determined by both genetic and environmental factors. The decrease in the diseases of the latter category may be regarded as dysgenic in the sense that selection against the genes is relaxed, but so long as the pattern of small family size continues the manifestation of those diseases would be prevented. The same transition should in the near future result in a significant reduction in consanguineous marriages and hence to some extent in a reduction in mortality and morbidity. Again, this effect would be dysgenic, because of the increase in the carriers of rare detrimental genes.

The variations in the distribution of family-planning practice, as well as variations in fertility among some social strata, are rapidly decreasing, but because of the lack of our knowledge on many important points, it is difficult to evaluate the extent of the assumed dysgenic effect upon some mental traits that are correlated with social stratification. The reduction in childhood mortality correlated with smaller family size must have resulted in relaxation of selection against most, if not all, of the detrimental genes affecting viability. The rate of infertility has been slowly decreasing, but the net intensity of selection due to infertility seems unlikely to undergo relaxation. The loss of variation in the number of children among potentially fertile women is obvious, but this reflects social conformity more than genetic homogeneity, so that the fecundity of future populations may scarcely be affected.

In the overall assessment of a benefit or harm for future generations, the possible effects upon the immediate or near future must be distinguished from those taking place in the long run. Within the limit of our present knowledge, the balance for the former appears to be far more in the direction of benefit, while for the latter it may be rather in the reverse. We may hope that the increased opportunities of using birth-control means for eugenic purpose, if guided by proper genetic counseling could counterbalance the presumed dysgenic effects.

References

1. *Standardized Vital Rates for All Japan: 1920–1960.* Research Series 155, Tokyo. Ministry of Health and Welfare, Institute of Population Problems, 1963, p. 17.
2. Penrose, L. S. Review of Court Brown, W. M., et al. *Abnormalities of the Sex Chromosome Complement in Man, Ann Hum Genet* **28** : 199–200 (Nov) 1964.
3. Lenz, W. Epidemiologie von Missbildungen, *Pädiatrie Pädologie* **1** : 38–50, 1965
4. Newcombe, H. B., and Tavendale, O. G. Maternal Age and Birth Order Correlations. Problems of Distinguishing Mutational From Environmental Components, *Mutat Res* **1** : 446–467, 1964.
5. Newcombe, H. B. Screening for Effects of Maternal Age and Birth Order in a Register of Handicapped Children, *Ann Hum Genet* **27** : 367–382, 1964.
6. Nei, M., and Imaizumi, Y. *Random Mating and Frequency of Consanguineous Marriages,* Annual Report of the National Institute of Radiological Sciences of Japan (1962), 1963, p. 48.
7. Dahlberg, G. Inbreeding in Man, *Genetics* **14** : 421–454, 1929.
8. *Summary of Eighth National Survey of Family Planning,* Population Problems Series 19, Tokyo: Mainichi Newspapers, Population Problems Research Council, 1965, pp. 32, 57–58.
9. *1960 Population Census of Japan,* Tokyo. Office of the Prime Minister, Bureau of Statistics, 1962, vol. 2, pp. 372–373.
10. Matsunaga, E. Measures Affecting Population Trends and Possible Genetic Consequences, Paper B. 12/I/E/22, read before the United Nations World Population Conference, Belgrade, Yugoslavia, Aug 31, 1965.

11. Schull, W. J., and Neel, J. V. *The Effect of Inbreeding on Japanese Children*, New York: Harper & Row, Publishers, Inc., 1965, pp. 92–96.
12. *Fifth Public Opinion Survey on Birth Control in Japan*, Population Problems Series 16, Tokyo. Mainichi Newspapers, Population Problems Research Council, 1959, p. 21.
13. *Sixth Opinion Survey on Family Planning and Birth Control: A Preliminary Report*, Population Problems Series 18, Tokyo. Mainichi Newspapers, Population Problems Research Council, 1962, pp. 25–26.
14. Aoki, H. Report of the Fourth Fertility Survey in 1962, *J. Population Problems* **90** : 1–54 (March) 1964.
15. King, J. L. The Effect of Litter Culling—or Family Planning—on the Rate of Natural Selection, *Genetics* **51** : 425–429 (March) 1965.
16. Crow, J. F. Some Possibilities for Measuring Selection Intensities in Man, *Hum Biol* **30** : 1–13 (Feb) 1958.
17. Spuhler, J. N. "Empirical Studies on Quantitative Human Genetics," in *Proceedings of the UN/WHO Seminar on the Use of Vital and Health Statistics for Genetic and Radiation Studies*, New York: United Nations, 1962, pp. 241–252.
18. Burgeois-Pichat, J. in discussion, Spuhler, J. N.[17]
19. Goodman, L. A. Some Possible Effects of Birth Control on the Incidence of Disorders and on the Influence of Birth Order, *Ann Hum Genet* **27** : 41–52, 1963.

25   Genetic Effects of Family Planning

John H. Graham, Bernice H. Cohen, and W. J. Schull

We have considered the genetic effects of the widespread use of family-planning services and have concluded that they will have short-run and long-run genetic effects which appear either to be beneficial or to be uncertain and to require further research.

1. SHORT-RUN EFFECTS WHICH RESULT FROM CHANGES IN AGE OF PARENTS, DIFFERENT SPACING OF PREGNANCIES, AND REDUCTION IN FAMILY SIZE. The introduction of family planning tends to decrease the average age of parents in industrialized countries and to increase the average age in less-developed countries. The changes in average age come about through a reduction in the number of children born to older parents in the more-developed countries and to younger parents in the less-developed countries. The genetic effects of this change appear entirely beneficial.

A committee of the American Society of Human Genetics was appointed by its president to advise the Committee on Population Growth and the American Future (established by congressional statute and chaired by John D. Rockefeller, III) on the genetic effects of family planning.

From the *American Journal of Human Genetics*, 24:350–351, 1972. Copyright © 1972 by the American Society of Human Genetics. Reprinted with permission of the author and The University of Chicago Press.

The older the parents, the more apt they are to have children with chromosomal abnormalities (e.g., Down's syndrome, Klinefelter's syndrome) or single gene defects (e.g., chondrodystrophy). Furthermore, it is well established that very young mothers are prone to have children who are more vulnerable to environmental insults, although the reasons for the latter are not clearly genetic.

The introduction of family planning may, but does not necessarily always, result in an increase in the interval between pregnancies. Lengthening the interval is beneficial since the available evidence suggests that there is a direct relation between short birth interval and high fetal, infant, child, and maternal mortality rates. There is also a suggestion that congenital malformations are more frequent when the interval between births is short, although again the genetic contribution to malformation is uncertain.

A third effect of increased use of family planning services is reduction in the number of pregnancies per family and in average family size. These changes have beneficial effects. Certain genetically-related problems which are magnified when there are numerous pregnancies, such as Rh sensitization in Rh-incompatible marriages, are alleviated by having fewer pregnancies per marriage. Consanguinity may be reduced when there is reduction in family size, and if it occurs, the frequency of homozygosity for deleterious autosomal recessive traits (e.g., phenylketonuria, galactosemia) will diminish.

In short, most indications are that widespread use of family-planning services will be genetically beneficial in the short run through reducing the incidence of certain hereditary disorders. The major effect to be expected is fewer children with chromosomal disorders such as Down's syndrome.

2. LONG-RUN CHANGES IN THE GENE POOL. The long-run genetic effects of family planning are much less predictable. The effect upon the frequencies of the various genes in the pool of genes available to the human species at some far distant point in time is a matter of considerable importance. It is not possible to make precise estimates about this important matter. In fact, it is not possible to know which of the genes in the gene pool will be of greatest importance at any distant point in time. Therefore, we regard it as important that the genetic variability of the species be maintained and protected against reduction to a narrow set of genetic possibilities. Narrowing of the range of possibilities might result if family-planning measures were unevenly applied throughout the planet. We believe that variability can best be assured by removing the impediments to use of family-planning services in all countries throughout the world. We cannot emphasize too strongly the importance of making family-planning services cheap, easy to obtain, and as widely available as possible.

3. THE EFFECTS OF CHEMICAL CONTRACEPTIVES. The issue about which we have the most concern is the genetic effect of the widespread use of chemical contraceptives. This is an area of complete ignorance but one which is susceptible of study. We

therefore urge that great attention be given to evaluating the short- and long-run effects of chemical contraceptives both on chromosomes and on the incidence of point mutations.

In summary, we feel that family planning has many genetic advantages in the short run, and that carefully organized research should provide the monitoring needed for protection of the species.

26   The Tragedy of the Commons

Garrett Hardin

At the end of a thoughtful article on the future of nuclear war, Wiesner and York (1) concluded that: "Both sides in the arms race are ... confronted by the dilemma of steadily increasing military power and steadily decreasing national security. *It is our considered professional judgment that this dilemma has no technical solution.* If the great powers continue to look for solutions in the area of science and technology only, the result will be to worsen the situation."

I would like to focus your attention not on the subject of the article (national security in a nuclear world) but on the kind of conclusion they reached, namely that there is no technical solution to the problem. An implicit and almost universal assumption of discussions published in professional and semipopular scientific journals is that the problem under discussion has a technical solution. A technical solution may be defined as one that requires a change only in the techniques of the natural sciences, demanding little or nothing in the way of change in human values or ideas of morality.

In our day (though not in earlier times) technical solutions are always welcome. Because of previous failures in prophecy, it takes courage to assert that a desired technical solution is not possible. Wiesner and York exhibited this courage; publishing in a science journal, they insisted that the solution to the problem was not to be found in the natural sciences. They cautiously qualified their statement with the phrase, "It is our considered professional judgment..." Whether they were right or not is not the concern of the present article. Rather, the concern here is with the important concept of a class of human problems which can be called "no technical solution problems," and, more specifically, with the identification and discussion of one of these.

It is easy to show that the class is not a null class. Recall the game of tick-tack-toe. Consider the problem, "How can I win the game of tick-tack-toe?" It is well known that I cannot, if I assume (in keeping with the conventions of game theory)

From *Science*, **162** (Dec. 13, 1968), 1243–1248. Copyright 1968 by the American Association for the Advancement of Science. Reprinted with permission of the author and the publisher.

that my opponent understands the game perfectly. Put another way, there is no "technical solution" to the problem. I can win only by giving a radical meaning to the word "win." I can hit my opponent over the head; or I can drug him; or I can falsify the records. Every way in which I "win" involves, in some sense, an abandonment of the game, as we intuitively understand it. (I can also, of course, openly abandon the game—refuse to play it. This is what most adults do.)

The class of "no technical solution problems" has members. My thesis is that the "population problem," as conveniently conceived, is a member of this class. How it is conventionally conceived needs some comment. It is fair to say that most people who anguish over the population problem are trying to find a way to avoid the evils of overpopulation without relinquishing any of the privileges they now enjoy. They think that farming the seas or developing new strains of wheat will solve the problem—technologically. I try to show here that the solution they seek cannot be found. The population problem cannot be solved in a technical way, any more than can the problem of winning the game of tick-tack-toe.

What Shall We Maximize?

Population, as Malthus said, naturally tends to grow "geometrically," or, as we would now say, exponentially. In a finite world this means that the per capita share of the world's goods must steadily decrease. Is ours a finite world?

A fair defense can be put forward for the view that the world is infinite; or that we do not know that it is not. But, in terms of the practical problems that we must face in the next few generations with the foreseeable technology, it is clear that we will greatly increase human misery if we do not, during the immediate future, assume that the world available to the terrestrial human population is finite. "Space" is no escape (2).

A finite world can support only a finite population; therefore, population growth must eventually equal zero. (The case of perpetual wide fluctuations above and below zero is a trivial variant that need not be discussed.) When this condition is met, what will be the situation of mankind? Specifically, can Bentham's goal of "the greatest good for the greatest number" be realized?

No—for two reasons, each sufficient by itself. The first is a theoretical one. It is not mathematically possible to maximize for two (or more) variables at the same time. This was clearly stated by von Neumann and Morgenstern (3), but the principle is implicit in the theory of partial differential equations, dating back at least to D'Alembert (1717–1783).

The second reason springs directly from biological facts. To live, any organism must have a source of energy (for example, food). This energy is utilized for two purposes: mere maintenance and work. For man, maintenance of life requires about 1600 kilo-calories a day ("maintenance calories"). Anything that he does over and above merely staying alive will be defined as work, and is supported by "work calories" which he takes in. Work calories are used not only for what we call work in common speech; they are also required for all forms of enjoyment, from swimming and automobile racing to playing music and writing poetry. If our goal is to maximize population it is obvious what we must do: We must make the work calories per person approach as close to zero as possible. No gourmet meals, no vacations, no sports, no music, no literature, no art. ... I think that everyone will grant, without argument or proof, that maximizing population does not maximize goods. Bentham's goal is impossible.

In reaching this conclusion I have made the usual assumption that it is the acquisition of energy that is the problem. The appearance of atomic energy has led some to question this assumption. However, given an infinite source of energy, population growth still produces an inescapable problem. The problem of the acquisition of energy is replaced by the problem of its dissipation, as J. H. Fremlin has so wittily shown (4). The arithmetic signs in the analysis are, as it were, reversed; but Bentham's goal is still unobtainable.

The optimum population is, then, less than the maximum. The difficulty of defining the optimum is enormous; so far as I know, no one has seriously tackled this problem. Reaching an acceptable and stable solution will surely require more than one generation of hard analytical work—and much persuasion.

We want the maximum good per person; but what is good? To one person it is wilderness, to another it is ski lodges for thousands. To one it is estuaries to nourish ducks for hunters to shoot; to another it is factory land. Comparing one good with another is, we usually say, impossible because goods are incommensurable. Incommensurables cannot be compared.

Theoretically this may be true; but in real life incommensurables *are* commensurable. Only a criterion of judgment and a system of weighting are needed. In nature the criterion is survival. Is it better for a species to be small and hideable, or large and powerful? Natural selection commensurates the incommensurables. The compromise achieved depends on a natural weighting of the values of the variables.

Man must imitate this process. There is no doubt that in fact he already does, but unconsciously. It is when the hidden decisions are made explicit that the arguments begin. The problem for the years ahead is to work out an acceptable theory of weighting. Synergistic effects, nonlinear variation, and difficulties in discounting the future make the intellectual problem difficult, but not (in principle) insoluble.

Has any cultural group solved this practical problem at the present time, even on an intuitive level? One simple fact proves that none has: there is no prosperous population in the world today that has, and has had for some time, a growth rate of zero. Any people that has intuitively identified its optimum point will soon reach it, after which its growth rate becomes and remains zero.

Of course, a positive growth rate might be taken as evidence that a population is below its optimum. However, by any reasonable standards, the most rapidly growing populations on earth today are (in general) the most miserable. This association (which need not be invariable) casts doubt on the optimistic assumption that the positive growth rate of a population is evidence that it has yet to reach its optimum.

We can make little progress in working toward optimum population size until we explicitly exorcize the spirit of Adam Smith in the field of practical demography. In economic affairs, *The Wealth of Nations* (1776) popularized the "invisible hand," the idea that an individual who "intends only his own gain," is, as it were, "led by an invisible hand to promote ... the public interest" (5). Adam Smith did not assert that this was invariably true, and perhaps neither did any of his followers. But he contributed to a dominant tendency of thought that has ever since interfered with positive action based on rational analysis, namely, the tendency to assume that decisions reached individually will, in fact, be the best decisions for an entire society. If this assumption is correct it justifies the continuance of our present policy of laissez-faire in reproduction. If it is correct we can assume that men will control their individual fecundity so as to produce the optimum population. If the assumption is not correct, we need to reexamine our individual freedoms to see which ones are defensible.

Tragedy of Freedom in a Commons

The rebuttal to the invisible hand in population control is to be found in a scenario first sketched in a little-known pamphlet (6) in 1833 by a mathematical amateur named William Forster Lloyd (1794–1852). We may well call it "the tragedy of the commons," using the word "tragedy" as the philosopher Whitehead used it (7): "The essence of dramatic tragedy is not unhappiness. It resides in the solemnity of the remorseless working of things." He then goes on to say, "This inevitableness of destiny can only be illustrated in terms of human life by incidents which in fact involve unhappiness. For it is only by them that the futility of escape can be made evident in the drama."

The tragedy of the commons develops in this way. Picture a pasture open to all. It is to be expected that each herdsman will try to keep as many cattle as possible on the commons. Such an arrangement may work reasonably satisfactorily for centuries because tribal wars, poaching, and disease keep the numbers of both man and beast well below the carrying capacity of the land. Finally, however, comes the day of reckoning, that is, the day when the long-desired goal of social stability becomes a reality. At this point, the inherent logic of the commons remorselessly generates tragedy.

As a rational being, each herdsman seeks to maximize his gain. Explicitly or implicitly, more or less consciously, he asks, "What is the utility *to me* of adding one more animal to my herd?" This utility has one negative and one positive component.

(1) The positive component is a function of the increment of one animal. Since the herdsman receives all the proceeds from the sale of the additional animal, the positive utility is nearly +1.

(2) The negative component is a function of the additional overgrazing created by one more animal. Since, however, the effects of overgrazing are shared by all the herdsmen, the negative utility for any particular decision-making herdsman is only a fraction of −1.

Adding together the component partial utilities, the rational herdsman concludes that the only sensible course for him to pursue is to add another animal to his herd.

And another; and another. . . . But this is the conclusion reached by each and every rational herdsman sharing a commons. Therein is the tragedy. Each man is locked into a system that compels him to increase his herd without limit—in a world that is limited. Ruin is the destination toward which all men rush, each pursuing his own best interest in a society that believes in the freedom of the commons. Freedom in a commons brings ruin to all.

Some would say that this is a platitude. Would that it were! In a sense, it was learned thousands of years ago, but natural selection favors the forces of psychological denial (8). The individual benefits as an individual from his ability to deny the truth even though society as a whole, of which he is a part, suffers. Education can counteract the natural tendency to do the wrong thing, but the inexorable succession of generations requires that the basis for this knowledge be constantly refreshed.

A simple incident that occurred a few years ago in Leominster, Massachusetts, shows how perishable the knowledge is. During the Christmas shopping season the parking meters downtown were covered with plastic bags that bore tags reading: "Do not open until after Christmas. Free parking courtesy of the mayor and city council." In other words, facing the prospect of an increased demand for already scarce space, the city fathers reinstituted the system of the commons. (Cynically, we suspect that they gained more votes than they lost by this retrogressive act.)

In an approximate way, the logic of the commons has been understood for a long time, perhaps since the discovery of agriculture or the invention of private property in real estate. But it is understood mostly only in special cases which are not sufficiently generalized. Even at this late date, cattlemen leasing national land on the western ranges demonstrate no more than an ambivalent understanding, in constantly pressuring federal authorities to increase the head count to the point where overgrazing produces erosion and weed-dominance. Likewise, the oceans of the world continue to suffer from the survival of the philosophy of the commons. Maritime nations still respond automatically to the shibboleth of the "freedom of the seas." Professing to believe in the "inexhaustible resources of the oceans," they bring species after species of fish and whales closer to extinction (9).

The National Parks present another instance of the working out of the tragedy of the commons. At present, they are open to all, without limit. The parks themselves are limited in extent—there is only one Yosemite Valley—whereas population seems to grow without limit. The values that visitors seek in the parks are steadily eroded. Plainly, we must soon cease to treat the parks as commons or they will be of no value to anyone.

What shall we do? We have several options. We might sell them off as private property. We might keep them as public property, but allocate the right to enter them. The allocation might be on the basis of wealth, by the use of an auction system. It might be on the basis of merit, as defined by some agreed-upon standards. It might be by lottery. Or it might be on a first-come, first-served basis, administered to long queues. These, I think, are all the reasonable possibilities. They are all objectionable. But we must choose—or acquiesce in the destruction of the commons that we call our National Parks.

Pollution

In a reverse way, the tragedy of the commons reappears in problems of pollution. Here it is not a question of taking something out of the commons, but of putting something in—sewage, or chemical, radioactive, and heat wastes into water; noxious and dangerous fumes into the air; and distracting and unpleasant advertising signs into the line of sight. The calculations of utility are much the same as before. The rational man finds that his share of the cost of the wastes he discharges into the commons is less than the cost of purifying his wastes before releasing them. Since this is true for everyone, we are locked into a system of "fouling our own nest," so long as we behave only as independent, rational, free-enterprisers.

The tragedy of the commons as a food basket is averted by private property, or something formally like it. But the air and waters surrounding us cannot readily be fenced, and so the tragedy of the commons as a cesspool must be prevented by different means, by coercive laws or taxing devices that make it cheaper for the polluter to treat his pollutants than to discharge them untreated. We have not progressed as far with the solution of this problem as we have with the first. Indeed, our particular concept of private property, which deters us from exhausting the positive resources of the earth, favors pollution. The owner of a factory on the bank of a stream—whose property extends to the middle of the stream—often has difficulty seeing why it is not his natural right to muddy the waters flowing past his door. The law, always behind the times, requires elaborate stitching and fitting to adapt it to this newly perceived aspect of the commons.

The pollution problem is a consequence of population. It did not much matter how a lonely American frontiersman disposed of his waste. "Flowing water purifies itself every 10 miles," my grandfather used to say, and the myth was near enough to the truth when he was a boy, for there were not too many people. But as population became denser, the natural chemical and biological recycling processes became overloaded, calling for a redefinition of property rights.

How to Legislate Temperance?

Analysis of the pollution problem as a function of population density uncovers a not generally recognized principle of morality, namely: *the morality of an act is a function of the state of the system at the time it is performed (10)*. Using the commons as a cesspool does not harm the general public under frontier conditions, because there is no public; the same behavior in a metropolis is unbearable. A hundred and fifty years ago a plainsman could kill an American bison, cut out only the tongue for his dinner, and discard the rest of the animal. He was not in any important sense being wasteful. Today, with only a few thousand bison left, we would be appalled at such behavior.

In passing, it is worth noting that the morality of an act cannot be determined from a photograph. One does not know whether a man killing an elephant or setting fire to the grassland is harming others until one knows the total system in which his act appears. "One picture is worth a thousand words," said an ancient Chinese;

but it may take 10,000 words to validate it. It is as tempting to ecologists as it is to reformers in general to try to persuade others by way of the photographic shortcut. But the essence of an argument cannot be photographed: it must be presented rationally—in words.

That morality is system-sensitive escaped the attention of most codifiers of ethics in the past. "Thou shalt not ..." is the form of traditional ethical directives which make no allowance for particular circumstances. The laws of our society follow the pattern of ancient ethics, and therefore are poorly suited to governing a complex, crowded, changeable world. Our epicyclic solution is to augment statutory law with administrative law. Since it is practically impossible to spell out all the conditions under which it is safe to burn trash in the back yard or to run an automobile without smog-control, by law we delegate the details to bureaus. The result is administrative law, which is rightly feared for an ancient reason—*Quis custodiet ipsos custodes?*— "Who shall watch the watchers themselves?" John Adams said that we must have "a government of laws and not men." Bureau administrators, trying to evaluate the morality of acts in the total system, are singularly liable to corruption, producing a government by men, not laws.

Prohibition is easy to legislate (though not necessarily to enforce); but how do we legislate temperance? Experience indicates that it can be accomplished best through the mediation of administrative law. We limit possibilities unnecessarily if we suppose that the sentiment of *Quis custodiet* denies us the use of administrative law. We should rather retain the phrase as a perpetual reminder of fearful dangers we cannot avoid. The great challenge facing us now is to invent the corrective feedbacks that are needed to keep custodians honest. We must find ways to legitimate the needed authority of both the custodians and the corrective feedbacks.

Freedom to Breed Is Intolerable

The tragedy of the commons is involved in population problems in another way. In a world governed solely by the principle of "dog eat dog"—if indeed there ever was such a world—how many children a family had would not be a matter of public concern. Parents who bred too exuberantly would leave fewer descendants, not more, because they would be unable to care adequately for their children. David Lack and others have found that such a negative feedback demonstrably controls the fecundity of birds (*11*). But men are not birds, and have not acted like them for millenniums, at least.

If each human family were dependent only on its own resources; *if* the children of improvident parents starved to death; *if*, thus, overbreeding brought its own "punishment" to the germ line—*then* there would be no public interest in controlling the breeding of families. But our society is deeply committed to the welfare state (*12*), and hence is confronted with another aspect of the tragedy of the commons.

In a welfare state, how shall we deal with the family, the religion, the race, or the class (or indeed any distinguishable and cohesive group) that adopts overbreeding as a policy to secure its own aggrandizement (*13*)? To couple the concept of freedom to breed with the belief that everyone born has an equal right to the commons is to lock the world into a tragic course of action.

Unfortunately this is just the course of action that is being pursued by the United Nations. In late 1967, some 30 nations agreed to the following (14):

> The Universal Declaration of Human Rights describes the family as the natural and fundamental unit of society. It follows that any choice and decision with regard to the size of the family must irrevocably rest with the family itself, and cannot be made by anyone else.

It is painful to have to deny categorically the validity of this right; denying it, one feels as uncomfortable as a resident of Salem, Massachusetts, who denied the reality of witches in the 17th century. At the present time, in liberal quarters, something like a taboo acts to inhibit criticism of the United Nations. There is a feeling that the United Nations is "our last and best hope," that we shouldn't find fault with it; we shouldn't play into the hands of the archconservatives. However, let us not forget what Robert Louis Stevenson said: "The truth that is suppressed by friends is the readiest weapon of the enemy." If we love the truth we must openly deny the validity of the Universal Declaration of Human Rights, even though it is promoted by the United Nations. We should also join with Kingsley Davis (15) in attempting to get Planned Parenthood-World Population to see the error of its ways in embracing the same tragic ideal.

Conscience Is Self-eliminating

It is a mistake to think that we can control the breeding of mankind in the long run by an appeal to conscience. Charles Galton Darwin made this point when he spoke on the centennial of the publication of his grandfather's great book. The argument is straightforward and Darwinian.

People vary. Confronted with appeals to limit breeding, some people will undoubtedly respond to the plea more than others. Those who have more children will produce a larger fraction of the next generation than those with more susceptible consciences. The difference will be accentuated, generation by generation.

In C. G. Darwin's words: "It may well be that it would take hundreds of generations for the progenitive instinct to develop in this way, but if it should do so, nature would have taken her revenge, and the variety *Homo contracipiens* would become extinct and would be replaced by the variety *Homo progenitivus*" (16).

The argument assumes that conscience or the desire for children (no matter which) is hereditary—but hereditary only in the most general formal sense. The result will be the same whether the attitude is transmitted through germ cells, or exosomatically, to use A. J. Lotka's term. (If one denies the latter possibility as well as the former, then what's the point of education?) The argument has here been stated in the context of the population problem, but it applies equally well to any instance in which society appeals to an individual exploiting a commons to restrain himself for the general good—by means of his conscience. To make such an appeal is to set up a selective system that works towards the elimination of conscience from the race.

Pathogenic Effects of Conscience

The long-term disadvantage of an appeal to conscience should be enough to condemn it; but it has serious short-term disadvantages as well. If we ask a man who is exploiting a commons to desist "in the name of conscience," what are we saying to him? What does he hear?—not only at the moment but also in the wee small hours of the night when, half asleep, he remembers not merely the words we used but also the nonverbal communication cues we gave him unawares? Sooner or later, consciously or subconsciously, he senses that he has received two communications, and that they are contradictory: (i) (intended communication) "If you don't do as we ask, we will openly condemn you for not acting like a responsible citizen"; (ii) (the unintended communication) "If you *do* behave as we ask, we will secretly condemn you for a simpleton who can be shamed into standing aside while the rest of us exploit the commons."

Everyman then is caught in what Bateson has called a "double bind." Bateson and his co-workers have made a plausible case for viewing the double bind as an important causative factor in the genesis of schizophrenia (17). The double bind may not always be so damaging, but it always endangers the mental health of anyone to whom it is applied. "A bad conscience," said Nietzche, "is a kind of illness."

To conjure up a conscience in others is tempting to anyone who wishes to extend his control beyond the legal limits. Leaders at the highest level succumb to this temptation. Has any President during the past generation failed to call on labor unions to moderate voluntarily their demands for higher wages, or to steel companies to honor voluntary guidelines on prices? I can recall none. The rhetoric used on such occasions is designed to produce feelings of guilt in noncooperators.

For centuries it was assumed without proof that guilt was a valuable, perhaps even an indispensable, ingredient of the civilized life. Now, in this post-Freudian world, we doubt it.

Paul Goodman speaks from the modern point of view when he says: "No good has ever come from feeling guilty, neither intelligence, policy, nor compassion. The guilty do not pay attention to the object but only to themselves, and not even to their own interests, which might make sense, but to their anxieties" (18).

One does not have to be a professional psychiatrist to see the consequences of anxiety. We in the Western world are just emerging from a dreadful two-centuries-long Dark Ages of Eros that was sustained partly by prohibition laws, but perhaps more effectively by the anxiety-generating mechanisms of education. Alex Comfort has told the story well in *The Anxiety Makers* (19); it is not a pretty one.

Since proof is difficult, we may even concede that the results of anxiety may sometimes, from certain points of view, be desirable. The larger question we should ask is whether, as a matter of policy, we should ever encourage the use of a technique the tendency (if not the intention) of which is psychologically pathogenic. We hear much talk these days of responsible parenthood; the coupled words are incorporated into the titles of some organizations devoted to birth control. Some people have proposed massive propaganda campaigns to instill responsibility into the nation's (or the world's) breeders. But what is the meaning of the word responsibility in this context? Is it not merely a synonym for the word conscience? When we use the word responsibility in the absence of substantial sanctions are we not trying to browbeat a free man in a commons into acting against his own interest? Responsibility is a verbal counterfeit for a substantial *quid pro quo*. It is an attempt to get something for nothing.

If the word responsibility is to be used at all, I suggest that it be in the sense Charles Frankel uses it (*20*). " Responsibility," says this philosopher, "is the product of definite social arrangements." Notice that Frankel calls for social arrangements—not propaganda.

Mutual Coercion Mutually Agreed Upon

The social arrangements that produce responsibility are arrangements that create coercion, of some sort. Consider bank-robbing. The man who takes money from a bank acts as if the bank were a commons. How do we prevent such action? Certainly not by trying to control his behavior solely by a verbal appeal to his sense of responsibility. Rather than rely on propaganda we follow Frankel's lead and insist that a bank is not a commons; we seek the definite social arrangements that will keep it from becoming a commons. That we thereby infringe on the freedom of would-be robbers we neither deny nor regret.

The morality of bank-robbing is particularly easy to understand because we accept complete prohibition of this activity. We are willing to say " Thou shalt not rob banks," without providing for exceptions. But temperance also can be created by coercion. Taxing is a good coercive device. To keep downtown shoppers temperate in their use of parking space we introduce parking meters for short periods, and traffic fines for longer ones. We need not actually forbid a citizen to park as long as he wants to; we need merely make it increasingly expensive for him to do so. Not prohibition, but carefully biased options are what we offer him. A Madison Avenue man might call this persuasion; I prefer the greater candor of the word coercion.

Coercion is a dirty word to most liberals now, but it need not forever be so. As with the four-letter words, its dirtiness can be cleansed away by exposure to the light, by saying it over and over without apology or embarrassment. To many, the word coercion implies arbitrary decisions of distant and irresponsible bureaucrats; but this is not a necessary part of its meaning. The only kind of coercion I recommend is mutual coercion, mutually agreed upon by the majority of the people affected.

To say that we mutually agree to coercion is not to say that we are required to enjoy it, or even to pretend we enjoy it. Who enjoys taxes? We all grumble about them. But we accept compulsory taxes because we recognize that voluntary taxes would favor the conscienceless. We institute and (grumblingly) support taxes and other coercive devices to escape the horror of the commons.

An alternative to the commons need not be perfectly just to be preferable. With real estate and other material goods, the alternative we have chosen is the institution of private property coupled with legal inheritance. Is this system perfectly just? As a genetically trained biologist I deny that it is. It seems to me that, if there are to be differences in individual inheritance, legal possession should be perfectly correlated with biological inheritance—that those who are biologically more fit to be the custodians of property and power should legally inherit more. But genetic recombination continually makes a mockery of the doctrine of "like father, like son" implicit in our laws of legal inheritance. An idiot can inherit millions, and a trust fund can keep his estate intact. We must admit that our legal system of private property plus inheritance is unjust—but we put up with it because we are not convinced, at the moment, that anyone has invented a better system. The alternative of the commons is too horrifying to contemplate. Injustice is preferable to total ruin.

It is one of the peculiarities of the warfare between reform and status quo that it is thoughtlessly governed by a double standard. Whenever a reform measure is proposed it is often defeated when its opponents triumphantly discover a flaw in it. As Kingsley Davis has pointed out (21), worshippers of the status quo sometimes imply that no reform is possible without unanimous agreement, an implication contrary to historical fact. As nearly as I can make out, automatic rejection of proposed reforms is based on one of two unconscious assumptions: (i) that the status quo is perfect; or (ii) that the choice we face is between reform and no action; if the proposed reform is imperfect, we presumably should take no action at all, while we wait for a perfect proposal.

But we can never do nothing. That which we have done for thousands of years is also action. It also produces evils. Once we are aware that the status quo is action, we can then compare its discoverable advantages and disadvantages with the predicted advantages and disadvantages of the proposed reform, discounting as best we can for our lack of experience. On the basis of such a comparison, we can make a rational decision which will not involve the unworkable assumption that only perfect systems are tolerable.

Recognition of Necessity

Perhaps the simplest summary of this analysis of man's population problems is this: the commons, if justifiable at all, is justifiable only under conditions of low-population density. As the human population has increased, the commons has had to be abandoned in one aspect after another.

First we abandoned the commons in food gathering, enclosing farm land and restricting pastures and hunting and fishing areas. These restrictions are still not complete throughout the world.

Somewhat later we saw that the commons as a place for waste disposal would also have to be abandoned. Restrictions on the disposal of domestic sewage are widely accepted in the Western world; we are still struggling to close the commons to pollution by automobiles, factories, insecticide sprayers, fertilizing operations, and atomic energy installations.

In a still more embryonic state is our recognition of the evils of the commons in matters of pleasure. There is almost no restriction on the propagation of sound waves in the public medium. The shopping public is assaulted with mindless music, without its consent. Our government is paying out billions of dollars to create supersonic transport which will disturb 50,000 people for every one person who is whisked from coast to coast 3 hours faster. Advertisers muddy the airwaves of radio and television and pollute the view of travelers. We are a long way from outlawing the commons in matters of pleasure. Is this because our Puritan inheritance makes us view pleasure as something of a sin, and pain (that is, the pollution of advertising) as the sign of virtue?

Every new enclosure of the commons involves the infringement of somebody's personal liberty. Infringements made in the distant past are accepted because no contemporary complains of a loss. It is the newly proposed infringements that we vigorously oppose; cries of "rights" and "freedom" fill the air. But what does "freedom" mean? When men mutually agreed to pass laws against robbing, mankind became more free, not less so. Individuals locked into the logic of the commons are free only to bring on universal ruin; once they see the necessity of mutual coercion, they become free to pursue other goals. I believe it was Hegel who said, "Freedom is the recognition of necessity."

The most important aspect of necessity that we must now recognize, is the necessity of abandoning the commons in breeding. No technical solution can rescue us from the misery of overpopulation. Freedom to breed will bring ruin to all. At the moment, to avoid hard decisions many of us are tempted to propagandize for conscience and responsible parenthood. The temptation must be resisted, because an appeal to independently acting consciences selects for the disappearance of all conscience in the long run, and an increase in anxiety in the short.

The only way we can preserve and nurture other and more precious freedoms is by relinquishing the freedom to breed, and that very soon. "Freedom is the recognition of necessity"—and it is the role of education to reveal to all the necessity of abandoning the freedom to breed. Only so, can we put an end to this aspect of the tragedy of the commons.

References

1. J. B. Wiesner and H. F. York. *Sci. Amer.* **211** (No. 4), 27 (1964).
2. G. Hardin. *J. Hered.* **50,** 68 (1959); S. von Hoernor, *Science* **137,** 18 (1962).
3. J. von Neumann and O. Morgenstern. *Theory of Games and Economic Behavior* (Princeton Univ. Press, Princeton, N.J., 1947), p. 11.
4. J. H. Fremlin, *New Sci.*, No. 415 (1964), p. 285.
5. A. Smith. *The Wealth of Nations* (Modern Library, New York, 1937), p. 423.
6. W. F. Lloyd. *Two Lectures on the Checks to Population* (Oxford Univ. Press, Oxford, England, 1833), reprinted (in part) in *Population, Evolution, and Birth Control*, G. Hardin, Ed. (Freeman, San Francisco, 1964), p. 37.
7. A. N. Whitehead. *Science and the Modern World* (Mentor, New York, 1948), p. 17.

8. G. Hardin, Ed. *Population, Evolution, and Birth Control* (Freeman, San Francisco, 1964), p. 56.
9. S. McVay. *Sci. Amer.* **216** (No. 8), 13 (1966).
10. J. Fletcher. *Situation Ethics* (Westminster, Philadelphia, 1966).
11. D. Lack. *The Natural Regulation of Animal Numbers* (Clarendon Press, Oxford, 1954).
12. H. Girvetz. *From Wealth to Welfare* (Stanford Univ. Press, Stanford, Calif., 1950).
13. G. Hardin. *Perspec. Biol. Med.* **6,** 366 (1963).
14. U. Thant. *Int. Planned Parenthood News,* No. 168 (February 1968), p. 3.
15. K. Davis. *Science* **158,** 730 (1967).
16. S. Tax, Ed. *Evolution after Darwin* (Univ. of Chicago Press, Chicago, 1960), vol. 2, p. 469.
17. G. Bateson, D. D. Jackson, J. Haley, J. Weakland. *Behav. Sci.* **1,** 251 (1956).
18. P. Goodman. *New York Rev. Books* **10** (8), 22 (23 May, 1968).
19. A. Comfort. *The Anxiety Makers* (Nelson, London, 1967).
20. C. Frankel. *The Case for Modern Man* (Harper, New York, 1955), p. 203.
21. J. D. Roslansky. *Genetics and the Future of Man* (Appleton-Century-Crofts, New York, 1966), p. 177.

27   The Tragedy of the Commons Revisited

Beryl L. Crowe

There has developed in the contemporary natural sciences a recognition that there is a subset of problems, such as population, atomic war, and environmental corruption, for which there are no technical solutions (*1, 2*). There is also an increasing recognition among contemporary social scientists that there is a subset of problems, such as population, atomic war, environmental corruption, and the recovery of a livable urban environment, for which there are no current political solutions (*3*). The thesis of this article is that the common area shared by these two subsets contains most of the critical problems that threaten the very existence of contemporary man.

The importance of this area has not been raised previously because of the very structure of modern society. This society, with its emphasis on differentiation and specialization, has led to the development of two insular scientific communities— the natural and the social—between which there is very little communication and a great deal of envy, suspicion, disdain, and competition for scarce resources. Indeed, these two communities more closely resemble tribes living in close geographic proximity on university campuses than they resemble the "scientific culture" that

From *Science,* **166** (Nov. 28, 1969), 1103–1107. Copyright 1969 by the American Association for the Advancement of Science. Reprinted with permission of the author and the publisher.

C. P. Snow placed in contrast to and opposition to the "humanistic culture" (4).

Perhaps the major problems of modern society have, in large part, been allowed to develop and intensify through this structure of insularity and specialization because it serves both psychological and professional functions for both scientific communities. Under such conditions, the natural sciences can recognize that some problems are not technically soluble and relegate them to the nether land of politics, while the social sciences recognize that some problems have no current political solutions and then postpone a search for solutions while they wait for new technologies with which to attack the problem. Both sciences can thus avoid responsibility and protect their respective myths of competence and relevance, while they avoid having to face the awesome and awful possibility that each has independently isolated the same subset of problems and given them different names. Thus, both never have to face the consequences of their respective findings. Meanwhile, due to the specialization and insularity of modern society, man's most critical problems lie in limbo, while the specialists in problem-solving go on to less critical problems for which they can find technical or political solutions.

In this circumstance, one psychologically brave, but professionally foolhardy soul, Garrett Hardin, has dared to cross the tribal boundaries in his article "The tragedy of the commons" (1). In it, he gives vivid proof of the insularity of the two scientific tribes in at least two respects: first, his "rediscovery" of the tragedy was in part wasted effort, for the knowledge of this tragedy is so common in the social sciences that it has generated some fairly sophisticated mathematical models (5); second, the recognition of the existence of a subset of problems for which science neither offers nor aspires to offer technical solutions is not likely, under the contemporary conditions of insularity, to gain wide currency in the social sciences. Like Hardin, I will attempt to avoid the psychological and professional benefits of this insularity by tracing some of the political and social implications of his proposed solution to the tragedy of the commons.

The commons is a fundamental social institution that has a history going back through our own colonial experience to a body of English common law which antedates the Roman conquest. That law recognized that in societies there are some environmental objects which have never been, and should never be, exclusively appropriated to any individual or group of individuals. In England the classic example of the commons is the pasturage set aside for public use, and the "tragedy of the commons" to which Hardin refers was a tragedy of overgrazing and lack of care and fertilization which resulted in erosion and underproduction so destructive that there developed in the late 19th century an enclosure movement. Hardin applies this social institution to other environmental objects such as water, atmosphere, and living space.

The cause of this tragedy is exposed by a very simple mathematical model, utilizing the concept of utility drawn from economics. Allowing the utilities to range between a positive value of 1, and a negative value of 1, we may ask, as did the individual English herdsman, what is the utility to me of adding one mory animal to my herd that grazes on the commons? His answer is that the positive utility is near 1 and the negative utility is only a fraction of minus 1. Adding together the component partial utilities, the herdsman concludes that it is rational for him to add another animal to his herd; then another, and so on. The tragedy to which Hardin refers develops because the same rational conclusion is reached by each and every herdsman sharing the commons.

Assumptions Necessary to Avoid the Tragedy

In passing the technically insoluble problems over to the political and social realm for solution, Hardin has made three critical assumptions: (i) that there exists, or can be developed, a "criterion of judgment and a system of weighting ..." that will "render the incommensurables ... commensurable ..." in real life; (ii) that, possessing this criterion of judgment, "coercion can be mutually agreed upon," and that the application of coercion to effect a solution to problems will be effective in modern society; and (iii) that the administrative system, supported by the criterion of judgment and access to coercion, can and will protect the commons from further desecration.

If all three of these assumptions were correct, the tragedy which Hardin has recognized would dissolve into a rather facile melodrama of setting up administrative agencies. I believe these three assumptions are so questionable in contemporary society that a tragedy remains in the full sense in which Hardin used the term. Under contemporary conditions, the subset of technically insoluble problems is also politically insoluble, and thus we witness a full-blown tragedy wherein "the essence of dramatic tragedy is not unhappiness. It resides in the remorseless working of things."

The remorseless working of things in modern society is the erosion of three social myths which form the basis for Hardin's assumptions, and this erosion is proceeding at such a swift rate that perhaps the myths can neither revitalize nor reformulate in time to prevent the "population bomb" from going off, or before an accelerating "pollution immersion," or perhaps even an "atomic fallout."

Eroding Myth of the Common Value System

Hardin is theoretically correct, from the point of view of the behavioral sciences, in his argument that "in real life incommensurables *are* commensurable." He is, moreover, on firm ground in his assertion that to fulfill this condition in real life one needs only "a criterion of judgment and a system of weighting." In real life, however, values are the criteria of judgment, and the system of weighting is dependent upon the ranging of a number of conflicting values in a hierarchy. That such a system of values exists beyond the confines of the nation-state is hardly tenable. At this point in time one is more likely to find such a system of values within the boundaries of the nation-state. Moreover, the nation-state is the only political unit of sufficient dimension to find and enforce political solutions to Hardin's subset of "technically insoluble problems." It is on this political unit that we will fix our attention.

In America there existed, until very recently, a set of conditions which perhaps made the solution to Hardin's problem subset possible: we lived with the myth that we were "one people, indivisible...." This myth postulated that we were the great "melting pot" of the world wherein the diverse cultural ores of Europe were poured into the crucible of the frontier experience to produce a new alloy—an American civilization. This new civilization was presumably united by a common value system that was democratic, equalitarian, and existing under universally enforceable rules contained in the Constitution and the Bill of Rights.

In the United States today, however, there is emerging a new set of behavior patterns which suggest that the myth is either dead or dying. Instead of believing and behaving in accordance with the myth, large sectors of the population are developing life-styles and value hierarchies that give contemporary Americans an appearance more closely analogous to the particularistic, primitive forms of "tribal" organizations living in geographic proximity than to that shining new alloy, the American civilization.

With respect to American politics, for example, it is increasingly evident that the 1960 election was the last election in the United States to be played out according to the rules of pluralistic politics in a two-party system. Certainly 1964 was, even in terms of voting behavior, a contest between the larger tribe that was still committed to the pluralistic model of compromise and accommodation within a winning coalition, and an emerging tribe that is best seen as a millennial revitalization movement directed against mass society—a movement so committed to the revitalization of old values that it would rather lose the election than compromise its values. Under such circumstances former real-life commensurables within the Republican Party suddenly became incommensurable.

In 1968 it was the Democratic Party's turn to suffer the degeneration of commensurables into incommensurables as both the Wallace tribe and the McCarthy tribe refused to play by the old rules of compromise, accommodation, and exchange of interests. Indeed, as one looks back on the 1968 election, there seems to be a common theme in both these camps—a theme of return to more simple and direct participation in decision-making that is only possible in the tribal setting. Yet, despite this similarity, both the Wallaceites and the McCarthyites responded with a value perspective that ruled out compromise and they both demanded a drastic change in the dimension in which politics is played. So firm were the value commitments in both of these tribes that neither (as was the case with the Goldwater forces in 1964) was willing to settle for a modicum of power that could accrue through the processes of compromise with the national party leadership.

Still another dimension of this radical change in behavior is to be seen in the black community where the main trend of the argument seems to be, not in the direction of accommodation, compromise, and integration, but rather in the direction of fragmentation from the larger community, intransigence in the areas where black values and black culture are concerned, and the structuring of a new community of like-minded and like-colored people. But to all appearances even the concept of color is not enough to sustain commensurables in their emerging community as it fragments into religious nationalism, secular nationalism, integrationists, separationists, and so forth. Thus those problems which were commensurable, both interracial and intraracial, in the area of integration become incommensurable in the era of Black Nationalism.

Nor can the growth of commensurable views be seen in the contemporary youth movements. On most of the American campuses today there are at least ten tribes involved in "tribal wars" among themselves and against the "imperialistic" power of those "over 30." Just to tick them off, without any attempt to be comprehensive, there are: the up-tight protectors of the status quo who are looking for middle-class union cards, the revitalization movements of the Young Americans for Freedom, the reformists of pluralism represented by the Young Democrats and the Young Republicans, those committed to New Politics, the Students for a Democratic Society, the Yippies, the Flower Children, the Black Students Union, and the Third World Liberation Front. The critical change in this instance is not the rise of new groups; this is expected within the pluralistic model of politics. What is new are value positions assumed by these groups which lead them to make demands, not as points for bargaining and compromise with the opposition, but rather as points which are "not negotiable." Hence, they consciously set the stage for either confrontation or surrender, but not for rendering incommensurables commensurable.

Moving out of formalized politics and off the campus, we see the remnants of the "hippie" movement which show clear-cut tribal overtones in their commune movements. This movement has, moreover, already fragmented into an urban tribe which can talk of guerrilla warfare against the city fathers, while another tribe finds accommodation to urban life untenable without sacrificing its values and therefore moves out to the "Hog Farm," "Morning Star," or "Big Sur." Both hippie tribes have reduced the commensurables with the dominant WASP tribe to the point at which one of the cities on the Monterey Peninsula felt sufficiently threatened to pass a city ordinance against sleeping in trees, and the city of San Francisco passed a law against sitting on sidewalks.

Even among those who still adhere to the pluralistic middle-class American image, we can observe an increasing demand for a change in the dimension of life and politics that has disrupted the elementary social processes: the demand for neighborhood (tribal?) schools, control over redevelopment projects, and autonomy in the setting and payment of rents to slumlords. All of these trends are more suggestive of tribalism than of the growth of the range of commensurables with respect to the commons.

We are, moreover, rediscovering other kinds of tribes in some very odd ways. For example, in the educational process, we have found that one of our best empirical measures in terms both of validity and reproducibility—the I.Q. test—is a much better measure of the existence of different linguistic tribes than it is a measure of "native intellect" (6). In the elementary school, the different languages and different values of these diverse tribal children have even rendered the commensurables that obtained in the educational system suddenly incommensurable.

Nor are the empirical contradictions of the common value myth as new as one might expect. For example, with respect to the urban environment, at least 7 years ago Scott Greer was arguing that the core city was sick and would remain sick until a basic sociological movement took place in our urban environment that would move all the middle classes to the suburbs and surrender the core city to the "... segregated, the insulted, and the injured" (7). This argument by Greer came at a time when most of us were still talking about compromise and accommodation of interests, and was based upon a perception that the life styles, values, and needs of these two groups were so disparate that a healthy, creative restructuring of life in the core city could not take place until pluralism had been replaced by what amounted to geographic or territorial tribalism; only when this occurred would urban incommensurables become commensurable.

Looking at a more recent analysis of the sickness of the core city, Wallace F. Smith has argued that the productive model of the city is no longer viable for the purposes of economic analysis (8). Instead, he develops a model of the city as a site for leisure consumption, and then seems to suggest that the nature of this model is such that the city cannot regain its health because it cannot make decisions, and that it cannot make decisions because the leisure demands are value-based and, hence, do not admit of compromise and accommodation; consequently there is no way of deciding among these various value-oriented demands that are being made on the core city.

In looking for the cause of the erosion of the myth of a common value system, it seems to me that so long as our perceptions and knowledge of other groups were formed largely through the written media of communication, the American myth that we were a giant melting pot of equalitarians could be sustained. In such a perceptual field it is tenable, if not obvious, that men are motivated by interests. Interests can always be compromised and accommodated without undermining our very being by sacrificing values. Under the impact of the electronic media, however, this psychological distance has broken down and we now discover that these people with whom we could formerly compromise on interests are not, after all, really motivated by interests but by values. Their behavior in our very living room betrays a set of values, moreover, that are incompatible with our own, and consequently the compromises that we make are not those of contract but of culture. While the former are acceptable, any form of compromise on the latter is not a form of rational behavior but is rather a clear case of either apostasy or heresy. Thus, we have arrived not at an age of accommodation but one of confrontation. In such an age "incommensurables" remain "incommensurable" in real life.

Erosion of the Myth of the Monopoly of Coercive Force

In the past, those who no longer subscribed to the values of the dominant culture were held in check by the myth that the state possessed a monopoly on coercive force. This myth has undergone continual erosion since the end of World War II owing to the success of the strategy of guerrilla warfare, as first revealed to the French in Indochina, and later conclusively demonstrated in Algeria. Suffering as we do from what Senator Fulbright has called "the arrogance of power," we have been extremely slow to learn the lesson in Vietnam, although we now realize that war is political and cannot be won by military means. It is apparent that the myth of the monopoly of coercive force as it was first qualified in the civil rights conflict in the South, then in our urban ghettos, next on the streets of Chicago, and now on our college campuses has lost its hold over the minds of Americans. The technology of guerrilla warfare has made it evident that, while the state can win battles, it cannot win wars of values. Coercive force which is centered in the modern state cannot be sustained in the face of the active resistance of some 10 percent of its population unless the state is willing to embark on a deliberate policy of genocide directed against the value dissident groups. The factor that sustained the myth of coercive force in the past was the acceptance of a common value system. Whether the latter exists is questionable in the modern nation-state. But, even if most members of the nation-state remain united around a common value system which makes incommensurables for the majority commensurable, that majority is incapable of enforcing its decisions upon the minority in the face of the diminished coercive power of the governing body of the nation-state.

Erosion of the Myth of Administrators of the Commons

Hardin's thesis that the administrative arm of the state is capable of legislating temperance accords with current administrative theory in political science and touches on one of the concerns of that body of theory when he suggests that the "... great challenge facing us now is to invent the corrective feedbacks that are needed to keep the custodians honest."

Our best empirical answers to the question—*Quis custodiet ipsos custodes?*—"Who shall watch the watchers themselves?"—have shown fairly conclusively (9) that the decisions, orders, hearings, and press releases of the custodians of the commons, such as the Federal Communications Commission, the Interstate Commerce Commission, the Federal Trade Commission, and even the Bureau of Internal Revenue, give the large but unorganized groups in American society symbolic satisfaction and assurances. Yet, the actual day-to-day decisions and operations of these administrative agencies contribute, foster, aid, and indeed legitimate the special claims of small but highly organized groups to differential access to tangible resources which are extracted from the commons. This has been so well documented in the social sciences that the best answer to the question of who watches over the custodians of the commons is the regulated interests that make incursions on the commons.

Indeed, the process has been so widely commented upon that one writer has postulated a common life cycle for all of the attempts to develop regulatory policies (10). This life cycle is launched by an outcry so widespread and demanding that it generates enough political force to bring about the establishment of a regulatory agency to insure the equitable, just, and rational distribution of the advantages among all holders of interest in the commons. This phase is followed by the symbolic reassurance of the offended as the agency goes into operation, developing a period of political quiescence among the great majority of those who hold a general but unorganized interest in the commons. Once this political quiescence has developed, the highly organized and specifically interested groups who wish to make incursions into the commons bring sufficient pressure to bear through other political processes to convert the agency to the protection and furthering of their interests. In the last phase even staffing of the regulating agency is accomplished by drawing the agency administrators from the ranks of the regulated.

Thus, it would seem that, even with the existence of a common value system accompanied by a viable myth of the monopoly of coercive force, the prospects are very dim for saving the commons from differential exploitation or spoilation by the administrative devices in which Hardin places his hope. This being the case, the natural sciences may absolve themselves of responsibility for meeting the environmental challenges of the contemporary world by relegating those problems for which there are no technical solutions to the political or social realm. This action will, however, make little contribution to the solution of the problem.

Are the Critical Problems of Modern Society Insoluble?

Earlier in this article I agreed that perhaps until very recently, there existed a set of conditions which made the solution of Hardin's problem subset possible; now I suggest that the concession is questionable. There is evidence of structural as well as value problems which make comprehensive solutions impossible and these conditions have been present for some time.

For example, Aaron Wildavsky, in a comprehensive study of the budgetary process, has found that in the absence of a calculus for resolving "intrapersonal comparison of utilities," the governmental budgetary process proceeds by a calculus that is sequential and incremental rather than comprehensive. This being the case "... if one looks at politics as a process by which the government mobilizes resources to meet pressing problems" (11) the budget is the focus of these problem responses and the responses to problems in contemporary America are not the sort of comprehensive responses required to bring order to a disordered environment. Another example of the operation of this type of rationality is the American involvement in Vietnam; for, what is the policy of escalation but the policy of sequential incrementalism given a new Madison Avenue euphemism? The question facing us all is the question of whether incremental rationality is sufficient to deal with 20th-century problems.

The operational requirements of modern institutions make incremental rationality the only viable form of decision-making, but this only raises the prior question of whether there are solutions to any of the major problems raised in modern society. It may well be that the emerging forms of tribal behavior noted in this article are the last hope of reducing political and social institutions to a level where incommensurables become commensurable in terms of values *and* in terms of comprehensive responses to problems. After all, in the history of man on earth we might well assume that the departure from the tribal experience is a short-run deviant experiment that failed. As we stand "on the eve of destruction," it may well be that the return to the face-to-face life in the small community unmediated by the electronic media is a very functional response in terms of the perpetuation of the species.

There is, I believe, a significant sense in which the human environment is directly in conflict with the source's of man's ascendancy among the other species of the earth. Man's evolutionary position hinges, not on specialization, but rather on generalized adaptability. Modern social and political institutions, however, hinge on specialized, sequential, incremental decision-making and not on generalized adaptability. This being the case, life in the nation-state will continue to require a singleness of purpose for success but in a very critical sense this singleness of purpose becomes a straightjacket that makes generalized adaptation impossible. Nowhere is this conflict more evident than in our urban centers where there has been a decline in the livability of the total environment that is almost directly proportionate to the rise of special purpose districts. Nowhere is this conflict between institutional singleness of purpose and the human dimension of the modern environment more evident than in the recent warning of S. Goran Lofroth, chairman of a committee studying pesticides for the Swedish National Research Council, that many breast-fed children ingest from their mother's milk "more than the recommended daily intake of DDT" (*12*) and should perhaps be switched to cow's milk because cows secrete only 2 to 10 percent of the DDT they ingest.

How Can Science Contribute to the Saving of the Commons?

It would seem that, despite the nearly remorseless workings of things, science has some interim contributions to make to the alleviation of those problems of the commons which Hardin has pointed out.

These contributions can come at two levels:

(1) Science can concentrate more of its attention on the development of technological responses which at once alleviate those problems and reward those people who no longer desecrate the commons. This approach would seem more likely to be successful than the "... fundamental extension in morality ..." by administrative law; the engagement of interest seems to be a more reliable and consistent motivator of advantage-seeking groups than does administrative wrist-slapping or constituency pressure from the general public.

(2) Science can perhaps, by using the widely proposed environmental monitoring systems, use them in such a way as to sustain a high level of "symbolic disassurance" among the holders of generalized interests in the commons—thus sustaining their political interest to a point where they would provide a constituency for the administrator other than those bent on denuding the commons. This latter approach would seem to be a first step towards the "... invention of the corrective feedbacks that are needed to keep custodians honest." This would require a major change in the behavior of science, however, for it could no longer rest content with development of the technology of monitoring and with turning the technology over to some new agency. Past administrative experience suggests that the use of technology to sustain a high level of "dis-assurance" among the general population would also require science to take up the role and the responsibility for maintaining, controlling, and disseminating the information.

Neither of these contributions to maintaining a habitable environment will be made by science unless there is a significant break in the insularity of the two scientific tribes. For, if science must, in its own insularity, embark on the independent discovery of "the tragedy of the commons," along with the parameters that produce the tragedy, it may be too slow a process to save us from the total destruction of the planet. Just as important, however, science will, by pursuing such a course, divert its attention from the production of technical tools, information, and solutions which will contribute to the political and social solutions for the problems of the commons.

Because I remain very suspicious of the success of either demands or pleas for fundamental extensions in morality, I would suggest that such a conscious turning by both the social and the natural sciences is, at this time, in their immediate self-interest. As Michael Polanyi has pointed out, "... encircled today between the crude utilitarianism of the philistine and the ideological utilitarianism of the modern revolutionary movement, the love of pure science may falter and die" (13). The sciences, both social and natural, can function only in a very special intellectual environment that is neither universal or unchanging, and that environment is in jeopardy. The questions of humanistic relevance raised by the students at M.I.T., Stanford Research Institute, Berkeley, and wherever the headlines may carry us tomorrow, pose serious threats to the maintenance of that intellectual environment. However ill-founded *some* of the questions raised by the new generation may be, it behoves us to be ready with at least some collective, tentative answers—if only to maintain an environment in which both sciences will be allowed and fostered. This will not be accomplished so long as the social sciences continue to defer the most critical problems that face mankind to future technical advances, while the natural sciences continue to defer those same problems which are about to overwhelm all mankind to false expectations in the political realm.

References and Notes

1. G. Hardin, *Science* **162,** 1243 (1968).
2. J. B. Wiesner and H. F. York, *Sci. Amer.* **211** (No. 4), 27 (1964).
3. C. Woodbury, *Amer. J. Public Health* **45,** 1 (1955); S. Marquis, *Amer. Behav. Sci.* **11,** 11 (1968); W. H. Ferry, *Center Mag.* **2,** 2 (1969).
4. C. P. Snow, *The Two Cultures and the Scientific Revolution* (Cambridge Univ. Press, New York, 1959).
5. M. Olson, Jr., *The Logic of Collective Action* (Harvard Univ. Press, Cambridge, Mass., 1965).
6. G. A. Harrison *et al., Human Biology* (Oxford Univ. Press, New York, 1964), p. 292; W. W. Charters, Jr. in *School Children in the Urban Slum* (Free Press, New York, 1967).
7. S. Greer, *Governing the Metropolis* (Wiley, New York, 1962), p. 148.
8. W. F. Smith, "The Class Struggle and the Disquieted City," a paper presented at the 1969 annual meeting of the Western Economic Association, Oregon State University, Corvallis.
9. M. Bernstein, *Regulating Business by Independent Commissions* (Princeton Univ. Press, Princeton, N. J., 1955); E. P. Herring, *Public Administration and the Public Interest* (McGraw-Hill, New York, 1936); E. M. Redford, *Administration of National Economic Control* (Macmillan, New York, 1952).

10. M. Edelman, *The Symbolic Uses of Politics* (Univ. of Illinois Press, Urbana, 1964).
11. A. Wildavsky, *The Politics of the Budgetary Process* (Little Brown, Boston, Mass., 1964).
12. Corvallis *Gazette-Times*, 6 May 1969, p. 6.
13. M. Polanyi, *Personal Knowledge* (Harper & Row, New York, 1964), p. 182.

28   Man's Evolutionary Future

Dudley Kirk

Garrett Hardin has addressed us in his usual artful and attractive style.* Much as I disagree with him, I must admit to enjoying his skillful presentation.

But we differ on many points. I will put them under two headings: first, the causes and nature of population trends in the United States (I believe we were talking primarily about the United States) and, second, the genetic implications of these changes.

First, the causes and nature of population trends in the United States: I feel that Garrett Hardin greatly underestimates man's ability to make rational adjustment to his environment and to adjust his family size accordingly. He makes this adjustment in part because of his own specific situation but also in line with the norms of the society at large. These he has absorbed in the process of socialization, often without realizing he has acquired them. Actually, most of the things we do we do because we think they are in some way right and proper, that is, in accord with the values of our society. People in their childbearing behavior, as in most other aspects of their behavior, tend to conform to the norms that exist both in the specific group to which they belong and also in the society at large. In other words, individuals react to their particular situation and they also react to the norms which are established by the collective responses of the society to its situation.

This relates to population trends in this way. Some of us may have the impression that in recent years we have been experiencing an astounding reduction in the birth rate and in fertility. What is not usually recognized is that we have been experiencing a continuous and rather rapid reduction in the birth rate since 1957. It started before the pill; it was well advanced before the more frenetic shouting by alarmists about population trends in the United States. Our society had already started to make its adjustment fourteen years ago. Individuals were participating in this not because they were concerned about national population problems—of course not— but because their particular life situation was such that they began to marry later, have children later, not so close together, and probably not so many. Both as individual couples and as a society we were pretty far along toward an adjustment toward zero population growth before the recent hullabaloo about the problem.

Excerpt from *Social Biology*, **19** (Dec. 1972), 362–366. Reprinted with permission from the author and the publisher.

* Dr. Kirk refers here to a 1972 paper by Hardin which covers the same topic as his paper in this text [Ed.].

The very success of the zero population growth idea is related to the fact that the people concerned were already "converted." The ZPG movement is upper-middle class, suburban, intellectual, and lily-white, not accidentally the group that experienced the greatest proportionate "baby boom" after World War II. Even more important, it is the group that has seen its former near-monopoly of the suburbs, the universities, the better beaches, the national parks, and prime recreation areas invaded by the more and more affluent middle and lower-middle-class masses. ZPG and the population alarmists are those whose near-monopoly of the better physical environment has been destroyed by growing affluence of the masses. This they see as a "population explosion."

It is only the inertia of past trends, reflected in our bulge of young adults, that will keep the total population growing for some years to come. The white population, and recently the total population, has reached a reproduction rate of one— that is, parents are having only enough children to reproduce themselves every generation. The only reason we continue to have population growth is that we have a disproportionate number of young people at the ages when people marry and have children. If present trends continue, we will shortly go below generation replacement, and we will in fact have zero population growth before a very long time. We will perhaps experience population decline in the not-too-distant future. This will pose quite a different set of problems not understood by alarmists about population growth.

Of course we aren't going to have zero population growth next year or in five or even ten years. In fact, I think it would be very unfortunate if we did. I could expand on that, but will simply say that distortions of the age structure might well be more deleterious than a somewhat larger population. We already are closing schools because we don't have enough primary school children and already have the opposite problems to those of baby-boom days for the youngest age groups. Such wide swings have many bad effects; but that is another subject.

Thus, I disagree with assertions that the state should in various ways pressure couples not to have children, and in particular not to have more than two children.

I am old enough to remember the 1930's, when there was much concern about our declining birth rate and the possible bad effects of a declining population. I have teased Charles Westoff, who is in charge of the staff preparing the report for the President's Commission on Population, that if they don't hurry up and get this report out they are going to have to change the whole focus. They may have to advise the President on methods of *raising* the birth rate. Obviously this is far-fetched now, but in fifteen or twenty years I view it as a real possibility.

Thus, the United States population was already making the adjustment before the shouters came on the scene, and the trend has not changed. Therefore, I am not concerned, as Dr. Hardin is, about a population control policy. There seems to me nothing on the scene to justify coercion or punitive measures to restrict childbearing.

I disagree with Dr. Hardin about present population trends and their implications. I also disagree with him on the genetic implications of what is being experienced today. He said in his paper, you will recall, that there is a misapprehension that over the long periods of declining fertility in the United States and other industrial countries the opportunities for natural selection decreased by a considerable amount. He is quite right, of course, in saying that, on the contrary, the index of opportunity for natural selection actually rose with the long downward trend in family size. This is because the index is determined by variance around the mean. It has no relation to the average itself. The variance around a two-child average can be just as important and even more important than the variance around an average of eight children, which women had in the colonial period.

But Dr. Hardin has perhaps not observed the more recent trends. These have not been in this direction. Actually, since the Second World War, the index for natural selection did in fact go down. This is documented in an article which appeared in the *Proceedings of the Academy of Sciences* (Kirk, 1968). In fact, this is just what ZPG advocates—that variance about a two-child average should be reduced, at least as regards more than two.

In any case, people voluntarily are moving to lower parities. This is just as true of minorities and disadvantaged as it is of the affluent white majority, and surveys indicate that the former don't in fact want larger families than the majority. Furthermore, we have been moving very rapidly in the direction of more and more voluntary control. More liberal laws on abortion should accelerate trends in that direction. If we may assume that the vast majority of children that will be born in the future will be wanted, I think we need to look at the genetic implications of *voluntary* childbearing as well. So there are two aspects: the efficiency with which birth control is practiced and the number of children couples really want to have.

Dr. Hardin is telling us that if you allow for differences in the efficiency with which people practice birth control you will promote something that as a biologist he calls *"homo progenitivus,"* who is going to outbreed *"homo contracipiens"* because by definition the people who practice contraception more effectively are weeding themselves out of the population. Does Dr. Hardin really believe this is something new? Reproductive differentials from differential effectiveness in use of birth control have existed in Western society for at least three hundred years. The striking fact is the extent to which these differentials have been *reduced* by the greater availability of effective methods of birth control. To take a specific example, men in *Who's Who* born a century ago and since have consistently had an average of two children. Native white women averaged 3.5 in 1875–79, but this has now fallen to about the level of men in *Who's Who* (Kirk, 1956). Modern Western society has been " weeding out " the planners, the cautious, the " conscientious " and the less impulsive through their more effective birth control for at least three centuries. But it is not at all clear that this has had much effect on the gene pool. Certainly social changes have been much more important than genetic change in shaping the personality and behavior of modern man. In any event, this factor is becoming *less* important as childbearing becomes more and more a voluntary matter subject to rational choice by birth control methods applied independent of the sexual act.

There is, of course, the assertion that people really want too many children for the social good. The fallacy of survey results on this subject has been clearly demonstrated (Ryder and Westoff, 1965). Respondents' " ideal " family size tends on the one hand to be retrospective (they are reluctant to give an " ideal " less than

271

the actual) and, on the other, to be too high because practical considerations often lead couples to have fewer children than their "ideal." In any event, the reported "ideal," like actual fertility, is now rapidly sinking to the level of generation replacement.

As a larger and larger percentage of births are wanted, the question becomes: *Who* wants children? Isn't it reasonable to suppose that those who do are probably better parents than those who for whatever reason decide not to have children? I would argue that many women—many couples—should not have any children. They are not equipped to provide good socialization for children. By contrast, I believe we can all think of examples of families with numerous children that have done a magnificent job of raising their children. To the extent that genetic factors are relevant, voluntary childbearing would be favoring by selection the kind of parents who would do the best by their children—give them the most love, have the most empathy, give them the best upbringing. Perhaps this is as much a speculation as what Dr. Hardin is saying, but isn't this a genetic result just as likely as selection of those who are more fecund or more careless in the use of birth control? Purely voluntary childbearing should have a very salutary result from a social point of view and perhaps even from a genetic point of view.

Now, let me return to one other point in Dr. Hardin's paper. He implies that coercive measures will be necessary, among other things, to limit parents to two children. I think this would be very distressing from a genetic point of view and certainly from a social point of view. Why should we strive for diversity in our ecology by demanding uniformity in our own behavior? Why should we not promote rather than inhibit a variety of situations in which children grow up? As I said earlier, there are important differences in the ability of parents to give the best upbringing to children. Because of this, isn't it much more desirable to have a diversity in size of families? Eighty-eight per cent of all females born who reached age 35 in 1966 had had children. I personally think this is too many mothers; I am sure that a lot of those shouldn't have had children. Given completely voluntary childbearing, I'm sure that many of them wouldn't have.

Thus, it seems to me that voluntary choice may very well have positive genetic influences; but I would add that the genetic impact of these things in any event is quite dubious when social changes are occurring so rapidly. We swing back and forth; differential fertility swings back and forth, and I think it is a nice question whether we can draw any major conclusions about the genetic implications of such short-range changes as have occurred in a single generation.

References

Kirk, Dudley. 1956. The fertility of a gifted group: A study of the number of children of men in *Who's Who*, p. 79–98. *In* The nature and transmission of the genetic and cultural characteristics of human populations. Milbank Memorial Fund, New York.

———. 1968. Patterns of survival and reproduction in the United States: Implications for selection. Proc. Nat. Acad. Sci. **59** : 662–670.

Ryder, Norman B., and Charles F. Westoff. 1969. Relationships among intended, expected, desired, and ideal family size: United States, 1965. Center for Population Research, NIH, Washington, D.C.

seven # Medical Genetics and Counseling

The paper by Crow which leads off this section sets the tone for those that follow. Crow raises important questions about where we are and where we are going in the field of medical genetics (and genetic engineering, a topic taken up in Section Nine). Crow reviews the symbiotic relation between genetics and medicine as it existed in the mid-1960s. Since then these disciplines have stimulated each other's growth, producing a vigorous off-shoot in the area of genetic counseling. This point is substantiated by Fraser's paper on genetic counseling in this section. But as Childs (1974) points out, medicine and the teaching of health education in schools are still largely innocent of understanding the practical consequences of genetic diversity in health and illness.

Areas of particular concern surrounding medical genetics include the legal processes involved in screening populations for genetic disease: Fletcher mentions these matters here and the paper by Gary further pursues the sociological implications of laws and tests on sickle-cell screening in Section Eight.

Amniocentesis is the technique of extracting fluid from the amniotic sac surrounding the fetus. The amniotic fluid is then tested for biochemical traits or, more often, the fetal cells in the fluid are cultured in the laboratory and subsequently characterized biochemically or chromosomally. Amniocentesis is a major component of the prenatal diagnosis of genetic defect. Prenatal diagnosis followed by elective abortion (or voluntary sterilization, or conscientious use of contraception) constitutes the core of genetic counseling. Neel in his paper here discusses the status of therapeutic abortion for genetic defects prior to the Supreme Court's decision in 1973 that abortion early in pregnancy is the prerogative of the mother. Since 1973 abortions have become widespread and, besides having facilitated greater effectiveness in genetic counseling, the court decision has engendered strong arguments pro and con about the morality of abortion in both the popular and scientific press. The paper by Ayala in this section presents a biologist's view on abortion, although Ayala's previous profession as a priest may not be unrelated to what he has to say here. The following paper, by Fletcher, carries the discussion on abortion forward to a discussion on the fate of newborns with severe genetic defects—raising provocative questions.

Several aspects of medical genetics with important social correlates are not represented by papers in this collection.

These include, among others, the genetics of cancer, immunogenetics, and pharmacogenetics. Cancer deserves special mention. Although the genetic basis of most forms of cancer is not understood, animal studies have provided us with important information on the development of cancer (Heston, 1974), and a small number of cancer categories fit genetic models very well (Knudson, 1975). The future looks promising for a clearer understanding of inherited susceptibility to cancer.

Bibliography

Ad Hoc Committee. Genetic counseling. *Amer. J. Hum. Genet.*, **27** : 240–242, 1975. States the conditions necessary for optimal genetic counseling.

Anonymous. A "fix" of life for a hemophiliac. *People Magazine*, July 22, 1974, pp. 58–59.

Bergsma, D., ed. *Ethical, Social and Legal Dimensions of Screening for Human Genetic Disease.* New York: Stratton/Symposia Specialists, 1974.

Boon, R. A., and D. F. Roberts. The social impact of haemophilia. *J. Biosocial Science*, **2** : 237–264, 1970. The everyday living problems of hemophiliacs are enormous.

Bylinsky, G. What science can do about hereditary disease. *Fortune*, Sept. 1974, pp. 148–160.

Callahan, D. Ethics, law, and genetic counseling. *Science* **176** : 197–200, 1972.

Childs, B. A place for genetics in health education, and vice versa. *Amer. J. Hum. Genet.*, **26** : 120–135, 1974.

Culliton, B. Abortion: Liberal laws do make abortion safer for women. *Science*, **188** : 1091–1092, 1975.

Culliton, B. Amniocentesis: HEW backs test for prenatal diagnosis of disease. *Science*, **190** : 537–540, 1975.

Culliton, B. Fetal research. *Science*, **187** : 237–241, 411–413, and 1175–1176, 1975. Describes the making, the nature, and the consequences of a new Massachusetts law restricting fetal research.

Culliton, B. Genetic screening. *Science*, **189** : 119–120, 1975, and **191** : 926–929, 1976.

Etzioni, A. *Genetic Fix.* New York: Macmillan, 1973. A sociologist's overview of the new genetics.

Fletcher, John. The brink: The parent-child bond in the genetic revolution. *Theological Studies*, **33** : 457–485, 1972. An ethicist studies the moral problems faced by couples undertaking amniocentesis.

Friedmann, T. Prenatal diagnosis and genetic disease. *Sci. Amer.*, **225** : 34–42, 1971.

Gayton, W. and L. Walker. Down's Syndrome: Informing the parents. *Amer. J. Diseases of Childhood*, **127** : 510–512, 1974. Many parents were told their child had Down's long after its birth, although parents suspected abnormality. The unnecessary delay caused parental anxieties.

Halacy, D. S., Jr. *Genetic Revolution*. New York: Harper & Row, 1974. A popular work on the new genetics.

Hamilton, M., ed. *The New Genetics and the Future of Man*. Grand Rapids, Mich. : W. B. Eerdmans, 1972.

Hemphill, M. Pretesting for Huntington's disease. *Hastings Center Report*, June 1973, pp. 12–13.

Heston, W. E. Genetics of cancer. *J. Heredity*, **65** : 262–272, 1974.

Hilton, B., D. Callahan, M. Harris, P. Condliffe, and B. Berkley. *Ethical Issues in Human Genetics*. New York: Plenum, 1973.

Knudson, A. G., Jr. Genetics of human cancer. *Genetics*, **79** : 305–316, 1975.

Leonard, C. O., G. A. Chase, and B. Childs. Genetic counseling: A consumer's view. *New Eng. J. Med.*, **287** : 433–439, 1972.

Lipkin, M., and P. T. Rowley, eds. *Genetic Responsibility on Choosing Our Children's Genes*. New York: Plenum, 1974.

McKusick, V. A. The growth and development of human genetics as a clinical discipline. *Amer. J. Hum. Genet.*, **27** : 261–273, 1975. The author considers human genetics to be solely medical.

Massie, R., and S. Massie. *Journey*. New York: Knopf, 1975. A couple with a hemophiliac son describe their life and the shortcomings of medical treatment in the United States.

Milunsky, A. *The Prevention of Genetic Disease and Mental Retardation*. Philadelphia: Saunders, 1975.

Milunsky, A., and G. J. Annas. *Genetics and the Law*. New York: Plenum, 1976.

Motulsky, A. G. Brave new world? *Science*, **185** : 653–663, 1974. The ethics of genetic counseling and related activities are discussed.

Murray, R. F. Genetic disease and human health. *Hastings Center Report*, Sept. 1974, pp. 4—6. Raises the question of what is genetic health.

Ramsey, P. Screening: An ethicist's view. In *Ethical Issues in Human Genetics*, B. Hilton et al., eds. New York: Plenum, 1973.

Restak, R. Genetic counseling for defective parents. *Psych. Today*, Sept. 1975, pp. 21–93 passim.

Reynolds, B. D., M. H. Puck, and A. Robinson. Genetic counseling: An appraisal. *Clinical Genetics*, **5** : 177–187, 1974.

Rhine, S. A., J. Cain, R. Cleary, C. Palmer, and J. Thompson. Prenatal sex detection with endocervical smears: Successful results utilizing Y-body fluorescence. *Amer. J. Obstet. Gyn.*, **122** : 155–160, 1975.

Schneider, E. L., E. J. Stanbridge, C. J. Epstein, M. Golbus, G. Abbo-Halbasch, and G. Rogers. Mycoplasma contamination of cultured amniotic fluid cells: Potential hazard to prenatal chromosomal diagnosis. *Science*, **184** : 477–480, 1974.

Shaw, M. W. Genetic counseling. *Science*, **184** : 751, 1974.

Sklar, J., and B. Berkov. Abortion, illegitimacy, and the American birth rate. *Science*, **185** : 909–915, 1974. Legalizing abortion dramatically decreased illegitimate births.

Stevenson, A. C., B. C. Clare Davison, and M. W. Oakes. *Genetic Counseling.* Philadelphia: Lippincott, 1970.

Sultz, H. A., E. R. Schlesinger, and J. Feldman. An epidemiologic justification for genetic counseling in family planning. *Amer. J. Public Health*, **62** : 1489–1492, 1972.

29   Genetics and Medicine

James F. Crow

The first half of the twentieth century was also the first half-century of Mendelism as a recognized science. During this period the major beneficiary of genetic knowledge has been agriculture. Selection, inbreeding, hybridization, artificial fertilization, and the use of pedigree information are now routine techniques of animal breeding. The plant breeder in addition can utilize haploidy, polyploidy, cytoplasmic factors, and various forms of asexual propagation. As a result, high-yielding, disease-resistant crop varieties are commonplace, and strains of plants and livestock have been made to order for particular environments and for specialized human needs.

In contrast, the major practical contribution of genetics in its second half-century is likely to concern man more directly, especially through its effects on medicine. The beginning of this period, from 1905 to 1965, has already produced some outstanding discoveries, and more are likely to follow at an accelerating pace.

That Mendel's discovery would ultimately have great influence on man's knowledge of and view toward himself was stated early by R. C. Punnett, among others. In the eleventh edition of the *Encyclopaedia Britannica*, which was published in 1910–11, he wrote:

> Increased knowledge of heredity means increased power of control over the living thing, and as we come to understand more and more the architecture of the plant or animal we realize what can and what cannot be done towards modification or improvement. The experiments of Biffen on the cereals have demonstrated what may be done with our present knowledge in establishing new, stable and more profitable varieties of wheat and barley, and it is impossible to doubt that as this knowledge becomes more widely disseminated it will lead to considerable improvements in the methods of breeding animals and plants.

> It is not, however, in the economic field, important as this may be, that Mendel's discovery is likely to have most meaning for us: rather it is in the new light in which man will come to view himself and his fellow creatures. Today we are almost entirely ignorant of the unit-characters that go to make the difference between one man and another. A few diseases, such as alcaptonuria and congenital cataract, a digital malformation, and probably eye colour, are as yet the only cases in which inheritance has been shown to run upon Mendelian lines. The complexity of the subject must render investigation at once difficult and slow; but the little that we

Paper number 1061 from the Genetics Laboratory, University of Wisconsin. Part of the work discussed in this paper was done with the support of the National Institutes of Health (GM-08217 and GM-07666).

know today offers the hope of a great extension in our knowledge at no very distant time. If this hope is borne out, if it is shown that the qualities of man, his body and his intellect, his immunities and his diseases, even his very virtues and vices, are dependent upon the ascertainable presence or absence of definite unit-characters whose mode of transmission follows fixed laws, and if also man decides that his life shall be ordered in the light of this knowledge, it is obvious that the social system will have to undergo considerable changes.

Soon after Mendel's work came to be generally recognized, there were reports of human characteristics showing Mendelian inheritance. Naturally, traits that were easily visible, such as those affecting the fingers, and that were expressed in the heterozygote would be most easily studied. It is not surprising, therefore, that the first convincing demonstration of Mendelian inheritance in man was Farabee's report on brachydactyly in 1905 (Stern 1960).

It was also recognized early by Garrod (1909) that alkaptonuria was inherited as if caused by homozygosity for a recessive gene. The evidence came from the inheritance pattern and the high frequency of consanguineous marriages among parents of affected persons. Soon after X-chromosome inheritance was understood, it was realized that color-blindness and hemophilia followed an X-linked pattern of inheritance.

The discovery of new inherited phenotypes has continued steadily for the past 50 years (see, for example, Becker 1964). The number of diseases known to be inherited as single gene differences has now grown to several hundred, the exact number depending on the rigor of the criteria used in admitting a disease to the list. Most of these are individually very rare, although collectively they add up to something like one per cent of all children born. A list of many such diseases, with their approximate incidences, has been published in a United Nations report (1958: pp. 197–200).

Human chromosome studies, until recently, were concerned mainly with establishing the chromosome number and identifying the X and Y chromosomes. The Y chromosome, being small, was overlooked in some early studies; but for two or three decades it has been generally recognized that there is a Y chromosome, identifiable during meiosis by the unequal synaptic pair that it makes with the X.

A most striking feature of human chromosome studies was the long period of time during which 48 was accepted as the correct diploid number. To the surprise of almost everyone the number was shown by Tjio and Levan in 1956 to be 46.

New techniques in chemistry and cytology plus new concepts from molecular biology and medicine have led to striking advances in the past decade and a half. Genetics has become one of the most active parts of medicine. I should like to discuss genetics and medicine from three viewpoints: genetics in the service of medicine, some contributions of medicine to genetics, and some genetic consequences of advances in medicine and public health. I want especially to emphasize the reciprocal relationship between genetics and medicine, the extent to which each has benefited from the other in the past, and the expectations for continued mutual benefit in the future.

Genetics in the Service of Medicine

One contribution genetics is making to medicine is in providing a deeper insight into disease-producing mechanisms. The recent developments in human cytogenetics furnish an example: Complete mystery suddenly changed to deep understanding with the discovery by Lejeune, Turpin, and Gautier (1959) that mongolism, or Down's syndrome, is caused by a chromosome anomaly. The previously known facts, familiar to students of the disease, were (1) mongoloid children practically always came from normal parents; (2) recurrence in a sibship was very rare; (3) concordance in one-egg twins was practically 100 per cent, but two-egg concordance was nearly zero; (4) on those rare occasions when a mongoloid person did reproduce, about half the children were affected; (5) the disease produced a number of seemingly unrelated anomalies; and (6) there was a striking increase in incidence with maternal age. The first five are immediately explained as predictable consequences of autosomal trisomy arising through nondisjunction. The sixth is not necessarily expected, but it is not particularly surprising; the D_1, 18, and XXY (Klinefelter's) types all have a maternal age effect. The original papers reporting these discoveries have been reprinted (Boyer 1963).

The hypothesis that trisomy is the cause of mongolism was suggested several times prior to 1959, but cytological techniques were then inadequate. In an early paper Haldane (1932) argued that nondisjunction and attached-X chromosomes exist in man. His reasoning about the latter was based on what we now realize was a wrong hypothesis of human sex determination; he was assuming, by analogy with *Drosophila*, that the Y chromosome was unimportant in sex determination. Speaking of *Drosophila*, Haldane said, " For example, primary nondisjunction both of the X and of the fourth chromosome occurs about once in 2000 individuals. We might therefore expect cytological abnormality to be reasonably frequent in man." Later in the same paper he said prophetically, " It seems possible that satisfactory mitoses might be observed in a culture of leukocytes. If so, the development of human cytology in relation to genetics will become possible."

As soon as human trisomy was discovered, it could be predicted that partial trisomy would be found, caused by half-translocations and other well-understood cytological mechanisms. One might also expect mosaicisms, caused by mitotic nondisjunction or chromosome loss. These predictions have been abundantly confirmed. In many instances cytological analysis has been useful in identifying otherwise normal persons with a high probability of having abnormal children.

It might be expected that there would be primary trisomic types for all the chromosomes, as in *Datura*. However, the phenotypes would not necessarily be predictable and many would probably be lethal. That many are indeed lethal is demonstrated by the finding of trisomic types among miscarried fetuses. Recently a specific deletion syndrome has been recognized, the " cri-du-chat " (Lejeune *et al.* 1963), and a somatic deletion is invariably associated with chronic myelogenous leukemia (Nowell and Hungerford 1960). Human cytogenetics is becoming an analytical tool comparable in precision to the cytogenetics of maize and is now a standard part of medical pathology. (For a recent review of techniques and some of the results, see Yunis 1965.)

Another area where genetic knowledge, this time together with biochemistry, has led to a deeper understanding of medical problems started in 1908 with Garrod's description of "inborn errors of metabolism." For many years only a few additional cases were discovered, but since 1950 there have been dozens of new types found (see Stanbury, Wyngaarden, and Fredrickson 1965). In contrast to the chromosomal anomalies, where the understanding is deep but where not much can be done to prevent or repair the damage, the inborn errors can sometimes be corrected by supplying or withholding the appropriate dietary element. A dramatic example is the prevention of much, if not all, of the brain damage in phenylketonuria by early initiation of a diet low in phenylalanine. Another example is galactosemia, where the substitution of glucose for lactose in the diet prevents the symptoms. Heterozygous carriers of a typical recessive disease caused by an enzyme deficiency may be expected to have the enzyme in reduced amount. Thus carrier detection is possible, once an appropriate method for assaying the enzyme is developed.

Detailed chemical knowledge is not always necessary for some of the methods of biochemical genetics to be applied. Beadle and Tatum exploited the possibilities of organ transplants in order to determine the mode of action of some genes affecting *Drosophila* eye pigments. Similarly, much has been learned about factors affecting blood coagulation from the patterns of repair when transfusions are given among various persons with superficially similar genetic coagulation defects (Biggs and Macfarlane 1962).

An area of medicine where genetics has had a major influence is immunology. Shortly after their discovery, the ABO blood groups were found to be inherited, although it was some time later when the exact inheritance was understood. An important landmark in immunogenetics was the association of Rh incompatibility with hemolytic disease of the newborn. Previously the inheritance of erythroblastosis fetalis had been obscure. For example, in 1941 Snyder suggested, "The condition is apparently dependent upon a recessive gene with variable expression including icterus gravis, generalized congenital hydrops with or without jaundice, and severe congenital anemia." At the time this was a reasonable hypothesis, since the disease frequently recurs in sibships but shows no obvious association between parent and child. Once again, an understanding of the mechanism brought order out of the previous confusion. The pattern of maternal-fetal incompatibility could now be sought in other systems, and the rapid discovery during the 1950's of new blood group systems was the result. (For a general review, see Race and Sanger 1962).

As early as 1916 Little and Tyzzer showed that the genotype has a great deal to do with acceptance or rejection of tissue transplants. They showed that a tumor from an inbred strain of mice could be transplanted to other members of the same strain and to the F_1 hybrids between this and other strains, but only very rarely to the F_2. Later these results were generalized to transplants of normal tissues in mice and by Loeb and Wright (1927) in guinea pigs. The rule is that a graft will not be accepted if the donor carries any histocompatibility gene not possessed by the recipient. From the frequency of acceptances by the F_2 generation of grafts from the original parents, from the F_1 generation, and from other F_2's, the minimum number of independent histocompatibility loci involved in the differences between mouse strains could be estimated at about a dozen. Roughly this number has now been found at the Jackson Laboratory.

The fundamental identity of graft rejection and antigen-antibody reaction was long suspected and finally proved in a number of ways. One kind of evidence, contributed by the study of a genetic disease in man, was the poor rejection of grafts by persons who had agammaglobulinemia, an X-linked recessive disorder manifested by a deficiency of plasma cells and poor antibody formation (Good and Zak 1956).

The study of blood groups in cattle by Irwin and his students led to another important development in immunogenetics. Out of this study grew Owen's discovery in 1945 that cattle twins whose blood vessels had interconnections during their prenatal life each had two kinds of blood cells, corresponding to the two kinds that the twins would have had separately had the mixture not occurred. Furthermore, both twins failed to produce antibodies to either kind of cell. This observation and later work by Medawar and others led to the discovery of immune tolerance, whereby under certain conditions an animal cannot make antibody against antigens to which he was exposed early in life.

The understanding of the genetics of histocompatibility has already led to the successful transplantation of organs between identical twins. Various methods of minimizing immune response have permitted partially or temporarily successful transplants between persons not of identical genotypes. Fuller understanding of the way tolerance can be achieved and antibodies selectively suppressed should make possible a wider range of successful transplants. The prediction that the surgery of the future can replace as well as repair may then be fulfilled. Donors might some day be from nonhuman primate strains which are bred for genetic homogeneity and to which the human recipients are made immunologically tolerant, as suggested by Lederberg (Wolstenholme 1963). Or perhaps, if the conditions leading to immunological tolerance in man were better understood, a child could be made tolerant to both parents, in which case he could receive grafts not only from either parent, but also from any of his sibs, since no child in the family would have genes not possessed by one of the parents.

It is interesting that in the history of immunogenetics, from Landsteiner's discovery of blood groups until quite recent times, the emphasis has been on the genetics of antigens. There are now a number of theories of antibody formation, most of them based on genetic models. Human medicine has once again provided some critical material in the form of Bence-Jones proteins from patients with multiple myeloma. Since each patient produces a single type of "antibody" protein, a full-scale chemical attack on each type is possible. By amino acid sequence analysis of proteins produced in this and similar disorders, one may hope to determine the chemical basis for the various antibody globulins. The future control or prevention of human disease through either stimulation or suppression of appropriate clones of antibody-producing cells is an exciting possibility.

A major medical trend of the past century has been the decrease in infectious disease, with the result that a larger fraction of medical problems are now constitutional diseases. Most of these probably have a substantial genetic component, so the importance of genetics in medical practice and public health is correspondingly increased. Genetic knowledge can aid the practicing physician and his patient in several ways, of which I shall discuss three.

GENETICS AS AN AID TO DIAGNOSIS. An inquiry into the family history is, of course, a standard part of medical diagnostic procedure. Whether the disease is genetic, environmental, or, as is frequently the case, both, the previous occurrence of the disease in a relative alerts the physician to the possibility of a genetic cause and increases the probability of an early and correct diagnosis.

If a disease is caused by an X-linked gene or by an autosomal dominant, the pattern of inheritance is usually clear if the pedigree is large enough. One can then determine which infants have a high risk of developing the condition. Separation of low-risk from high-risk individuals in the population may permit the employment of complicated or expensive screening procedures that would be impractical on a wide scale.

On the other hand, with a recessive disease the value of pedigree information is far less, for there are usually no ancestors or collateral relatives (except sibs) who show the condition. Once a child with a recessive disease has been born, all subsequent sibs are suspect and should be carefully watched. Consanguineous marriages, although extremely rare in the general population, are by no means unusual among the parents of persons with rare recessive diseases; therefore any instance of a consanguineous marriage should alert the physician to the possibility of a recessive disease in children. Similarly, some population isolates may have a high frequency of some gene, often one that is virtually unknown elsewhere.

PREVENTION OF DISEASE. The function of genetics in disease prevention is not completely distinct from that in disease diagnosis; the value of finding persons with a high probability of developing the disease is the same, whether it be for preventive or curative measures. If the disease is understood and can be predicted, it may be possible to create an environment in which the genetic damage is minimized or the limited genetic potential is maximized. An example is the special diet for galactosemia, mentioned before.

GENETIC COUNSELING. Genetic counseling has been practiced on a small scale for several decades (Hammons 1959). Now it is becoming a more widely recognized part of medicine. The rapid acquisition of new knowledge means an increase in the number of known inherited diseases. Most of these are rare, but collectively they account for a great deal of human misery. The United Nations (1958) report, previously mentioned, lists more than a hundred supposedly monogenic, severe traits with a total incidence of somewhat more than one per cent of all births.

In some instances genetic knowledge is very firm. For a dominant gene with full penetrance the geneticist can give an exact risk figure, usually 50 per cent. The recurrence risk—the risk that a second child will be affected—with a recessive disease from normal parents is safely taken as 25 per cent in most cases, since the

285

probability that the first child resulted from a new mutation is very small. Improved methods of carrier detection, such as that for galactosemia, will eventually make possible the identification of carriers of many diseases before they produce affected children.

In some cases, as mentioned, valuable information is obtained from biochemical analysis. But I should like to emphasize also that an important source of information of a more routine type is the study of pedigrees. A great deal of knowledge about the inheritance of eye defects, for example, has been gathered mainly through pedigree studies (François 1961).

Another area of great progress in the past few years has been in cytogenetic analysis. Persons who carry chromosomes with balanced rearrangements (usually translocations) have a high probability of producing unbalanced gametes which, when fertilized, become nonviable or abnormal zygotes. For example, some instances of mongolism are known to be caused by a translocation. A cytogeneticist can sometimes identify those sibs or other relatives of the affected person who are carriers of the translocation and therefore have a high risk of producing abnormal children.

As knowledge increases, the number of diseases about which the counselor can give precise and useful information will increase correspondingly. At present the number is regrettably small, but fortunately accurate information is accruing rapidly.

It seems to me that one aspect of counseling has lagged behind. There are many conditions for which the heredity is not simple and not understood—indeed, it may not even be known whether the disease is primarily genetically caused—but for which there is a substantial clustering within families or pedigrees. For many such conditions empirical risk figures could be obtained. Parents who have had a severely impaired child would often like to know the risk of recurrence. They do not care whether a 25 per cent risk is based on genetic knowledge or past experience, as long as both risk figures are equally reliable.

With more attention to preventive medicine, better systems of medical record keeping, and more precise vital statistics, we could advise better than we now do without waiting for improvement in fundamental knowledge. We need not only more accurate but more specialized risk figures. For example, the recurrence risk following the birth of a child with cleft palate is usually given as about 5 per cent. But this is a composite value, undoubtedly an average of a quite heterogeneous collection. It might well be that the recurrence risk differs greatly depending on whether there are half a dozen previous normal sibs or whether the affected child is the only child. It certainly changes if one of the parents is affected of if there were two affected children rather than one. By analogy with life insurance rate calculations, we could have risks adjusted for age, sex, other relatives affected, relevant environmental factors, maternal age, race, or anything else that has a measurable modifying effect—whether the basis for this modification is understood or not.

A particularly useful summary of empirical risks was prepared by Fraser (1954). But this is a small beginning. With better systems of reporting and recording diseases and with computer systems for collating and analyzing, the genetic counselor could provide useful information in far more cases than he now can. Risks can be made more accurate and refined as additional risk-modifying factors are discovered. Collection of risk data has not yet attracted the amount of serious work that it might have. New fundamental knowledge is always welcome, of course, and in the long run it is far more desirable. But my point here is that much could be done now, with what is already known or obtainable, to make genetic counseling a more precise art.

As a single example, the information given by a genetic counselor regarding mongolism prior to 1959 might seem now to be hopelessly outdated. But as far as risk information is concerned, it would have been quite good. The great majority of cases are ordinary trisomy; the recurrence risk is very low; the maternal age effect is striking enough to take seriously; and the probability of its occurrence in the rare offspring of a mongoloid female is very high. The number of mistakes made by reliance on empirical information would have been relatively few. Now, of course, a good cytological analysis can help to detect the few cases that are the result of chromosome breakage. Those relatives who carry balanced rearrangements, and thus have a higher risk of producing abnormal children, can be identified. It must be stressed, however, that even in such cases the actual risk figures can only be empirical; indeed, the risk may differ in the two sexes. It has been found that females with balanced heterozygous translocations transmit unbalanced gametes far more frequently than do males. It is true that experiments with *Drosophila* and various plants have shown that alternate segregation from translocation heterozygotes is frequently higher in one sex than the other, but this information is not sufficient to permit predictions about the situation in man. Furthermore, the mechanism for the unequal transmission may be of other, less likely sorts; for example, possibly not all of the sperms function regularly, as may be the case in *Drosophila*, or there may possibly be gamete selection. In any case one has to rely on the empirical information.

These instances in which genetics has made a major impact on medicine are only a small sample, and a biased one. In particular, I have said nothing about the ramifications of the new knowledge of the gene and its action coming from molecular genetics and the study of microorganisms. Most of this is too new to have made its great impact, but that it eventually will is as certain as any future prediction can be.

There is a possibility that genetic defects can be repaired at more fundamental levels, cellular or intracelluar, rather than at levels more distantly removed from the primary gene action. One can hope to repair agammaglobulinemia by stimulating or introducing competent plasma cells, or to cause the endogenous production of antihemophilic globulin, or to repair the defective enzyme in phenylketonuria rather than prescribing a diet that is unsatisfactory in many respects. A large part of the developmental engineering that is being discussed as a future possibility (e.g., Wolstenholme 1963) can be expected in the realm of repair of gene deficiencies.

I should like to end this section by saying that to me it seems the greatest contribution genetics can make to medicine is to increase the ever-deepening chemical and biological knowledge of the human organism on which rational prevention, diagnosis, and treatment of disease must rest.

Some Contributions of Medicine to Genetics

Contributions of medicine to genetics can be conveniently divided into two categories. One contribution is to the formal genetics of man—the extending to man of the kinds of genetic knowledge that already exist for such organisms as *Drosophila*. The second is more basic: It is the contribution of medicine to our understanding of genetic principles. There are so many ways that man is unsuited for genetic research that it is something of a surprise and a testimony to the ingenuity of the various workers that so much has been learned.

THE FORMAL GENETICS OF MAN. Genetic analysis depends on gene markers, and most of man's phenotypic differences on which genetic research depends have been learned from medicine. As already mentioned, a large number of inherited traits and their mode of inheritance are known. In distinction to the mouse or *Drosophila*, in which most of the known genes are recessive, most of the known monogenic traits in man are dominant. This difference is readily understandable, for inbreeding is a part of the routine methodology of mouse and *Drosophila* genetics. But it means that most of the recessive genes in man are yet to be discovered; many may come to expression only very rarely, or never, because of elimination through heterozygous selection before they have a chance to become homozygous.

The human linkage map is still in a primitive state. Most known genes are very rare, so the probability that two of them will be found segregating in the same family is vanishingly small. As a consequence, the only known linkages depend on associations with a few markers with high frequencies, such as blood group genes.

The seeming synapsis of the X and Y chromosomes during spermatogenesis appeared to provide a possibility for crossing over between the X and Y, as was known to occur in some fishes. Haldane (1936) pointed out that this crossover might be taking place and that traits determined by genes located on the homologous parts of the X and Y chromosomes would have a characteristic pattern of inheritance. For example, the sex of a child with a dominant trait inherited through the father should be correlated with that of the affected paternal grandparent. Any exceptions would be attributable to crossing over between the gene and the end of the homologous region. Thus the geneticist would have a possible basis not only for identifying genes in this region but also for mapping their location without having to find families segregating simultaneously for two rare genes. For recessive genes similar, but more indirect, procedures were given.

On the basis of such studies, several maps of the human X and Y chromosomes were made (e.g., Snyder 1941). The situation soon become one of an embarrassment of riches, as too many traits seemed to satisfy Haldane's criteria. It gradually came to be realized that various sex-biases in the disease manifestation could stimulate this mode of inheritance, and none of the cases has since stood up under rigorous analysis. So it now appears that Haldane's suggestion was an ingenious failure.

Another way in which genes may be identified as being on a particular chromosome is by the examination of phenotypic frequencies in a population trisomic for this chromosome. Such an approach has been applied to the analysis of blood group phenotypes in individuals with mongolism, and a possible association of the ABO locus with the mongolism chromosome was reported (Shaw and Gershowitz 1963). As new protein polymorphisms are discovered, this procedure could be applied to them. It is likely that, with electrophoretic or gel-diffusion methods, two protein types (say, A and B) could be resolved into three phenotypes. These could be AA, BB, and AB in diploids and AAA, BBB, AAB, and ABB in trisomics. Even if the last two cannot be distinguished, one could easily compute the expected frequencies of $AAB + ABB$ in terms of the gene frequencies, p_A and p_B, and compare them with the expectations in diploids. If the gene is near the centromere and non-disjunction is at the second meiotic division or is postzygotic, the proportion of heterozygotes will be the same in triploids as in diploids, namely $2p_A p_B$. This result is formally equivalent to endosperm inheritance in maize. On the other hand, if the nondisjunction is at the first meiotic division, the frequency of $AAB + ABB$ types can be shown to be $3p_A p_B$, or 50 per cent greater than that in diploids. Finally, if the gene is far from the centromere, the frequency approaches $8p_A p_B/3$ with either first or second division nondisjunction, a 33 per cent increase.

Thus, with enough data, one might hope to identify a gene as being on the mongolism chromosome, except for the unlikely possibility that the gene is very close to the centromere and nondisjunction hardly ever occurs at the first meiotic division. Conversely, a 50 per cent increase would show not only that the gene is near the centromere but also that, as in *Drosophila* (Merriam and Frost 1964), meiotic nondisjunction is mainly, if not exclusively, at the first division.

Both of these methods are applicable only to the chromosomes that produce viable trisomic types, and are thus very restricted. Deletion mapping has been useful in *Drosophila* and even more successful, though without the cytological correlation, in microorganisms. Some possible instances of autosomal hemizygosity in individuals carrying a recessive gene on one chromosome and a deletion including the corresponding locus on the homologous chromosome have been reported (see Elmore *et al.* 1966). By the accumulation of such information one may hope, in cytologically favorable cases, to associate known genes with specific chromosome regions.

One of the products of the study of human disease and abnormalities has been the elucidation of the sex-determining mechanism in man. For many years it had been assumed that man would follow the same rules as *Drosophila* and that the Y chromosome would be unimportant; although if human geneticists had taken the silk moth as a model, the role of the Y chromosome would have been appreciated. Not only were the discoveries of the XXY (Klinefelter) type and the XO (Turner) type important medically as explanations of the etiology of these two well-known but little-understood syndromes; they also showed unambiguously the role of the Y chromosome in human sex determination.

The discovery of the XO mouse and later the XXY mouse and XXY cat suggested that the Y chromosomes may have general importance in the mammals. The XO mouse is a phenotypically normal and fertile female; the XXY male is normal, though sterile. Thus the male sex seems to be determined entirely by the presence or absence of the Y chromosome. That this is not the whole story in man is clear from the fact that the XO and XXY types are not fully normal females and males.

A surprise has been the discovery of persons with more than three sex-chromosomes. Individuals with as many as five have been found, with various combinations of X's and Y's. In the first place it is not clear what mechanism leads to such whole- ̄ae increases in chromosome number for the sex chromosomes. In the second place, ̄c was surprising that such extreme examples of aneuploidy would be viable.

A reason for their viability was suggested by the discovery of the single active X principle (Lyon 1962). The observed fact that the maximum number of Barr (chromatin) bodies in the human interphase nucleus was always one less than the number of X chromosomes led to the proposal that the Barr bodies represent inactive X chromosomes. It has now been convincingly demonstrated that, within a particular clone of somatic cells, only one X chromosome is fully functioning. This provides not only an explanation of how a person with genotype XXX, XXXX, or XXXXX can be viable but also a totally new explanation for the phenomenon of dosage compensation (Muller 1950). Muller has presented evidence for the existence in *Drosophila* of genes whose function is to adjust the action of X-chromosome genes such that the phenotype is the same with one X chromosome as with two. The single active X in mammals accomplishes exactly the same purpose, namely, assuring that the male and female have the same phenotype for sex-linked but not sex-related characters. The mechanism, however, is totally different from that in *Drosophila*.

CONTRIBUTIONS TO BASIC GENETICS. The single active X hypothesis is of interest as an example of formal human genetics. It is of far greater genetic interest, however, as a case where the study of man (plus some very important related findings in the mouse) has demonstrated a new mechanism of dosage compensation. It provides an excellent example of how medicine has made an important contribution to basic genetics.

There are others. One is the association between genes and enzymes implied by Garrod's early studies on inborn errors of metabolism, already mentioned. In this case a deep insight into the nature of gene action was gained by the study of persons with diseases resulting from metabolic errors. Another example is agammaglobulinemia. This X-linked disease has shown the importance of genetic control over the development of plasma cells and the synthesis of gamma globulins, and the study of this "experiment of nature" has aided the understanding of immune mechanisms (Good and Zak 1956).

One of the greatest single contributions of medicine to basic genetics has been through the study of diseases caused by abnormal hemoglobins. That sickle cell anemia is an inherited disease has been known for some time. Pauling's suggestion that there was a defect in the hemoglobin molecule led eventually to the demonstration by Ingram that the difference between S-hemoglobin and normal hemoglobin lay in the substitution for valine of glutamic acid at a specific position in the molecule. This first demonstration that a mutation can lead to a single amino acid substitution, at a time when the nature of the gene was beginning to be understood, provided a key part in the story of the role of the genes in protein synthesis (see Ingram 1963).

Many more hemoglobin variants are now known that depend on an amino acid substitution in the alpha or the beta chain. With this chemical knowledge comes a greater understanding of the wide variety of hemoglobinopathies.

With improved methods for analysis of the amino acid sequence of proteins, it is possible to study in detail the hemoglobins and other proteins from various animals. Thus the comparative molecular structure of related species can be studied, and for the first time the rate of evolution of the amino acid composition of proteins can be estimated. Studies of hemoglobin, cytochrome-C, and ribonuclease have shown that the rate of change is roughly one amino acid change per ten million years (for a recent review, see Epstein and Motulsky 1965). Most changes are consistent with there having been a single nucleotide substitution, so the rate probably represents only a slight underestimate of the number of mutations that have been incorporated into the species.

We can try to relate this estimate to other estimates of the rate of evolution given by population genetics theory. Haldane (1957) showed that the number of gene substitutions that a species can make with a certain total intensity of selection depends almost entirely on the initial frequency of the mutant and hardly at all on the selective advantage of the mutant. Using this principle and making what he considered reasonable assumptions about the relevant numerical quantities, he concluded that a species could probably make a gene substitution every 100 to 1000 generations. If there are 10,000 genes per genome, then the number of changes incorporated per gene would be one per million to ten million generations. So the rates of evolution of hemoglobins and cytochrome are not unexpected, if Haldane's principle is an important factor in determining evolutionary rates.

One great advantage of man as a subject for genetic research is that we know a great deal about human morphology, physiology, and behavior, and therefore small differences are noticed. Another advantage, of particular importance for epidemiological and population research, is that many relevant records are already available. We can expect still more information in the future from better medical record-keeping systems, computerization, record linkage, and census records. Excellent material for combined medical and genetic study has been isolated populations, such as the Amish and Hutterites.

One way in which available data can be used takes advantage of the fact that a person's surname is inherited as if the appropriate gene were carried on the Y chromosome with a delay of one generation in its expression. By exploiting this, one can estimate the inbreeding coefficient from the proportion of marriages of persons with the same name. There are a number of obvious pitfalls, but in some circumstances the method may be useful (Crow and Mange 1965).

Through the study of gene frequencies in different populations, medical genetics has helped bring about a better understanding of the way in which genetic variability is maintained. The best understood human polymorphism is sickle cell anemia, where the S gene has been maintained in some parts of the world by the heterozygote's advantage of a greater resistance to falciparum malaria, as established by both epidemiological and physiological studies. The interrelationships of the hemoglobins involved are complicated, but it is already clear that a number of other hemoglobin abnormalities have also been maintained because of resistance to malaria. Haldane was the first to emphasize the importance of disease resistance as a factor in human evolution. Recent findings, such as those above, have apparently borne out Haldane's thesis and, furthermore, suggest that some contemporary polymorphisms are relics of disease-resistance mechanisms that were once important but are no longer relevant because of the virtual eradication of the disease.

Two decades ago it was fashionable to explain the frequencies of seemingly neutral genes (such as blood group factors) in different populations as determined primarily by random drift. However, after Fisher (1939) had analyzed Nabours' data on grouse locusts, he and Ford were inclined to attribute most such differences to differences in selection intensities in different populations. There have been a number of recent reports of very strong selection at a number of loci, particularly those concerned with the hemoglobins and ABO compatibility. Although intense selection at a few loci is possible, every animal breeder knows that he cannot select intensively for a great many characters at the same time. Thus the number of strongly selected polymorphisms is limited.

It has been apparent for some time that medical studies of large populations living under primitive conditions should provide some of the best information on the way in which selection is acting in man and some insights as to how it has operated in the past. Morton (unpublished) has recently completed a study of over 1000 Brazilian families chosen from a group characterized by high fertility, genetic variability, consanguinity, and infant mortality. Such a population maximizes the possibility that evidence for selection will be found. Included in the study were analyses of all the blood groups for which suitable sera are available, haptoglobins, transferrins, and the Gm and Inv factors, as well as tests of phenylthiocarbamide tasting. The general result was that no evidence of intense selection was found for any of these factors, except the previously known selection against incompatible ABO types. Thus it would appear that most of these polymorphisms at present are not associated with large selective differences, although differences too small to be detected even with this large population sample could still be responsible for maintenance of polymorphism through heterozygote advantage, selection for rarity, or other known mechanisms.

Origin of new genes by duplication has been recognized by *Drosophila* workers for many decades; in particular the phenomenon was emphasized by Bridges, who noticed many "repeats" in salivary gland chromosomes. Smithies, Connell, and Dixon (1962) have provided convincing chemical evidence that in human haptoglobins the Hp^2 allele is an almost complete duplication, combining Hp^{1F} and Hp^{1S} which differ from each other probably by only a single amino acid. The analogy with the Bar duplication is close, and the predicted triplication that would arise from unequal synapsis in an Hp^2 homozygote has been found. Presumably this triplication is selectively disadvantageous, for it is very rare. Nance (1963) has suggested a mechanism for maintenance of the haptoglobin polymorphism that differs from the usual assumption of heterozygote advantage. He suggests that, for some unknown reason, Hp^2 has a slight selective advantage, but that the two Hp^1 alleles are recurrently produced by unequal synapsis and crossing over in Hp^2 homozygotes, an event complementary to that which produces the triplication. Such recombination is probably not extremely rare; by analogy with Bar reversion, it would be expected to occur occasionally although it has not yet been directly demonstrated. However, if selection is very slight at the haptoglobin loci, unequal synapsis and crossing over might be sufficient to account for the presence of Hp^{1F} and Hp^{1S} in the population. If this hypothesis turns out to be correct, it will be an instance in which human genetics has demonstrated a new type of polymorphism-maintaining mechanism.

More examples could be given, but those mentioned are illustrative of the fact that medical genetics, in addition to its role in medical research and practice, has made substantial contributions to basic genetics. As cell and molecular genetics continue to develop new techniques, more and more of these will be applicable to the study of human genetics, and we can expect the interaction between basic genetics and medical genetics to become stronger.

Some Genetic Consequences of Medical Advances

The path connecting genetics and medicine, once narrow and rarely traveled, is now a busy highway. It is probably inevitable that there are some collisions.

One problem, which has long been foreseen but which becomes greater in proportion to the success of medicine and other forms of environmental improvement, is the perpetuation of diseases and abnormalities that were once ruthlessly eliminated by a rigorous environment. Clearly many persons who in an earlier period would have died or been so incapacitated as to lower their fertility are surviving and having children. To the extent that these conditions are heritable, they are being transmitted to future generations, adding to the already heavy burden of disease.

From studies of consanguineous marriages, it has been estimated that the average person carries some four or five hidden recessive genes that, if made homozygous, would produce recognizable and reasonably serious disease. Despite the fact that the average person carries several such genes, the number of different kinds is so great, and the individual mutant types are so rare, that it is very unlikely that two unrelated persons have even one severe recessive gene in common. As carrier detection becomes cheaper and more accurate, such rare occurrences when two persons carry the same hidden gene can be discovered. Two persons who know that they carry the same gene may, depending on the nature of the disease and many other factors, choose not to marry or not to have children.

If such marriages are avoided, how much will it restrict the range of marriage partners? From the incidence of any given disease the carrier frequency can be estimated. With random mating, if p^2 is the incidence, then $2p(1 - p)$ is the carrier frequency, and the probability that two unrelated persons share the gene is the square of the carrier frequency, or about $4p^2$. This is four times the incidence. Summing the probabilities for all known recessive genes gives a value of only one or two per cent; so the restriction on marriage partners would be quite small. But if such a practice only altered the choice of mate and not the average number of children, there would be a corresponding additional contribution of harmful genes to future generations.

How may we assess the extent to which the gene frequency may be changed by avoidance of marriage in certain cases? To make the discussion concrete, I shall use phenylketonuria (PKU) as an example. With low phenylalanine diets many persons with PKU will grow up normally and may be expected to have children of their own. (I shall ignore the complications that may ensure when a mother who cannot metabolize phenylalanine has a child. It may well be that special diets will be required during pregnancy, or that wholly unforeseen difficulties may appear.)

No one knows the factors responsible for the present incidence of PKU, which is roughly one per 10,000 live births. It may be that the gene is completely recessive, in which case a mutation rate of 10^{-4} would be required to account for the present incidence, if it is assumed that the population is at equilibrium and that the fitness of homozygotes is zero. If the amount of inbreeding in the past has been appreciable or if there has been some selection against heterozygotes, a higher rate must be postulated. On the other hand, there may be or may have been in the past some selection in favor of heterozygotes. To account for the present incidence by heterozygote advantage would require only a one per cent selective advantage, a value too small to be measured. However, we can still assess the impact on the next few generations of relaxed selection against homozygotes without knowing how the gene frequency is maintained.

Let us assume the successful repair of all homozygotes for the harmful gene. This means that a fraction of the population, equal to p^2 which would otherwise not reproduce now does so, and therefore p^2 recessive genes are contributed to the next generation above the number that would be there otherwise. If p' is the frequency of mutant genes that would otherwise be present in the next generation, the proportion will be approximately $p' + p^2$. If p and p' are nearly the same, and there is no reason to think that the gene frequency is changing rapidly, the frequency has been increased by a fraction p; that is, it is $p(1 + p)$ instead of p. If p is $1/100$, as it is for PKU, the gene frequency in the next generation, if complete repair has resulted in normal fertility of the present generation, would be increased by one per cent of its former value.

An increase of one per cent in the gene frequency means that the frequency of homozygotes would be multiplied by a factor $(1.01)^2$, which is about a two per cent increase. With a two per cent increase per generation, it would require about 40 generations, or 12 centuries, to double the incidence. In comparison with the rate at which knowledge is changing, that is a long time.

In general, if p is the gene frequency and s is the selective disadvantage of the homozygote before treatment, a completely successful treatment will cause an increase in gene frequency sp and, with random mating, the genotype will increase by about $2sp$. With less than a complete restoration of fertility, the increase would, of course, be correspondingly less. This formula also makes it clear that the rarer the trait (i.e., the smaller p is) the less is the proportional increase in incidence in future generations.

I conclude that the successful treatment of rare recessive genes will not cause any great problem for the foreseeable future. Compared with the great alleviation of suffering in this generation by prevention or cure, the cost to the next few generations (who presumably will have the same or better cures available) seems small.

With a dominant gene the story is quite different. If I again let s stand for the amount by which the fitness is impaired in the untreated condition, a complete repair or at least restoration of normal fertility will increase the incidence of the disease in the next generation by a fraction s. For example, a dominant disease that reduces the expectation of children by 35 per cent would increase in incidence by about this amount in the next generation if all cases were to become fully fertile. This principle is easily seen for the case of a dominant gene that is lethal or sterilizing ($s = 1$). For such a gene, the entire incidence in this generation is the result of new mutations. If the disease is repaired and full fertility restored, the mutations from this generation are added to those that occur next generation, so the incidence would be doubled.

An example of a disease whose heredity is not known exactly, but which depends at least in part on one or more dominant genes, is pyloric stenosis. Prior to World War I this condition was nearly always fatal, but with the development of Rammstedt's operation many cases were repaired surgically. These persons have in many cases reached the age of reproduction and have now produced a number of children. About one in five of the sons of a parent who has had the operation is affected. Thus the incidence is increased by this amount beyond those cases that would have occurred anyhow.

Thus, we see that the cure of a dominant disease has a far greater effect on the next generation than the cure or prevention of a recessive disease. On the other hand, the consequences for society may not be so greatly different, at least if the person so treated is aware of the heredity of his condition. A person with a dominant disease that has been successfully treated knows that, if he has children, each of them has a 50 per cent chance of having the same condition. If the cure is complete, nontraumatic, and inexpensive, there is little reason to fear the disorder, and the parent would probably have little hesitation in subjecting his children to the risk. However, if the disease is severe, and if the treatment is only partially satisfactory, he would probably prefer not to subject his children to the 50 per cent risk. So the situation for the next generation would have a certain tendency to adjust itself to the perpetuation mainly of those diseases that are most effectively treated. The interest of society in lowering the incidence of disease is matched by the deep interest of the parent in his own children.

Again, a person who has been successfully treated for a recessive disease has little reason to fear for his own children. For example, a man who has been treated for PKU would have about one chance in 50 of marrying a woman who carried the same gene. Thus, unlike the one who has survived a dominant disease, he has little reason to be concerned over his own children. The concern is primarily one of society for an increased average incidence. But as we have seen, the percentage increase for recessive diseases is very slight unless the gene is common.

Thus the situation does not call for immediate grave concern. The repair of dominant conditions creates a problem only to the extent that parents are uninformed or choose to have children subject to the same risk that they themselves had. With rare recessives there is only a small increase, though a permanent and cumulative one.

But what about common conditions—weak hearts, diabetes, various constitutional disorders, mental disease? Here the answers are far harder. We need to know at least the following for each condition: (1) Is the repair or environmental advance affecting the reproductive pattern? (2) How complete is the repair? And, hardest, (3) How heritable is the trait? At present only the crudest guesses can be made.

One thing is abundantly clear: An environmental improvement must be permanent. Any return to the old conditions ordinarily brings back the original incidence, plus any increase gained by the perpetuation of genes that would otherwise have been eliminated. So we are certain to have to devote an increasing fraction of our economic resources to making up for one another's genetic deficiencies.

The question then becomes, how soon and to what extent should man start to intervene in his genetic future?

There is a wide measure of agreement that prospective parents are entitled to the best information available about the risk of their having diseased or defective children. Considerable human suffering would thereby be eliminated, but it is probably a small fraction of the total.

There seems to be no hesitancy on the part of society to improve the environment. This may mean cultural and educational improvements, as emphasized previously by Beadle (Ch. 17). It also can mean the most sophisticated kinds of developmental engineering as envisaged in the word euphenics (Wolstenholme 1963).

A recent symposium (Sonneborn 1965) presented several possible uses of molecular and cell biology to influence heredity. There were a number of suggestions for removal, addition, or replacement of genes based mainly on procedures developed for microorganisms. Indeed the potential possibilities are enormous. However, all such suggestions have one thing in common: They are not now practical, and one doesn't yet have to worry about the possible consequences of their misuse or accept any responsibility for having advocated them.

H. J. Muller, writing in that same symposium (Sonneborn 1965), fears that too much discussion of such possibilities as cell transplants, DNA transformations, directed mutation, or suitably designed episomes may result in escapism and postponement of practical genetic manipulation. In his words, " It would be intellectually dishonest and morally reprehensible of us to exploit the hope of mankind's eventual success in this enterprise as an excuse for not giving our support to the great re-educational process that could make possible, by means now physically available, a most significant advance in the genetic constitution of our species." He is referring, of course, to artificial insemination.

Genetic engineering, if it becomes practical, will probably be most effective for single gene traits. But the hereditary components of many of the most important human traits, intelligence, general health, emotional stability, unselfishness—to the extent that they are genetic—are likely to be polygenic. The principle that like begets like is the best guide to prediction for some time to come, and selection the most effective means of change. If we really want to change the human population by genetic means, Muller's method is the most likely to succeed.

But it is much less clear what society or the individuals comprising it want to do. There is no groundswell of public opinion in favor of doing anything by way of positive eugenics.

The rate of increase in knowledge is high and accelerating. To the next generation many of our beliefs of today will seem quite primitive, and any attempts at genetic manipulation of the population may seem similarly crude. On the other hand, to wait for complete knowledge is to do nothing forever.

Will future generations regard our generation somewhat as we do the pioneers who destroyed our forests and wildlife—as geneticists without the wisdom and courage to look to the future? Or, on the other hand, will they regard this generation as one which prudently refrained from rushing to act too soon in ignorance?

I have one conviction; it is high time that the social implications of our expanding genetic knowledge be discussed. Early eugenics was crude, oversimplified, and got confused in various dubious (and in some cases disastrous) political movements. I hope we are ready for a more mature consideration of eugenics and euphenics as complementary possibilities. It may well be that the second century of Mendelism will mark the beginning of a serious and informed consideration of the extent to which man can and should influence his biological future, with full deliberation on both the opportunities and the risks.

Literature Cited

Becker, P. E., 1964. *Humangenetik*. Georg Thieme Verlag, Stuttgart.

Biggs, R., and R. G. Macfarlane, 1962. *Human Blood Coagulation and Its Disorders*. F. A. Davis, Philadelphia.

Boyer, S. H., 1963. *Papers on Human Genetics*. Prentice-Hall, Inc., Englewood Cliffs, New Jersey.

Crow, J. F., and A. P. Mange, 1965. Measurement of inbreeding from the frequency of marriages between persons of the same surname. Eugen. Quart. **12** : 199–203.

Elmore, S. M., W. E. Nance, B. J. McGee, M. Engel-de Montmollin, and E. Engel, 1966. Pycnodysostosis, with a familial chromosome anomaly. Amer. J. Med. **40** : 273–282.

Epstein, C. J., and A. G. Motulsky, 1965. Evolutionary origins of human proteins. Progr. Med. Genet. **4** : 85–127.

Fisher, R. A., 1939. Selective forces in wild populations of *Paratettix texanus*. Ann. Eugen. **9** : 109–122.

François, J., 1961. *Heredity in Ophthalmalogy*. C. V. Mosby Co., St. Louis.

Fraser, F. C., 1954. Medical genetics in pediatrics. J. Pediat. **44** : 85–103.

Garrod, A. E., 1909. *Inborn Errors of Metabolism*. Reprinted 1963, Oxford University Press, Oxford.

Good, R. A., and S. J. Zak, 1956. Disturbances in gamma globulin synthesis as "experiments of nature." Pediatrics **18** : 109–149.

Haldane, J. B. S., 1932. Genetical evidence for a cytological abnormality in man. J. Genet. **26** : 341–344.

————, 1936. A search for incomplete sex-linkage in man. Ann. Eugen. **7** : 28–57.

————, 1957. The cost of natural selection. J. Genet. **55** : 511–524.

Hammons, H. G. (ed.), 1959. *Heredity Counseling*. Hoeber-Harper, New York.

Ingram, V. M., 1963. *The Hemoglobins in Genetics and Evolution*. Columbia Univ. Press, New York.

Lejeune, J., J. Lafourcade, R. Berger, J. Vialatte, M. Boeswillwald, P. Seringe, and R. Turpin, 1963. Trois cas de délétion partielle du bras court d'un chromosome 5. Compt. Rend. **257** : 3098–3102.

Lejeune, J., R. Turpin, and M. Gautier, 1959. Le mongolisme, premier example d'aberration autosomique humaine. Ann. Génét. Hum. **1** : 41–49.

Little, C. C., and E. E. Tyzzer, 1916. Further experimental studies on the inheritance of susceptibility of a transplantable tumor. J. Med. Res. **33** : 393–453.

Loeb, L., and S. Wright, 1927. Transplantation and individuality differentials in inbred families of guinea pigs. Amer. J. Pathol. **3** : 251–283.

Lyon, M. F., 1962. Sex chromatin and gene action in the mammalian X-chromosome. Amer. J. Hum. Genet. **14** : 135–148.

Merriam, J. R., and J. N. Frost, 1964. Exchange and nondisjunction of the X chromosomes in female *Drosophila melanogaster*. Genetics **49** : 109–122.

Muller, H. J., 1950. Evidence for the precision of genetic adaptation. Harvey Lect. **43** : 165–229.

Nance, W. E., 1963. Genetic control of hemoglobin synthesis. Science **141** : 123–130.

Nowell, P. C., and D. A. Hungerford, 1960. A minute chromosome in human chronic granulocytic leukemia. Science **132** : 1497.

Owen, R. D., 1945. Immunogenetic consequences of vascular anastomosis between bovine twins. Science **102** : 400–401.

Punnett, R. C., 1910–11. Mendelism. *Encyclopaedia Britannica* (11th ed.).

Race, R. R., and Ruth Sanger, 1962. *Blood Groups in Man* (4th ed.). F. A. Davis Company, Philadelphia.

Shaw, M. W., and H. Gershowitz, 1963. Blood group frequencies in mongols. Amer. J. Hum. Genet. **15** : 495–496.

Smithies, O., G. E. Connell, and G. H. Dixon, 1962. Chromosomal rearrangements and the evolution of haptoglobin genes. Nature **196** : 232–236.

Snyder, L. H., 1941. *Medical Genetics*. Duke Univ. Press, Durham, North Carolina.

Sonneborn, T. M. (ed.), 1965. *The Control of Human Heredity and Evolution*. Macmillan, New York.

Stanbury, J. B., J. B. Wyngaarden, and D. S. Frederickson, 1965. *The Metabolic Basis of Inherited Disease*. McGraw-Hill, New York.

Stern, C., 1960. *Principles of Human Genetics* (2nd ed.). W. H. Freeman and Co., San Francisco.

Tjio, J. H., and A. Levan, 1956. The chromosome number of man. Hereditas **42** : 1–6.

United Nations, 1958. Report of the United Nations Scientific Committee on the Effects of Atomic Radiation. General Assembly, Official Records, 13th Session, Supplement 17 (A/3838).

Wolstenholme, G. (ed.) 1963. *Man and His Future*. Little, Brown, Boston.

Yunis, J. J., 1965. *Human Chromosome Methodology*. Academic Press, New York.

30 Genetic Counseling

F. C. Fraser

Recent dramatic advances in medical genetics have resulted in a comparable increase of interest in genetic counseling. These advances have included rapid growth in knowledge of the ways in which a large number of diseases are inherited, improvements in the ability to examine human chromosomes and detect chromosomal abnormalities, the ability to diagnose certain diseases in the second trimester of pregnancy, and the advent of screening programs for certain diseases in high-risk populations. The increased interest in genetic counseling has created two major effects: (1) a demand for more and better counseling services, and (2) a growing realization that we know very little about the present extent of, and need for, counseling services and the optimal methods of delivering counseling. A workshop was therefore organized under the sponsorship of the National Genetics Foundation, Inc., which was designed to evaluate and make recommendations about the status of genetic counseling, its goals, nature, achievements, and needs.

Definition

It was agreed that

genetic counseling is a communication process which deals with the human problems associated with the occurrence, or the risk of occurrence, of a genetic disorder in a family. This process involves an attempt by one or more appropriately trained persons to help the individual or family (1) comprehend the medical facts, including the diagnosis, the probable course of the disorder, and the available management; (2) appreciate the way heredity contributes to the disorder, and the risk of recurrence in specified relatives; (3) understand the options for dealing with the risk of recurrence; (4) choose the course of action which seems appropriate to them in view of their risk and their family goals and act in accordance with that decision; and (5) make the best possible adjustment to the disorder in an affected family member and/or to the risk of recurrence of that disorder.

A report based on a Workshop on Genetic Counseling sponsored by the National Genetics Foundation, Inc., held in Washington, D.C., December 10–13, 1972. Support for the workshop and for publication of this report came from the National Institute of General Medical Sciences, grant no. GM 19997. With the assistance of R. M. Antley, R. Y. Berini, B. Childs, J. F. Crow, C. J. Epstein, R. D. Freeman, J. G. Hall, L. L. Heston, M. M. Kaback, R. L. Masland, W. J. Mellman, A. G. Motulsky, J. V. Neel, D. Rimoin, M. Rivas, L. E. Rosenberg, G. S. Sachs, W. S. Sly, J. R. Sorensen, A. G. Steinberg, and M. W. Thompson.

Description

Medical genetics units often provide diagnostic services such as karyotyping, dermatoglyphic analysis, syndrome recognition, and, in some cases, biochemical tests related to the inborn errors of metabolism. They also, in some cases, take part in the management of children with rare inborn errors of metabolism. Third, they provide counseling, and it is implicit in the above definition that counseling does not include the diagnosis or clinical management of the patient's disease. For instance, amniocentesis is an obstetrical procedure, the biochemical or cytogenetic tests performed on the resulting specimen are diagnostic, the advice given to the parents is genetic counseling, and the termination of the pregnancy (if such occurs) is management. Screening populations for hereditary conditions such as phenylketonuria or sickle cell trait is diagnostic, and counseling is provided for the high-risk individuals so identified. It is impossible to draw a hard and fast line as to where in the genetic workup diagnosis stops and counseling begins, but one might consider that diagnostic procedures not requiring any knowledge of genetics are not part of genetic counseling, while taking the family history of a counselee is.

Most counseling involves the occurrence of a particular disease in a child and the concern of the parents as to whether their future children might be similarly affected. Parents may also want to know about the risk for the affected child's children or the children of unaffected sibs. Other situations in which counseling is requested may involve, for instance, a person contemplating marriage who is concerned about some aspect of the future spouse's family history. There may be a specific disease in a near relative, ancestors from a different racial group, or the future spouse may be a relative. Occasionally the question may involve the presence of a disease or racial admixture in the family of a child being considered for adoption. Recently the advent of screening programs for certain diseases in high-risk populations has introduced prospective counseling—where the counselees have been identified as high-risk couples before the disease has ever occurred in the family (such as carriers of the Tay-Sachs gene).

In many cases the family doctor is the most appropriate person to do the counseling since he knows the family, its attitudes, and the socioeconomic background better than a consultant. However, he may have neither the genetic knowledge nor the time to devote to what may be a series of interviews. And frequently the family does not have a family doctor. In any case, some counseling problems are so complex or require sufficiently specialized tests that the services of a professional genetic counselor are required. The present report will deal primarily with cases that, for various reasons, have been referred to a medical genetics unit for counseling.

The process of genetic workup and counseling may be considered in several stages: validation of the diagnosis, obtaining the family history, estimation of the risk of recurrence, helping the family to reach a decision and take appropriate action, and follow-up.

VALIDATION OF DIAGNOSIS. The patient and/or family may be referred to the medical genetics unit for diagnosis by biochemical or cytogenetic tests or in the hope that the geneticist will recognize a syndrome not generally known to the medical profession. In the latter case the genetics unit will need the services of someone skilled in "syndromology," who may in turn need to obtain further tests such as radiographs or electrocardiograms. If the problem is biochemical, it may involve specialized diagnostic tests that are not available in most laboratories, such as enzyme assays, so that specimens may have to be sent to one of the few laboratories in the country that does the assay. This is often done on an informal basis, but to meet the increasing demand for, and complexity of, referral needs, several voluntary agencies have set up more formal organizations whereby patients suspected of having a genetic disease which requires an enzyme assay or other test not generally available can be referred to the nearest center where the test is available; the appropriate specimens (or sometimes the patient) can be sent to the center for diagnostic workup.

Often the diagnosis already has been made by a physician before referral to a medical genetics unit, and the counselee wants to be informed about the implications of the disease for the family. Sometimes the situation is clear-cut; for example, the patient has a well-documented case of phenylketonuria, cystic fibrosis of the pancreas, cleft lip, or some other disease diagnosed by a competent physician. The counselor must be aware of genetic heterogeneity and when he should require further diagnostic procedures. For instance, is the sporadic case of "achondroplasia" really diastrophic dwarfism? does the "phenylketonuric" child actually have benign hyperphenylalaninemia? and does the boy referred as having sex-linked Duchenne muscular dystrophy actually have the recessive limb-girdle type?

No genetic counselor, however brilliant and well trained, can be expected to be so expert in all branches of medicine that he can provide the necessary diagnostic expertise to deal with all the types of cases referred for genetic counseling. Therefore *he must have access to expert consultants* in pediatrics, neurology, cardiology, ophthalmology, radiology, orthopedics, and other branches of medicine; and he must have the judgment to know when to call upon them.

GENETIC WORKUP AND ESTIMATE OF RISK. *Evaluating needs of counselee.* It should go without saying (but sometimes does not) that the counseling must be tailored to the needs of the counselee. The central question usually concerns the probabilitd that a particular disease that has already occurred in a family will recur in specifiey relatives, most often in the sibs of an affected child. Occasionally the question takes some other form: for instance, the effects of consanguineous marriage, the significance of racial admixture in the ancestry, or the deleterious genetic effects of exposure to radiation or other mutagens. The concerns vary widely from case to case. The counselees may have come out of idle curiosity or may not really know why they have come, having been referred by a well-meaning doctor without adequate explanation. At the other end of the spectrum are parents who may be stunned by the birth of a defective child, who may consider it an act of retribution or an expression of their own imperfection, who may be torn with guilt or feel that they are doomed to have imperfect children. Exploring these feelings may be far

301

more important than providing a statistical estimate of the risk, and somewhere during the counseling process there should be an opportunity to do so, but in practice this aspect of counseling tends to be neglected.

Follow-up studies suggest that many parents need counseling because they have overestimated the magnitude of the risk and therefore need reassuring. Others, perhaps, need to have their anxieties heightened, in view of an underestimated high risk! A preliminary interview in which the parents state their needs may be useful in bringing such problems to the surface, but the counselor needs to be aware of them throughout the interview(s).

Evaluating family history. A detailed family history is taken from the counselees (who we shall assume, for the present discussion, are the parents of a defective or diseased child) and a pedigree is constructed. Some centers are developing methods of storing pedigree data by computer for possible linkage with other families [1–5].

It is generally useful to record "cultural" characteristics of the family, including racial background, religion, occupation, and grandparents' birthplace. Age and state of health of first-, second-, and third-degree relatives are recorded. Miscarriages and stillbirths are recorded for the near relatives. Unless there are specific indications, such as a dominant or sex-linked recessive pattern of inheritance or consanguinity, it is usually not worthwhile to go beyond third-degree relatives, since the information will usually be irrelevant as well as unreliable. Names are recorded (including married names of female relatives), and when a relative is reported to have some condition of possible relevance, the address should be noted as well as names of hospitals. The pedigree may reveal a specific pattern of inheritance and make a risk estimate possible even if the diagnosis is in doubt. The pedigree may also help in making the diagnosis—for instance, to distinguish between Hunter (sex linked) and Hurler (autosomal recessive) types of mucopolysaccharidosis. An opportunity should be provided for the interviewer to see the parents separately, in case there are factors or attitudes that might not be revaled in front of the spouse.

Sometimes the family history is straightforward and needs no further work, but not infrequently relevant facts will need to be corroborated or clarified by correspondence with relatives, hospitals, or doctors. A consent for release of information should be obtained from the parents. This usually has no legal authority except for records of the parents themselves or their children, but nevertheless it often seems to satisfy hospital record librarians in the case of more distant relatives. Record searching and correspondence can be a very time-consuming part of the counseling process.

Estimating risk. The next step is to apply the relevant knowledge from the literature to the problem. If the disease is well known to the counselor, he may have the necessary information at hand. Even with relatively well known diseases, many questions may come up besides the risk of recurrence. How long will a child with Down syndrome live? What is the chance that a child who inherits the gene for Marfan syndrome will have a dissecting aortic aneurysm? What proportion of children with neurofibromatosis are mentally retarded? It may be difficult, if not impossible, to find good answers to these and many other questions about genetic diseases. *Extensive research is needed to provide better information on the clinical risks associated with many genetic diseases.*

Not infrequently the disease is not well known to the counselor, and a literature search is necessary. McKusick's catalogue [6] is a good starting point for Mendelian disorders. The National Foundation's *Clinical Delineation of Birth Defects* series contains a wealth of information, and there are now several catalogues of syndromes that can be useful [7–11]. But much of the relevant information must be gleaned from the original literature, and new information, possibly significant, is constantly appearing. Keeping up with this constant outflow, so that answers may be as precise as possible, is a persistent challenge for the counselor.

Assuming that the counselor can classify the disease in question as Mendelian, he can apply the appropriate Mendelian law to the particular family situation and estimate the probability of recurrence. In some cases the application of Bayesian algebra can refine the estimate, depending on the number of unaffected children or, in the case of diseases of later onset, on the age of relevant family members [12, 13]. For chromosomal and multifactorial diseases or those of unknown etiology, empirical risk estimates from the literature must be applied.

COMMUNICATION OF RISK TO COUNSELEE(S). *First counseling session.* Now the recurrence risk must be conveyed to the counselees, with all its implications. This may be done at the first interview if the situation is straightforward or at the second if the estimation of risk has required some investigation. The risk may vary from quite small (less than 5%) to large (25%, 50%, or occasionally 100%). Whatever it is, it can be presented against the background of the 2%–3% risk that any child runs of having a major defect of some kind. Conveying the risk can be simple or very complex. At one end of the spectrum is a well-balanced couple with an optimistic bent, whose child has a condition with a low recurrence risk. They have no "hang-ups" about having transmitted the condition to the child, they receive the news of the low recurrence risk happily, and leave the interview reassured. At the other end is a guilt-ridden couple who may be subjected to all sorts of family and social pressures, and who are not able to comprehend the concept of statistical odds. They may indeed still be so involved with working out the impact of the present child's illness that they have not even thought of the question of future children.

The first couple will probably do well without any further attention from the counselor. The second may require a series of interviews over a period of months or years.

Probably a substantial portion of counselees fall into the first category and are reasonably well satisfied with a single interview. In spite of the current emphasis on the second category [14], it must not be assumed that all counselees are riddled with complexes and torn by family pressures. The problem is to tell one from the other. It does not depend on the magnitude of the risk. Some parents counseled with a low recurrence risk are still torn with uncertainties and need long-term support. Some with a high risk of recurrence of serious disease take a decision, such as sterilization, with a minimum of internal or external conflict and live happily with it. The psychodynamics of counseling is discussed in more detail below.

The question of the risk for more distant family members may come up. For instance, parents of an affected child may worry about their normal children transmitting the disease to their children. For them the risk may be negligible, and it is advisable to point this out and to suggest that the parents bring this up with the relatives, who may be embarrassed to ask. The case of high-risk relatives is discussed under Family Follow-up.

Follow-up. It is useful to follow up the counseling interview with a letter to the parents setting forth the facts and a letter to the referring physician, of course, to inform him of what has been done. This not only keeps him in touch with the situation but also may improve his understanding of the role of genetic counseling. At least one center tapes the interview and gives a duplicate tape to the parents for future reference.

For certain diseases such as Down syndrome, cystic fibrosis of the pancreas, and hemophilia, there may be pamphlets describing the disease and its causes in lay terms which the parents may find useful.

A follow-up call to the parents a week or two after the counseling session may be used to determine whether the counselees have grasped the information provided, to reinforce it, to give them a chance to air any second thoughts, and to arrange for further interviews where indicated.

Finally, the counselees may need one or more follow-up interviews during which they can grasp the facts of the situation more fully and work out the implications for their particular situation. Their attitudes and/or their situation may change with time, and new medical developments may change the available options, as in the case of amniocentesis, for instance [15]. Follow-up for assessing the effects of the counseling will be discussed later.

DECISION AND APPROPRIATE ACTION. As we have said, the counselees may make a decision easily and quickly, before the end of the counseling interview, or may do so only with much difficulty, over a long period of time, and perhaps with a number of interviews with the counselor or with additional specialized help.

The decision may be simply to go ahead with having a family as usual or to limit the number of children to some degree. Contraception may be adopted for a time to see how things work out with the present child. Some couples decide to adopt a child (increasingly difficult to arrange with the falling birthrate) and then take a chance with another child of their own. Others seek sterilization of one of the parents. Artificial insemination may be a solution in the case of autosomal recessive diseases or when the husband carries the mutant gene, but this option is rarely chosen for genetic problems. In some cases the wife may have accidentally become pregnant, so that there is no time for protracted decision making.

In the past few years, several developments have changed the situation quite radically with respect to avoidance or termination of pregnancy. The advent of contraceptive pills has made it a good deal easier for those who wish to avoid having children to do so. Diaphragms, condoms, and (even more so) "rhythm" were sufficiently unreliable to make the prospect of another potentially diseased baby a constant menace. The pill provided increased security but, on the other hand, the prospect of years "on the pill" and the fear of a single forgetfulness resulting in a high-risk pregnancy can be formidable.

Second, there is the radical change in social and legal attitudes to abortion from a generally proscriptive to a generally permissive one. In some societies the mere fact that a pregnancy is unwanted is sufficient ground for termination; in others there must be danger to the life or health of the mother, and in some there are no grounds for taking the life of an unborn human being. The counselor must, of course, respect the religious and moral attitudes of the parents, but these may sometimes conflict with the laws of the church or state. The parents, in good conscience, may wish to have a pregnancy terminated but may find it difficult to do so because of legal restrictions. The situation has improved markedly in many areas, but there are still regions where the law does not allow the parents to take the necessary steps to prevent the birth of a child with a high risk of disease.

Where prenatal diagnosis is possible, amniocentesis is a useful new option, provided that the parents are willing to accept abortion of the fetus if it is shown to be abnormal. More than 30 genetic diseases can now be diagnosed by study of cells obtained from the embryo by amniocentesis [16], and others can be detected by study of the amniotic fluid. Thus amniocentesis is indicated where the parents are at risk for having children with a detectable disease or a chromosomal aberration. The latter group includes translocation carriers, older mothers (over 35 or 40 depending on the center), or mothers who have had a previous child with a chromosomal aberration. Most couples who want subsequent children will accept this approach. A more difficult counseling situation is presented by X-linked recessive diseases where the disease cannot be positively diagnosed in the fetus, but the sex of the fetus can be determined and the pregnancy terminated if it is a boy or allowed to continue if it is a girl, with virtually no risk. Understandably, parents find it much more difficult to face the prospect of abortion if the fetus has a 50:50 chance of being normal than if it is certain to be defective.

It must be strongly emphasized that the couple should be thoroughly counseled *before* the amniocentesis is done, and arrangements should be made for the necessary diagnostic workup of the cells and/or fluid. All too often the genetics unit is presented with a specimen of amniotic fluid for diagnosis without having been advised of what the problem is or having had the opportunity to make certain that the parents understand the situation. If the problem is chromosomal, some centers routinely obtain karyotypes of the parents *before* the amniocentesis, since an unusual chromosome, suspected of being abnormal, might be interpreted as a normal variant if present in one of the parents.

Little is known about the psychodynamics of the genetic counseling process, although there is a growing literature on the impact on parents and family with respect to a number of specific diseases [17–24]. *We need to know a great deal more about parental attitudes; how they feel about the fact of having had a defective child, about themselves for having possibly transmitted the defect to the child, and about the information given by the counselor.* This is discussed further under Psychodynamics of Counseling.

FAMILY FOLLOW-UP. A facet of counseling that tends to be neglected is the *extension of counseling to other members of the family*. This may be because the necessary effort will overtax the resources of the genetics unit or because seeking out high-risk individuals without their consent raises problems of invasion of privacy. Family follow-up may be useful for two reasons: (1) to counsel high-risk (and reassure low-risk) relatives, and (2) to employ preventive measures.

Counseling high-risk individuals. In the case of dominant or sex-linked recessive diseases or chromosomal rearrangements, there may be many family members at high risk of having affected children, who could benefit from counseling. Well-known examples are hemophilia, Duchenne muscular dystrophy, and Huntington chorea. There may also be low-risk individuals who are unnecessarily worried about having affected children. Bringing counseling to the former individuals may be difficult—confronting people with the news that they are likely to have defective children requires some type of introduction. This may be arranged through the counselee, who may approach the relatives and find out if they are interested in counseling, or through the family doctor, if any. Calling a family conference may work in some cases. In a family with dominantly inherited spinocerebellar ataxia, for instance, a family gathering of 95 members was arranged, and 29 early cases of the disease were detected before these individuals had begun their families (*Time*, Jan. 25, 1971).

In some cases the risk of being a carrier can only be estimated from Mendelian laws, while in others the heterozygous carrier can be identified by biochemical tests, with varying degree of reliability. For instance, some types of hemophilia carriers may be diagnosed reliably with modern immunoassay methods, while for Duchenne muscular dystrophy about 70%–80% of the carriers can be identified. The counselor must be aware of the possible serious psychological effects that may result from the discovery that one is a carrier of a gene that could produce serious disease in one's children. Even more difficult is the situation where diagnosis of heterozygosity means that the carrier will develop a serious or fatal disease himself. It is necessary to be sure that the family member really wants to know, before doing the test, and to be willing to provide appropriate support if the test is positive. In the case of Huntington chorea, for instance, some counselors believe it would be better not to have a test that would detect the disease before its overt onset. Others feel that those high-risk individuals who would want to know whether they are carriers should have the option of doing so, if available, particularly since they may wish to avoid the possibility of passing the mutant gene on to their children. In a limited survey, about half the potential carriers thought they would like to know and half would not (A. G. Motulsky and G. S. Omenn, personal communication). Much more information is needed on the psychological effects of finding out that one carries a deleterious gene.

Employing preventive measures. Follow-up of high-risk relatives is particularly important in the case of diseases where measures are available to protect the individual against the harmful effects of the mutant gene.

Arrangements should be made, for instance, to check subsequent sibs of children with inborn errors of metabolism in which treatment must be initiated early, such as the genetic types of cretinism, hemophilia, the adrenogenital syndromes, phenyl-ketonuria, and galactosemia. There are also the pharmacogenetic diseases, such as G6PD deficiency, pseudocholinesterase deficiency, the porphyrias, and malignant hyperthermia, where the mutant individual must be protected against specific environmental agents. Patients with α_1-antitrypsin deficiency should be warned against smoking, and those with hypercholesterolemia and hypertriglyceridemia can be put on regimes protective of the cardiovascular system. Patients with multiple polyposis can be protected from carcinoma of the colon by colectomy if the disease is discovered in time. In these and many other cases, follow-up of the relatives can be a useful exercise in preventive medicine.

RANGE OF PROBLEMS BROUGHT FOR COUNSELING. *Individual cases.* The variety of problems a counselor must be prepared to deal with is enormous. They can roughly be broken down into diseases showing Mendelian inheritance, those with a chromosomal basis, multifactorial disorders, nonfamilial conditions, those where there is no information, and a miscellaneous group which includes the genetic hazards of consanguineous marriages, racial admixture, or exposure to possible mutagens, paternity exclusion, and screening of donors for artificial insemination. Adoption agencies may want to know the relevance of a child's family history of disease, or racial ancestry, to its adoptability. The proportions of the various types will vary among centers depending on what the center is best known for, the population served, and whether there are specialty clinics for certain common diseases (cystic fibrosis of the pancreas, diabetes, hemophilia) at which the counseling is done by the staff of the clinic. In this situation the quality of the counseling depends in part on the genetic knowledge and counseling expertise available in the clinic, which unfortunately may not always be top quality. Consultation with medical geneticists is recommended to evaluate such situations.

Over 1,000 conditions caused by single mutant genes in man have been identified [6], and although no one counselor will meet them all, he must be prepared to deal with a great variety. In one series of 349 consecutive cases, there were 164 Mendelian disorders among which 79 different diseases were represented, 45 of them by only one case each—that is, diseases with which the counselor had had no previous experience. Among 139 cases of multifactorial and miscellaneous diseases, there were 44 different conditions [16]. Obviously the counselor must have broad knowledge as well as access to the extensive literature which is constantly growing.

Special groups. In addition to individual cases seen for counseling, there is an increasing demand for counseling services directed to groups. We have already mentioned the need for counseling extensive families in which a dominant or sex-linked disease is segregating. There is also a growing demand for counseling services in conjunction with screening programs directed towards populations at high risk for specific diseases. These include sickle cell anemia in populations of African descent, G6PD deficiency in Mediterranean and African populations, thalessemia in Greeks, and Tay-Sachs disease in Ashkenazi Jews. Of these, two have some definitive management to offer. Individuals with G6PD deficiency can be warned to avoid certain drugs and other noxious agents. Couples who carry the Tay-Sachs disease gene can be offered monitoring of the pregnancy by amniocentesis and abortion of affected fetuses. Neither sickle cell disease nor thalassemia can be diagnosed prenatally because there is no acceptable technique for obtaining fetal blood, although techniques have been developed for diagnosis of the condition on minute amounts of fetal blood, once obtained. It may not be long before satisfactory techniques for drawing blood from the fetus are perfected [25]. Until then, all that can be offered heterozygotes is counseling regarding the possible consequences of marrying other heterozygotes, namely, the 25% probability of having a severely diseased child.

Experience is rapidly accumulating, but *much needs to be learned about the psychological effects of learning that one carries a mutant gene, not harmful to the carrier, but which can be very harmful when two carriers marry* [26]. Extensive screening programs are going on, but we know little about such vital questions as the optimal age for screening, how to counsel with minimum damage and maximum benefit, the psychological effects of being labeled a carrier of a harmful gene, and the social pressures that may be brought to bear on carriers. Will they become social pariahs in their own minds or the eyes of society? Will pressure be brought to bear against their having children, if they choose to take the risk? These and other questions urgently need answers and have aroused so much concern that there is a growing feeling (but by no means unanimous) that such screening programs should not be implemented unless there is something more to offer mutant gene carriers than simply genetic counseling. The positive effects of counseling must be weighed against the negative effects. We do not have the data to evaluate either adequately. There is a need for pilot studies in which these factors can be measured before we embark on large-scale screening programs.

METHODOLOGY. The term "genetic counseling" was coined by Sheldon Reed to replace the term "genetic hygiene" which had unpleasant eugenic implications. For a long time the typical counseling process consisted of a single interview in which the pedigree was taken and a prediction of recurrence risk given. The literature on genetic counseling dealt mainly with methods of arriving at recurrence risk estimates. Gradually, however, the psychological complexities involved in the counseling situation began to be recognized, and now an increasing number of articles on the impact of genetic diseases on the family and the techniques, philosophy, psychodynamics, and ethics of counseling are appearing [15, 27–35].

The methodology of counseling is in a highly experimental stage, and it will no doubt be some time (if ever) before there is any general agreement on optimal procedures.

In some centers the interviewing and counseling are still done by one person; in others, the counseling approach is a multiple interview, multidisciplinary process of varying degrees of complexity. Very often the counselor has no formal training in the techniques of counseling, although such training would presumably be an advantage. Having the same person take the family history and do the counseling has the advantage that the original interview gives the counselor some opportunity to get to know the counselees and something of the family background and to establish some degree of rapport before entering the counseling phase. The multi-disciplinary approach may provide the benefits of having several skills represented. For instance, a nurse or other trained worker (e.g., a genetic associate) may take the family history and collate the relevant medical information. Some centers find it useful to send out a family history form to be prepared by the family before the interview. A social worker may also do a preliminary interview to evaluate the social situation and needs of the family. A trained human or medical geneticist then studies the data, reviews the literature if necessary, orders appropriate tests or consultations to validate the diagnosis, and estimates the recurrence risk. Finally, there is a counseling session in which all three workers meet with the parents (and perhaps other family members) and discuss the situation, with the nurse acting as the family advocate to make sure they understand the terminology and implications of the information given and have had a chance to ask questions. There may be a follow-up interview with one member of the team to reinforce the information given and allow for more questions from the parents. A letter may be sent to the family and the referring physician summarizing the information given. This serves as a permanent record for future reference and discussion. The family may be sent a questionnaire to make sure they have understood the counseling figures.

Several centers tape the interview, and at least one presents the parents with a duplicate tape for future reference. In another center, the counselees are given a battery of psychological tests to evaluate their mental and emotional status, the family history is then taken by a trained worker, and a medically trained geneticist then sees them for counseling. Further psychological tests are given at follow-up; while this is done partly for research to evaluate the effects of counseling, it can also improve the efficiency of the counseling process. In at least one center a psychiatrist sits in on the interview and interacts with both counselor and counselee. Several centers use a one-way window so that group members and students can observe and evaluate.

Mention should also be made of a more informal kind of counseling that goes on in hallways and at the bedside. In hospitals that have a genetic counselor on staff, his visits to the wards serve to increase interest in genetics, particularly if he takes time to ask searching questions of the residents and to increase the number of referrals for formal counseling.

A number of centers are experimenting with counseling in groups, but it is too early to formulate any definite conclusions about their relative advantages and disadvantages. Group counseling may be effective when used in group screening programs for high-risk populations, provided counseling is available on an individual basis for those who need it. Group sessions may also be helpful for parents of genetically defective children, *after* the initial counseling interview, when they may aid in working out some of the psychological and practical problems created by the presence of a defective gene or chromosome. These should presumably be directed by someone with appropriate experience, both psychiatric and genetic.

One example is the Little People's Parents Auxiliary, in which both dwarfs and tall parents take part and learn from each other. Other groups arrange visits to parents who have just had a defective child, so that parents who have previously had a child with the same defect can provide support and practical information. The pros and cons concerning such groups need further exploration.

Thus there is *no consensus as to what constitutes an optimal counseling approach.* It seems clear, at least to some, that some of the more extensive routines outlined above are much more elaborate than a good many families need. On the other hand, one or two interviews with a single counselor will not meet the needs of some families. Perhaps we must learn how to evaluate counselees initially and modify the counseling routine according to progress.

Areas of Ignorance and Unmet Needs

It should now be clear that there are many unanswered questions and areas of ignorance concerning genetic counseling.

SUPPLY AND DEMAND FOR COUNSELING SERVICES. There has recently been a quantum jump in the interaction between genetic services and the public, partly because of screening programs such as those for Tay-Sachs disease or amniocentesis for mothers over 35. It is clear that we still have very little idea of how great the need for genetic counseling is in comparison to the supply of counselors. In one survey of 130 couples, it appeared that 40% had waited until one or more children had been born subsequent to the affected one before seeking counseling, 15% had had a second affected child, and 39% had waited more than 2 years after the affected child before seeking counseling (S. Wright, personal communication). Obviously not all those who would benefit by counseling are getting it. But what proportion? Since the need (if not the demand) is greater than the supply, *what priorities should determine who receives the services?*

Ideally one might say it should be all who could benefit by it, but many who would benefit are not even aware of the existence of the services or perhaps even of their need. Thus there is *a need for public education.*

Perhaps, then, it should be all who are concerned enough to ask—both those with a high risk who need to be warned and those with a low risk who need to be reassured. In less than 2 years, the National Genetics Foundation has received over 13,000 inquiries about genetic diseases and malformations. This, and the number of examples of misinformation from inexpert sources that genetic counselors hear about from their counselees, suggest that the demand for expert counseling services far exceeds the supply. In modern pediatric hospitals, from 10%–25% of the admissions are for cases of clearly genetic etiology, but only a small proportion of these receive counseling—10% in one survey (J. G. Hall, personal communication) and presumably much less in hospitals where there is no genetics unit.

Since the demand far exceeds the supply, perhaps efforts should be made to provide counseling at least for all those with a high recurrence risk. To achieve this on an individual basis, it would be necessary that the medical profession be so familiar with medical genetics that they could do much of the counseling themselves, which seems unlikely.

Groups that seem particularly in need of counseling services include families with psychiatric diseases, schools for handicapped children (the blind, in particular), high-risk individuals identified by population screening programs, and individuals far from medical centers. The latter group would require either funds for travel to centers or traveling clinics provided by the medical centers. Several centers are experimenting with traveling clinics, and this facility requires further evaluation. It is probably the only practical method for reaching out to the whole population with the manpower available. In view of the trend towards community center medical care, we also need some guidelines as to how far genetic services should be decentralized. Perhaps there should be a hierarchy, with simple cases handled at the local community level and the most complex referred to the medical center.

OPTIMAL TIME FOR COUNSELING. Another question about which we have very little knowledge is the optimal time at which counseling should begin. No doubt it varies from case to case. If counseling is offered too soon after the birth of a defective child, the parents may be too involved with working out the problems posed by the present child to worry about the question of future children and may not be able to take in what the counselor is saying. On the other hand, if counseling is delayed too long, other children may have already been born, some of them affected, or the parents may have refrained unnecessarily from having children or from sexual relations, when the risk was in fact low. It would be difficult to draw up specific guidelines, but at least the counselee(s) could be offered counseling reasonably soon after the event in such a way that they would know it was available whenever they were ready for it. There is need for education here.

DIRECTIVE OR NONDIRECTIVE COUNSELING? Opinions differ widely on how directive genetic counseling should be. Some counselors (a minority) believe that counseling should stop at the point where an estimate of risk is given and that the parents should make up their own minds what to do, without benefit (or otherwise) of further advice from the counselor. At the other extreme is the authoritative father-figure type (more often a physician than a genetic counselor) who tells the parents not to have more children, or to go ahead, as the case may be. " Genetic counseling offers the only help. As a result, doctors have advised ... relatives who are in line for the disease not to have children ..." (*Time*, Jan. 25, 1971).

Most counselors take neither extreme. Many counselees need help in the first place in understanding the meaning of a statistical probability. They may also appreciate the opportunity of discussing the pros and cons of the situation with someone who is not only sympathetic but well informed. Most counselors would refrain from directly telling the counselee what to do, but most would at least talk the matter over and point out the various factors that might be considered. Often the counselee will, after discussion, ask what the counselor would do if he were in the same situation. Many counselors would go so far as to say that although it is impossible to extrapolate himself entirely into the counselee's situation, since he is not the counselee, he thinks he would probably take a certain course of action. Although this has been opprobriously termed " behavior control " by some ethicists, the counselor is no more controlling behavior than is a surgeon who recommends an elective operation, a stockbroker who advises buying a particular stock, or a professor who recommends a particular course to a student. Many counselees come for advice and are disappointed if they do not get it. The counselor must of course avoid projecting his own personality defects into the situation, but in general he will not be of much help to the counselee if he maintains an Olympian detachment, concerned only with the statistical probability and not the unique combination of factors entering into the counselee's personal situation. Some counselors seem to have a natural flair for establishing empathy with the counselee, but others do not. Thus *training in the general techniques of counseling would be a useful part of the training of a genetic counselor.*

PSYCHODYNAMICS OF COUNSELING. Too little is known about the psychodynamics of genetic counseling. There may be various obstacles that prevent the counselee from being able to apply risk figures rationally to his particular situation. Denial, guilt, and anger may need to be reduced to manageable proportions before planning for the future begins [36]. Mothers may feel guilty because of smoking, excessive weight gain, or not wanting the baby; or some may have unsuccessfully tried to abort the pregnancy. Fathers may worry about a possible relation to previous venereal disease. One parent may blame the other, or the other's family, to the detriment of marital relationships. Parents may feel that they are defective because they carry a deleterious mutant gene and feel guilty for having passed it on to the child. It may be useful to emphasize that we all probably carry several deleterious genes and that (in the case of recessives) they were just unlucky that they both happened to carry the same one. When the mutant gene comes from a known carrier, this argument does not apply, and the parent may need help in accepting responsibility for the child's disease.

Studies are now appearing on the impact of specific diseases on the family and on the subsequent reproductive behavior of the parents of defective children [17–24, 27, 28, 33]. Some of these are widely discrepant. For instance, one study claimed that following the birth of a child with Down syndrome, reproduction virtually ceased among the sibs of the parents of the affected child [37]; whereas no evidence of such an effect was found by another group [38].

Parents also differ widely in their reaction to a given risk. The same figure may look formidable to one person and negligible to another, and the same defect may look trivial to one person and tragic to another [39].

To provide optimal counseling, we need to know much more about the effects of a defective child on the life styles and reproductive behavior of families and about the effects of counseling thereon.

Some studies of this kind are being done, but few data are available so far. One survey produced results that were rather discouraging in terms of the degree of understanding and retention of the information given [33], but those counseled were drawn from clinics for specific diseases and had not specifically sought counseling. Other studies have shown that counselees do, in general, make "reasonable" decisions about reproduction according to the risks they face [27].

A special problem is presented by large families with a high incidence of a genetically determined disease, particularly those involving an autosomal dominant gene with adult onset, such as Huntington chorea, multiple polyposis, or spinocerebellar dystrophy. Some congenital conditions fit here, too, for example, ectodermal dystrophy, aniridia, or diabetes insipidus. What are the psychodynamic effects of this menace hanging over the family? And, in particular, what prompts these high-risk individuals to keep on producing high-risk children? What is the role and what are the effects of counseling here?

EVALUATION OF RESULTS. A number of centers are now making attempts to evaluate the results of their counseling procedures. These range from simple follow-up questionnaires to intensive psychometric evaluations before, immediately after, and some time after the counseling. It is difficult to quantify results in any meaningful terms. For one thing, it is virtually impossible to obtain control observations, that is, the attitudes and reproductive behavior of other comparable families who were not counseled. Furthermore, it would be difficult to avoid observer effects—the fact of undergoing a psychological examination directed toward the parental attitudes before counseling might well change these attitudes.

A useful technique for evaluating doctor-patient communication which might be adapted for the present purpose is the health belief model [40]. This model suggests that people will accept preventive, and probably other forms, of health care provided that they perceive their own susceptibility to the condition being discussed, its seriousness, and the benefit to be had in accepting the care. Personal experiences, attitudes, and psychological characteristics may becloud these perceptions and constitute barriers to acceptance of the health care. On the other hand, certain cues, which may consist of special social experiences or which may emanate from the mass media, may be required to trigger action.

This seems a useful way in which to assess the effectiveness of genetic counseling. The counselee is more likely to receive and to act upon the counseling if he perceives the burden the disease is likely to impose upon himself or his family, grasps the fact and extent of his own susceptibility, and perceives that the counselor is the source of information and support which will allow him to use the information to make rational decisions about reproduction. The counselor might test the attitudes of the counselee, looking for barriers which might prevent him from perceiving or grasping the counseling. These might consist of religious attitudes, insufficient knowledge, difficulty in accepting the disease, guilt, superstition, and so on. The counselor might look for specific forms of information or social cues which might persuade the counselee to attend and to understand the message of the counseling. Similarly, the effectiveness of the counseling might be studied by an assessment of the perceptions of the counselee and by exploration of possible barriers and lack of cues to promote action.

Attempts to standardize post-counseling evaluations so that data from different centers could be compared in a systematic way would probably be fruitless because of the multiplicity of variables involved. However there is *room for collaborative studies on selected diseases*. For instance, a number of centers might collect data in a standardized way on a particular disease such as Huntington chorea so that data could be pooled and significant numbers accumulated beyond what would be possible for any one center to collect. Such efforts to enhance understanding of family attitudes to particular diseases should be encouraged. A working party to draw up specific plans would be useful.

Organization of Genetic Counseling Services

The need for counseling services is far in excess of the supply. It has been estimated that by the mid-1980s, in North America, one person with training in medical genetics will be needed for every 200,000 persons.

It should be obvious from the report so far that genetic counseling is a team affair. No individual could set himself up as a counselor and provide satisfactory service without the support of a cytogenetics laboratory, a biochemistry department geared to do screening for a variety of genetic diseases, the diagnostic facilities of a modern hospital, and the clinical expertise of a variety of specialists. Thus it is generally agreed that a genetic counseling service should be provided by a group and should be situated in or associated with a large medical center, preferably a university hospital.

The relation of the genetics unit to other hospital services varies widely among centers. Some specialty clinics (e.g., cystic fibrosis, hemophilia, diabetes) may do their own counseling and refer only difficult or disgruntled patients to geneticists. Some services (e.g., maxillofacial, hearing, ophthalmology) may refer patients as a regular part of the workup. Sometimes the genetics department may provide a geneticist to regularly attend a given clinic, such as a muscular dystrophy clinic. The medical genetics unit may run a regular clinic where patients are seen by various members of the team. Other units see referred patients by appointment.

Since counseling is vitally dependent upon research on the identification and delineation of new syndromes, the improvement of data on penetrance and expressivity, and the refinement of recurrence risk estimates, it is desirable that *the counseling service be associated with an ongoing research program*.

The personnel of the group should include an M.D. trained in genetics to take responsibility for the medical acts performed by the group and, depending on the size of the group, a number of others with either M.D. or Ph.D. degrees, or both, and training in the techniques of genetic counseling. There must be a cytogeneticist available. A variety of auxiliary personnel such as public health nurses, social workers, or genetic associates can provide invaluable service in interviewing, searching files and literature sources, collating information, and following up families.

Perhaps the optimal system would be a multitiered one such as exists in other branches of medicine. Simple cases might be counseled by the primary physician at the local level—when primary physicians have acquired the knowledge to handle the simple cases and to screen out those that are not so simple. These more complex cases might be seen by a professional genetics counselor, either at the medical center if the family lives near one or at the local community level by means of traveling clinics. Individuals needing sophisticated biochemical tests and children with esoteric syndromes or other complex problems could be referred to the genetics unit at the medical center.

Finally, the voluntary health agencies can aid the cause by providing appropriate literature, helping people who need counseling to find it, and supporting counseling services. Societies devoted to specific inherited diseases such as cystic fibrosis of the pancreas, hemophilia, or muscular dystrophy often have literature outlining the modes of inheritance of these diseases and perhaps recommending genetic counseling. The National Foundation provides a list of those who state that they provide genetic counseling services [41], sponsors courses on various aspects of medical genetics, and maintains over 50 counseling clinics. The National Genetic Foundation provides a referral service whereby anyone with a counseling problem can apply to the foundation and be referred to the nearest appropriate genetic counseling center. They also support a reference network, a group of centers spread across the continent that provide special diagnostic tests for genetic diseases that are so rare that the necessary tests are not maintained routinely. A doctor who has a patient suspected of having one of these rare diseases can apply to the foundation and will be put in touch with the nearest appropriate center. Thus voluntary health agencies can play an important role in improving the quality of genetic counseling services available to the public.

References

1. Emery, A. E. H.: The prevention of genetic disease in the population. *Int. J. Environ. Studies* 3 : 37–41, 1972.
2. Kimberling, W. J.: Computers and gene localization, in *Perspectives in Cytogenetics: The Next Decade*, edited by Wright, S. W., Crandall, B. F., Boyer, L., Springfield, Ill., Thomas, 1972, pp. 131–147.
3. Krush, A. J., Sharp, E. A., Lynch, H. T., Friden, F.: A computer-based system of coding for genetic studies of large kindreds (abstr.). *Am. J. Hum. Genet.* 22 : 21a, 1970.
4. McKusick, V. A.: Family-oriented follow-up. *J. Chronic Dis.* 22: 1–7, 1969.
5. Smith, C., Holloway, S., Emery, A. E. H.: Individuals at risk in families with genetic disease. *J. Med. Genet.* 8 : 453–469, 1971.
6. McKusick, V. A.: *Mendelian Inheritance in Man*, 3d ed. Baltimore, Johns Hopkins Press, 1971.
7. Bergsma, D. (ed.): *Birth Defects Atlas and Compendium*. Baltimore, Williams & Wilkins, 1973.
8. Gellis, S. S., Feingold, M. (eds.): *Atlas of Mental Retardation Syndromes*. U.S. Department of Health, Education, and Welfare, Washington, D.C., Government Printing Office, 1968.
9. Gorlin, R. J., Pinborg, J. J.: *Syndromes of the Head and Neck*, 2d ed. New York, McGraw-Hill, 1971.
10. Holmes, L. B., Moser, H. W., Halldorsen, S., Mack, C., Pant, S. S., Matzilevich, B.: *Mental Retardation: An Atlas of Diseases with Associated Physical Abnormalities*. New York, Macmillan, 1972.
11. Smith, D. W.: *Recognizable Patterns of Human Malformation*. Philadelphia, Saunders, 1970.

12. Murphy, E. A., Mutalik, G. S.: The application of Bayesian methods in genetic counseling. *Hum. Hered.* 19 : 126–151, 1969.
13. Murphy, E. A.: Probabilities in genetic counseling. *Birth Defects: Orig. Art. Ser.* 9(4) : 19–33, 1973.
14. Tips, R. L., Smith, G. S., Lynch, H. T., McNutt, C. W.: The "whole family" concept in clinical genetics. *Am. J. Dis. Child.* 107 : 67–76, 1964.
15. Sly, W. S.: What is genetic counseling? *Birth Defects: Orig. Art. Ser.* 9(4) : 5–18, 1973.
16. Clow, C. L., Fraser, F. C., Laberge, C., Scriver, C. R.: On the application of knowledge to the patient with genetic disease. *Prog. Med. Genet.* 9 : 159–213, 1973.
17. Carr, E. F., Oppé, T. E.: The birth of an abnormal child: telling the parents. *Lancet* 2 : 1075–1077, 1971.
18. Clifford, E., Crocker, E. C.: Maternal responses: the birth of a normal child as compared to the birth of a child with a cleft. *Cleft Palate J.* 8 : 298–306, 1971.
19. Emery, A. E. H., Watt, M. S., Clack, E. R.: The effects of genetic counselling in Duchenne muscular dystrophy. *Clin. Genet.* 3 : 147–150, 1972.
20. Freeman, R. D.: The crisis of diagnosis: need for intervention. *J. Special Educ.* 5 : 389–414, 1971.
21. Maddison, D., Raphael, B.: Social and psychological consequences of chronic disease in childhood. *Med. J. Aust.* 2 : 1265–1270, 1971.
22. McCollum, A. T., Gibson, L. E.: Family adaptation to the child with cystic fibrosis. *J. Pediatr.* 77 : 571–578, 1970.
23. Reiss, J. A., Menasthe, V. D.: Genetic counseling and congenital heart disease. *J. Pediatr.* 80 : 655–656, 1972.
24. Richards, I. D., McIntosh, H. T.: Spina bifida survivors and their parents: a study of problems and services. *Dev. Med. Child Neurol.* 15 : 292–304, 1973.
25. Valenti, C.: Antenatal detection of hemoglobinopathies. *Am. J. Obstet. Gynecol.* 115 : 851–853, 1973.
26. Murray, R.: Screening: a practitioner's view, in *Ethical Issues in Human Genetics*, edited by Hilton, B., Callahan, D., Harris, M., Condliffe, P., Berkley, B., New York, Plenum, 1973, pp. 121–130.
27. Carter, C. O., Roberts, J. A. F., Evans, K. A., Buck, A. R.: Genetic clinic: a follow-up. *Lancet* 1 : 281–285, 1971.
28. Emery, A. E. H., Watt, M. S., Clack, E. R.: Social effects of genetic counselling. *Br. Med. J.* 1 : 724–726, 1973.
29. Fraser, F. C.: Counseling in genetics: its intent and scope. *Birth Defects: Orig. Art. Ser.* 6(1) : 7–12, 1970.
30. Fraser, F. C.: Genetic counseling. *Hosp. Practice* 6(1) : 49–56, 1971.
31. Fraser, F. C.: Survey of counseling practices, in *Ethical Issues in Human Genetics*, edited by Hilton, B., Callahan, D., Harris, M., Condliffe, P., Berkely, B., New York, Plenum, 1973, pp. 7–13.
32. Hsia, Y. E.: Choosing my children's genes: genetic counseling. *Yale Med.* 8(1) : 14–17, 1973.
33. Leonard, C. O., Chase, G. A., Childs, B.: Genetic counseling: a consumer's view. *N. Engl. J. Med.* 287 : 433–439, 1972.
34. Nance, W. E.: Genetic counseling for the hearing impaired. *Audiology* 10 : 222–233, 1971.

35. Thompson, M. W.: Genetic counseling in clinical pediatrics. *Clin. Pediatr.* 6 : 199–209, 1967.
36. Hecht, F., Holmes, L. B.: What we don't know about genetic counseling. *N. Engl. J. Med.* 287 : 464–465, 1972.
37. Tips, R. L., Smith, G. S., Perkins, A. L., Bergman, E., Meyer, D. L.: Genetic counseling problems associated with trisomy 21, Down's disorder. *Am. J. Ment. Defic.* 68 : 334–339, 1963.
38. Fraser, F. C., Latour, A.: Birth rates in families following birth of a child with mongolism. *Am. J. Ment. Defic.* 72 : 883–886, 1968.
39. Fraser, F. C.: Heredity counseling: the darker side. *Eugen. Q.* 3 : 45–51, 1956.
40. Rosenstock, I. M.: Why people use health services. *Milbank Mem. Fund Q.* 44 : 94–127, 1966.
41. Lynch, H. T.: *International Directory of Genetic Services*, 4th ed. White Plains, N.Y., The National Foundation, 1974.
42. Epstein, C. J.: Who should do genetic counseling, and under what circumstances? *Birth Defects: Orig. Art. Ser.* 9(4) : 39–48, 1973.

31   Some Genetic Aspects of Therapeutic Abortion

James V. Neel

This presentation discusses the genetic indications for interruption of pregnancy. To those expecting simple, clear-cut formulas, I must apologize. The issues are such that the field of obstetrics and gynecology deals with only part of an increasingly complex problem, a problem which raises some of the most difficult biomedical questions that our society is trying to answer. The problem cannot be settled readily, but certain courses of action can be suggested.

For some 26 years, the University of Michigan has maintained a Heredity Clinic. The limitations on staff, time, and space being what they are, we have not attempted to see all the people referred to the University Hospital with potential problems in genetic counseling, but rather, both for service and training purposes, have tried to consult each year with a limited number of people with representative questions requiring such counseling. One of the issues arising from time to time has been whether a given probability of congenital abnormality in a child is an indication to interrupt a pregnancy. Each time this question has fallen to me, I have been troubled by the nearly total absence of guide lines.

For present purposes, we will define a possible " genetic indication " for terminating a pregnancy as an appreciable probability that the unborn child will, either at birth or subsequently, be found to have a major abnormality in which genetic factors

From *Obstetrics and Gynecology*, **30**, no. 4 (Oct. 1967), 493–497. Copyright © 1967 by American College of Obstetricians and Gynecologists. Reprinted with permission of the author and the publisher.

are of paramount importance. The most common situation involves serious congenital defect. It must be recognized that the cause of congenital defects is extremely complex, involving both environmental and genetic factors of various types. Setting aside for the moment questions of legality of abortion on genetic grounds (see below), let us consider the wide range of recurrence risks with which one may be confronted in genetic counseling. Five examples, any one of which might prompt the question of the propriety of terminating a pregnancy, will suffice:

1. The first child had anencephaly, but no other instance of congenital defect was known in the family, and the mother is again pregnant. Here the recurrence risk for this or other severe defect of the central nervous system is about 4%.[2]

2. Severe central nervous system defects were found in 2 children; the immediate family history was otherwise negative for congenital defects, and the mother is again pregnant. Here the recurrence risk is 12%.[5]

3. Parents are second cousins, the first child was normal, but the second child developed infantile amaurotic idiocy at 6 months. On the well-justified assumption of simple recessive inheritance, the recurrence risk is 25%.

4. For the fourth example, we turn to a problem seen 14 years ago. The first child of normal-appearing parents had died of a complex cardiac malformation, the second had been found to have cystic fibrosis of the pancreas. Moreover, a rising anti-Rh titer was encountered during that second pregnancy, although there was no clinical erythroblastosis. The mother was Rh-negative and the father was found very probably to be $R_1 R_2$, so that all subsequent children would be Rh incompatible. Our calculation of the probability of normal outcome, should the mother again become pregnant, was 36%, computed as the product of 3 probabilities: normality for congenital heart disease (95%); normality for cystic fibrosis (75%); and nonoccurence of Rh disease (50%). This example was chosen to illustrate not only the complexity sometimes encountered in counseling, but also the rapid evolution of knowledge. Successive developments in the anticipation and treatment of erythroblastosis would now greatly lessen the likelihood of severe disease from Rh incompatibility, while the prophylactic use of an anti-D serum during the pregnancies of predisposed women has the possibility of preventing the disease altogether.[4, 7]

5. The final example concerns pregnancy in a woman with the dominantly inherited disorder termed Huntington's chorea. The child has a 50% chance of inheriting the disease, and there is the additional consideration of the inability of the mother to care for the child. While Huntington's chorea and the other dominantly inherited degenerative diseases of the nervous system can scarcely be termed "congenital defect," we must recognize the narrow margin which sometimes separates "inherited predisposition" and congenital defect.

At this point we should take note of the emerging possibilities for converting these a priori probabilities to figures which present a more satisfying prospect for meaningful intervention. In the case of a woman who has a family history of hemophilia and has produced a hemophilic son, half of her future sons but none of her daughters can be expected to be affected. The risk, regardless of sex, is 1 in 4. If sex-chromatin studies on cells obtained by amniocentesis reveal the fetus to be a girl, the woman can be completely reassured (unless her husband is hemophilic), but if the fetus is a boy, the risk is now 50%. One can visualise an even greater change in the odds, to certainty, in a case where one could apply newly described technics for the culture of cells obtained by amniocentesis[11] to a pregnant woman with a translocation which predisposes her offspring to Down's syndrome. That is, given that this woman's abnormal karyotype had become known because of a previous child with mongolism, one could hope to detect the presence of mongolism in future pregnancies at a very early stage. However, amniocentesis during the first trimester is not without its risks, nor is it at all certain that the specimen will contain the cells necessary for diagnosis or culture. But assuming the necessary technical advances, it seems clear we can or shortly will be able to define a spectrum of risk to an unborn child ranging from a low percentage to virtual certainty. Incidentally, there is an important philosophic difference between these genetic risks and the rubella risks, in that the latter are nonrecurrent.

We must also take note of developments which could make the distinction between contraception and abortion somewhat academic. Hardin, in a thought-provoking argument for a virtually complete abrogation of existing legislation regarding the termination of early pregnancies, has particularly emphasized how the development of the so-called " morning after " pill, capable either of interfering with implantation or causing sloughing of the early embryo, will bring to a focus the moral issues involved in *any* termination of pregnancy. But a point which he did *not* make was that the intrauterine device, which apparently acts by preventing implantation of the blastocyst, and which is one of the most effective means of control of fertility, involves the very same issue, the only difference being that the IUD requires (hopefully) one decision, the " morning after " pill many.

The foregoing, then, are examples of the kinds of situations in which the question of inducing an abortion on genetic grounds is apt to arise. They will obviously stem in many instances from the failure of family planning; for the other possible genetic consequences of such planning the reader is referred to the recent excellent paper of Matsunaga. Let us turn now to the practicalities of therapeutic abortion. There are at present only 2 states (Colorado and North Carolina) in which the prospect of an abnormal child is recognized as a legal indication for abortion. On the other hand, there are many states in which (mostly because of data on the adverse effects of maternal rubella on the fetus during the first trimester, with risks to the fetus no greater than some of the genetic risks just considered) there has been tacit recognition that such a prospect constitutes sufficient grounds for an abortion. But at a time when in one state physicians are being threatened with legal action because they did induce an abortion but in another state are being sued because they failed to abort, in each instance for virtually the same indications, it is clear that some clarification of the indications for inducing an abortion is urgently needed.

Under the circumstances, I have become interested in the section on abortion in the Model Penal Code drawn up by the American Law Institute. This section, usually with some modifications, is now being considered for adoption by several states. The pertinent sentence reads as follows:

A licensed physician is justified in terminating a pregnancy if he believes there is a substantial risk that continuance of the pregnancy would gravely impair the

319

physical or mental health of the mother or that the child would be born with grave physical or mental defect, or that the pregnancy resulted from rape, incest, or other felonious intercourse.

I assume, although the section does not make it explicitly clear, that the consent of the individual involved is a prerequisite. With this reservation, I favor general adoption of such a code, which would achieve an important clarification of the present confusion. However, there is an ambiguity to that key sentence which generates a certain uneasiness. How great must a risk be to be termed substantial, and where does one draw the dividing line between grave and nongrave? In drawing up this suggestion, the Institute was undoubtedly concerned with not infringing on the judgment of the physician. However, in view of the temper of the times, we may anticipate that enactment of this legislation will be followed by a whole series of precedent-setting legal decisions. Specifically, when a pregnancy terminates in an abnormal child, will the physician now be vulnerable to legal harassment because where there was reason to suspect such an outcome, he failed to advise the mother accordingly, or sufficiently strongly?

It seems appropriate at this time to suggest that the ACOG*set up a committee to establish guidelines on these indications. Such a committee could delineate a series of situations which in its collective judgment involved "substantial" risk of "grave"defect. These guidelines would have no legal basis, but yet could not fail to influence both medical practice and court opinion. The Royal Medico-Psychological Association of Britain, in its recent policy statement on therapeutic abortion,[9] felt that no "spelling out in detail" was necessary. While in general I agree with this viewpoint, from the genetic standpoint the spectrum of risk is so wide, and genetic knowledge even today so poorly disseminated, that I believe the practicing physician would welcome some guidelines. Since a value judgment is clearly involved, it would seem appropriate that any committee which might be established should reach well beyond the medical and allied professions, to include legal experts, theologians, and social scientists. Although the physician certainly bears a special responsibility in these matters, in my opinion he will do himself a disservice, and only delay the clarification we seek, by failing to recognize the need to enlist all the best thinking on this subject.

Among the many difficult value judgments to be made, one in particular has impressed itself on us over the years. This is the question of weighing immediate versus delayed impact of a genetic defect. The nature of medical practice being what it is, the immediate seems more pressing and has a way of pre-empting our attention. But consider, for example, the woman who has given birth to 2 anencephalic children. There is a 1 in 8 chance that her next child will be a frightful monster, but the monster will almost surely die shortly after birth and 7 normal fetuses will be aborted for every one such birth prevented by abortion. Consider, on the other hand, the apparently normal young woman whose father has Huntington's chorea. There is a 1 in 2 chance she has inherited the gene herself, and if so, another 1 in 2 chance she will transmit it to any child. There is thus a 1 in 4 probability any child to which

* American College of Obstetricians and Gynecologists.

she gives birth will some 30 to 40 years later develop progressive degenerative disease of the central nervous system, with prolonged emotional trauma to the family and a long and expensive hospitalization. Or consider a marriage in which both parents have either the sickle-cell trait or thalassemia minor, with a 1 in 4 probability that any child will have a chronic, debilitating anemia, with death at a relatively early age. Any one of us can document instances in which the prolonged care of such individuals as these just mentioned has entailed exorbitant cost to the family, or to some agency or institution, with the ultimate result a very marginal performance by the individual in our complex society. While I do not for one moment wish to place a price tag on a human life, I cannot help wondering how that same sum spent on normal children might advance the interests of society. Which of the 4 foregoing situations, if any, best justifies an interruption of pregnancy? For these as well as our 5 earlier examples, where should the dividing line be drawn?

The Model Code also groups pregnancies resulting from rape and incest in a juxtaposition of apparent equivalence. Although both are morally abhorrent, from the biologic standpoint incest is in a very different category from rape. The child of rape faces the biologic risk of any illegitimate child, but the child of incest faces the added risk of high inbreeding. It has been very difficult to obtain an accurate evaluation of these risks. Recently we published the results of a small series of such children.[1] Among 18 children of father-daughter or brother-sister relationships in which the mother's pregnancy became known prior to the birth of the child, there were 6 children who, on follow-up 6 months after birth, either had died or had a major defect, whereas among 18 children born to a similarly ascertained, matched series of unwed mothers, none had died and only one had a major congenital defect. Carter[3] has recently reported an even smaller series, in which the results were quite similar. Taken at face value, these biological effects are greater than would be predicted from the results of first-cousin marriage,[10] and perhaps reflect the nonrandom selection that enters into this situation, but whatever the reason, the risk would seem to qualify as "substantial."

It is appropriate in a discussion of this type to recognize that except in those instances where a parent has manifest disease, or one or both parents has a defined genetic carrier state, we usually cannot identify high-risk families until after the birth of a defective child. Since the great majority of defective children are single events within the sibship, and are born to normal parents, even a rather liberal attitude towards therapeutic abortion on genetic grounds could result in only a fractional decrease in the frequency of congenital defect.

Conclusions

At the outset, we recognized that this general problem was only one aspect of a much larger issue. The term "genetic engineering" is frequently heard these days. Although I have been unable to determine with certainty who first coined the term, it apparently stems from the theoretical possibility of direct intervention into the genetic material, created by our increasing knowledge of DNA. Where such intervention activates or represses the expression of genes in somatic tissue, it is in effect not too different from, for example, hormone therapy. But where it actually alters genes, especially in germinal material, a new principle would be introduced. If directed mutation is involved, then there are implications for the gene pool of the next generation, and in many respects "genetic engineering" becomes one approach to what has commonly been termed "eugenics." The issues raised by the prospect of eugenic measures are many and controversial. It seems only appropriate to point out that the decision to induce an abortion on genetic grounds also has implications for the gene pool of the next generation, and so falls under the broad category of eugenic measures. Until we understand the genetic structure of human populations far better than is the case at present, and until we have a modest experience with legalized abortion, it is fitting to proceed cautiously and conservatively.

References

1. Adams, M. S., and Neel, J. V. Children of incest. *Pediatrics 40*: 55–62, 1967.
2. Carter, C. O. "The Inheritance of Common Congenital Malformation." In *Progress in Medical Genetics* (Vol. 4). Steinberg, A. G., and Bearn, A. G., Eds. Grune, New York, 1965, p. 59.
3. Carter, C. O. Risk to offspring of incest. *Lancet 1*: 436, 1967.
4. Clarke, C. A. "The Prevention of Rh-immunization." In *Proceedings Third International Congress of Human Genetics*. Crow, J. F., and Neel, J. V., Eds. Johns Hopkins Press, Baltimore, 1967.
5. Fraser, Roberts, J. A. Personal communication.
6. Hardin, G. The history and future of birth control. *Perspect. Biol. Med. 10*: 1, 1966.
7. McConnell, R. B. The prevention of Rh-haemolytic disease. *Ann. Rev. Med. 17*: 291, 1966.
8. Matsunaga, E. Possible genetic consequences of family planning. *JAMA 198*: 533, 1966.
9. Royal Medico-Psychological Association. Statement on therapeutic abortion. *JAMA 199*: 167, 1967.
10. Schull, W. J., and Neel, J. V. *The Effects of Inbreeding on Japanese Children.* Harper, New York, 1965.
11. Steel, M. W., and Breg, W. R. Chromosome analysis of human amniotic-fluid cells. *Lancet 1*: 383, 1966.

32   The Question of Abortion

Francisco J. Ayala

Amniocentesis—taking a sample of the amniotic fluid surrounding the fetus for clinical examination—has become common practice in U.S. hospitals. Some routine tests of the fetal skin cells obtained with the fluid permit detection of any one of a number of severe ailments determined by the genetic constitution of the future child. Some ailments are relatively common. Down's syndrome (also called mongoloid idiocy) is caused by the presence of an extra chromosome-21; about one out of every 600 live births is affected. About one out of every 400 liveborn males suffers from Klinefelter's syndrome, a condition caused by the presence of one or more extra X-chromosomes. Although generally less severe than mongoloid idiocy, Klinefelter's accounts for more than one per cent of institutionalized mentally defective males.

Down's, Klinefelter's, and a host of other serious genetic conditions can be detected by amniocentesis. A pregnant woman informed that her future child will be a mongoloid idiot may choose, under the present law, to have a therapeutic abortion, or may decide to continue her pregnancy and have the baby. In the latter case, the child, severely handicapped mentally and physically, will become an emotional and financial burden not only to the mother, but to society. One billion eight hundred million dollars are spent in the United States each year for the social care of mongoloid idiots.

Abortion is practiced not only in cases of abnormal genetic constitution of the fetus, but for a variety of other reasons, for instance, because continuation of pregnancy poses physical or mental risks to the mother. In any case, abortion raises a plethora of legal and ethical issues.

The legal state of the question can be briefly summarized. On January 22, 1973, the U.S. Supreme Court overruled as unconstitutional all state laws that prohibit or restrict in any way a woman's right to obtain an abortion during the first three months of pregnancy. The states can regulate abortion during the last six months of pregnancy, but they can *prohibit* it only during the last ten weeks. Therefore, according to the U.S. Supreme Court, any legislation prohibiting abortion during the first six and a half months of pregnancy infringes upon the constitutional rights of pregnant women. However, for the last ten weeks of pregnancy, when the fetus is judged to be " viable," that is, capable of surviving outside the mother's uterus, the states can establish legislation precluding abortions.

There are reasons why the legalization of abortion should be supported, regardless of whether one considers it an immoral act. The state must operate within a framework of popular consent. Ours is a pluralistic society where many respectable citizens believe that abortion is morally acceptable, while others equally respectable are convinced of the opposite. Do the latter have a right to impose their ethical beliefs upon the former? If the law does not prohibit abortion, each citizen can follow the dictates of his or her own conscience. As the Jesuit, Father Austin J. Fagothey has recently observed: "The state cannot make all its citizens adopt the same philosophy of rights, the same respect for human life, and the same judgment on the personality of the human fetus. To attempt it would involve a grave violation of the freedom of its citizens."[1]

But what about the *morality* of abortion? Legality and morality overlap, but each has areas of competence where the other does not apply; some immoral acts may not be prohibited by the law because it does not concern the society's interests and some actions are regulated by law although they are *per se* morally neutral. A very strong argument for the immorality of abortion is the following. Among a person's rights, the right to life is the most fundamental. It is not derived from civil law, but rather it is an inalienable right of all human beings. A fetus has incipient human life that is destroyed by abortion, which thus infringes upon the most basic of all human rights.

Is this reasoning correct? Let us examine it. The argument is correct to state that all persons have inherent rights, and the most basic of them is the right to life. The argument states, correctly also, that there is incipient human life in a fetus. But is "human life' synonymous with "human being" or "person"? Clearly not. There is human life in my hand, in my eyes, and in the other parts of my body. If a surgeon removes one of my kidneys, no sane person will accuse him of having killed a human being, although he has destroyed, in a sense, human life. I kill several cells of my body, and thus destroy a particle of human life, when I prick my finger with a needle to obtain a blood sample. Such "destruction" of human life may be morally justified and indeed called for by the circumstances. The right to allow my finger to be pricked, or my kidney to be extirpated, belongs to me; the finger and the kidney have life but are not human beings and, thus, have no rights. Therefore, the question is not whether a fetus has life (it does), but whether it is a human being.

It may be argued that the question of whether the fetus is a person from the moment of conception is not decisive, since everybody agrees that a fetus has the *potentiality* to become a human being. But to be or to have something potentially is not the same, neither legally nor morally, as to be or to have it in fact. Every American-born citizen has the potentiality to become President of the United States; but if a child is murdered the slayer is not accused of presidential murder. Every human spermatozoon and ovum have the potentiality to become human beings, but although many die, they are not mourned as human deaths. About 3000 million spermatozoa are delivered in each human ejaculate; all of them, or all but one, die each time a man ejaculates. Each woman loses between 400 and 500 ova during her lifetime.

A fertilized egg is not a human being although it has the potentiality to become one. The fertilized egg is a single miscroscopic cell containing the genetic information to direct the development of the individual. To have the complete set of genetic instructions for an individual is not the same as to be an individual. An architect's plans for a building has the instructions for the construction of the building, but the plans are not the building. There are ten trillion (one followed by 13 zeros) cells in the body of an adult human being; each one of them (or at least most of them) has the same set of genetic instructions in its nucleus as the fertilized egg from which the individual developed. Millions of cells die each day in every individual.

Christian theologians have always recognized that a fertilized egg, or even a fetus during early stages of development is not a human being. St. Thomas Aquinas studied this matter carefully and concluded that the fetus during the first few weeks of pregnancy is *incapable* of having a human soul. There is no single point in time at which the fetus, or the infant, suddenly becomes a person. The development of the human being occurs gradually starting from conception. As development progresses, "personhood" gradually develops. The personhood does not exist separately from the human being, nor does the soul. Modern Christian theologians, like Piet Schoonenberg, believe that the soul is not created at a given moment by God, but rather is created gradually with the human body as the human being develops. To quote Father Schoonenberg: "The creation of the human soul is neither more nor less than the beginning of a new person in a whole world, which is constantly created by God as a world in which there is an increase in human persons."[2] This notion is fully compatible with the Biblical view of man, as cogently shown by Robert Francoeur, a theologian who happens to be also a biologist.[3]

Corrective surgery, like amputation of a leg or excision of a kidney, is morally justified when commensurate reasons exist. The greater the destruction of human life, the more serious the reasons to justify it need to be. Similarly, an abortion justified by nothing but caprice would be immoral. In my opinion, however, when adequate reasons exist abortion is morally justified; but the more advanced the pregnancy, the more compelling the reasons required for its moral justification. Many people believe that the death penalty is morally acceptable. What the commensurate reasons are to justify abortion at various stages of pregnancy is not for me to determine. Ultimately, this, like all other moral decisions, must be made by the responsible person.

[1] The quotation of Father Fagothey is from his article, "The abortion question" in *Santa Clara Today*, December 1973. [2] The quotation of Piet Schoonenberg is from *God's World in the Making* (Duquesne Univ. Press: Pittsburgh, 1965). [3] The reference to Robert Francoeur is to his book, *Evolving World, Converging Man* (Holt, Rinehart and Winston: New York, 1970).

33   Abortion, Euthanasia, and Care of Defective Newborns

John Fletcher

ABSTRACT. *Growing use of abortion to prevent births of infants with unfavorable prenatal diagnoses raises ethical questions about active euthanasia for newborn infants with similar impairments. Two opposing ethical arguments are those of Paul Ramsey, who equates genetically indicated abortion with infanticide disapprovingly, and of Joseph Fletcher, who equates the morality of abortion with selective euthanasia approvingly. Though radically different, these arguments treat the ethical aspects of the prenatal and postnatal situations as essentially similar. There are, however, different moral features between the two situations, in that the postnatal situation is characterized by the independent physical existence of the infant, the possibility of treatment, and the formation of parental loyalty to the infant. Thus, a decision for abortion after prenatal diagnosis does not necessarily commit parents to euthanasia in the management of a seriously damaged infant.*

The medical literature has increasingly dealt with the utility and versatility of prenatal diagnosis for the detection of a variety of inborn errors of metabolism,[1-6] chromosomal abnormalities and variants[7-12] and polygenic conditions (e.g., spina bifida and anencephaly).[13, 14] Most of these diagnostic procedures entail amniocentesis, but some, such as the recent maternal serum test for indexes of neural-tube abnormalities,[15, 16] require confirmation by invasive technics only if the test is positive. In every instance, the information obtained about the fetus affords the parents and attending physician data that they may use to decide whether or not to abort the fetus.

Although the proper use of these diagnostic findings in deciding on elective abortion is controversial,[17, 18] my purpose here is to consider a different issue: does the ethical reasoning that is applied to prenatal management bear any relation to decision-making about survival after the birth of the infant? The ethical problem facing the medical profession is simply this: how should physicians and parents now understand their obligation to care for the newborn defective infant in the light of arguments for genetically indicated abortion after amniocentesis? To be consistent, a person might ask whether the arguments that support abortion after prenatal diagnosis of genetic disease also support euthanasia of the same infants who slip through that screen and are born. The debate in ethics is at present polarized between a disapproving view, which tends to equate genetically indicated abortion with infanticide,[19-21] and an approving view, which tends toward equating the morality of abortion and selective euthanasia of the defective newborn.[22, 23]

A third position, for which I argue, accentuates parental freedom to participate in

Supported in part by a grant (GM 19922-02) from the National Institutes of Health.

From the *New England Journal of Medicine*, 292 (Jan. 9, 1975), 75–78. Copyright 1975 by the Massachusetts Medical Society. Reprinted with permission of the author and the publisher.

life-and-death decisions independently in both the prenatal and postnatal situations, accepts abortion of a seriously defective fetus, but disapproves euthanasia of defective newborns.

Paul Ramsey, an exponent of the first view,[20] rejects arguments for abortion that are based upon a positive prenatal diagnosis of a severe fetal disease and the socioeconomic harm that will be done to the family, because he holds that the same arguments might be used under similar circumstances to justify "infanticide." Infanticide is his term for deliberately bringing about the death of a newborn defective infant. His position is built upon the presupposition that there are no clear-cut moral differences between abortion and infanticide in the same disease. In this perspective, abortion for Lesch–Nyhan syndrome, Tay–Sachs disease, or other lethal diseases would be invalid since these justifications could be used for killing the same infant born without benefit of prenatal diagnosis.

In an earlier essay, Ramsey explained that his method of testing right and wrong action in such cases was fashioned upon the ethical measure of "universalizability." This test asks, "what would be the case if everyone in a morally relevant, like situation did as I am doing (whether there is any tendency for them to do so or not)?"[24] Ramsey is primarily interested in using this method for ethical appraisal of genetically indicated abortions. He is not saying that we will necessarily begin to commit infanticide because we do abortions to prevent genetic disorders. He is saying that if we would not do the infanticides, we should not be doing the abortions. In short, there are means that are in themselves wrong regardless of the good ends that may be desired through them.

Joseph Fletcher's arguments, leading the other side of the debate, not only support abortion after prenatal diagnosis, but also advocate setting aside traditional restraints against euthanasia for defective infants. In an article on ethics and euthanasia,[25] Fletcher reasons that a decision to abort a defective fetus, which is "subhuman life," is logically of the same order as a decision to end a "subhuman life in extremis" in old age. Discussing euthanasia, he abhors moral distinctions between acts of commission and omission ("allowing to die"), whether concerned with the newborn or the terminally ill older person. His thought clearly relates the ethical arguments for genetically indicated abortions to justifications for euthanasia of newborn defective infants:

> If we are morally obliged to put an end to a pregnancy when an amniocentesis reveals a terribly defective fetus, we are equally obliged to put an end to a patient's hopeless misery when a brain scan reveals that a patient with cancer has advanced brain metastases.
>
> Furthermore ... it is morally evasive and disingenuous to suppose that we can condemn or disapprove positive acts of care and compassion but in spite of that approve negative strategies to achieve exactly the same purpose. This contradiction has equal force whether the euthanasia comes at the fetal point on life's spectrum or at some terminal point post-natally.[25]

The bearing of Fletcher's thought on the problem here, in contrast to Ramsey's, is that if we would do abortions based on prenatal diagnosis, we should be active in ending the suffering of infants born with the same condition. To do less—to allow

327

a newborn with severe disease to die by withholding support—is in this view hypocritical, since we would be active in ending the same life in utero.

He comes to his conclusions in ethics through the reasoning of consequentialism, for which "only the end (a proportionate good) makes sense of what we do (the means)."[22] He does not mean that any end justifies any means, and he so states earlier in the passage just quoted. When a fetus or newborn is seriously incapacitated, however, the human harm prevented and suffering relieved by abortion and euthanasia justifies these actions.

In spite of the polar difference between the two positions, the positions exhibit an interesting similarity. Each is based in part on the assumption that the prenatal and postnatal situations are so similar that ethical behavior in the former determines ethical guidance in the latter. Each ethicist, in his own terms, sees the fetus on the same level of value and dignity as he sees the newborn infant. Ramsey is unequivocal about the status of a fetus as a fellow human being whose irreducible dignity derives from God.[19, 24] Fletcher's reflections on "indicators of humanhood" lead him to delay conferral of human status upon the fetus or the infant until qualitative and quantitative measures can be made.[26] In the former view, humanhood is a free gift of God; in the latter view, humanhood is a human choice. In both views, moral guidance is given to be consistent in actions before and after birth.

My task is to show that there are morally relevant differences between abortion and euthanasia, even when one considers them for the same depth of impairment. The task is complex because an understanding of human development and growth as a process militates against lifting up one sequence of development too far out of relation with another. The defective newborn infant is the same being, but at a different stage of development, as a fetal candidate for prenatal diagnosis. Yet the post-natal situation in which parents and physicians face life-and-death decisions for extremely ill infants has features of a kind different from decision making about abortion.

What are the differences? One is that the separate physical existence of the infant, apart from the mother, confronts parents, physicians and legal institutions with independent moral claims for care and support. A newborn infant is clearly a patient. The movement of the fetus prepares the parents emotionally for the acceptance of the infant as a separate individual, but, before extra-uterine viability the well-being of the fetus should not be considered independently from the mother's condition. H. T. Englehardt has argued persuasively on the difference between a fetus and an infant, noting that "as soon as the fetus actualizes its potential viability, it can play a full social role and can be understood as a person."[27] A moderate ethical stance will avoid the one extreme of regarding every fetus as already a human being with rights and the other extreme of withholding human status until quality-of-life standards have been passed. The former extreme provides no rational grounds for the legitimate interests of parents, family and society to be expressed and guided in abortion decisions. The latter provides no rational ground for the interest of the newborn infant to be expressed in medical decision making. The decision of the United States Supreme Court on abortion responds to the moral imperative that new human life requires a social protection wider than prenatal and medical care, but it observes that before viability, claims for protection cannot be made compellingly without violating the privacy of the mother.[28] In law, the source of the difference between a wider parental freedom before fetal viability and a larger social responsibility after viability lies in the growing moral claims that the new human life makes upon society.

The second major difference is the fact that after birth the disease in the infant is more available to physicians for palliation or perhaps even cure. Confrontation with disease in an independently existing life requires physicians to respond within their obligations to heal and to relieve suffering. The most noteworthy disease at present that can be successfully treated in utero is erythroblastosis fetalis (Rh disease). The moral claim to relieve suffering in a diseased fetus may be more answerable in the future when genetic therapies are possible. For the present, however, the real situation for parents and physicians is that they must wait until birth to respond to the specificity of a disease with decisions to treat or not to treat.[29]

Thirdly, parental acceptance of the infant as a real person is much more developed at birth than in the earlier stages of pregnancy. The medical literature amply describes pregnancy as a crisis bringing about in the parents, especially in the mother, a series of behavioral changes that prepare them for caring for the infant.[30-33] We should expect loyalty to the developing life to grow, change, and moderate the ambivalence about the fetus usually present in the parents. An example of the depth of parental loyalty can be seen when parents of a defective newborn " mourn " the loss of the expected healthy child and reconsider acceptance of the child with a defect.[34] Since acceptance of the new life undergoes a development of its own, it should be readily apparent that increasing parental loyalty to the infant constitutes a major difference when one compares abortion of a fetus to euthanasia of an infant.

The effect of these three differences is to establish the newborn infant, even with a serious defect, as a fellow human being who deserves protection on both a legal and an ethical basis, and thus each of the differences contributes to an argument against euthanasia.

Two additional reasons round out the case for opposition to euthanasia. The first is the potential brutalization of those who participate in it. This point has been persuasively made by Bok in a discussion of infanticide,[35] and it is the source of Lorber's rejection of euthanasia as " an extremely dangerous weapon in the hands of unscrupulous individuals."[36] The second reason is the destructive social consequences of changing the ethical ambience of the birth of infants from one of thorough caring for life to one in which the public accepted a policy of euthanasia and supported its legalization. I see no way of making this social change that would not undermine the optimal moral condition for the beginning of life: the experience of trust. Erikson described the "basic trust" required between mother and infant necessary for the first task of forming healthy personality.[37] His thesis was that the task of mothering is strengthened, among other things, by a world view based on confidence that life is good, even though death and tragedy are part of life.

A society that supports acceptance of defective newborns, where reasonably possible, does more to nurture patterns of acceptance in parents and thus reinforce the child's basic trust in the world's trustworthiness.

329

A brief comment here is appropriate regarding the fact that physicians, parents, and ethicists are confronted with real cases of terribly damaged newborns for whom death is the desirable outcome when therapy either is not available or will only prolong the ordeal without definite ground for hope.[38] In such cases, if we would reject euthanasia because of a negative ethical assessment of the action and its consequences, what should be done? Allowing the infant to die by withholding support while relieving pain is a decision, in my view, that can be ethically justified for reasons of mercy to the infant and relief of meaningless suffering of the parents and medical team. If death is understood as a good outcome in such cases, however difficult the emotional acceptance of the infant's death, parents and physicians do not "do harm" to the infant by the decision, assuming that every reasonable therapeutic step has been taken or evaluated negatively. The crucial difference between euthanasia and allowing to die is that the self-restraint imposed by the latter choice is more consistent with ethical and legal norms that physicians and parents do no harm to the infant.

I have argued here that a decision for abortion after prenatal diagnosis does not necessarily commit parents to one course of action in the care of an infant born with the same degree of illness. The structure of my argument is based upon three differences between the fetus and the infant that are sufficiently grounded in human experience to be verified by observation. A defective newborn is a separate individual, whose disease is available for treatment, and whose parents are prepared by the process of a typical pregnancy to accept the infant. When these human experiences interact with the beliefs and values of the religious and humanistic communities that provide our culture with visions of the ultimately desirable, this interaction produces strong ethical and theologic backing for caring for the infant. If we choose to be shaped by Judeo-Christian visions of the "createdness" of life within which every creature bears the image of God, we ought to care for the defective newborn as if our relation with the Creator depended on the outcome. If we choose to be shaped by visions of the inherent dignity of each member of the human family, no matter what his or her predicament, we ought to care for this defenseless person as if the basis of our own dignity depended on the outcome. We are in a very hazardous situation ethically if we allow abortion for medical reasons in utero and then attempt to usher new life into a thoroughly caring context at birth, unless we are fully responsive to the imperatives to treat disease in the newborn when it appears.

References

1. Milunsky, A., Littlefield, J. W., Kanfer, J. N., et al: Prenatal genetic diagnosis. N. Engl. J. Med. 283 : 1370–1381, 1441–1447, 1498–1504, 1970.
2. Gerbie, A. B., Nadler, H. L., Gerbie, M. V.: Amniocentesis in genetic counseling: safety and reliability in early pregnancy. Am. J. Obstet. Gynecol. 109 : 765–770, 1971.
3. Epstein, C. J., Scheider, E. L., Conte, F. A., et al: Prenatal detection of genetic disorders. Am. J. Hum. Genet. 24 : 214–226, 1972.
4. Valenti, C.: Antenatal detection of hemoglobinopathies: a preliminary report. Am. J. Obstet. Gynecol. 115 : 851–853, 1973.

5. Antenatal Diagnosis. Edited by A. Dorfman. Chicago, University of Chicago Press, 1972.
6. Antenatal Diagnosis of Genetic Disease. Edited by A. E. H. Emery. Edinburgh, Churchill Livingstone, 1973.
7. Steele, M. W., Breg, W. R., Jr.: Chromosome analysis of human amniotic-fluid cells. Lancet 1 : 383–385, 1966.
8. Jacobson, C. B., Barter, R. H.: Intrauterine diagnosis and management of genetic defects. Am. J. Obstet. Gynecol. 99 : 796–807, 1967.
9. Nadler, H. L.: Antenatal detection of hereditary disorders. Pediatrics 42 : 912–918, 1968.
10. Nadler, H. L., Gerbie, A.: Present status of amniocentesis in intrauterine diagnosis of genetic defects. Obstet. Gynecol. 38 : 789–799, 1971.
11. Milunsky, A., Atkins, L., Littlefield, J. W.: Polyploidy in prenatal genetic diagnosis. J. Pediatr. 79 : 303–305, 1971.
12. *Idem:* Amniocentesis for prenatal genetic studies. Obstet. Gynecol. 40 : 104–108, 1972.
13. Emery, A. E. H., Eccleston, D., Scrimgeour, J. B., et al: Amniotic fluid composition in malformations of the central nervous system. J. Obstet. Gynaecol. Brit. Commonw. 79 : 154–158, 1972.
14. Allan, L. D., Ferguson-Smith, M. A., Donald, I., et al: Amniotic-fluid alphafetoprotein in the antenatal diagnosis of spina bifida. Lancet 2 : 522–525, 1973.
15. Brock, D. J. H., Bolton, A. E., Monaghan, J. M.: Prenatal diagnosis of anencephaly through maternal serum-alphafetoprotein measurement. Lancet 2 : 923–924, 1973.
16. Brock, D. J. H., Bolton, A. E., Scrimgeour, J. B.: Prenatal diagnosis of spina bifida and anencephaly through maternal plasma alphafetoprotein measurement. Lancet 1 : 767–769, 1974.
17. Gustafson, J. M.: Genetic counseling and the uses of genetic knowledge—an ethical overview. Ethical Issues in Human Genetics. Edited by B. Hilton, D. Callahan, et al. New York, Plenum Press, 1973, pp. 101–112.
18. Ramsey, P.: Screening: an ethicist's view. Ethical Issues in Human Genetics. Edited by B. Hilton, D. Callahan, et al. New York, Plenum Press, 1973, pp. 147–161.
19. *Idem:* Feticide/Infanticide upon request. Religion in Life 39 : 170–186, 1970.
20. *Idem:* Abortion. Thomist 37 : 174–226, 1973.
21. Dyck, A. J.: Perplexities for the would-be liberal in abortion. J. Reprod. Med. 8 : 351–354, 1972.
22. Fletcher, J.: The Ethics of Genetic Control. Garden City, New York, Doubleday, 1974, pp. 121–123, 152–154, 185–187.
23. Tooley, M.: Abortion and infanticide. Phil. Pub. Affairs 2 : 37–65, 1972.
24. Ramsey, P.: Reference points in deciding about abortion. The Morality of Abortion. Edited by J. T. Noonan Jr. Cambridge, Harvard University Press, 1970, pp. 60–100.
25. Fletcher, J.: Ethics and euthanasia. Am. J. Nursing 73 : 670–675, 1973.
26. Fletcher, J.: Indicators of humanhood: a tentative profile of man. Hastings Center Rep. 2(5) : 1–4, 1972.
27. Englehardt, H. T., Jr.: Viability, abortion, and the difference between a fetus and an infant. Am. J. Obstet. Gynecol. 116 : 429–434, 1973.
28. Roe vs Wade: 410 US 113 (1973).

29. Duff, R. S., Campbell, A. G. M.: Moral and ethical dilemmas in the special-care nursery. N. Engl. J. Med. 289 : 890–894, 1973.
30. Kennell, J. H., Klaus, M. H.: Care of the mother of the high-risk infant. Clin. Obstet. Gynecol. 14 : 926–954, 1971.
31. Bibring, G. L.: Some considerations of the psychological processes in pregnancy. Psychoanal. Study Child. 14 : 113–121, 1959.
32. Caplan, G.: Emotional Implication of Pregnancy and Influences on Family Relationships in the Healthy Child. Cambridge, Harvard University Press, 1960.
33. Nadelson, C.: " Normal " and " special " aspects of pregnancy. Obstet. Gynecol. 41 : 611–620, 1973.
34. Solnit, A. J., Stark, M. H.: Mourning and the birth of a defective child. Psychoanal. Study Child. 16 : 523–537, 1961.
35. Bok, S.: Ethical problems and abortion. Hastings Center Studies 2 : 33–52, 1974.
36. Lorber, J.: Selective treatment of myelomeningocele: to treat or not to treat? Pediatrics 53 : 307–308, 1974.
37. Erikson, E.: The healthy personality. Psychol. Issues 1 : 56–65, 1959.
38. McCormick, R. A.: To save or let die: the dilemma of modern medicine. JAMA 229 : 172–176, 1974.

eight Human Evolution, Race, and Intelligence

A variety of writings on past, present, and future human evolution is presented in this book. The paper by Allison in this section illuminates genetic forces operative in human survival in the past and present, particularly with respect to resistance to disease, notably malaria. As J. B. S. Haldane surmised quite a while ago, probably much of selection during human evolution has involved resistance to infectious diseases.

Allison uses a common shorthand in discussing red blood cell hemoglobin (Hb) types. Hemoglobin A refers to the predominant hemoglobin in humans, which contains two alpha and two beta chains. Hemoglobin S refers to the mutant form which causes red blood cells to deform (sickle) and which differs from A only in one amino acid in the beta chain. Hemoglobins C and N refer to mutant hemoglobins with other simple differences from Hb A. Hemoglobin F refers to normal fetal hemoglobin, which differs from Hb A enormously, having two gamma chains instead of the two beta chains found in Hb A; and T refers to a thalassemic condition (Mediterranean anemia) where one or another kind of hemoglobin chain is absent or rare in an individual, producing abnormal hemoglobin molecules. The paper by Allison refers to the Hb A–Hb S heterozygote as "sickle-cell trait" or "sickle-cell trait carrier."

Allison briefly mentions the Rh–ABO interaction in mother-child incompatibility leading to hemolytic disease of the newborn (erythroblastosis fetalis). The ABO blood group system has a decided effect in mitigating the intensity and frequency of erythroblastosis fetalis. But the theory that ABO is an influential system in disease resistance, supported by Allison (and by Wood, 1974), is disputed by Wiener (1970).

The importance of the sickle-cell gene in human evolution in Africa has been amplified by Wiesenfeld (1967). He gives a plausible explanation of the correlation between sickle-cell gene frequency and agricultural development in tropical Africa. He concludes that the replacement of nomadic hunting–gathering by settled agriculture in the rain forest allowed the mosquito vector of malaria to concentrate around settlements, which selectively increased the frequency of Hb AS heterozygotes. This in turn stabilized malarial infections at a relatively low level in the population —on balance, a beneficial feedback effect. It now appears that in addition to the sickle-cell gene, West Africans have also had for centuries a native plant which when eaten is effective in ameliorating sickle-cell anemia (Isaacs-Sodeye et al., 1975). Different genetic polymorphisms involving malaria resistance in other regions are now being investigated (e.g., Baer et al., 1976).

The paper by Gary in this section reports on current problems involving sickle-cell anemia. He discusses the sociopolitical aspects of sickle cell in the American black community (see also Cerami and Peterson, 1975; Kellon and Beutler, 1974; Reilly, 1973).

The paper by Loomis indicates the critical adaptation of skin color to sunlight intensity in different latitudes (but see Deol, 1975). Loomis' conclusions disagree with the rationale for skin color given in the Hindu epic, the *Mahabharata*, written long before the birth of Christ, which states that "Brahmins [priests, high caste] are fair, and sudras [peasants, servants] are black." How the European idea of race, or caste, evolved is given in Eiseley (1968). Perhaps another human adaptation to climate is body build; differences in surface-area/volume relations in different climes may be important in thermoregulation (Schreider, 1964).

Also in this section are papers dealing with human intelligence, a topic that has evolutionary, political, and social significance (Mead et al., 1968; Cronbach, 1975; Daniels, 1973; Kamin, 1974; Richardson and Spears, 1972). The plain fact is that intelligence is defined as one pleases *quot capita tot sententia* (there are as many opinions as there are heads—The Genetic Basis of Opinions, Rule 1). A broad genetical perspective on intelligence is provided by Stern here. Some parts of Matsunaga's paper, presented in Section Six, are related to what Stern has to say. Preceding Stern's paper, and using an adversary format, the views of Lewontin on a paper by Jensen and Jensen's reply are put on display (see also Bodmer and Cavalli-Sforza, 1970; Lerner, 1972; Tizard, 1974; Thoday, 1969; Willerman et al., 1974). The paper by Goldsby points out other factors in the "IQ equation" that often are overlooked in technical reviews on the subject.

Since intellectual abilities are the product of a gene "pool of abilities" and such social factors as economic and educational opportunity, I personally favor Pettigrew's (1964) outlook that what we should be looking at in children is not differences in performance but differences in motivation and emotion (including fear) in the face of unequal incentives for success and unequal chances of success. With respect to black–white differences in IQ scores, Pettigrew says, "The final, definitive research must await a racially integrated America in which opportunities are the same for both races. But, ironically, by that future time the question of racial difference in intelligence will have lost its salience; scholars will wonder why we generated so much heat over such an irrelevant topic."

Bibliography

Baer, A., L. E. Lie-Injo, Q. B. Welch, and A. N. Lewis. Genetic factors and malaria in the Teuman. *Amer. J. Hum. Genet.*, **28** : 179–188, 1976.

Bajema, C. J., ed. *Natural Selection in Human Populations.* New York: Wiley, 1971.

Bodmer, W. F., and L. Cavalli-Sforza. Intelligence and race. *Sci. Amer.*, **223** : 19–29, 1970. See also comments in *Nature*, **229** : 71–72, 1971, and in *Sci. Amer.*, **224** : 6–8, 1971.

Cavalli-Sforza, L. The genetics of human populations. *Sci. Amer.*, **231** : 81–89, 1974.

Cerami, A., and C. M. Peterson. Cyanate and sickle cell disease. *Sci. Amer.*, **232** : 45–50, 1975. Cyanate is one treatment of this anemia.

Cronbach, L. J. Five decades of public controversy over mental testing. *Amer. Psychologist*, **30** : 1–14, 1975.

Culliton, B. J. Sickle cell anemia: The route from obscurity to prominence: National program raises problems as well as hopes. *Science*, **178** : 138–142 and 283–286, 1972.

Daniels, N. The smart white man's burden. *Harper's Magazine*, Oct. 1973, pp. 24–40.

Deol, M. S. Racial differences in pigmentation and natural selection. *Ann. Hum. Genet.*, **38** : 501–503, 1975.

Dobzhansky, T. *Genetic Diversity and Human Equality.* New York: Basic Books, 1973.

Dobzhansky, T. Is genetic diversity compatible with human equality? *Soc. Biol.*, **20** : 280–288, 1973.

Ehrlich, P. R., and J. P. Holdren. The "lost genius" debate. *Saturday Review*, **54**, May 1, 1971, p. 61.

Eiseley, L. Race: The reflections of a biological historian. In *Science and the Concept of Race*, M. Mead et al., eds. New York: Columbia University Press, 1968, pp. 80–87.

Goldsby, R. A. Human races: Their reality and significance. *Sci. Teacher*, **40** : 14–18, 1973.

Harrison, G. A., and A. J. Boyce. *The Structure of Human Populations.* New York: Oxford University Press, 1972.

Harrison, G. A., and J. J. Owen. Studies in the inheritance of human skin colour. *Ann. Hum. Genet.*, **28** : 27–37, 1964.

Isaacs-Sodeye, W. A., E. A. Sofowora, A. O. Williams, V. O. Marquis, A. A. Adekunle, and C. O. Anderson. Extract of

Fagara zanthoxyloides root in sickle cell anemia. *Acta Haematologica*, **53** : 158–164, 1975.

Jencks, C. What color is IQ? Intelligence and race. *New Republic*, Sept. 13, 1969, pp. 25–29. Further comments are printed in the Sept. 27 and Oct. 25, 1969, issues.

Jensen, A. *Educability and Group Differences*. New York: Harper & Row, 1973.

Kamin, L. J. *The Science and Politics of IQ*. New York: Wiley/Halstead, 1974.

Kellon, D. G., and E. Beutler. Physician attitudes about sickle cell disease and sickle cell trait. *J. Amer. Med. Assoc.*, **227** : 71–72, 1974.

Layzer, D. Heritability analyses of IQ scores; science or numerology? *Science*, **183** : 1259–1266, 1974. See also letters, *Science*, **188** : 1125–1130, 1975.

Lerner, I. M. Polygenic inheritance and human intelligence. In *Evolutionary Biology*, Vol. 6, T. Dobzhansky, M. Hecht, and W. Steere, eds. New York: Appleton, 1972, pp. 399–414.

Lewontin, R. C. Further remarks on race and the genetics of intelligence. *Bull. Atomic Sci.*, 1970, pp. 23–25. This is a rebuttal to Jensen's statement given in this section.

Loehlin, J. C., G. Lindzey, and J. N. Spuhler. *Race Differences in Intelligence*. San Francisco: Freeman, 1975.

Mead, M., T. Dobzhansky, E. Tobach, and R. E. Light, eds. *Science and the Concept of Race*. New York: Columbia University Press, 1968.

Nei, M., and A. K. Roychoudhury. Gene differences between Caucasian, Negro, and Japanese populations. *Science*, **177** : 434–436, 1972.

Nichols, P. L., and V. E. Anderson. Intellectual performance, race, and socioeconomic status. *Soc. Biol.*, **20** : 367–374, 1973.

Pettigrew, T. Race, mental illness, and intelligence: A social psychological view. *Eugen. Quart.*, **11** : 189–215, 1964.

Powledge, T. M. The new ghetto hustle. *Sat. Review Sciences*, **1** : 38–47, 1973. Reports on shortcomings of sickle cell programs in the U.S.

Provine, W. B. Geneticists and the biology of race crossing. *Science*, **182** : 790–796, 1973. Some closeted skeletons are examined.

Ramot, B., ed. *Genetic Polymorphisms and Disease in Man* New York: Academic, 1974.

Reilly, P. Sickle cell anemia legislation. *J. Legal Med.*, **1** No. 4 : 39–48 and No. 5 : 36–40, 1973.

Richardson, K., and D. Spears, eds. *Race and Intelligence*. Baltimore: Penguin, 1972. A nontechnical paperback.

Scarr-Salapatek, S. Race, social class, and IQ. *Science*, **174** : 1285–1295, 1971.

Scarr-Salapatek, S. Some myths about heritability and IQ. *Nature*, **251** : 463–464, 1974.

Schreider, E. Ecological rules, body-heat regulation, and human evolution. *Evolution*, **18** : 1–9, 1964.

Stern, C. Model estimates of the number of gene pairs involved in pigmentation variability of the Negro-American. *Hum. Hered.*, **20** : 165–168, 1970.

Thoday, J. M. Limitations to genetic comparison of populations. *J. Biosoc. Sci. Suppl.*, **1** : 3–14, 1969.

Tizard, B. IQ and race. *Nature*, **247** : 316, 1974.

Wiener, A. S. Blood groups and disease. *Amer. J. Hum. Genet.*, **22** : 476–483, 1970.

Wiesenfeld, S. L. Sickle-cell trait in human biological and cultural evolution. *Science*, **157** : 1134–1140, 1967.

Willerman, L., A. F. Naylor, and N. C. Myrianthopoulos. Intellectual development of children from interracial matings: Performance in infancy and at four years. *Behavior Genetics*, **4** : 83–90, 1974.

Wood, C. S. Preferential feeding of *Anopheles gambiae* mosquitoes on human subjects of blood group O: A relationship between the ABO polymorphism and malaria vectors. *Hum. Biol.*, **46** : 385–404, 1974.

Young, L. B., ed. *Evolution of Man.* New York: Oxford University Press, 1970.

34 Polymorphism and Natural Selection in Human Populations

A. C. Allison

I should like to begin by considering the abnormal hemoglobins, which provide much the best evidence of polymorphism and natural selection in man. At the Cold Spring Harbor Symposium in 1955 I put forward the following propositions:

(1) The homozygous sickle-cell condition is virtually lethal in Africa. In view of the high frequencies of the sickle-cell gene in many regions, the rate of elimination of the gene could not be compensated by recurrent mutation.

(2) Balanced polymorphism has resulted because the sickle-cell heterozygote is at an advantage, mainly as a consequence of protection against falciparum malaria.

(3) Malaria exerts its selective effect mainly through differential viability of subjects with and without the sickle-cell gene between birth and reproductive age, and to a much lesser extent through differential fertility.

(4) High frequencies of the sickle-cell gene are found only in regions where falciparum malaria is, or was until recently, endemic.

(5) In most New World Negro populations, frequencies of the sickle-cell gene are lower than would be expected from dilution of the African gene pool by racial admixture. This is probably the result of elimination of sickle-cell genes without counterbalancing heterozygous advantage.

(6) In regions where two genes for abnormal hemoglobins co-exist, and interact in such a way that individuals possessing both genes are at a disadvantage (such as the $Hb_\beta{}^S$ and $Hb_\beta{}^C$ or $Hb_\beta{}^T$), then as a result of selection these genes will tend to be mutually exclusive in populations.

In the nine years that have elapsed most of these statements have been challenged, but enough evidence has now accumulated to establish beyond reasonable doubt that, insofar as these statements are testable, they are substantially correct. The evidence will be summarized briefly and somewhat dogmatically; a more extended discussion with full documentation is to be found in a book on *Polymorphism in Man* (Allison and Blumberg, 1965).

Much recent information on abnormal hemoglobins has been summarized at a meeting held earlier this year at Ibadan, Nigeria, under the auspices of the Council for International Organizations of Medical Sciences, the proceedings of which will shortly be published.

From *Cold Spring Harbor Symposia on Quantitative Biology*, **29** : 137–149, 1964. Copyright 1964 by Cold Spring Harbor Laboratory, Cold Spring Harbor, N.Y. Reprinted with permission of the author and the publisher.

Lethality of the Homozygous Sickle-Cell Condition in Africa

It is now generally agreed by workers in Africa that sickle-cell disease is common in childhood, occurring at about the expected frequency of homozygotes in the populations concerned (see Vandepitte and Stijns, 1964; Watson-Williams and Weatherall, 1964). The data do not support the suggestion of Nance (1963) that the incidence of sickle-cell disease might be much less than expected from the heterozygote frequency because of Hb_β^{AS} duplications. Although such duplications may occur, they must be uncommon, as the absence of Hb–A in many studied cases of genotype Hb_β^S/Hb_β^C is sufficient to illustrate.

There is also agreement that most African sickle-cell homozygotes die in childhood (see Vandepitte and Stijns, 1964). A few subjects with S + F hemoglobin patterns (mainly genetic variants of sickle-cell disease) survive to adulthood and reproduce with difficulty (see Fullerton, Hendrickse and Watson-Williams, 1964). Nevertheless, it can be said that the fitness of the sickle-cell homozygote under African conditions is close to zero. Even the upper limits of mutation at the Hb_β locus estimated by Vandepitte et al. (1955) and Frota-Pessoa and Wajntal (1963)—which are certainly in excess of the true figure—are far too low to replace the loss of sickle-cell genes from populations.

Malaria in Sickle-Cell Heterozygotes

The considerable amount of work that has been carried out on this subject has been widely misinterpreted, as the quotations in some textbooks (e.g., Harris, 1959; Wagner and Mitchell, 1964) will show. At the beginning of the investigation (Allison, 1954a) it was pointed out that only results on young children have any validity because of the powerful effects of acquired immunity in children of school age and adults. Before this point was appreciated, attempts to confirm the observations on schoolchildren and adult populations failed, and these results have been passed from review to review and so fossilized in the literature. Eventually the point sank in, and series of observations on the right age groups were made, which tell the story plainly enough for all to see.

In Table 1 all the published observations that have been made on malaria in susceptible African children with and without the sickle-cell trait are collected together. The only data that have been omitted are those in older children where the authors themselves give reasons for concluding that acquired immunity is having a powerful effect (i.e., in Nigerian children and children from Northern Ghana over five years of age). Where individual groups in the same investigation were small (groups A and B of Foy, Brass, Moore, Timms, Kondi, and Olouch, 1955; three groups of Walters and Chwatt, 1956; and two groups of preschool children of Garlick, 1960), the figures have been pooled, which simplifies the analysis but does not affect the overall result. The observations have been analyzed by the method of Woolf (1955) which allows comparisons to be made between populations with different gene frequencies and attack rates.

In column 7 of Table 1 the relative incidence of *P. falciparum* infections in child-ren with and without the sickling trait is expressed relative to an incidence of unity

Table 1. *P. falciparum* Parasite Rates in African Children.

Authors	Subjects and Age in Years	Sickle-Cell Trait		Non-Sickle-Cell Trait		Relative Incidence (1)	Weight	Woolf χ^2	Probability
		Falcip-arum	Total	Falcip-arum	Total				
(1) Allison (1954a)	Uganda, <6	12	43	113	247	2.18	7.58	4.60	0.05 > p > 0.02
(2) Foy et al. (1955)	Kenya, <6	131	241	154	241	1.49	28.81	4.53	0.05 > p > 0.02
(3) Raper (1955)	Uganda, <10	73	191	494	1,009	1.55	38.26	7.36	0.01 > p > 0.001
(4) Colbourne and Edington (1956)	S. Ghana	42	173	270	842	1.47	27.10	4.05	0.05 > p > 0.02
(5) Colbourne and Edington (1956)	N. Ghana, <5	11	15	165	177	5.00	2.32	6.01	0.02 > p > 0.01
(6) Walters and Chwatt (1956)	Nigeria, <5	162	213	680	890	1.02	31.24	0.01	p > 0.99
(7) Edington and Laing (1957)	N. Ghana, <4	13	19	109	127	2.79	3.24	3.42	0.10 > p > 0.50
(8) Garlick (1960)	Nigeria, <6	51	91	245	342	1.98	16.95	7.93	0.01 > p > 0.001
(9) Allison and Clyde (1961)	Tanganyika, <5	77	136	272	407	1.54	24.38	4.00	0.05 > p > 0.02
(10) Thompson (1962, 1963)	S. Ghana	34	123	176	593	1.10	20.52	0.20	p > 0.50

(1). Incidence of *P. falciparum* infections in non-sickle-cell trait groups relative to unity in corresponding sickle-cell groups.
Weighted mean relative incidence = 1.46.
Difference from unity, χ^2 = 29.2 for 1 d.f., p < 0.001.
Heterogeneity between groups χ^2 = 13.5 for 9 d.f., 0.20 > p > 0.10.

341

Table 2. Incidence of Heavy *P. falciparum* Infections in African Children.

Authors	Classification of Infection	Sickle-Cell Heavy infections	Sickle-Cell Total	Non-Sickle-Cell Heavy infections	Non-Sickle-Cell Total	Relative Incidence (1)	Weight	Woolf χ^2	Probability
(1) Allison (1954a)	Group 2 or 3	4	43	70	247	3.86	3.38	6.16	$0.02 > p > 0.01$
(2) Foy et al. (1955)	Heavy	21	241	38	241	1.96	11.99	5.43	$0.02 > p > 0.01$
(3) Raper (1955)	$>1000/\mu l$	35	191	374	1,009	2.63	25.49	23.74	$p < 0.001$
(4) Colbourne and Edington (1956)	$>1000/\mu l$	3	173	57	842	4.11	2.79	5.59	$0.02 > p > 0.01$
(5) Colbourne and Edington (1956)	$>1000/\mu l$	5	15	75	177	1.47	3.07	0.46	$p > 0.50$
(8) Garlick (1960)	$>1000/\mu l$	25	91	147	342	1.99	14.91	7.06	$0.01 > p > 0.001$
(9) Allison and Clyde (1961)	$>1000/\mu l$	36	136	152	407	1.66	20.71	5.27	$0.05 > p > 0.02$
(10) Thompson (1962, 1963)	$>5630/\mu l$	3	123	42	593	3.05	2.72	3.38	$0.10 > p > 0.05$

(1) Incidence of heavy *P. falciparum* infections in non-sickle cell trait groups relative to unity in corresponding sickle-cell trait groups.
Weighted mean relative incidence = 2.17.
Difference from unity $\chi^2 = 51.379$ for 1 d.f., $p < 0.001$.
Heterogeneity between groups $\chi^2 = 5.719$ for 7 d.f., $0.7 > p > 0.5$.

in the sickling children. It will be clear that the relative incidence is always above unity, the weighted mean being 1.46. The next column but one has the χ^2 values of the differences, which are significant at the 5% level in 8 out of the 10 groups. There is a probability of much less than 1 in 1,000 that the total difference from unity would occur by chance; and the heterogeneity between groups, despite the differences in conditions under which the observations were made, is no more than would be expected by chance.

In Table 2 the incidence of heavy *P. falciparum* infections is compared in sickling and nonsickling subjects. The differences are even more striking: the weighted mean relative incidence is 2.17 and the difference from unity is very highly significant ($\chi^2 = 51.379$ for 1 d.f.), with only very slight heterogeneity between groups. Other data that cannot be analyzed in the same way also show considerable protection by the sickle-cell trait (e.g., the difference in parasite levels above and below $1000/\mu l$ in Congolese children studied by Vandepitte, 1959, gives $\chi^2 = 18.9$ for 1 d.f.; $p < 0.01$). Since there is evidence both from Asia and Africa (Figures 34–1 and 34–2) that mortality from falciparum malaria is related to parasite counts, the results in Table 2 provide powerful evidence that sickle-cell trait carriers are more likely to survive in malarious environments than nonsickling children. Direct evidence that this is so comes from available data on malarial mortality summarized in Table 3. The probabilities of obtaining the observed results by chance are calculated from the binomial distribution and combined by the method of Fisher (1959), giving $\chi^2 = 46.4$ ($p < 0.001$). It is difficult to see what further proof any reasonable person could expect.

Figure 34-1. Number of fatal cases of falciparum malaria in relation to the parasite count in peripheral blood before treatment (data of Field, 1949).

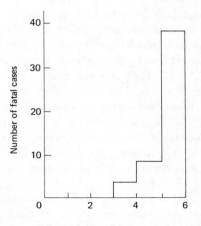

Log parasites/mm³ before treatment

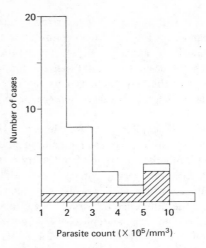

Figure 34-2. Total number of Congolese children with parasite counts $\times 10^5$ per cu. mm. observed by Vandepitte and Delaisse (1957) and number of fatal cases (hatched). There is a progressive increase in the proportion of fatal cases as the parasite count increases.

Differential Viability Vs. Differential Fertility

Resistance against malaria might favor the sickle-cell heterozygote either by increasing the viability of this genotype between birth and reproductive age or by increasing the relative fertility of sickle-cell trait carriers. Because it is difficult to demonstrate any effect of the sickle-cell trait on malaria in adults, Allison (1954a) concluded that the main selective effect was exerted by differential mortality in young children before appreciable immunity against malaria developed.

Allison (1956) obtained direct evidence in support of this interpretation. In the Musoma district of Tanganyika the frequency of sickle-cell heterozygotes in the adult population (38.4%) was significantly higher than that in the corresponding population of young infants. A number of other workers have found higher frequencies of sickle-cell trait carriers in older than younger age groups. Available data were summarized by Allison (1964): The unweighted mean ratio of trait incidences in schoolchildren and adults to young children is 1.17 to 1, the difference from unity being highly significant. Although these results have to be treated with some reservation because of various possible sources of bias, they provide strongly suggestive evidence in support of the differential viability hypothesis. In an extensive study of the progeny of 4,700 Congolese families, Burke, de Bock, and de Wolf (1958) also found highly significant evidence of a lower mortality among children carrying the sickle-cell trait than other children.

The case for differential fertility among women as a result of placental malarial infection has been argued by Livingstone (1957). As Rucknagel and Neel (1961) have pointed out, in a population with 35% sickle-cell trait carriers equilibrium can

Table 3. Deaths from Malaria in Relation to the Sickle-Cell Trait in African Children.

Author	Subjects	No. of Deaths	No. with Sickle-Cell Trait	Incidence of Sickle-Cell Trait in Population	Probability
Raper (1956)	Uganda (Kampala)	16	0	0.20	0.028148
J. and C. Lambotte-Legrand (1958)	Congo (Leopoldville)	23	0	0.235	0.0021095
Vandepitte (1959)	Congo (Luluaborg)	23	1	0.25	0.115938
Edington and Watson-Williams (1964)	Ghana (Accra)	13	0	0.08	0.33826
Edington and Watson-Williams (1963)	Nigeria (Ibadan)	29	0	0.24	0.00034953

$\chi^2 = 46.4$ (10 d.f.), $p < 0.001$

be attained if the fertility of sickle-cell trait carriers (both sexes) is 1.97 times that of normal subjects and fertilities are additive in the sexes. If the excess fertility applies only to females, the difference has to be increased accordingly. Quite considerable data from Africa, summarized by Allison (1964) shows that nowhere is there any finding of enough increased female fertility of sickle-cell heterozygotes to account for the persistence of the gene. The largest series investigated (Burke et al., 1958) showed that the fecundity of sickling mothers is not superior to the mean in the Congo.

Among the Black Caribs of British Honduras, Firschein (1961) found a fertility ratio of sickle-cell trait to normal mothers of about 1.45 to 1, which is sufficient to maintain the sickle-cell gene even in the absence of differential mortality due to malaria; very few adult males were included in this study, and no attempt was made to estimate differential viability.

The most reasonable conclusion is that differential female fecundity is at most a minor contributory factor to the persistence of the sickle-cell gene in Africa, but may be more important where the malaria transmission rate is lower, as in British Honduras.

Distribution of the Sickle-Cell Gene

One of the original arguments in support of the malaria hypothesis was the distribution of the sickle-cell gene in Africa and elsewhere (Allison, 1954b). The extensive surveys that have since been carried out have all supported this view, with the exception of one preliminary claim (Foy, Brass, and Kondi, 1956) which seems to have been based on a technical error. Thus, in the Congo (Hiernaux, 1962), Tripolitania (Modica, Levadiotti, and Sorrenti, 1960), Greece (Barnicot, Allison, Blumberg,

Deliyannis, Krimbas, and Ballas, 1963), Arabia (Lehmann, Maranjian, and Maurant, 1963) and India (see Shukla and Solanki, 1958; Mital, Parok, Sukumaran, Sharma, and Dave, 1962), high frequencies of sickling are confined to regions where malaria is or was endemic.

Frequencies of the Sickle-Cell Gene in the New World

Interpretation of the results of selection in New World populations is complicated by their uncertain origin. Slaves were obtained between 1520 and 1800 from areas widely scattered along the West Coast of Africa, from Cape Verde to Angola. Historical evidence indicates that the greatest number came from the neighborhood of Ghana, and this is supported by the relatively high frequencies of Hb-C in New World Negro populations. Over much of West Africa the incidence of the sickle-cell trait is of the order of 20%, and it is unlikely that the frequencies in the populations transported to the New World were less than this. Lower rates are found in Liberia and that neighborhood, but it is known that few slaves were taken from these regions; and this is confirmed by the exceedingly low frequencies of Hb-N in the Americas.

The second source of uncertainty is the degree of non-Negro admixture in the New World populations. This problem is more easily resolved by the use of suitable genetic markers, as in the recent study of Workman, Blumberg, and Cooper (1963). A comparison was made of 15 polymorphic traits in American White and Negro populations living in a rural Southern United States community with frequencies observed in West African Negroes and other American Negro and White populations. Estimating the total amount of gene migration, m, from the Whites to the American Negroes (assuming no selection), the polymorphic traits fell into two groups. In the larger group, including several red-cell blood-group antigens, the estimates of m (0.1 to 0.2) are consistent with the hypothesis that migration alone can account for the differences in gene frequencies between the West African and American Negro populations. In the second group, significantly higher estimates of gene migration were obtained (0.46 to 0.69 in the case of the sickle-cell gene). It was concluded that these resulted from both gene migration and different adaptive values of the traits in West African and American environments. Watson-Williams and Weatherall (1964) have also concluded that the frequencies of the sickle-cell gene in American Negroes are too low to be accounted for by non-Negro admixture. These results are of considerable interest because they suggest that a selective change in the genetic structure of a human population, that is to say, evolution, can take place within the short historical span of some three hundred years.

The frequencies of the sickle-cell gene in different West Indian populations are summarized in Table 4. Jonxis (1959) drew attention to the higher frequencies of the sickle-cell trait in Surinam (which is malarious) than in Curaçao (which is not). The sickle-cell trait is also very common in the Djukas of Surinam and the Black Caribs of British Honduras, both living in highly malarious environments. J. Parker-Williams (unpublished) has carried out comparisons of populations living in different West Indian islands. Significant, higher frequencies of the sickle-cell gene were

Table 4. Incidence of Sickle-Cell and Hemoglobin-C Genes in West Indian Populations.

Population	No. Tested	% Hb-S Trait	% Hb-C Trait	Reference
Honduras, Black Caribs	705	23.3	2.6	Firschein (1961)
Curaçao	1502	7.2	8.0	Jonxis (1959)
Surinam, Kabel	519	16.8	5.2	Jonxis (1959)
Surinam, Moengo	172	20.3	4.7	Jonxis (1959)
Surinam, Stoelman's Island	275	11.3	3.3	Jonxis (1959)
Surinam, Djukas	343	15	3	Liachowitz et al. (1958)
Jamaica	1018	10.9	3.1	Went (1957)
Jamaica		11.5	1.8	J. Parker-Williams (unpublished)
St. Vincent	748	8.7	2.7	Same
Dominiq	664	9.5	1.5	Same
Barbados	912	7.0	4.6	Same
St. Lucia	825	14.0	3.8	Same

found in St. Lucia and Jamaica (malarious) than Barbados (nonmalarious). In St. Lucia significantly higher frequencies of sickle-cells were found in Gros Islet, which was highly malarious, than in the Soufriere district, with a low incidence of malaria. The frequencies of Hb-C were not significantly different in all the regions studied. Although these results must be interpreted with caution because of the uncertain origin of the populations concerned, they are consistent with the view that malaria has played a substantial part in maintaining high frequencies of the sickle-cell gene in some parts of the New World.

Interaction of Allele Genes for Hemoglobin Variants

Allison (1955) drew attention to the special case of two alleles, such as Hb_β^S and Hb_β^C, which occur in the same region and interact to produce a heterozygote Hb_β^S/Hb_β^C of lowered fitness. There has been some doubt about the fitness of this genotype, and one approach to this problem is examination of large numbers of adults to see whether the corresponding phenotype occurs at frequencies different from Hardy-Weinberg expectation. Watson-Williams and Weatherall (1964) reported the results of hemoglobin electrophoresis of specimens from 12,387 Yoruba blood donors from Ibadan, Nigeria. Only a slight deficiency of the S + C type was observed; but, as Allison (1964) pointed out, these results must be interpreted with caution, since the blood donors were nearly all male and were, to a considerable extent, selected through relationship to anemic subjects. When an unselected panel of donors was analyzed, a considerable deficiency of S + C was found, and since it is quite clear that Hb^S/Hb^C females are unduly prone to complications in pregnancy (Fullerton et al., 1964), there is no doubt that this genotype is less fit than the mean. From theoretical considerations Allison (1955) predicted that the Hb_β^S and Hb_β^C genes ought to be mutually exclusive in West African populations. Observations

347

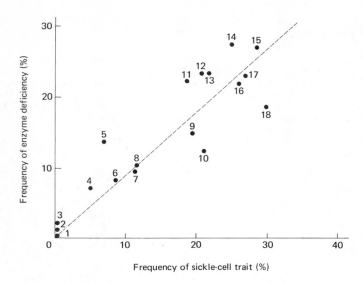

Figure 34-3. Frequencies of the sickle-cell and glucose-6 phosphate dehydrogenase deficiency traits in African popula tions (combined data of Motulsky, 1960, and Allison, 1960).

made shortly afterwards by Edington and Lehmann (1956) and Allison (1956) showed that this was so: high frequencies of $Hb_\beta{}^C$ were found in Northern Ghana, but low frequencies of $Hb_\beta{}^S$; to the South and East frequencies of $Hb_\beta{}^S$ increased while those of $Hb_\beta{}^C$ fell. Allison (1956) concluded that most populations could be in or near a state of equilibrium resulting from selection. However, Neel, Hiernaux, Linhard, Robinson, Zuelzer, and Livingstone (1956) found that some populations living in Liberia and Guinea have low incidences of both Hb-S and Hb-C, and could not be in equilibrium, and the interpretation was further complicated by the fact that in the latter regions β-thalassemia is also relatively common (Oleson, Oleson, Livingstone, Cohen, Zuelzer, Robinson, and Neel, 1959; Neel, Robinson, Zuelzer, Livingstone, and Sutton, 1961).

In general, comparison between Figure 34–4 and Figures 34–3 and 34–6 shows a striking difference. Were there no unfit interaction heterozygotes, positive correlations with frequencies of different abnormal hemoglobins and glucose-6-phosphate dehydrogenase (G-6-PD) deficiency might be expected. In contrast, the data for Hb-S and Hb-C remain strongly suggestive of mutual exclusion. The populations with low frequencies of both S and C are anthropologically distinct from other West African populations (Livingstone, 1958; Cabannes, 1964), and it is possible that selection has not yet raised the frequencies of abnormal hemoglobin genes to near equilibrium values in these cases.

Similar problems are posed by the presence of both sickling and β-thalassemia in Greece. Here again the interaction heterozygote $(Hb_\beta{}^S/Hb_\beta{}^T)$ is at a disadvantage, and it is found that where sickling frequencies are high, thalassemia frequencies are

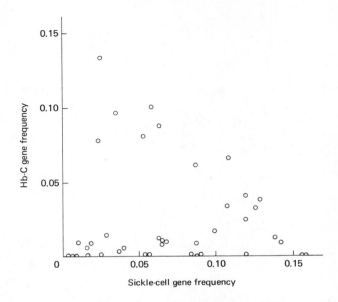

Figure 34-4. Frequencies of sickle-cell and hemoglobin C genes in West African populations.

low, and vice versa (Figure 34–5). In South-east Asian populations Hb-E and β-thalassemia are relatively common and interact to produce a somewhat disadvantageous heterozygote, and it would again be of interest to have accurate population and fitness data; the data of Flatz and Sundaragiati (1964) again suggest mutual exclusion.

Blood Groups and Natural Selection

At the 1955 meeting I pointed out that we still have no explanation for the persistence of the Rh blood-group polymorphism in the face of selection against the heterozygote, and that the inverse relationship of ABO incompatibility and Rh immunization could provide a mechanism which would help to stabilize one of these loci if the other were independently stabilized. These points remain valid, and nothing done during the past nine years has thrown any light on the matter.

Gorman (1964) has recently suggested that selection against the Rh-negative gene may be operating indirectly through malaria. His argument is that there are genetically-controlled differences in the capacity to form antibodies and that peoples living in malarious environments may be, in general, superior antibody formers. Hence a higher proportion of Rh-incompatible matings would result in hemolytic disease of the newborn, and there would be more intense selection against the Rh-negative gene. Unfortunately, the facts are in exactly the opposite direction. Hemolytic disease due to Rh incompatibility is exceedingly rare in tropical countries such as Nigeria, but somewhat more common in African populations living in nonmalarious

349

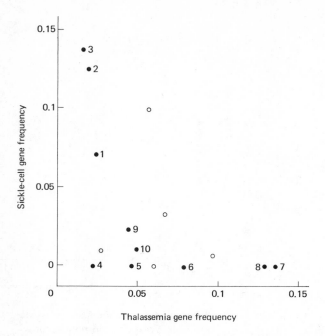

Figure 34-5. Frequencies of the sickle-cell gene plotted against frequencies of the β-thalassemia gene in different Greek populations (Barnicot et al., 1963 ●; Stamatoyannopoulos and Fessas, 1964 ○).

regions, such as South Africa. The argument for distribution is also weak. Very low frequencies of Rh-negatives are found in Chinese, Japanese, Eskimos, etc., and Gorman's argument that these may never have had the Rh-negative gene is begging the question.

Further data have accumulated establishing beyond reasonable doubt that possession of particular ABO blood groups predisposes individuals to certain diseases (see Fraser Roberts, 1959; Clarke, 1961; Allison and Blumberg, 1965). Unfortunately, the biochemical factors underlying these susceptibility differences are unknown, and the associations tell us very little about the action of selection on the blood-group genes. This is so because in most cases the diseases take their toll after reproductive age (see Reed, 1960), and because it is not known whether the associations are different for homozygous and heterozygous A or B subjects.

One of the problems that has attracted attention is the relationship of blood-group status to susceptibility to infection by micro-organisms. The higher susceptibility of nonsecretors of ABO, and subjects of blood group A, to rheumatic fever and its complications, may be due to their higher susceptibility to virulent streptococcal infection (Van Ryjsewijk and Gosling, 1963). A remarkably clear-cut result was obtained with influenza A2 virus by McDonald and Zuckerman (1962): those infected with the virus, which was new to the population of Air Force recruits

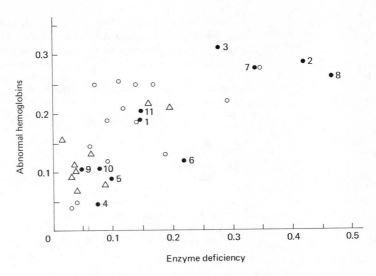

Figure 34-6. Frequency of glucose-6-phosphate dehydrogenase deficiency plotted against combined phenotype frequency of sickling and β-thalassemia in various Sardinian and Greek populations (data of Bernini et al., 1960 ○; Allison et al., 1963 ●; Stamatoyannopoulos and Fessas, 1964 △).

studied, showed a highly significant relative excess of group O and corresponding deficiency of group A, as compared with controls in all three regions (Table 5). This was not seen with other strains of influenza, but the results were complicated by prior exposure and immunity to those infections. The marked difference of susceptibility to influenza A2 according to blood group status is of interest because influenza, complicated by bacterial infection, could certainly have brought about selection during reproductive age in the past.

It has been known for many years that certain bacteria, protozoa, and helminths have antigens that are related to blood-group substances. These facts have suggested to a number of workers that particular organisms might be resisted unequally by subjects possessing different blood groups, either because of " naturally occurring " antibodies or because of inability to produce antibodies to antigens resembling those possessed by the host as a result of immunological tolerance. This approach has been taken to extremes by Pettenkofer and his colleagues (Pettenkofer and Bickerich, 1960; Vogel, Pettenkofer, and Helmbold, 1960) who have suggested that the present day distribution of ABO blood groups in human populations can be explained largely through selection exerted by epidemics of plague and smallpox. The arguments are too involved to discuss here; they are considered in detail by Allison and Blumberg (1965). Suffice it to say that the claims are altogether unjustified by the scanty and unsatisfactory evidence provided.

This does not, of course, invalidate the whole approach, which has much to commend it, as well as some support from animal experiments. Thus Rowley and Jenkin

Table 5. Relative Incidence of Blood Groups O and A in Patients with Influenza A2 as Compared with Controls by Geographical Region (McDonald and Zuckerman, 1962).

Region	No. in Disease Series	Relative Incidence O : A	χ^2
1	313	1.699	17.929
2	316	1.315	4.926
3	72	1.463	2.130
Total	701		24.985 (p < 0.00001)

χ^2	Difference from unity D.F. = 1	22.869
	Heterogeneity D.F. = 2	2.196

(1962) found that "opsonins" which increase the uptake by mouse phagocytic cells, and probably intracellular destruction, of *Salmonella typhimurium* CS are present only at low concentration in mouse serum but at much higher concentrations in rat and pig sera. This was thought to be due to the fact that the bacterium and mouse tissues share antigenic components, so that the mice are unable, because of immunological tolerance, to produce the appropriate opsonins. Some experimental evidence in support of this view was obtained. Other examples of heterophile immunological reactions, occurring apparently nonspecifically between antigens and antibodies, are reviewed by Jenkin (1963) and Boyden (1963), who emphasize that sharing of antigens between hosts and parasites may affect susceptibility to disease.

In general, we need have no doubt that selection *is* operating on human blood-group genes, and observations in animals indicate that these effects may be quite strong. Thus, several small inbred lines of chickens have remained polymorphic for one or more blood-group systems. All of the highly inbred White Leghorn lines studied by Gilmour (1962) still show segregation of at least one to four blood-group loci. The levels of inbreeding reached are the result of from 21 to 26 generations of full-sib matings, corresponding to computed coefficients of 98.9 to 99.6%. Briles (1960) sampled 73 different closed populations (inbred up to coefficients of 86%) and found segregation at the B locus in all but two of them; and most were polymorphic for other blood-group systems as well. This type of evidence argues powerfully for heterozygous advantage, because in the absence of selection many of the populations should have become homozygous for the blood-group alleles. Briles and Gilmour have produced detailed evidence that in a number of instances heterozygotes are at an advantage in terms of hatchability and viability.

In this connection, Morton et al. (1965) have obtained evidence that several transferrin genotypes in chickens have different probabilities of surviving. The reasons for the difference are unknown, but the findings point up the weakness of the claim by Martin and Jandl (1959) that selection may operate through inhibition by transferrin of the multiplication of micro-organisms. This inhibition is an example of the iron-chelating effect previously described by Schade (see 1961) and no convincing evidence of a difference between effects of normal and variant forms of transferrin in this respect has been forthcoming.

Regarding the possible effects of selection on blood-group antigens at the gametic level, Edwards, Ferguson, and Coombs (1964) were unable to demonstrate A or B antigens on human spermatozoa of nonsecretors by the mixed agglutination and mixed antiglobulin reactions, and Holborow, Braun, Glynn, Hawes, Gresham, O'Brien, and Coombs (1960), and Allison and Bishop (unpublished) found none by the fluorescent antiglobulin technique. Hence the claims of Güllbring (1957) and Shahani and Southam (1962) that the A and B antigens segregate on separate spermatozoa must be treated with reservation. Matsunaga and Hiraizumi (1962) presented data indicating that heterozygous AO and BO fathers transmitted more than the expected 50% of O-bearing sperm to their children. Since neither sperm incompatibility nor reproductive compensation appeared to account for the observations, it was concluded that they demonstrate prezygotic selection operating in the ABO blood groups of man, probably due to meiotic drive or sperm competition. This study has been criticized by Novitski (1962) and Reed (1963), and the criticisms only partially answered by Hiraizumi (1963). The recent evidence of Hiraizumi (1964) presented in this volume, shows no indication of prezygotic selection.

Selection and the Sex Ratio

As everybody knows, in man equal numbers of homogametic and heterogametic zygotes would be expected, but vital statistics from many countries shows a preponderance of male births, usually about 105 males to 100 females. Over millions of births this excess cannot be due to chance, and its significance must be considered. The proportion has not been constant, as Parkes (1963) has reminded us. The sex ratios of births in England and Wales from 1841 to the present time are summarized in Figure 34–7. The decrease in proportion of males during the first 30 years of this period is not understood; perhaps registration was incomplete. For the next 40 years the ratio hovered around 104 males per 100 females and then at the time of the First World War rose to about 105. A subsequent decrease was followed by another sharp peak coincident with the Second World War.

The sex ratio changes in another way. In the middle of the last century, the excess of males at birth disappeared by the age of 5 years because of the differential mortality of males over this period, and thereafter females were in excess. By the turn of the century mortality had decreased and sex ratios were about equal up to the age group 15–19 years; thereafter, owing to differential mortality and emigration, falling sharply to below 90 males per 100 females in the 25–29 age group (Figure 34–8). At the present time, with the further decrease in child mortality, the excess of males found at birth is being preserved much longer, up to the age group 25–29. Even now, females exceed males from the age group 35–39 onwards, and the decrease in relative numbers of males becomes very sharp after the age 55. The sex ratio among

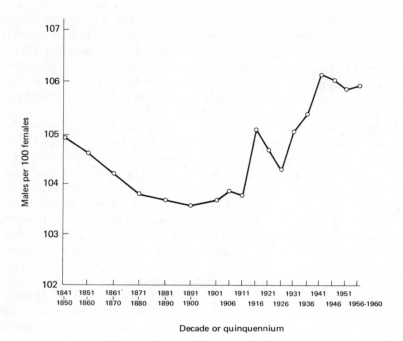

Decade or quinquennium

Figure 34-7. Sex ratio for live births in England and Wales, 1841–1960 (from Parkes, 1963).

births is also inversely correlated with parental age, and the low age of parents during the two World Wars probably explains the coincident peaks of male birth frequencies.

Thus there is abundant evidence that mortality after birth is higher among males than females. Until recently, there was an even more striking difference among stillbirths, and it has been widely accepted that this is true also of abortuses (Crew, 1952). However, Tietze (1948) and McKeown and Lowe (1951) have pointed out that it is difficult to sex abortuses reliably by morphology, so that the evidence for a very high primary sex ratio in man is inadequate. There is scope here for cytological study, since human sex chromatin is detectable at 10–12 days gestation (Park, 1957).

Evidence from animals indicates the existence of a high primary sex ratio and proportionately greater loss of male fetuses during pregnancy. Crew (1952) estimated that the sex ratio of pigs at conception was about 150. Nuclear sexing has shown a high sex ratio before implantation in the golden hamster (Lindahl and Sundell, 1958), which implies that there is a considerable differential mortality *in utero* since the sex ratio at birth is about unity.

The sex ratio represents, in fact, a most interesting polymorphism determined by a simple genetic switch mechanism and subject to selection. The substantial excess of males at conception found in some species (and possibly present in man) might

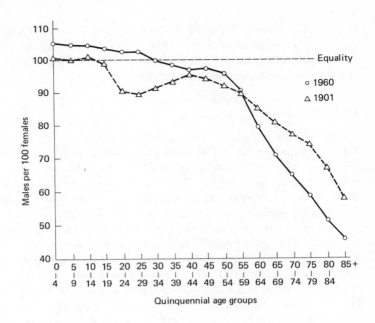

Figure 34-8. Sex ratio according to age (England and Wales, from Parkes, 1963).

be due to formation of a higher proportion of functional Y-bearing than X-bearing spermatozoa, to some advantage of the former in the female genital tract, or to some other more obscure mechanism. The higher mortality of male than female fetuses is presumably the expression of sex-linked or sex-limited factors reducing viability. Some social consequences of the current trend in sex ratios are discussed by Parkes (1963).

There are other reports of deviation of the sex ratio. The most important is the indication that the sex ratio of offspring is altered by parental exposure to X-irradiation (Lejeune and Turpin, 1957; Schull and Neel, 1958; Tanaka and Okhura, 1961). The ratio was increased after paternal, and decreased after maternal exposure, which suggests that some daughters of exposed fathers and some sons of exposed mothers failed to develop because of damage to genes on the parental X-chromosomes. This is the main indication of a genetic—as opposed to a somatic—effect of X-irradiation in man.

The second series of reports concerns departures from randomness of the sexes of successive children. Renkonen, Mäkelä, and Lehtovaara (1961) reported that the birth of a boy renders a woman less likely to give birth to further boys in the future. Reviewing Finnish and French data, Edwards (1962) states that there is good evidence for a correlation between the sexes of successive children. One interpretation of the findings of Renkonen et al., if substantiated, would be that bearing a male fetus could immunize a woman against a male (Y) antigen, which might harm subsequent male fetuses or inactivate a proportion of Y-bearing sperm before fertilization. Such immunization can be achieved in homografting experiments. However, McLaren (1962) found no change in the sex ratio among offspring of inbred mice deliberately immunized against the male antigen.

General Comment

The evidence for selection acting on the sickle-cell gene is an order of magnitude better than that for selection operating on other abnormal hemoglobins and G-6-PD deficiency, which is, in turn, a good deal better than that acting on most other human polymorphic systems. We can, however, be quite confident that selection *is* acting on blood groups, the sex ratio, and probably other polymorphisms, although of the underlying mechanisms we are sadly ignorant. The last ten years have seen posed the problems and provided some of the solutions, but progress in this field is not furthered by the uncontrolled speculations that have become common. Observations on animals provide powerful evidence for selection acting on blood-group and transferrin polymorphisms, and by analogy we may suppose that similar effects are operating in man. But well-controlled, precise evidence on human populations is needed before this can be accepted as true. The existence of human polymorphisms controlled by selection is established, but much more work is needed to establish how widespread these effects are and the magnitude of the selective forces involved.

References

Allison, A. C. 1954a. Protection by the sickle-cell trait against subtertian malarial infection. Brit. Med. J., *i*: 290.

————. 1954b. The distribution of the sickle-cell trait in East Africa and elsewhere and its apparent relationship to the incidence of subtertian malaria. Trans. Roy. Soc. Trop. Med. Hyg. *48*: 312.

————. 1955. Aspects of polymorphism in Man. Cold Spring Harbor Symp. Quant. Biol. *20*: 239.

————. 1956. The sickle-cell and haemoglobin C genes in some African populations. Ann. Hum. Genet. *21*: 67.

————. 1960. Glucose-6-phosphate dehydrogenase deficiency in red blood cells of East Africans. Nature *186*: 431–432.

————. 1965. Population genetics of abnormal haemoglobins. *In* J. H. P. Jonxis, ed., *Abnormal Haemoglobins in Africa*. Blackwell, Oxford.

Allison, A. C., B. A. Askonas, N. A. Barnicot, B. S. Blumberg, and C. Krimbas. 1963. Deficiency of glucose-6-phosphate dehydrogenase in Greek populations Ann. Hum. Genet. *26*: 237.

Allison, A. C., and B. S. Blumberg. 1965. *Polymorphism in Man*. Little, Brown and Co., Boston.

Allison, A. C., and D. F. Clyde. 1961. Malaria in African children with deficient glucose-6-phosphate dehydrogenase. Brit. Med. J. *i*: 1346–1348.

Barnicot, N. A., A. C. Allison, B. S. Blumberg, G. Deliyannis, C. Krimbas, and A. Ballas. 1963. Haemoglobin types in Greek populations. Ann. Hum. Genet. *26*: 229–236.

Bernini, R. N., V. Carcassi, B. Latte, A. G. Motulsky, L. Romei, and M. Siniscalco. 1960. Indagini genetische sulla predispositione al favismo. III. Distribuzione della frequenze geniche per il locus gd oin Sardegna. Interazoine con la malaria e la talassemia il livello populazionistico. Recond. Acad. Natl. dei Lincei 199: 1–18.

Boyden, S. V. 1963. Cellular recognition of foreign matter. International Review of Experimental Pathology 2: 311. Academic Press, New York.

Briles, W. E. 1960. Blood groups in chickens, their nature and utilization. World's Poultry Sci. 16: 223.

Burke, J., C. de Bock, and C. de Wolf. 1958. La drépanocytémie simple et l'anémie drépanocytaire au Kwango (Congo Belge). Memoires in 8° de l'Acad. Roy. Sciences Colon. Nouv. Ser. 7: 1.

Carbannes, R. 1965. Distribution of haemoglobin variants. In J. H. P. Jonxis, ed., Abnormal Haemoglobins in Africa. Blackwell, Oxford.

Clarke, C. A. 1961. Blood groups and disease. Progr. Med. Genet. 1: 81.

Colbourne, M. J., and G. M. Edington. 1956. Sickling and malaria in the Gold Coast. Brit. Med. J. i: 784.

Crew, F. A. E. 1952. The factors which determine sex, p. 741–792. In Parkes (ed.), Marshall's Physiology of Reproduction. Longmans, London.

Edington, G. M., and W. N. Laing. 1957. Relationship between haemoglobins S and C and malaria in Ghana. Brit. Med. J. ii: 143.

Edington, G. M., and H. Lehmann. 1956. The distribution of haemoglobin C in West Africa. Man. 56: 34.

Edington, G. M., and G. J. Watson-Williams. 1965. Sickling, haemoglobin C, glucose-6-phosphate dehydrogenase deficiency and malaria in Western Nigeria. In J. H. P. Jonxis, ed., Abnormal Haemoglobins in Africa. Blackwell, Oxford.

Edwards, A. W. F. 1962. Genetics and the human sex ratio. Adv. Genet. 11: 239.

Edwards, R. G., L. C. Ferguson, and R. R. A. Coombs. 1964. Blood group antigens in human spermatozoa. J. Reprod. Fertil. 17: 153–161.

Field, J. W. 1949. Blood examination and prognosis in acute falciparum malaria. Trans. Roy. Soc. Trop. Med. Hyg. 43: 33.

Firschein, I. L. 1961. Population dynamics of the sickle-cell trait in the Black Caribs of British Honduras. Am. J. Hum. Genet. 13: 233.

Fisher, R. A. 1959. Statistical Methods for Research Workers. Oliver and Boyd, Edinburgh, p. 99.

Flatz, G., and B. Sundaragiati. 1964. Malaria and haemoglobin E in Thailand. Lancet ii: 385.

Foy, H., W. Brass, R. A. Moore, G. Timms, A. Kondi, and T. Olouch. 1955. Two surveys to investigate the relation of the sickle-cell trait and malaria. Brit. Med. J. ii: 1116.

Foy, H., W. Brass, and A. Kondi. 1956. Sickling and malaria. Brit. Med. J. i: 289.

Frota-Pessoa, O., and A. Wajntal. 1963. Mutation rates of the abnormal hemoglobin genes. Amer. J. Hum. Genet. 15: 123.

Fullerton, W. T., J. P. de V. Hendrickse, and E. J. Watson-Williams. 1965. Haemoglobin SC disease in pregnancy. In J. H. P. Jonxis, ed., Abnormal Haemoglobins in Africa. Blackwell, Oxford.

Garlick, J. P. 1960. Sickling and malaria in South West Nigeria. Trans. Roy. Soc. Trop. Med. Hyg. 54: 146.

Gilmour, D. G. 1962. Blood groups in chickens. Ann. N.Y. Acad. Sci. 97: 166.

Gorman, J. G. 1964. Selection against the Rh-negative gene by malaria. Nature *202*: 676.

Güllbring, B. 1957. Investigation on the occurrence of blood group antigens in spermatozoa from man, and serological demonstration of the segregation of characters. Acta Med. Scand. *159*: 169.

Harris, H. 1959. *Introduction to Human Biochemical Genetics*. 2nd ed. University Press, Cambridge, 310 p.

Hiernaux, J. 1962. Donnes génétiques sur six populations de la République du Congo. Ann. Soc. Belge Med. Trop. *40*: 339.

Hiraizumi, Y. 1963. Assumption in tests for meiotic drive. Science *139*: 406.

Holborow, E. J., P. C. Braun, L. E. Glynn, M. D. Hawes, G. A. Gresham, T. F. O'Brien, and R. R. A. Coombs. 1960. The distribution of the blood group A antigen in human tissues. Brit. J. Exp. Path. *41*: 430.

Jenkin, C. R. 1963. Heterophile antigens and their significance in the host-parasite relationship. Adv. Immunol. *3*: 351.

Jonxis, J. H. P. 1965. The frequency of haemoglobin S and C carriers in Curaçao and Surinam. *In* J. H. P. Jonxis, ed., *Abnormal Haemoglobins in Africa*. Blackwell, Oxford.

Lambotte-Legrand, J., and C. Lambotte-Legrand. 1958. Notes complémentaires sur la drépanocytose, 11. Sicklémie et malaria. Ann. Soc. Belge. Med. Trop. *38*: 45.

Lehmann, H., G. Maranjian, and A. E. Mourant. 1963. Distribution of sickle-cell haemoglobin in Saudi Arabia. Nature *198*: 492.

Lejeune, J., and R. Turpin. 1957. Mutations radioinduites chez l'homme et dose de doublement. Sur la validité d'une estimation directe. Compt. Rend. *244*: 2425.

Liachowitz, C., J. Elderlein, I. Gilchrist, H. W. Brown, and H. M. Ranney. 1958. Abnormal hemoglobins in the Negroes of Surinam. Amer. J. Med. *24*: 19.

Lindahl, P. E., and G. Sundell. 1958. Sex ratio of the golden hamster before uterine implantation. Nature *182*: 139.

Livingstone, F. B. 1957. Sickling and malaria. Brit. Med. J. *i*: 762.

———. 1958. Anthropological implications of sickle-cell gene distribution in West Africa. Amer. Anthrop. *60*: 533.

McDonald, J. C., and A. J. Zuckerman. 1962. ABO blood groups and acute respiratory virus disease. Brit. Med. J. *ii*: 89.

McKeown, T., and C. R. Lowe. 1951. The sex ratio of stillbirths related to cause and duration of stillbirths. An investigation of 7,066 stillbirths. Human Biol. *23*: 41.

McLaren, A. 1962. Does maternal immunity to male antigen affect the sex ratio of the young? Nature *195*: 1323.

Martin, C. M., and J. H. Jandl. 1959. Inhibition of virus multiplication by transferrin. J. Clin. Invest. *38*: 1024.

Matsunaga, E., and Y. Hiraizumi. 1962. Prezygotic selection in the ABO blood groups. Science *135*: 432.

Mital, M. S., J. G. Parok, P. K. Sukumaran, R. S. Sharma, and P. J. Dave. 1962. A focus of sickle-cell gene near Bombay. Acta Haemat. *27*: 257.

Modica, H., M. Levadiotti, and A. M. Sorrenti. 1960. Incidenza della sicklemia, progressa malaria e distribuzione razziale dell'oasi costiera di Tauroga. Arch. Ital. Sci. Med. Trop. Parasitol. *41*: 595–603.

Morton, J. R., D. G. Gilmour, E. M. McDermid, and A. L. Ogden. 1965. Association of blood group and protein polymorphisms with embryonic mortality in the chicken. Genetics *51*: (1), in press.

Motulsky, A. G. 1960. Metabolic polymorphisms and the role of infectious diseases in human evolution. Human Biol. *32*: 28.

Nance, W. E. 1963. Genetic control of hemoglobin synthesis. Science *141*: 123.

Neel, J. V., J. Hiernaux, J. Linhard, D. D. Robinson, W. W. Zuelzer, and F. B. Livingstone. 1956. Data on the occurrence of hemoglobin C and other abnormal hemoglobins in some African populations. Amer. J. Hum. Genet. *8*: 138.

Neel, J. V., A. R. Robinson, W. W. Zuelzer, F. B. Livingstone, and H. E. Sutton. 1961. The frequency of elevations in the A₂ and fetal hemoglobin fractions in the natives of Liberia and adjacent regions with data on haptoglobin and transferrin types. Amer. J. Hum. Genet. *13*: 263.

Novitski, E. 1962. Meiotic drive. Science *137*: 861.

Olesen, E. B., K. Olesen, F. B. Livingstone, F. Cohen, W. W. Zuelzer, A. R. Robinson, and J. V. Neel. 1959. Thalassemia in Liberia. Brit. Med. J. *i*: 1385.

Park, W. W. 1957. The occurrence of sex chromatin in early human and macaque embryos. J. Anat. *91*: 369.

Parkes, A. S. 1963. The sex ratio of human populations, p. 91. *In* Wolstenholme, ed., *CIBA Foundation Symposium on Man and His Future*. Churchill, London.

Pettenkofer, H. J., and R. Bickerich. 1960. Uber Antigen-gemeinschaften zwischen den menschlichen Blutgruppen ABO und den erregern gemeingefährlicher Krankheiten. J. Bakt. Parasitol. *179*: 433.

Raper, A. B. 1955. Malaria and the sickling trait. Brit. Med. J. *i*: 1186–1189.

———. 1956. Sickling in relation to morbidity from malaria and other diseases. Brit. Med. J. *i*: 965.

———. 1959. Further observations on sickling and malaria. Trans. Roy. Soc. Trop. Med. Hyg. *53*: 110.

Reed, T. E. 1960. Polymorphism and natural selection in blood groups. *In Genetic Polymorphisms and Geographic Variations in Disease*. U.S. Public Health Service, Washington.

———. 1963. Assumptions in tests for meiotic drive. Science *139*: 408.

Renkonen, K. O., O. Mäkelä, and R. Lehtovaara. 1961. Factors affecting the human sex ratio. Ann. Med. Exp. Biol. Fenn. *39*: 173.

Roberts, J. A. Fraser. 1959. Some associations between blood groups and disease. Brit. Med. Bull. *15*: 129.

Rowley, D., and C. R. Jenkin. 1962. Antigenic cross-reaction between host and parasite. Nature *193*: 151.

Rucknagel, D. L., and J. V. Neel. 1961. The hemoglobinopathies. Progr. Med. Genet. *i*: 158.

Schade, A. L. 1961. The microbiological activity of siderophilin, p. 261. *In Protides of Biological Fluids*, Elsevier, Amsterdam.

Schull, J., and J. V. Neel. 1958. Radiation and the sex ratio in man. Science *128*: 343.

Shahani, S., and A. L. Southam. 1962. Immunofluorescent study of the ABO blood group antigens in human spermatozoa. Amer. J. Obstet. Gynec. *84*: 660.

Shukla, R. N., and B. R. Solanik. 1958. Sickle-cell trait in Central India. Lancet *i*: 297.

Stamatoyannopoulos, G., and P. H. Fessas. 1964. Thalassemia and glucose-6-phosphate dehydrogenase, sickling and malarial endemicity in Greece. A study of five areas. Brit. Med. J. *i*: 875–879.

Tanaka, K., and K. Okhura. 1961. Genetic effects of radiation in man: a study on the sex ratio in the offspring of radiological technicians. Proc. 2nd Int. Conf. Human Genetics, Rome.

Thompson, G. R. 1962. Significance of haemoglobins S and C in Ghana. Brit. Med. J. *i*: 682.

————. 1963. Malaria and stress in relation to haemoglobins S and C. Brit. Med. J. *ii*: 976.

Tietze, C. 1948. A note on the sex ratio of abortions. Human Biol. *20*: 156.

Vandepitte, J. 1965. The incidence of haemoglobinoses in the Belgian Congo. *In* J. H. P. Jonxis, ed., *Abnormal Haemoglobins in Africa*. Blackwell, Oxford.

Vandepitte, J., and J. Delaisse. 1957. Sicklémia et paludisme. Ann. Soc. Belge Med. Trop. *37*: 703.

Vandepitte, J., and J. Stijns. 1965. Haemoglobinopathies in the Congo (Leopoldville) and the Rwanda Burundi. *In* J. H. P. Jonxis, ed., *Abnormal Haemoglobins in Africa*, Blackwell, Oxford.

Vandepitte, J., W. W. Zuelzer, J. V. Neel, and J. Colaert. 1955. Evidence concerning the inadequcy of mutation as an explanation of the frequency of the sickle-cell gene in the Belgian Congo. Blood *10*: 341.

Van Rijsewijk, M. G. H., and N. E. O. Goslings. 1963. Secretor status of streptococcus pyogenes group A carriers and patients with rheumatic heart disease or acute glomerulonephritis. Brit. Med. J. *ii*: 542.

Vogel, J., H. J. Pettenkofer, and W. Helmbold. 1960. Über die Populationsgenetik der ABO Blutgruppen. Acta Genet. *10*: 267.

Wagner, R. P., and H. K. Mitchell. 1964. *Genetics and Metabolism*, 2nd ed. John Wiley, New York.

Walters, J. H., and L. J. Bruce-Chwatt. 1956. Sickle-cell anemia and falciparum malaria. Trans. Roy. Soc. Trop. Med. Hyg. *50*: 51.

Watson-Williams, G. J., and D. J. Weatherall. 1965. The laboratory characterization of haemoglobin variants. *In* J. H. P. Jonxis, ed., *Abnormal Haemoglobins in Africa*. Blackwell, Oxford.

Went, L. N. 1957. Incidence of abnormal haemoglobins in Jamaica. Nature *180*: 1131.

Wilson, T. 1961. Malaria and glucose-6-phosphate dehydrogenase. Brit. Med. J. *ii*: 246 and 895.

Woolf, B. 1955. On estimating the relation between blood group and disease. Ann. Hum. Genet. *19*: 251.

Workman, P. L., B. S. Blumberg, and A. J. Cooper. 1963. Selection, gene migration and polymorphic stability in a U.S. White and Negro population. Amer. J. Hum. Genet. *15*: 429.

35   The Sickle Cell Controversy

Lawrence E. Gary

Among the health policies facing the nation, sickle cell anemia has become a major issue, especially in the black community. During the past three years, various black professional associations have discussed and debated the problem.[1] Black political organizations have taken specific positions on designing programs to deal with it.[2]

After President Nixon's 1971 health message to Congress gave the first significant public support to the study and treatment of sickle cell anemia and after Congress had introduced several bills providing support for sickle cell programs, medical journals, popular magazines, and newspapers began to publish articles about its clinical, psychosocial, and political aspects.[3] With widely diverse groups supporting research, fund-raising and informational programs, it is not surprising that the disease became a subject of major controversy in the black community, that myths and conflicts became associated with it, and that many implications of the programs initiated were not at first apparent.[4] As noted by Michaelson:

> The current enthusiasm for sickle cell anemia may in the end reveal less about this killing disease than about contemporary American medical politics. It may be part of the larger effort to "save" not black children but an absolute and elitist system of medical care which has oppressed patients of all races and classes for a century.[5]

The primary objectives of this article are to identify and examine critically the major issues in the sickle cell controversy from a sociopolitical perspective and to discuss the potential role of social work in dealing with the problem.

Definition of the Disease

According to Lin-Fu, the comprehensive term sickle cell disease includes sickle cell anemia, sickle cell trait, and sickle cell variants. All are hereditary disorders that have clinical, hematologic, and pathological features related to the presence of sickle hemoglobin (hemoglobin S) in the red blood cells.[6] What is important to note is that sickle cell anemia is an *inherited* defect of the red blood cells.[7]

Herrick first recognized and described the abnormal sickle cell in 1910.[8] Then for almost forty years medical research added little to the knowledge of the disease. In 1949, Pauling et al., using chemical and electrophoretic techniques, demonstrated

The author wishes to thank Delores Duncan, Sharon Prather, and Theodis Thompson for their assistance.

Reprinted with permission of the author and the National Association of Social Workers, from *Social Work*, Vol. **19**, No. 3 (May 1974), pp. 263–272.

the molecular structure of sickle cell anemia. In the same year, working independently, Neel and Beet developed the heterozygous-homozygous hypothesis about how the disease is inherited.[9] Scott and Kessler pointed out the following differences in the red blood cells of a normal person and one with sickle cell anemia:

> Normal red blood cells are disk shaped. In sickle cell anemia, the red blood cells assume a crescent or sickle shape when their supply of oxygen is low. [While] the life span of normal red blood cells is about 120 days, [the life span of the red cells of sickle cell anemia patients] is often considerably less than 60 days. The person with sickle cell anemia can produce new cells, often at a rapid rate, but he becomes anemic because production cannot keep pace with the rapid destruction.[10]

Although there is no cure for sickle cell anemia, scientists have identified several substances such as urea and cyanate, that physicians have found somewhat successful in helping patients cope with the sickle cell crisis—that is, the periodic outbreaks of severe pain.[11]

It is crucial to distinguish between sickle cell anemia and the sickle cell trait, which Scott and Kessler define as "a condition in which a small percentage of the red blood cells have the sickle shape but behave in all other respects like normal red blood cells."[12] Until recently it had been assumed, for the most part, that the sickle cell trait was a benign condition with no clinical manifestations. However, an increasing number of reports show that persons with the trait may occasionally experience a sickle cell crisis or have other abnormal conditions related to sickling. Precipitating factors include flight at high altitudes, infection, underwater swimming, and alcoholic intoxication.[13]

Since clinical manifestations are absent or infrequent, specific tests are necessary to discover the presence of the trait. Several techniques, such as stained-blood-smear examination, sickle-cell-slide preparation, sickle-turbidity tubule test, and hemoglobin electrophoresis, have been devised to identify it.[14] It is estimated that in the United States at least fifty thousand blacks have sickle cell anemia and that two million blacks have the trait.[15] These figures suggest that this health problem affects a large proportion of the black population. However, from a sociopolitical perspective, the significant fact is that—there being no known cure for sickle cell anemia—at the present time the only way to prevent its spread is to control the mating of persons who have the trait.[16]

Major Sociopolitical Issues

With the advances in biological and medical science, man is developing a variety of techniques for controlling the quality and quantity of the human population.[17] Developments in genetics have many implications for manipulating the composition of the population in an industrial, technological society. As Sorenson observed:

> These developments pose complex questions of a moral, ethical, political, psychological, or economic nature. For instance, what genetic attributes or constitutions are desirable? Who is to decide? Should genetic anomalies be reduced in a population? Who shall say how or when?[18]

Since sickle cell anemia is a genetic disorder, the black community is understandably concerned about the social consequences of strategies of treatment that will have a direct impact on the future of a large number of black people. The concern is compounded because of the significant shift in this country toward limiting social programs that are designed to broaden opportunities for the disadvantaged.[19] Those who defend the shift have not done so solely on biological or genetic grounds, but the net effect seems similar to that of programs dealing with the sickle cell problem: a tendency to blame the victim. Given this reality, it is not surprising that black groups are beginning to raise questions about governmental interest in financing sickle cell programs.

Priorities

One factor contributing to the sickle cell controversy is the relationship between priorities of health care and the publicity given to sickle cell anemia. As Scott noted:

> In 1967 there were an estimated 1,155 new cases of sickle cell anemia, 1,206 of cystic fibrosis, 813 of muscular dystrophy, and 350 of phenylketonuria. Yet volunteer organizations raised $1.9 million for cystic fibrosis, $7.9 million for muscular dystrophy, but less than $100,000 for sickle cell anemia.[20]

These figures have changed, especially in the past two years, but still sickle cell anemia has not received proportionate private financial support for research and education. Until recently the federal government, through research grants of the National Institutes of Health (NIH), did not provide the level of support for sickle cell anemia suggested by its incidence.

Now that both the federal government and private foundations are providing greater financial support for research and education on sickle cell anemia, many blacks are beginning to question the importance and attention given the disease. The Center for Black Education issued the following statement:

> How important is this disease to black people? The answer is that while sickle cell anemia is a crippler and a killer well-deserving of a cure, there are many things which more commonly afflict black people in America. There is lead poisoning which millions of black children living in run-down slum homes contract from paint peeling off the walls. There is malnutrition from inadequate diets of many black people in the rural southern part of the country as well as in the cities. There are dope addiction and hypertension which are common in the crowded cities of Chicago, New York, and Los Angeles.[21]

Governmental statistics tend to support these assertions. For example, maternal mortality is three times higher for blacks than for whites; infant mortality is also greater for blacks.[22] It is generally known that the death rate from hypertension is higher for blacks than whites. This disease killed approximately 13,500 blacks last year while sickle cell anemia caused 340 deaths among blacks. It has been estimated that 80–85 percent of the victims of lead poisoning are nonwhite.[23]

Commenting on the differential in the mortality rates for specific diseases in the black and the white communities, Knowles and Prewitt pointed out:

> By and large, medical science has not made a dramatic reduction in mortality due to malignant disease even in the white community so the lack of medical care does not affect the course of these diseases markedly from one racial group to the next. Medical science, however, found cures and controls for most infectious diseases, which have greatly reduced their mortality rates in the white community. But the health institutions have failed to effectively extend this significant medical progress to the black community.[24]

One implication is that medical knowledge is available for dealing effectively with many health problems facing the black community, but for various reasons the health care system is not delivering the needed services. Some blacks argue that the sickle cell program is diverting attention from the major causes of black illness. One aspect of the sickle cell anemia controversy relates directly to priorities of health care perceived by individuals and groups in the black community.

Myths and Misconceptions

Inadequate information for the public has led to several myths and misconceptions about the sickle cell disease. Television, radio, and the press have widely publicized the disorder, but in several instances, publicity has tended to exaggerate its hazards. One nationally recognized private social service agency referred to the disease as the "black scourge." It has also been called the "killer disease."[25]

Medical literature has sometimes added to the confusion. In several instances medical publications have stated that few sickle cell anemia patients live beyond their 40s, but this statement is based on rather limited studies; many live well beyond that age.[26] What is significant with respect to the longevity of the sickle cell anemia patient is that malnutrition and generally poor living conditions can influence the severity of the disease. The impact of such factors is clearly related to the social structure of our society and the delivery of health services to the black community.[27]

Brochures issued by NIH have equated the sickle cell trait with sickle cell anemia.[28] The National Sickle Cell Anemia Control Act has also confused the terms. For example, the first line of the law reads, "... sickle cell anemia is a debilitating, inheritable disease that affects approximately two million American citizens and has been largely neglected."[29] This error is most unfortunate in that the law sets national policy for the sickle cell problem. Some state laws have also tended to confuse the sickle cell trait and sickle cell anemia. As a result of this, it is charged that jobs and insurance have been denied to persons who carry the trait. Apparently, there is little or no evidence that trait carriers have a higher risk of contracting infectious disease, living less than the normal life span, or being unproductive on the job. Culliton concluded, "Careful, controlled studies of carriers of the trait are few ... so the matter is clouded by a slew of impressions and ... erroneous notions."[30]

Misinformation or lack of information has also led to the fallacy that only blacks have sickle cell disease. Using literature published by leading private and public agencies, United Klans of America has circulated material that attempts to associate black people with monkeys and apes, arguing that sickle cell anemia is common to blacks and to these animals.[31] United Klans implies that no other race is subject to sickle cell anemia. However, Lin-Fu has noted that both the sickle cell trait and sickle cell anemia are found not only in Africans and their descendants, but also in Greeks, Italians, Arabs, Southern Iranians, Asiatic Indians, some Amerindians, and some Mexicans.[32] Because of racial mixing, it has been estimated that 20 percent of the whites in the United States have African genes. Therefore, it follows that some whites have sickle cell anemia or the sickle cell trait.

The stigmatizing of black people for a genetic disorder must be looked at in relation to the negative way blacks have long been viewed in Western society. In the light of the foregoing, many blacks are naturally sensitive to labeling sickle cell anemia as primarily a black disease. At the 1971 annual convention of the National Association for the Advancement of Colored People, delegates deleted the words in a sickle cell resolution that described the disease as one largely affecting blacks.[33]

Laws on Testing

With more money from the federal government and the new techniques revealing abnormal hemoglobin in a small blood sample, there has been a burgeoning of mass screening for sickle cell anemia and the sickle cell trait. These screening programs have aroused great concern in the black community. To provide support for them, states have passed sickle testing laws. Most are voluntary, but according to Bowman, at least fifteen states have compulsory sickle cell testing laws.[34] Only twenty-two states have no sickle cell laws.[35] The Virginia law, which includes mandatory sickle cell examination for inmates of correctional and mental institutions, lists the following requirement for testing children:

> Every child ... shall on or within 30 days after beginning attendance in a public kindergarten, elementary or secondary ... furnish a certificate from a duly licensed physician that the child has been tested ... or that such tests are not, in his professional judgment and discretion, deemed necessary.[36]

The last sentence suggests that white children will be excluded from mandatory testing.

The Massachusetts law reads as follows:

> Every child which the commissioner of public health, by rule or regulation, may determine is susceptible to the disease known as sickle cell trait or sickle cell anemia, shall be required to have a blood test to determine whether or not he has such disease before being admitted to a public school, or in the first year of attendance in a public school.[37]

In this law, it is clear that the terms sickle cell anemia and sickle cell trait are used interchangeably. Both the Virginia and the Massachusetts laws are directed at children, who are a captive audience. As Powledge suggests:

> If the purpose of the screening is to find sickle cell disease (anemia) it is probably too late. Almost all children with the disease have undergone crises and have been

diagnosed long before this. If the purpose of the screening is to find carriers and offer them genetic counseling, it is far too early. A child of seven ... should not be expected to deal with the notion that he might give his future children a fatal disease unless he selects a marriage partner who is not a carrier.[38]

There are indications that the Massachusetts law has not been fully enforced. Some state officials have advanced substitute legislative proposals that provide voluntary screening for the sickle cell disease and other genetic diseases, flexibility in adding other diseases to the list, and proper counseling and treatment. Nonetheless, the impact of the compulsory sickle cell law has had a negative effect on the black community.

Sickle cell laws present other problems. For instance, in Washington, D.C., to pass the law the city council had to declare sickle cell a communicable disease— merely adding to the sickle cell confusion.[39] Many sickle cell laws require screening tests before marriage and in most cases these laws merely amend the regulations for venereal disease.[40] Given the high rate of venereal disease in the black community, labeling sickle cell disease as a black disorder and relating the two disorders can exaggerate the social psychological impact of the sickle cell problem and easily be misinterpreted by racist forces in our society. But Powledge identified the significant issue when she concluded that "the genetic screening provision of marriage license laws represents the opening wedge for governmental involvement in genetic criteria for procreation."[41]

It is believed that a compulsory sickle cell law drawn along racial lines might be unconstitutional. For this reason, many legislators have been reluctant to use the words *black* or *Negro* in the law. However, the New York marriage license law states that testing is required for "each applicant for a marriage license who is not of Caucasian, Indian, or Oriental race."[42] In general, most states have avoided the color question and indirectly the eugenic implications by shifting the decision as to who should be screened to state health officials, physicians, or marriage license clerks. By neglecting these problems, legislators have helped intensify the conflict in the black community. Ironically, black legislators introduced many compulsory laws.[43] They are encountering considerable pressures to work for the repeal of mandatory screening laws.

Genetic Counseling

Genetic counseling is also being widely debated by black professionals as well as community groups, since many believe that either compulsory or voluntary testing will lead to inadequate counseling. What role should genetic counseling play in the overall management of this disorder? Some blacks believe that in genetic counseling there may be a sophisticated attempt to limit the black population. Some have called it "professionally sanctioned genocide."

If testing reveals that a person has sickle cell anemia or the sickle cell trait, some argue that counseling should in the main provide information rather than advice about family planning. Experts on the sickle cell disease do not agree on the proper role of family planning in genetic counseling. Further, the function of birth control in helping solve social problems is a sensitive issue in the black community. Militants and some other black groups feel that it is a scheme to eliminate the black race in the United States. Thus linking family planning with efforts to control a disease intensifies conflict within the community. Some physicians favor aggressive genetic counseling that includes sterilization and abortion. For the most part, literature on genetic counseling is not well developed. In a casual review of the genetic literature, the author found virtually no articles on counseling for sickle cell disease.[44]

Feinberg has expressed great concern about widespread testing without adequate legal safeguards for persons tested.[45] He raised the following questions about possible violation of civil liberties in mass sickle cell testing programs:

> Is sickle cell testing of children or others who cannot give informed consent to their participation a civil liberties problem? Is such testing (voluntary or involuntary) a civil liberties problem if strict protection of confidentiality is not maintained? Should involuntary testing be prohibited if more carriers would be detected than information facilities can handle?[46]

In the light of such questions and issues, several groups have attempted to set forth principles and guidelines for the operation of mass screening programs that would offer adequate protection to individuals being tested. Among these groups are the Institute of Society, Ethics, and Life Sciences and the National Conference on the Mental Health Aspects of Sickle Cell Anemia held in 1972 at Meharry Medical College, Nashville, Tennessee.

Reports from both of these groups recommend, in different words and with different emphases, many of the same procedures and safeguards: voluntary participation, confidentiality, well-planned programs of education and treatment, effective testing procedures, and follow-up counseling. In addition, the institute's guidelines call for community involvement and provision of equal access. The Meharry report goes a step further and calls for total funding of educational, screening, and counseling programs.[47] Whatever guidelines are followed for sickle cell programs, a central problem is implementing them. Nonetheless, the recommendations of these two groups provide a basic frame of reference for mass screening programs for sickle cell disease. In addition, the guidelines might be adapted and applied to screening programs for other genetic disorders.

Role of Social Work

In this analysis, the author has identified major issues surrounding the programs for sickle cell disease. But each day new factors enter into the politics of this controversial health problem. What is the role of social work in dealing with the problem? How can social workers help resolve the controversy that, more and more, is intensifying conflict within the black community? How can they help dispel the myths and clarify the confusion? What should be their approach?

It is crucial for social workers to understand the dynamics of the issues involved. Moreover, they must critically analyze the philosophy and strategies of social scientists who have attempted to relate genetic principles and theories to problems of blacks.

The issues surrounding the sickle cell disease have been viewed in this article from a sociopolitical perspective. In addition, rather than relying fully on a medical model, social workers should examine the problem from a systems perspective. Various elements in the social system help to bring about most difficulties that plague the black community. Too often, however, social workers either neglect these elements or focus attention on only one or two of them when planning a program to ameliorate a situation.[48] Social workers alone cannot resolve all the controversial issues, the procedural questions, and the conflicts associated with programs for sickle cell disease. There must be a team effort. Physicians, psychiatrists, psychologists, geneticists, lawyers, civil rights leaders, legislators, educators, journalists, publicists, and other community leaders—these are among the persons who should be on the team.

If social workers are to be effective members of the team, it is imperative that they improve their knowledge of human genetics, especially with respect to this particular disorder. In most schools of social work, human development courses do not include the study of genetics. Perhaps the first task of professional social workers should be to try to persuade schools of social work to offer courses providing more substantive information about the principles of genetics and the impact that genetic disorders have on families. Since many social workers assist clients with genetic disorders, this proposed addition to the curriculum would tend to produce better trained professionals.[49] Efforts to introduce or expand the study of genetics in schools of social work should be viewed as a priority since more and more professionals are being employed in health settings.

Direct and Indirect Services

What specific services can social workers provide directly to victims of sickle cell anemia and their families? New types of service are opening up, in addition to the traditional ones. With training, social workers can become genetic counselors who function along with other professionals in sickle cell disease centers. For example, social workers are on the staff at the Sickle Cell Anemia Center of Howard University. The primary role of social workers at such a center is to educate and be supportive of patients with the disease. Among their responsibilities are the following:

1. Find out what social and medical services are available to persons suffering from the disease.
2. Mobilize family and community support systems.
3. Help assuage the "guilt complex" of parents whose children have the disease.
4. Act as liaison between school-age patients and the school system.
5. Assist in developing vocational training opportunities.
6. Interpret hospital or clinic procedures to patients and clients.
7. Help educate the general public concerning this disorder.
8. Organize families affected into treatment groups.

Social workers can play a variety of roles in providing indirect services to clients. There are positions at the national level for social workers who have expertise in sickle cell disease, positions with either the federal government or private organizations. For instance, a social worker is on the Advisory Committee for Sickle Cell Disease at the Department of Health, Education, and Welfare. Also, a social worker serves as the public education specialist on the National Sickle Cell Disease staff. There are of course more opportunities at the local level.

Since many black social workers are viewed as leaders in their respective communities, it is not surprising that many of them have been asked to serve on advisory committees for sickle cell disease centers and for screening, fund-raising, and public education programs. To serve effectively and judiciously, they must be knowledgeable about the physiological problems, psychological effects, and sociopolitical issues related to sickle cell disease. They must be aware of the available educational, screening, and counseling programs as well as the programs of prevention and treatment. They should be able to recognize the actual and potential impact—in some instances constructive, in others destructive—that the programs have on the community, especially the black community.

In the past year a number of voluntary groups have been organized that devote their efforts to programs for sickle cell disease. These organizations provide a variety of services that range from general public information to technical assistance on procedures for detection, educational and counseling programs for victims of the disease and for their families, and training for genetic counselors. If more federal and private funds continue to be made available and if at the same time funds for some other health and welfare programs continue to dry up, the new groups can be expected to compete strongly for funds.

The Black Panthers recently attacked several black organizations for exploiting the sickle cell situation as follows:

There are so many of these rip-off sickle cell organizations around that they defy listing . . . [they] are lining their pockets on one hand, and preaching how "evil" sickle cell disease is on the other. Receiving vast sums of money, these phoney foundations have done nothing at all in regard to sickle cell anemia, save perhaps putting out a few leaflets and posters, telling the people the need for more funds.[50]

While these are strong words, they suggest the level of conflict that exists in the black community regarding this disease. There are reports that the fly-by-night and dishonest sickle cell groups and their money-raising events are becoming a major problem in several cities.[51] Moreover, it is believed that some legitimate local voluntary groups may not be able to get their share of the funds they need to fight the disease because they do not have the "slick" approaches of the fly-by-night and dishonest groups. Nor do they have the credentials, influence, "track records," and grantsmanship skills that the more traditional institutions have developed.

If the voluntary grass-roots black organizations do not get their share of the funds, one can expect a constant struggle between them and the traditional organizations, as well as a continuing struggle between the legitimate and the dishonest organizations. Instead of presenting a united front on this common problem, the black community is likely to continue to be split by intensive conflict, which is sure to have other pernicious effects on community development.

Social workers must be aware of these and other problems associated with the sickle cell disease. Given our training and experience in interpersonal relationships, adult education, mental health, and other fields, many of us in the social work profession—if we have the suggested training in genetics—should be able to develop the skills necessary to offer needed support to persons with sickle cell disease and to their families.

Given our competence in community organization, social action, and social planning, we should be able to provide surer leadership and clearer direction in helping communities implement principles set forth in the two reports cited.[52] Implemented, these principles hold the promise for effective programs of education, screening, counseling, and treatment—programs that would guarantee the rights of the individuals participating and would help resolve the sickle cell controversy in the black community.

Notes and References

1. Examples are the National Medical Association and the National Association of Black Social Workers.
2. Examples are the Urban League, the National Association for the Advancement of Colored People (NAACP), and the Black Panthers.
3. *See*, for example, Howard A. Pearson, "Progress in Early Diagnosis of Sickle Cell Anemia," *Children* 18 (December 1971), pp. 222–226; Roland B. Scott, "A Commentary on Sickle Cell Disease," *National Medical Association Journal*, 63 (January 1971), pp. 1, 2, 60; John C. Lane and Robert B. Scott, "Awareness of Sickle Cell Anemia Among Negroes in Richmond, Va.," *Public Health Report*, 84 (November 1969), pp. 949–953; "Counterattack on a Killer," *Ebony*, 25 (October 1971), pp. 84–93; and "Sickle Cell Anemia" *Black Enterprise*, 2 (June 1972), pp. 23–25. For a comprehensive list of articles, *see* Charlotte Kenton, *Sickle Cell Anemia: Diagnosis, Pathology, Complications and Therapy*, January 1969–April 1972 (339 citations) and Kenton, *Sickle Cell Anemia: Epidemiologic, Genetic, Social, Legal and Ethical Aspects, January 1969 through May 1972* (116 citations) (Bethesda, Md.: Literature Search Program, National Library of Medicine).
4. Tabitha M. Powledge, "New Ghetto Hustle," *Saturday Review of Sciences*, 1 (February 1973), pp. 38–40, 45–49.
5. Michael G. Michaelson, "Sickle Cell Anemia: An 'Interesting Pathology,'" *Ramparts*, 10 (October 1971), p. 53.
6. Jane S. Lin-Fu, *Sickle Cell Anemia: A Review of the Literature* (Rockville, Md.: Public Health Service, U.S. Department of Health, Education & Welfare, 1965), p. 2. *Sickle cell variants* may be defined as "conditions wherein other abnormal hemoglobins occur in combination with sickle hemoglobin and will lead to the sickling state." *See National Sickle Cell Disease Programs—Answers to Common Questions about Sickle Cell Disease*, Department of Health, Education & Welfare Publications, NIH No. 73–364 (Washington, D.C.: U.S. Government Printing Office, 1973), p. 2.

7. For a discussion of the inherited nature of this disease, *see* Roland B. Scott and Althea D. Kessler, *Sickle Cell Anemia and Your Child* (Washington, D.C.: Howard University College of Medicine, 1971), pp. 7–8; M. Splaine, E. B. Haynes and G. P. Barclay, "Calculation for Changes in Sickle-Cell Trait Rates," *American Journal of Human Genetics*, 23 (July 1971), pp. 368–374.

8. James B. Herrick, "Peculiar Elongated and Sickle-Shaped Red Corpuscles in a Case of Severe Anemia," *Archives of Internal Medicine*, 6 (1910), p. 517.

9. Linus Pauling et al., "Sickle Cell Anemia: A Molecular Disease," and James V. Neel, "Inheritance of Sickle Cell Anemia," *Science*, 110 (1949), pp. 543 and 64, respectively; and E. A. Beet, "Genetics of Sickle Cell Trait in Bantu Tribe," *Eugenics*, 14 (1949), p. 279.

10. Scott and Kessler, op. cit., pp. 6, 7.

11. *See* Jessyca R. Gaver, *Sickle Cell Disease: Its Tragedy and Its Treatment* (New York: Lancer Books, 1972), pp. 75–83.

12. Scott and Kessler, op. cit., p. 7.

13. *See*, for example, Lin-Fu, op. cit., p. 12; Stephen R. Jones et al., "Sudden Death in Sickle Cell Trait," *New England Journal of Medicine*, 282 (February 5, 1970), pp. 323–325; William A. Miller et al., "Perirenal Hematoma in Association with Renal Infarction in Sickle Cell Trait," *Radiology*, 92 (1969), pp. 351–352; Sherman D. Nichols, "Splenic and Pulmonary Infarction in Negro Athletes," *Rocky Mountain Medical Journal*, 65 (1968), pp. 49–50; and Royal Rotter et al., "Splenic Infarction in Sicklemia During Airplane Flight: Pathogenesis, Hemoglobin Analysis and Clinical Features of Six Cases," *Annals of Internal Medicine*, 44 (1956), p. 257.

14. Jessyca R. Gaver and Carl Pochedly, "Sickle Cell Anemia: Recognition and Management," *American Journal of Nursing*, 71 (October 1971), pp. 1950–1951.

15. Robert B. Scott, "Health Care Priority and Sickle Cell Anemia," *Journal of the American Medical Association*, 214 (October 26, 1970), p. 731.

16. Gaver, op. cit.; and "Sickle Cell Anemia," *Urban Health*, 2, Special Issue (December 1973).

17. Amitai Etzioni, "Health Care and Self-Care: The Genetic Fix," *Society*, 10 (September/October 1973), pp. 28–32.

18. James R. Sorenson, *Social Aspects of Human Genetics* (New York: Russell Sage Foundation, 1971), p. 1.

19. S. M. Miller and Ronnie S. Ratner, "The American Resignation: The New Assault on Equality," *Social Policy* (May/June 1972), pp. 5–15. For a discussion on studies that question the usefulness of social programs, *see* Richard Hernstein, "IQ," *Atlantic Monthly* (September 1971), pp. 43–64; Arthur Jensen, "How Much Can We Boost IQ and Scholastic Achievement," *Harvard Education Review*, 39 (Winter 1969), pp. 1–123; Edward C. Banfield, *The Unheavenly City* (Boston: Little Brown & Co., 1968); and Christopher Jencks et al., *Inequality: A Reassessment of the Effects of Family and Schooling in America* (New York: Basic Books, 1972).

20. Robert B. Scott, op. cit., p. 731.

21. Center for Black Education, "Why Sickle Cell Anemia Should Not Be Placed on Blacks' List of Ills," *Muhammad Speaks* (January 21, 1972), p. 4.

22. U.S. Dept. of Commerce, Bureau of the Census, *The Social and Economic Status of Negroes in the United States*, 1970, BLS Report No. 394 (Washington, D.C.: U.S. Government Printing Office, 1971), p. 98.

23. *Fact Sheets on Institutional Racism* (New York: Foundation for Change, 1971), p. 3.
24. Louis L. Knowles and Kenneth Prewitt, *Institutional Racism in America* (Englewood Cliffs, N.J.: Prentice-Hall, 1969), p. 98.
25. "The Row over Sickle Cell," *Newsweek* (February 12, 1973), p. 63.
26. Shirley Linde, *Sickle Cell: A Guide to Prevention and Treatment* (New York: Pavilion Publishing Co., 1972), p. 88.
27. James E. Bowman, "Medicolegal, Social and Ethical Aspects of Sickle Cell Programs," and Stanley Smith, "The Socio-psychological Aspects of Sickle Cell Anemia," *Meadowbrook Staff Journal* (Spring 1973), pp. 33, 34, 39 and 4–10, respectively; and Michaelson, op. cit., p. 57.
28. Bowman, op. cit., p. 34.
29. *Public Law 92–294* (92nd Congress, S 2676, May 16, 1972), p. 1.
30. Barbara Culliton, "Sickle Cell Anemia: The Route from Obscurity to Prominence," *Science*, 178 (October 13, 1972), p. 142; and Powledge, op. cit., p. 45.
31. John 3X, "Klan Literature Suggests Motivation of Government's Sickle Cell Interest," *Muhammad Speaks* (April 14, 1972), p. 10.
32. Lin-Fu, op. cit.; and Frank B. Livingstone, *Abnormal Hemoglobins in Human Populations* (Chicago: Aldine Publishing Co., 1967).
33. *Washington Post*, November 15, 1972.
34. Bowman, op. cit., p. 35.
35. Powledge, op. cit., p. 40.
36. Commonwealth of Virginia, *An Act to Amend the Code of Virginia by Adding in Title 32 a Chapter Number 32–112.10 through 32–112.19, Relating Generally to the Detection and Control of Sickle Cell Trait*, April 10, 1972.
37. Commonwealth of Massachusetts, *Chapter 491 of Acts and Resolves, 1971*, S15A.
38. Powledge, op. cit., pp. 40, 45.
39. District of Columbia City Council, *Regulation Number 72–9*, May 3, 1972.
40. Powledge, op. cit., p. 45.
41. Ibid., p. 45.
42. Ibid., p. 46.
43. Bowman, op. cit., p. 35.
44. For an exception, *see* V. Heading, and J. Fielding, "Guidelines for Counseling Young Adults with Sickle Cell Traits," *American Journal of Public Health* (in press).
45. Irwin Feinberg, "Medicine and Liberty," *Civil Liberties*, 288 (July 1972), p. 6.
46. Irwin Feinberg, "Sickle Cell Tests," *Civil Liberties*, 291 (December 1972), p. 4.
47. "Ethical and Social Issues in Screening for Genetic Disease—A Group Report," *The New England Journal of Medicine*, 286 (May 25, 1972), pp. 1129–1132; and *Sickle Cell Screening—Medical, Legal, Ethical, Psychological and Social Problems: A Sickle Cell Crisis, Proceedings*, National Conference on the Mental Health Aspects of Sickle Cell Anemia held at Meharry College, Nashville, Tennessee, June 27–28, 1972 (Nashville, Tenn., in press).
48. Lawrence Gary, "Educating Blacks for the 1970s: The Role of Black Colleges," *Proceedings of the Fourth Annual Conference of the National Association of Black Social Workers, April 4–9, 1972* (Nashville, Tenn.), pp. 31, 32.
49. Amelia Schultz, "The Impact of Genetic Disorders," *Social Work*, 11 (April 1966), pp. 29–34.
50. "The Sickle Cell 'Game,'" *Black Panthers* (May 27, 1972), p. 11.

51. Bowman, op. cit., pp. 39, 40; and *Washington Post*, November 15, 1972.
52. *See* "Ethical and Social Issues in Screening for Genetic Disease—A Group Report." *See also Sickle Cell Screening—Medical, Legal, Ethical, Psychological and Social Problems: A Sickle Cell Crisis.*

36   Skin-Pigment Regulation of Vitamin-D Biosynthesis in Man

W. Farnsworth Loomis

Vitamin D mediates the absorption of calcium from the intestine and the deposition of inorganic minerals in growing bone; this "sunshine vitamin" is produced in the skin, where solar rays from the far-ultraviolet region of the spectrum (wavelength, 290 to 320 millimicrons) convert the provitamin 7-dehydrocholesterol into natural vitamin D (*1*) (Figure 36–1).

Unlike other vitamins, this essential calcification factor is not present in significant amounts in the normal diet; it occurs in the liver oils of bony fishes and, in very small amounts, in a few foodstuffs in the summer (see Table 1). Almost none is present in foodstuffs in winter.

Chemical elucidation of the nature of vitamin D has made it possible to eradicate rickets from the modern world through artificial fortification of milk and other foods with this essential factor. Before this century, however, mankind resembled the living plant in being dependent on sunshine for his health and well-being, a regulated amount of vitamin-D synthesis being essential if he were to avoid the twin dangers of rickets on the one hand and an excess of vitamin D on the other.

Unlike the water-soluble vitamins, too much vitamin D causes disease just as too little does, for the calcification process must be regulated and controlled much

Figure 36-1. Chemical structures of 7-dehydrocholesterol and vitamin D₃.

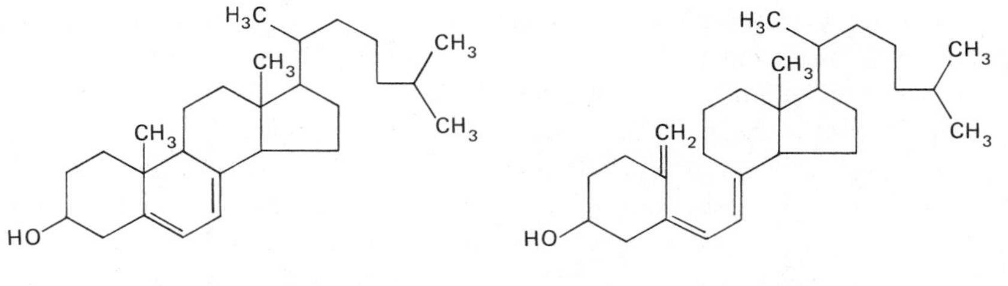

7-Dehydrocholesterol   Vitamin D₃

From *Science*, **157** (Aug. 4, 1967), 501–506. Copyright 1967 by the American Association for the Advancement of Science. Reprinted with permission of the author and the publisher.

Table 1. Vitamin-D Content of Two Fish-Liver Oils and of the Only Foodstuffs Known to Contain Vitamin D. [From K. H. Coward, *The Biological Standardization of the Vitamins* (Wood, Baltimore, 1938), p. 223].

Fish-Liver Oil or Foodstuff	Vitamin-D Content (I.U./gram)
Halibut-liver oil	2000–4000
Cod-liver oil	60–300
Milk	0.1
Butter	0.0–4.0
Cream	0.5
Egg yolk	1.5–5.0
Calf liver	0.0
Olive oil	0.0

as metabolism is regulated by the thyroid hormone. The term *vitamin D* is, in fact, almost a misnomer, for this factor resembles the hormones more closely than it resembles the dietary vitamins in that it is not normally ingested but is synthesized in the body by one organ—the skin—and then distributed by the blood stream for action elsewhere in the body. As in the case of hormones, moreover, the rate of synthesis of vitamin D must be regulated within definite limits if both failure of calcification and pathological calcifications are to be avoided.

Synthesis of too little vitamin D results in the bowlegs, knock-knees, and twisted spines (scoliosis) associated with rickets in infants whose bones are growing rapidly. Similar defects in ossification appear in older children and women deprived of this vitamin; puberty, pregnancy, and lactation predispose the individual toward osteomalacia, which is essentially adult rickets. In osteomalacia the bones become soft and pliable, a condition which often leads to pelvic deformities that create serious hazards during childbirth. Such deformities were common, for example, among the women of India who followed the custom of purdah, which demands that they live secluded within doors and away from the calcifying power of the sun's rays (2). Cod-liver oil or other source of vitamin D is a specific for rickets and osteomalacia, the usual recommended daily dosage being 10 micrograms of 400 international units (1 I.U. = 0.025 microgram of vitamin D).

Ingestion of vitamin D in amounts above about 100,000 I.U. (2.5 milligrams) per day produces the condition known as hypervitaminosis D, in which the blood levels of both calcium and phosphorus are markedly elevated and multiple calcifications of the soft tissues of the body appear. Ultimate death usually follows renal disease secondary to the appearance of kidney stones (3). Although this condition has been described only in patients given overdoses of vitamin D by mouth, similarly toxic results would probably follow the natural synthesis of equal doses of vitamin D by unpigmented skin exposed to excessive solar radiation. The body appears to have no power to regulate the amount of vitamin D absorbed from food and no power to selectively destroy toxic doses once they have been absorbed. These facts suggest that the physiological means of regulating the concentration of vitamin D in the body is through control of the rate of photochemical synthesis of vitamin D in the skin.

It is the thesis of this article that the rate of vitamin-D synthesis in the stratum granulosum of the skin is regulated by the twin processes of pigmentation and keratinization of the overlying stratum corneum, which allow only regulated amounts of solar ultraviolet radiation to penetrate the outer layer of skin and reach the region where vitamin D is synthesized. According to this view, different types of skin—white (depigmented and dekeratinized), yellow (mainly keratinized), and black (mainly pigmented)—are adaptations of the stratum corneum which maximize ultraviolet penetration in northern latitudes and minimize it in southern latitudes, so that the rate of vitamin-D synthesis is maintained within physiological limits (0.01 to 2.5 milligrams of vitamin D per day) throughout man's worldwide habitat.

Figure 36–2 provides evidence in support of this view, for it is apparent that there is a marked correlation between skin pigmentation and equatorial latitudes. In addition, the reversible summer pigmentation and keratinization activated by ultraviolet radiation and known as suntan represents a means of maintaining physiologically constant rates of vitamin-D synthesis despite the great seasonal variation in solar ultraviolet radiation in the northern latitudes.

Ultraviolet Transmission and Vitamin-D Synthesis

In 1958 Beckemeier (4) reported that 1 square centimeter of white human skin synthesized up to 18 I.U. of vitamin D in 3 hours. Using this figure, we calculate that an antirachitic preventive dose of 400 I.U. per day can be synthesized by daily exposure of an area of skin approximately equal to that of the nearly transparent pink cheeks of European infants (about 20 square centimeters). Perhaps this explains why mothers in northern climates customarily put their infants out of doors for "some fresh air and sunshine" even in the middle of winter.

From this high rate of synthesis by only a small area of thin unpigmented skin, one can calculate the daily amount of vitamin D that would be synthesized at the equator by the skin of adults who exposed almost all their $1\frac{1}{2}$ square meters (22,500 square centimeters) of body surface during the whole of a tropical day. Such a calculation shows that the skin of such individuals would synthesize up to 800,000 I.U. of vitamin D in a 6-hour period if the stratum corneum contained no pigment capable of filtering out the intense solar ultraviolet radiation.

Direct evidence that pigmented skin is an effective ultraviolet filter was provided by Macht, Anderson, and Bell (5), who used a spectrographic method to show that excised specimens of whole skin from Negroes prevented the transmission of ultraviolet radiation of wavelengths below 436 millimicrons, while excised specimens of white skin allowed radiation from both the 405- and the 356-millimicron bands of the mercury spectrum to pass through.

These early studies with whole skin were refined by Thomson (6), who used isolated stratum corneum obtained by blistering the skin with cantharides. He found that the average percentage of solar radiation of 300- to 400-millimicron wavelength transmitted by the stratum corneum of 22 Europeans were 64 percent, while the average for 29 Africans was only 18 percent. There was no overlapping of values for the two groups (Figure 36–3), but there was considerable variation within each

Figure 36-2. Distribution of human skin color before 1492. [Adapted from Brace and Montague, *Man's Evolution* (Macmillan, New York, 1965), p. 272.]

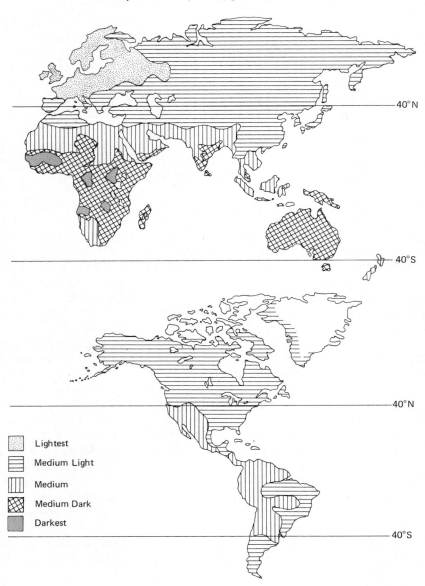

Lightest

Medium Light

Medium

Medium Dark

Darkest

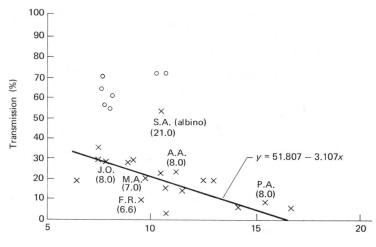

Figure 36-3. Variation of transmission of solar ultraviolet light (3000 to 4000 angstroms) through the stratum corneum, plotted against thickness of this layer. ○, Europeans; ×, Africans. The numbers in parentheses after initials are percentages for reflectance of blue light on the forearm. [From M. L. Thomson (6).]

group, the values for the Europeans varying from 53 to as high as 72 percent and those for the Africans (who were mainly Ibos but also included men from most of the Nigerian tribes) varying from 36 to as low as 3 percent.

In his careful studies, Thomson measured skin thickness as well as pigmentation and found that the former was a minor variable. Studies on the degree of blackness of the various African specimens were made by skin-reflectance measurements. These showed that the darker the skin is, the lower is the percentage of ultraviolet radiation transmitted. One specimen from an albino African showed transmission of 53 percent—a value within the range for the European group. Thomson concluded from these studies that skin pigmentation was mainly responsible for protecting the African from excessive solar ultraviolet radiation, the thickness of the horny layer in Africans playing only a minor role. Thomson did not mention the fact that skin pigmentation and thickening of the horny layer in Africans would protect against excessive vitamin-D synthesis as well as sunburn.

Thomson's results indicate that African stratum corneum filters out solar ultraviolet radiation equivalent to between 50 and 95 percent of that which reaches the vitamin-D synthesizing region of the skin of Europeans. This explains "the fact agreed to by all, that of all races the Negro is most susceptible to rickets" (7). It is clear from Thomson's figures that exposure of the face of Negro infants to winter sunlight in Scandinavia would result in synthesis of too little vitamin D to meet the infant's body requirements.

It was Hess who first proved that sunlight could cure rickets (8). Seeking experimental proof of a relationship between skin pigmentation and rickets, he took six white and six black rats and placed them on a rachitogenic diet containing low amounts of phosphorus. Exposing both groups to a critical amount of ultraviolet light, he found that all the white rats remained healthy while all the black rats developed rickets. He concluded (9), "It is manifest that the protective rays were rendered inert by the integumentary pigment."

To return now to Thomson's results and consider their bearing on hypervitaminosis, they explain why deeply pigmented Africans living near the equator and exposing almost all their body surface to the ultraviolet of the tropical sun do not suffer from kidney stones and other evidences of hypervitaminosis. Under conditions where untanned Europeans would synthesize up to 800,000 I.U. per day, deeply pigmented Africans would synthesize 5 to 10 percent as much; thus their daily production would fall within the acceptable range.

In this connection it is significant that Reinertson and Wheatley (10) found that the 7-dehydrocholesterol content of human skin does not vary significantly between Negroes and Whites. Skin from the back, abdomen, and thigh of adults of both races averaged 3.8 percent (standard deviation, 0.8 percent), the lowest result in their series being obtained in a specimen of the epidermis of the sole, an area that receives no radiation at all, while the highest result among adults was from a Negro. The highest content of all was found in a specimen from a 2-week-old infant that showed 8.8 percent of the provitamin, a fact that correlates well with the especially high need for vitamin D during the first 2 years of life.

In their paper and ensuing discussion, the above workers emphasize that 7-dehydrocholesterol is found almost entirely beneath the stratum corneum, thus establishing the fact that in man it is not present in the secretions of the sebaceous glands as it is in birds and some northern fur-covered animals which, respectively, obtain their vitamin D by preening or by licking their fur after the provitamin has been converted into the vitamin on the surface of the body. It would appear that vitamin D is made in man solely by the irradiation of the provitamin in the layers underneath the stratum corneum, a mechanism that would allow efficient regulation of the biosynthesis of this essential factor by varying the degree of ultraviolet penetration through differing amounts of pigmentation in the overlying stratum corneum.

Origin of White Skin

Having originated in the tropics where too much sunlight rather than too little was the danger, the first hominids had no difficulty in obtaining sufficient amounts of vitamin D until they extended their range north of the Mediterranean Sea and latitude 40°N (Figure 36–2), where the winter sun is less than 20 degrees above the horizon (11) and most of the needed ultraviolet is removed from the sun's rays by the powerful filtering action of the atmosphere through which the slanting rays have to pass. Before the present century, for example, there was a very high incidence of rickets among infants in London and Glasgow, because in these latitudes the midday sun is less than 35 degrees from the horizon for 5 and 6 months, respectively, of the year; in Jamaica and other southern localities, on the other hand, the sun's midday altitude is never less than 50 degrees and rickets is almost unknown (2). The farther north one goes, the more severe becomes this effect of latitude on the availability of winter ultraviolet radiation, an effect compounded by cloudy winter skies.

Having evolved in the tropics, early hominids were probably deeply pigmented and covered with fur, as are most other tropical primates. The first adaptation one might expect therefore to lowered availability of ultraviolet light as they moved north of the Mediterranean would be a reduction of fur, for Cruikshank and Kodicek have shown (12) that shaved rats synthesize four times more vitamin D then normal rats do.

As early hominids moved farther and farther north, their more deeply pigmented infants must have been especially likely to develop the grossly bent legs and twisted spines characteristic of rickets, deformities which would cripple their ability to hunt game when they were adults. In this connection, Carleton Coon has written (13), " Up to the present century, if black skinned people were incorporated into any population living either north or south of the fortieth degree of latitude, their descendants would eventually have been selected for skin color on the basis of this vitamin factor alone." Howells agrees (14): " This variety of outer color has all the earmarks of an adaptation, of a trait responding to the force of sunlight by natural selection." The skin, he continues " admits limited amounts of ultraviolet, which is needed to form vitamin D, but presumably diminishes or diffuses dangerous doses by a screen of pigment granules."

Even in 1934 Murray clearly recognized the implications of these facts (15): "As primordial man proceeded northwards into less sunlit regions, a disease, rickets, accomplished the extinction of the darker, more pigmented elements of the population as parents and preserved the whiter, less pigmented to reproduce their kind and by progressive selection through prehistoric times, developed and established the white race in far northern Europe as it appears in historic times; its most extreme blond types inhabiting the interior of the northern-most Scandinavian peninsula."

It is a curious fact that Murray's thesis is almost unknown to the general public, including physiologists, biochemists, and physicians, and that it is not generally accepted by anthropologists, with the exception of Coon and Howells, quoted above, even though it fits the facts of Figure 36-2. Both in Europe and China, skin pigmentation becomes lighter as one goes north, and it is lighter in young children; in almost all races the skin is lighter in the newborn infant (16) and gradually darkens as the individual matures, a change that parallels the declining need for vitamin D.

When Did European Hominids Begin to Turn White?

On the basis of the conclusion that white skin is an adaptation to northern latitudes because of the lowered availability of winter ultraviolet radiation, it appears probable that the early hominids inhabiting western Europe had lost much of their body hair and skin pigmentation even half a million years ago. Anthropological evidence indicates that early hominids such as the Heidelberg, Swanscomb, Steinheim, Fontechevade, and Neanderthal men lived north of the Mediterranean Sea —particularly during warm interglacial periods (*17*). It is important to recognize that the effect of latitude on the availability of ultraviolet light in winter is not related to climate but operates steadily and at all times, through glacial and interglacial period alike.

Hand axes and other early stone tools have been found throughout the tropics of the Old World and also in Europe as far north as the 50th degree of latitude (Figure 36-4). The presence of such stone tools as far north as England and France shows that some early hominids must already have adapted to the lowered level of ultraviolet radiation and consequent danger of rickets by partial loss of body hair and skin pigmentation, for without such adaptation they would have probably been unable to survive this far north.

England is at the same latitude as the Aleutian Islands, and no stone tools such as those found in southern England and France have ever been found in other areas at this latitude—for example, Mongolia and Manchuria. The unique combination of temperate climate and low levels of winter ultraviolet radiation in England and France is due to the powerful warming effect of the Gulf Stream on this particular northern area, which is unique in the world in this respect, for the Japan current in the Pacific is not as powerful as the Gulf Stream and warms only the Aleutian Islands, where no hominids existed until very recently.

Figure 36-4. Distribution of early stone tools throughout the tropics of the Old World and in Europe as far north as 50°N. (From Brace and Montague, *Man's Evolution* (Macmillan, New York, 1965), p. 231.]

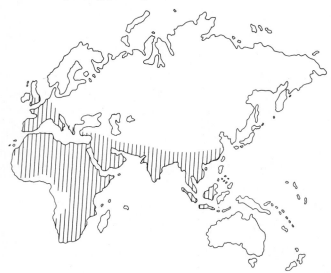

Occupation of northern Europe and even Scandinavia up to the Arctic Circle seems to have taken place during the Upper Paleolithic, when presumably partial depigmented men already adapted to latitude 50°N lost nearly all their ability to synthesize melanin and so produced the blond-haired, blue-eyed, fair-skinned peoples who inhabit the interior of the northernmost part of the Scandinavian peninsula.

It has been held that the abundant appearance of stone scrapers in the Upper Paleolithic indicates that this far-northern extension of man's habitat followed his use of animal skins for clothing, a change that would select powerfully for infants with nearly transparent skin on their cheeks, who were thus still able to synthesize a minimum antirachitic dose of vitamin D even when fully clothed during the Scandinavian winter. Certainly the pink-to-red cheeks of northern European children are uniquely transparent; their color is due to the high visibility of the blood that circulates in the subepidermal region.

The one exception to the correlation between latitude and skin color in the Old World is the Eskimo; his skin is medium dark and yet he remains completely free of rickets (18) during the long dark arctic winters. Murray noted long ago that the Eskimo's diet of fish oil and meat contains several times the minimum preventive dose of vitamin D, concluding (15), " Because of his diet of antirachitic fats, it has been unnecessary for the Eskimo to evolve a white skin in the sunless frigid zone. He has not needed to have his skin bleached by centuries of evolution to admit more antirachitic sunlight. He probably has the same pigmented skin with which he arrived in the far north ages ago." Similar considerations would apply in the case of any coastal peoples of Europe and Asia, who would have been able to expand northward without depigmentation as long as they obtained sufficient vitamin D from a diet of fish; only when they ventured into the interior would antirachitic selection for blond types, as in Scandinavia, presumably have taken place.

Yellow, Brown, and Black Adaptation

Human skin has two adaptive mechanisms for resisting the penetration of solar ultraviolet: melanin-granule production in the Malpighian layer and keratohyaline-granule production in the stratum granulosum. Melanin granules are black, whereas the keratohyaline granules produce keratin (from which nails, claws, horns, and hoofs are formed), which has a yellowish tinge. Particles of both types migrate toward the horny external layer, where they impart a black (melanin), yellow (keratin), or brown (melanin and keratin) tinge to the skin.

Thomson has shown (6) that, in Negroes, melanization of the stratum corneum plays the major role in filtering out excessive ultraviolet radiation, keratinization of the horny layer playing only a minor part. Mongoloids on the other hand have yellowish skin, since their stratum corneum is packed with disks of keratin (13) that allow them to live within 20 degrees of the equator even though their skin contains only small amounts of melanin (Figure 36–2). On the equator itself, however, even Mongoloid-derived peoples acquire pigmentation—for example, the previously medium-light-skinned Mongoloids who entered the Americas over the Bering Straits at latitude 66°N as recently as 20,000 to 10,000 years ago (Figure 36-2).

Even white-skinned peoples have to protect themselves against excessive doses of solar ultraviolet radiation in summer, for, as Blum has pointed out (*19*), on 21 June the solar ultraviolet is as intense in Newfoundland as it is at the equator, since at that time the two regions are at the same distance from the Tropic of Cancer (at 23°27'N). (At the equator, the solar ultraviolet is never less than on this date, while in Newfoundland it is never more.) In other words, adaptation to the variable intensities of solar ultraviolet in the north requires not only winter depigmentation but also the evolution of a reversible mechanism of summer repigmentation to keep the rate of vitamin-D synthesis constant throughout the year. It is significant that both the keratinization and melanization components of suntan are initiated by the same wavelengths which synthesize vitamin D, for it would be difficult to design a more perfect defense against excessive doses of vitamin D than this reversible response to ultraviolet light of these particular wavelengths—a pigmentation response that is further protected by the painful alarm bell of sunburn, which guarantees extreme caution against overexposure to solar ultraviolet in untanned individuals suddenly encountering a tropical sun.

Defenses against production of too much vitamin D therefore range from (i) reversible suntanning, as in Europeans, through (ii) constitutive keratinization, as in the Mongoloids of Asia and the Americas, to (iii) constitutive melanization, as in African and other truly equatorial peoples. The physiological superiority of melanization as a means of protection against ultraviolet was demonstrated by the ability, historically documented, of imported Nigerian slaves to outwork the recently adapted American Indians in the sun-drenched cane fields and plantations of the Caribbean and related tropical areas.

Additional evidence for the view that melanization of the stratum corneum is primarily a defense against the oversynthesis of vitamin D from solar ultraviolet is provided by the fact that the palms and soles of Negroes are as white as those of Europeans; *only* the palms and soles possess a thickly keratinized stratum lucidum under the external stratum corneum, which renders melanization of the latter unnecessary. The same reasoning explains the failure of the palms and soles of whites to sunburn during the summer.

Coon has written (*13*), "We cannot yet demonstrate why natural selection favors the prevalence of very dark skins among otherwise unrelated populations living in the wet tropics, but the answer may not be far away." Since overdoses of vitamin D administered orally are known to result in prompt and serious consequences, such as calcifications in the aorta and other soft tissues of the body, kidney stones, secondary renal disease, and death, it would appear that oversynthesis of vitamin D is sufficiently detrimental in young and old to favor the gradual selection for deeply pigmented skin near the equator, as seen, for example, in the repigmentation that has taken place among the equatorial American Indians during the last 10,000 years (Figure 36-2).

Secondary Results of Pigmentation and Depigmentation

It is known that black skin absorbs more heat than white skin; the studies of Weiner and his associates (*20*) show that black Yoruba skin reflects only 24 percent of incident light whereas untanned European skin reflects as much as 64 percent. Of themselves, these facts would lead one to expect that reflective white skin would be found near the equator while heat-absorbing black skin would be found in cold northern climates.

Since the exact opposite is true around the world, it seems clear that man has adapted his epidermis in response to varying levels of ultraviolet radiation despite the price he has had to pay in being badly adapted from the standpoint of heat absorbance and reflectance of visible and near-infrared wavelengths. Similar considerations naturally apply to summer pigmentation due to suntan; ultraviolet regulation rather than heat regulation explains why Caucasians are white in the winter but pigmented in the summer.

In addition to being badly adapted for maximum heat absorbance, white-skinned northern peoples are known to be particularly susceptible to skin cancer (*21*) and such skin diseases as psoriasis and acne. Therefore, only some powerful other advantage, such as relative freedom from rickets, would explain the worldwide correlation between high latitudes and white skin, for without some such factor it would seem that black or yellow skin would be the superior integument.

From this and other evidence, such as the fact that lion cubs and the young of other tropical animals develop rickets in northern zoos unless given cod-liver oil (*2*), it appears probable that depigmentation occurred north of latitude 40°N (a line marked by the Mediterranean Sea, the Great Wall of China, and the Mason-Dixon line) as an adaptation that allowed an increased penetration of winter ultraviolet radiation and consequent freedom from rickets. Certainly no other essential function of solar ultraviolet is known for man besides the synthesis of vitamin D.

Summary

The known correlation between the color of human skin and latitude (Figure 36–2) is explainable in terms of two opposing positive adaptations to solar ultraviolet radiation, weak in northern latitudes in winter yet powerful the year around near the equator. In northern latitudes there is selection for white skins that allow maximum photoactivation of 7-dehydrocholesterol into vitamin D at low intensities of ultraviolet radiation. In southern latitudes, on the other hand, there is selection for black skins able to prevent up to 95 percent of the incident ultraviolet from reaching the deeper layers of the skin where vitamin D is synthesized. Selection against the twin dangers of rickets on the one hand and toxic doses of vitamin D on the other would thus explain the world-wide correlation observed between skin pigmentation and nearness to the equator.

Since intermediate degrees of pigmentation occur at intermediate latitudes, at well as seasonal fluctuation in pigmentation (through reversible suntanning), is appears that different skin colors in man are adaptations of the stratum corneum which regulate the transmission of solar ultraviolet to the underlying stratum granulosum, so that vitamin-D photosynthesis is maintained within physiological limits throughout the year at all latitudes.

References and Notes

1. A. White, P. Handler, S. L. Smith, *Principles of Biochemistry* (McGraw-Hill, New York, ed. 3, 1964), p. 981.
2. C. H. Best and N. B. Taylor, *The Physiological Basis of Medical Practice* (Williams and Wilkins, Baltimore, ed. 3, 1943), pp. 1102, 1105.
3. F. Bicknell and F. Prescott, *The Vitamins in Medicine* (Grune and Stratton, New York, ed. 3, 1953), p. 578.
4. H. Beckemeier, *Acta Biol. Med. Ger.* **1,** 756 (1958); ——— and G. Pfennigsdorf, *J. Physiol. Chem.* **214,** 120 (1959).
5. D. I. Macht, W. T. Anderson, F. K. Bell, *J. Amer. Med. Assoc.* **90,** 161 (1928); W. T. Anderson and D. I. Macht, *Amer. J. Physiol.* **86,** 320 (1928).
6. M. L. Thomson, *J. Physiol. London* **127,** 236 (1955).
7. A. F. Hess and L. J. Unger, *J. Amer. Med. Assoc.* **69,** 1583 (1917).
8. ———, *ibid.* **78,** 1177 (1922).
9. A. F. Hess, *ibid.*, p. 1177.
10. R. P. Reinertson and V. R. Wheatley, *J. Invest. Dermatol.* **32,** 49 (1959).
11. F. Daniels, Jr., in *Handbook of Physiology*, D. B. Dill, E. F. Adolph, C. G. Wilber, Eds. (American Physiological Society, Washington, D.C., 1964), pp. 969–88.
12. E. M. Cruikshank and E. Kodicek, *Proc. Nutr. Soc. Engl. Scot.* **14,** viii (1955).
13. C. Coon, *The Living Races of Man* (Knopf, New York, 1965), pp. 232, 234.
14. W. W. Howells, *Mankind in the Making* (Doubleday, New York, 1959), p. 270.
15. F. G. Murray, *Amer. Anthropol.* **36,** 438 (1934).
16. E. A. Hooton, *Up from the Ape* (Macmillan, New York, 1946), p. 466.
17. It is possible that the most northern or "classic" Neanderthal died out some 35,000 years ago in western Europe because of rickets which became severe when the arctic weather of the last glaciation made it necessary for him to dress his infants warmly in animal skins during the winter months, a change that would drastically reduce the area of their skin exposed to solar ultraviolet.
18. W. A. Thomas, *J. Amer. Med. Assoc.* **88,** 1559 (1927).
19. H. F. Blum, *Quart. Rev. Biol.* **36,** 50 (1961).
20. J. S. Weiner, G. A. Harrison, R. Singer, R. Harris, W. Jopp, *Human Biol.* **36,** 294 (1964).
21. H. F. Blum, in *Radiation Biology*, A. Hollaender, Ed. (McGraw-Hill, New York, 1955), vol. 2, pp. 487, 509, 529.
22. The work discussed here was partially supported by grant E–443 of the American Cancer Society to Brandeis University, Waltham, Massachusetts. This article is Graduate Department of Biochemistry Publication No. 505.

37   Race and Intelligence

Richard C. Lewontin

In the Spring of 1653 Pope Innocent X condemned a pernicious heresy which espoused the doctrines of "total depravity, irresistible grace, lack of free will, predestination and limited atonement." That heresy was Jansenism and its author was Cornelius Jansen, Bishop of Ypres.

In the winter of 1968 the same doctrine appeared in the "Harvard Educational Review." That doctrine is now called "jensenism" by the "New York Times Magazine" and its author is Arthur R. Jensen, professor of educational psychology at the University of California at Berkeley. It is a doctrine as erroneous in the twentieth century as it was in the seventeenth. I shall try to play the Innocent.

Jensen's article, "How Much Can We Boost I.Q. and Scholastic Achievement?" created such a furor that the "Review" reprinted it along with critiques by psychologists, theorists of education and a population geneticist under the title "Environment, Heredity and Intelligence." The article first came to my attention when, at no little expense, it was sent to every member of the National Academy of Sciences by the eminent white Anglo-Saxon inventor, William Shockley, as part of his continuing campaign to have the Academy study the effects of inter-racial mating. It is little wonder that the "New York Times" found the matter newsworthy, and that Professor Jensen has surely become the most discussed and least read essayist since Karl Marx. I shall try, in this article, to display Professor Jensen's argument, to show how the structure of his argument is designed to make his point and to reveal what appear to be deeply embedded assumptions derived from a particular world view, leading him to erroneous conclusions. I shall say little or nothing about the critiques of Jensen's article, which would require even more space to criticize than the original article itself.

The Position

Jensen's argument consists essentially of an elaboration on two incontrovertible facts, a causative explanation and a programmatic conclusion. The two facts are that black people perform, on the average, more poorly than whites on standard I.Q. tests, and that special programs of compensatory education so far tried have not had much success in removing this difference. His causative explanation for these facts is that I.Q. is highly heritable, with most of the variation among individuals arising from genetic rather than environmental sources. His programmatic conclusion is that there is no use in trying to remove the difference in I.Q. by education since it arises chiefly from genetic causes and the best thing that can be done

From the *Bulletin of the Atomic Scientists* (March 1970), 2–8. Reprinted by permission of the author and *Science and Public Affairs*, the *Bulletin of the Atomic Scientists*. Copyright © 1970 by the Educational Foundation for Nuclear Sciences. (Illustrations omitted.)

for black children is to capitalize on those skills for which they are biologically adapted. Such a conclusion is so clearly at variance with the present egalitarian consensus and so clearly smacks of a racist elitism, whatever its merit or motivation, that a very careful analysis of the argument is in order.

The article begins with the pronouncement: "Compensatory education has been tried and it apparently has failed." A documentation of that failure and a definition of compensatory education are left to the end of the article for good logical and pedagogical reasons. Having caught our attention by whacking us over the head with a two-by-four, like that famous trainer of mules, Jensen then asks:

> What has gone wrong? In other fields, when bridges do not stand, when aircraft do not fly, when machines do not work, when treatments do not cure, despite all the conscientious efforts on the part of many persons to make them do so, one begins to question the basic assumptions, principles, theories, and hypotheses that guide one's efforts. Is it time to follow suit in education?

Who can help but answer that last rhetorical question with a resounding " Yes "? What thoughtful and intelligent person can avoid being struck by the intellectual and empirical bankruptcy of educational psychology as it is practiced in our mass educational systems? The innocent reader will immediately fall into close sympathy with Professor Jensen, who, it seems, is about to dissect educational psychology and show it up as a pre-scientific jumble without theoretic coherence or prescriptive competence. But the innocent reader will be wrong. For the rest of Jensen's article puts the blame for the failure of his science not on the scientists but on the children. According to him, it is not that his science and its practitioners have failed utterly to understand human motivation, behavior and development but simply that the damn kids are ineducable.

The unconscious irony of his metaphor of bridges, airplanes and machines has apparently been lost on him. The fact is that in the twentieth century bridges do stand, machines do work, and airplanes do fly, because they are built on clearly understood mechanical and hydrodynamic principles which even moderately careful and intelligent engineers can put into practice. In the seventeenth century that was not the case, and the general opinion was that men would never succeed in their attempts to fly because flying was impossible. Jensen proposes that we take the same view of education and that, in the terms of his metaphor, fallen bridges be taken as evidence of the unbridgeability of rivers. The alternative explanation, that educational psychology is still in the seventeenth century, is apparently not part of his philosophy.

This view of technological failure as arising from ontological rather than epistemological sources is a common form of apology at many levels of practice. Anyone who has dealt with plumbers will appreciate how many things " can't be fixed " or " weren't meant to be used like that." Physicists tell me that their failure to formulate an elegant general theory of fundamental particles is a result of there not being any underlying regularity to be discerned. How often men, in their overweening pride, blame nature for their own failures. This professionalist bias, that if a problem were soluble it would have been solved, lies at the basis of Jensen's thesis which can only be appreciated when seen in this light.

Having begun with the assumption that I.Q. cannot be equalized, Jensen now goes on to why not. He begins his investigation with a discussion of the "nature of intelligence," by which he means the way in which intelligence is defined by testing and the correlation of intelligence test scores with scholastic and occupational performance. A very strong point is made that I.Q. testing was developed in a western industrialized society specifically as a prognostication of success in that society by the generally accepted criteria. He makes a special point of noting that psychologists' notions of status and success have a high correlation with those of society at large, so that it is entirely reasonable that tests created by psychologists will correlate highly with conventional measures of success. One might think that this argument, that I.Q. testing is "culture bound," would militate against Jensen's general thesis of the biological and specifically genetical basis of I.Q. differences. Indeed, it is an argument often used against I.Q. testing for so-called "deprived" children, since it is supposed that they have developed in a subculture that does not prepare them for such tests. What role does this "environmentalist" argument play in Jensen's thesis? Is it simply evidence of his total fairness and objectivity? No. Jensen has seen, more clearly than most, that the argument of the specific cultural origins of I.Q. testing and especially the high correlation of these tests with occupational status cuts both ways. For if the poorer performance of blacks on I.Q. tests has largely genetic rather than environmental causes, then it follows that blacks are also genetically handicapped for other high status components of Western culture. That is, what Jensen is arguing is that differences between cultures are in large part genetically determined and that I.Q. testing is simply one manifestation of those differences.

In this light we can also understand his argument concerning the existence of "general intelligence" as measured by I.Q. tests. Jensen is at some pains to convince his readers that there is a single factor, g, which, in factor analysis of various intelligence tests, accounts for a large fraction of the variance of scores. The existence of such a factor, while not critical to the argument, obviously simplifies it, for then I.Q. tests would really be testing for "something" rather than just being correlated with scholastic and occupational performance. While Jensen denies that intelligence should be reified, he comes perilously close to doing so in his discussion of g.

Without going into factor analysis at any length, I will point out only that factor analysis does not give a unique result for any given set of data. Rather, it gives an infinity of possible results among which the investigator chooses according to his tastes and preconceptions of the models he is fitting. One strategy in factor analysis is to pack as much weight as possible into one factor, while another is to distribute the weights over as many factors as possible as equally as possible. Whether one chooses one of these or some other depends upon one's model, the numerical analysis only providing the weights appropriate for each model. Thus, the impression left by Jensen that factor analysis somehow naturally or ineluctably isolates one factor with high weight is wrong.

387

" True Merit "?

In the welter of psychological metaphysics involving concepts of "crystallized" as against "fluid" intelligence, "generalized" intelligence, "intelligence" as opposed to "mental ability," there is some danger of losing sight of Jensen's main point: I.Q. tests are culture bound and there is good reason that they should be, because they are predictors of culture bound activities and values. What is further implied, of course, is that those who do not perform well on these tests are less well suited for high status and must paint barns rather than pictures. We read that "We have to face it: the assortment of persons into occupational roles simply is not 'fair' in any absolute sense. The best we can hope for is that true merit, given equality of opportunity, act as a basis for the natural assorting process." What a world view is there revealed! The most rewarding places in society shall go to those with "true merit" and that is the best we can hope for. Of course, Professor Jensen is safe since, despite the abject failure of educational psychology to solve the problems it has set itself, that failure does not arise from lack of "true merit" on the part of psychologists but from the natural intransigence of their human subjects.

Having established that there are differences among men in the degree to which they are adapted to higher status and high satisfaction roles in Western society, and having stated that education has not succeeded in removing these differences, Jensen now moves on to their cause. He raises the question of "fixed" intelligence and quite rightly dismisses it as misleading. He introduces us here to what he regards as the two real issues. "The first issue concerns the genetic basis of individual differences in intelligence; the second concerns the stability or constancy of the I.Q. through the individual's lifetime." Jensen devotes some three-quarters of his essay to an attempt to demonstrate that I.Q. is developmentally rather stable, being to all intents and purposes fixed after the age of eight, and that most of the variation in I.Q. among individuals in the population has a genetic rather than environmental basis. Before looking in detail at some of these arguments, we must again ask where he is headed. While Jensen argues strongly that I.Q. is "culture bound," he wishes to argue that it is not environmentally determined. This is a vital distinction. I.Q. is "culture bound" in the sense that it is related to performance in a Western industrial society. But the determination of the ability to perform culturally defined tasks might itself be entirely genetic. For example, a person suffering from a genetically caused deaf-mutism is handicapped to different extents in cultures requiring different degrees of verbal performance, yet his disorder did not have an environmental origin.

Jensen first dispenses with the question of developmental stability of I.Q. Citing Benjamin Bloom's survey of the literature, he concludes that the correlation between test scores of an individual at different ages is close to unity after the age of eight. The inference to be drawn from this fact is, I suppose, that it is not worth trying to change I.Q. by training after that age. But such an inference cannot be made. All that can be said is that, given the usual progression of educational experience to which most children are exposed, there is sufficient consistency not to cause any remarkable changes in I.Q. That is, a child whose educational experience (in the broad sense) may have ruined his capacity to perform by the age of eight is not likely to experience an environment in his later years that will do much to alter those capacities. Indeed, given the present state of educational theory and practice, there is likely to be a considerable reinforcement of early performance. To say that children do not change their I.Q. is not the same as saying they cannot. Moreover, Jensen is curiously silent on the lower correlation and apparent plasticity of I.Q. at younger ages, which is after all the chief point of Bloom's work.

The Genetic Argument

The heart of Jensen's paper is contained in his long discussion of the distribution and inheritance of intelligence. Clearly he feels that here his main point is to be established. The failure of compensatory education, the developmental stability of I.Q., the obvious difference between the performance of blacks and whites can be best understood, he believes, when the full impact of the findings of genetics is felt. In his view, insufficient attention has been given by social scientists to the findings of geneticists, and I must agree with him. Although there are exceptions, there has been a strong professional bias toward the assumption that human behavior is infinitely plastic, a bias natural enough in men whose professional commitment is to changing behavior. It is as a reaction to this tradition, and as a natural outcome of his confrontation with the failure of educational psychology, that Jensen's own opposite bias flows, as I have already claimed.

The first step in his genetical argument is the demonstration that I.Q. scores are normally distributed or nearly so. I am unable to find in his paper any explicit statement of why he regards this point as so important. From repeated references to Sir Francis Galton, filial regression, mutant genes, a few major genes for exceptional ability and assortative mating, it gradually emerges that an underlying normality of the distribution appears to Jensen as an important consequence of genetic control of I.Q. He asks: "... is intelligence itself—not just our measurements of it—really normally distributed?" Apparently he believes that if intelligence, quite aside from measurement, were really normally distributed, this would demonstrate its biological and genetical status. Aside from a serious epistemological error involved in the question, the basis for his concern is itself erroneous. There is nothing in genetic theory that requires or even suggests that a phenotypic character should be normally distributed, even when it is completely determined genetically. Depending upon the degree of dominance of genes, interaction between them, frequencies of alternative alleles at the various gene loci in the population as allometric growth relations between various parts of the organism transforming primary gene effects, a character may have almost any uni-modal distribution and under some circumstances even a multi-modal one.

After establishing the near-normality of the curve of I.Q. scores, Jensen goes directly to a discussion of the genetics of continuously varying characters. He begins by quoting with approbation E. L. Thorndike's maxim: "In the actual race of life, which is not to get ahead, but to get ahead of somebody, the chief determining factor is heredity." This quotation along with many others used by Jensen shows a style of argument that is not congenial to natural scientists, however it may be a part of other disciplines. There is a great deal of appeal to authority and the acceptance of the empirically unsubstantiated opinions of eminent authorities as a kind of relevant evidence. We hear of "three eminent geneticists," or "the most distinguished exponent [of genetical methods], Sir Cyril Burt." The irrelevance of this kind of argument is illustrated precisely by the appeal to E. L. Thorndike, who, despite his eminence in the history of psychology, made the statement quoted by Jensen in 1905, when nothing was known about genetics outside of attempts to confirm Mendel's paper. Whatever the eventual truth of his statement turns out to be, Thorndike made it out of his utter ignorance of the genetics of human behavior, and it can only be ascribed to the sheer prejudice of a Methodist Yankee.

Heritability

To understand the main genetical argument of Jensen, we must dwell, as he does, on the concept of heritability. We cannot speak of a trait being molded by heredity, as opposed to environment. Every character of an organism is the result of a unique interaction between the genetic information and the sequence of environments through which the organism has passed during its development. For some traits the variations in environment have little effect, so that once the genotype is known, the eventual form of the organism is pretty well specified. For other traits, specification of the genetic makeup may be a very poor predictor of the eventual phenotype because even the smallest environmental effects may affect the trait greatly. But for all traits there is a many-many relationship between gene and character and between environment and character. Only by a specification of both the genotype and the environmental sequence can the character be predicted. Nevertheless, traits do vary in the degree of their genetic determination and this degree can be expressed, among other ways, by their heritabilities.

The distribution of character values, say I.Q. scores, in a population arises from a mixture of a large number of genotypes. Each genotype in the population does not have a unique phenotype corresponding to it because the different individuals of that genotype have undergone somewhat different environmental sequences in their development. Thus, each genotype has a distribution of I.Q. scores associated with it. Some genotypes are more common in the population so their distributions contribute heavily to determining the over-all distribution, while others are rare and make little contribution. The total variation in the population, as measured by the variance, results from the variation between the mean I.Q. scores of the different genotypes and the variation around each genotypic mean. The heritability of a measurement is defined as the ratio of the variance due to the differences between the genotypes to the total variance in the population. If this heritability were 1.0, it would mean that all the variation in the population resulted from differences between genotypes but that there was no environmentally caused variation around each genotype mean. On the other hand a heritability of 0.0 would mean that there was no genetic variation because all individuals were effectively identical in their genes, and that all the variation in the population arose from environmental differences in the development of the different individuals.

Defined in this way, heritability is not a concept that can be applied to a trait in general, but only to a trait in a particular population, in a particular set of environments. Thus, different populations may have more or less genetic variation for the same character. Moreover, a character may be relatively insensitive to environment in a particular environmental range, but be extremely sensitive outside this range. Many such characters are known, and it is the commonest kind of relation between character and environment. Finally, some genotypes are more sensitive to environmental fluctuation than others so that two populations with the same genetic variance but different genotypes, and living in the same environments, may still have different heritabilities for a trait.

The estimation of heritability of a trait in a population depends on measuring individuals of known degrees of relationship to each other and comparing the observed correlation in the trait between relatives with the theoretical correlation from genetic theory. There are two difficulties that arise in such a procedure. First, the exact theoretical correlation between relatives, except for identical twins, cannot be specified unless there is detailed knowledge of the mode of inheritance of the character. A first order approximation is possible, however, based upon some simplifying assumptions, and it is unusual for this approximation to be badly off.

A much more serious difficulty arises because relatives are correlated not only in their heredities but also in their environments. Two sibs are much more alike in the sequence of environments in which they developed than are two cousins or two unrelated persons. As a result, there will be an overestimate of the heritability of a character, arising from the added correlation between relatives from environmental similarities. There is no easy way to get around this bias in general so that great weight must be put on peculiar situations in which the ordinary environmental correlations are disturbed. That is why so much emphasis is placed, in human genetics, on the handful of cases of identical twins raised apart from birth, and the much more numerous cases of totally unrelated children raised in the same family. Neither of these cases is completely reliable, however, since twins separated from birth are nevertheless likely to be raised in families belonging to the same socioeconomic, racial, religious and ethnic categories, while unrelated children raised in the same family may easily be treated rather more differently than biological sibs. Despite these difficulties, the weight of evidence from a variety of correlations between relatives puts the heritability estimates for I.Q. in various human populations between .6 and .8. For reasons of his argument, Jensen prefers the higher value but it is not worth quibbling over. Volumes could be written on the evaluation of heritability estimates for I.Q. and one can find a number of faults with Jensen's treatment of the published data. However, it is irrelevant to questions of race and intelligence, and to questions of the failure of compensatory education, whether the heritability of I.Q. is .4 or .8, so I shall accept Jensen's rather high estimate without serious argument.

The description I have given of heritability, its application to a specific population in a specific set of environments and the difficulties in its accurate estimation are all discussed by Jensen. While the emphasis he gives to various points differs from mine, and his estimate of heritability is on the high side, he appears to have said in one way or another just about everything that a judicious man can say. The very judiciousness of his argument has been disarming to geneticists especially, and they have failed to note the extraordinary conclusions that are drawn from these reasonable premises. Indeed, the logical and empirical hiatus between the conclusions and the premises is especially striking and thought-provoking in view of Jensen's apparent understanding of the technical issues.

The first conclusion concerns the cause of the difference between the I.Q. distributions of blacks and whites. On the average, over a number of studies, blacks have a distribution of I.Q. scores whose mean is about 15 points—about 1 standard deviation—below whites. Taking into account the lower variance of scores among blacks than among whites, this difference means that about 11 per cent of blacks have I.Q. scores above the mean white score (as compared with 50 per cent of whites) while 18 percent of whites score below the mean black score (again as compared to 50 percent of blacks). If, according to Jensen, "gross socio-economic factors" are equalized between the tested groups, the difference in means is reduced to 11 points. It is hard to know what to say about overlap between the groups after this correction, since the standard deviations of such equalized populations will be lower. From these and related observations, and the estimate of .8 for the heritability of I.Q. (in white populations, no reliable estimate existing for blacks), Jensen concludes that:

> ... all we are left with are various lines of evidence, no one of which is definitive alone, but which, viewed altogether, make it a not unreasonable hypothesis that genetic factors are strongly implicated in the average Negro-White intelligence difference. The preponderance of evidence is, in my opinion, less consistent with a strictly environmental hypothesis than with a genetic hypothesis, which, of course, does not exclude the influence of environment on its interaction with genetic factors.

Anyone not familiar with the standard litany of academic disclaimers ("not unreasonable hypothesis," "does not exclude," "in my opinion") will, taking this statement at face value, find nothing to disagree with since it says nothing. To contrast a "strictly environmental hypothesis" with "a genetic hypothesis which ... does not exclude the influence of the environment" is to be guilty of the utmost triviality. If that is the only conclusion he means to come to, Jensen has just wasted a great deal of space in the "Harvard Educational Review." But, of course, like all cant, the special language of the social scientist needs to be translated into common English. What Jensen is saying is: "It is pretty clear, although not absolutely proved, that most of the difference in I.Q. between blacks and whites is genetical." This, at least, is not a trivial conclusion. Indeed, it may even be true. However, the evidence offered by Jensen is irrelevant.

Is It Likely?

How can that be? We have admitted the high heritability of I.Q. and the reality of the difference between the black and the white distributions. Moreover, we have seen that adjustment for gross socio-economic level still leaves a large difference.

Is it not then likely that the difference is genetic? No. It is neither likely nor unlikely. There is no evidence. The fundamental error of Jensen's argument is to confuse heritability of a character within a population with heritability of the difference between two populations. Indeed, between two populations, the concept of heritability of their difference is meaningless. This is because a variance based upon two measurements has only one degree of freedom and so cannot be partitioned into genetic and environmental components. The genetic basis of the difference between two populations bears no logical or empirical relation to the heritability within populations and cannot be inferred from it, as I will show in a simple but realistic example. In addition, the notion that eliminating what appear a priori to be major environmental variables will serve to eliminate a large part of the environmentally caused difference between the populations is biologically naive. In the context of I.Q. testing, it assumes that educational psychologists know what the major sources of environmental difference between black and white performances are. Thus, Jensen compares blacks with American Indians whom he regards as far more environmentally disadvantaged. But a priori judgments of the importance of different aspects of the environment are valueless, as every ecologist and plant physiologist knows. My example will speak to that point as well.

Let us take two completely inbred lines of corn. Because they are completely inbred by self-fertilization, there is no genetic variation in either line, but the two lines will be genetically different from each other. Let us now plant seeds of these two inbred lines in flower pots with ordinary potting soil, one seed of each line to a pot. After they have germinated and grown for a few weeks we will measure the height of each plant. We will discover variation in height from plant to plant. Because each line is completely inbred, the variation in height within lines must be entirely environmental, a result of variation in potting conditions from pot to pot. Then the heritability of plant height in both lines is 0.0. But there will be an average difference in plant height between lines that arises entirely from the fact that the two lines are genetically different. Thus the difference between lines is entirely genetical even though the heritability of height is 0!

Now let us do the opposite experiment. We will take two handfuls from a sack containing seed of an open-pollinated variety of corn. Such a variety has lots of genetic variation in it. Instead of using potting soil, however, we will grow the seed in vermiculite watered with a carefully made-up nutrient, Knop's solution, used by plant physiologists for controlled growth experiments. One batch of seed will be grown on complete Knop's solution, but the other will have the concentration of nitrates cut in half and in addition, we will leave out the minute trace of zinc salt that is part of the necessary trace elements (30 parts per billion). After several weeks we will measure the plants. Now we will find variation within seed lots which is entirely genetical since no environmental variation within lots was allowed. Thus heritability will be 1.0. However, there will be a radical difference between seed lots which is ascribable entirely to the difference in nutrient levels. Thus we have a case where heritability within populations is complete, yet the difference between populations is entirely environmental!

But let us carry our experiment to the end. Suppose we do not know about the difference in the nutrient solutions because it was really the carelessness of our assistant that was involved. We call in a friend who is a very careful chemist and ask him to look into the matter for us. He analyzes the nutrient solutions and discovers the obvious—only half as much nitrate in the case of the stunted plants. So we add the missing nitrates and do the experiment again. This time our second batch of plants will grow a little larger but not much, and we will conclude that the difference between the lots is genetic since equalizing the large difference in nitrate level had so little effect. But, of course, we would be wrong for it is the missing trace of zinc that is the real culprit. Finally, it should be pointed out that it took many years before the importance of minute trace elements in plant physiology was worked out because ordinary laboratory glassware will leach out enough of many trace elements to let plants grow normally. Should educational psychologists study plant physiology?

Having disposed, I hope, of Jensen's conclusion that the high heritability of I.Q. and the lack of effect of correction for gross socio-economic class are presumptive evidence for the genetic basis of the difference between blacks and whites, I will turn to his second erroneous conclusion. The article under discussion began with the observation, which he documents, that compensatory education for the disadvantaged (blacks, chiefly) has failed. The explanation offered for the failure is that I.Q. has a high heritability and that therefore the difference between the races is also mostly genetical. Given that the racial difference is genetical, then environmental change and educational effort cannot make much difference and cannot close the gap very much between blacks and whites. I have already argued that there is no evidence one way or the other about the genetics of inter-racial I.Q. differences. To understand Jensen's second error, however, we will suppose that the difference is indeed genetical. Does this mean that compensatory education, having failed, must fail? The supposition that it must arises from a misapprehension about the fixity of genetically determined traits. It was thought at one time that genetic disorders, because they were genetic, were incurable. Yet we now know that inborn errors of metabolism are indeed curable if their biochemistry is sufficiently well understood and if deficient metabolic products can be supplied exogenously. Yet in the normal range of environments, these inborn errors manifest themselves irrespective of the usual environmental variables. That is, even though no environment in the normal range has an effect on the character, there may be special environments, created in response to our knowledge of the underlying biology of a character, which are effective in altering it.

But we do not need recourse to abnormalities of development to see this point. Jensen says that "there is no reason to believe that the I.Q.'s of deprived children, given an environment of abundance, would rise to a higher level than the already privileged children's I.Q.'s." It is empirically wrong to argue that if the richest environmental experience we can conceive does not raise I.Q. substantially, that we have exhausted the environmental possibilities. In the seventeenth century the infant mortality rates were many times their present level at all socio-economic levels. Using what was then the normal range of environments, the infant mortality rate of the highest socio-economic class would have been regarded as the limit below which one could not reasonably expect to reduce the death rate. But changes in sanitation, public health and disease control—changes which are commonplace to us now but would have seemed incredible to a man of the seventeenth century— have reduced the infant mortality rates of "disadvantaged" urban Americans well below those of even the richest members of seventeenth century society. The argument that compensatory education is hopeless is equivalent to saying that changing the form of the seventeenth century gutter would not have a pronounced effect on public sanitation. What compensatory education will be able to accomplish when the study of human behavior finally emerges from its pre-scientific era is anyone's guess. It will be most extraordinary if it stands as the sole exception to the rule that technological progress exceeds by manyfold what even the most optimistic might have imagined.

The real issue in compensatory education does not lie in the heritability of I.Q. or in the possible limits of educational technology. On the reasonable assumption that ways of significantly altering mental capacities can be developed if it is important enough to do so, the real issue is what the goals of our society will be. Do we want to foster a society in which the "race of life" is "to get ahead of somebody" and in which "true merit," be it genetically or environmentally determined, will be the criterion for men's earthly reward? Or do we want a society in which every man can aspire to the fullest measure of psychic and material fulfillment that social activity can produce? Professor Jensen has made it fairly clear to me what sort of society he wants.

I oppose him.

38   Race and the Genetics of Intelligence: A Reply to Lewontin

Arthur R. Jensen

Professor Lewontin (Bulletin, March 1970) has likened my article, "How Much Can We Boost IQ and Scholastic Achievement?" ("Harvard Educational Review," Winter, 1969) to the "pernicious heresy ... of total depravity, irresistible grace, lack of free will, predestination and limited atonement" attributed to Bishop Jansen in the seventeenth century. Lewontin goes on to claim that the same doctrine is now called "jensenism" (a term coined by the "Wall Street Journal"), and that "jensenism" is "as erroneous in the twentieth century as it was in the seventeenth." Lewontin proposes to play the role of Pope Innocent X (who denounced Bishop Jansen in 1653) by holding up and condemning his own version, incomplete and distorted, of "jensenism."

From the *Bulletin of the Atomic Scientists* (May 1970) 17–23. Reprinted by permission of the author and *Science and Public Affairs*, the *Bulletin of the Atomic Scientists*. Copyright © 1970 by the Educational Foundation for Nuclear Science. (Illustrations omitted.)

Thus Lewontin sets the stage for the ad hominem flavor of the rest of his paper. His role may resemble that of Pope Innocent's in trying to put down what he perceives as a heresy, but readers of Lewontin's piece may be reminded of a closer ecclesiastical parallel in Bishop Wilberforce, who, in debating evolution with T. H. Huxley, resorted to commenting that Darwin's physiognomy bore a simian resemblance; and he begged to know of Huxley, "was it through his grandfather or grandmother that he claimed his descent from a monkey?" Thus we see Lewontin, albeit in a milder vein, referring to Edward L. Thorndike (probably America's greatest psychologist and a pioneer in twin studies of the heritability of intelligence) as a "Methodist Yankee" and to William Shockley (a Nobel Laureate in physics, author of some three hundred scientific articles, and winner of numerous scientific awards and distinctions) as "the eminent white Anglo-Saxon inventor." (True, Shockley has 85 patented inventions, including the junction transistor.) If Lewontin is trying to be uncomplimentary, it is interesting to see the labels he picks for this.

In connection with Lewontin's reference to Shockley, an error of fact calls for correction. Shockley has not urged the Academy to study "the effects of interracial mating." This is a distortion of Shockley's aim, which is to see the Academy openly encourage scientific enquiry into the genetics of human abilities and pro-clivities, including their racial aspects. Lewontin's approach makes it appear to me that he views the problems of criticizing my article as that of making a case for the "good guys" versus the "bad guys," and he wants there to be no doubt in the reader's mind that he is very much one of the "good guys." Thus he finally makes it perfectly clear in the last few sentences of his article that he opposes me mainly for ideological reasons and not on scientific or technical grounds.

A Persistent Question

Lewontin's statement that "Jensen has made it fairly clear to me what sort of society he wants" is not based on knowledge that Lewontin has of my social or political philosophy. It is a subjective surmise reflecting Lewontin's antipathy for anyone who would raise the question of genetic racial intelligence differences in an obviously non-political, scholarly context. The question of whether the observed racial differences in mental abilities and scholastic performance involve genetic as well as environmental factors is indeed tabooed. Nevertheless, it is a persistent question. My belief is that scientists in the appropriate disciplines must face the question and not repeatedly sweep it back under the rug. In the long run, the safest and sanest thing we can urge is intensive, no-holds-barred inquiry in the best tradition of science.

Before proceeding with comments on specific technical points in Lewontin's paper, it would be well to put them in proper perspective by giving a capsule summary of what my article was about.

Survey Findings

First, I reviewed some of the evidence and the conclusions of a nationwide survey and evaluation of the large, federally-funded compensatory education programs made by the U.S. Commission on Civil Rights, which concluded that these special programs had produced no significant improvement in the measured intelligence or scholastic performance of the disadvantaged children whose educational achievements these programs were specifically intended to improve. The massive evidence presented by the Civil Rights Commission suggests to me that merely applying more of the same approach to compensatory education on a still larger scale is not at all likely to lead to the desired results, namely, increasing the benefits of public education to the disadvantaged. The well-documented fruitlessness of these well-intentioned compensatory programs indicates the importance of now questioning the assumptions, theories and practices on which they were based.

I agree with Lewontin that these assumptions, theories and practices—espoused over the past decade by the majority of educators, social and behavioral scientists —are bankrupt. I do not blame the children who fail to benefit from these programs, as Lewontin would have his readers think. A large part of the failure, I believe, has resulted from the failure and reluctance of the vast majority of the educational establishment, aided and abetted by social scientists, to take seriously the problems of individual differences in developmental rates, patterns of ability, and learning styles. The prevailing philosophy has been that all children are basically very much alike—they are all "average children"—in their mental development and capabilities, and that the only causes of the vast differences that show up as they go through school are due to cultural factors and home influences that mold the child even before he enters kindergarten. By providing the culturally disadvantaged with some of the cultural amenities enjoyed by middle-class children for a period of a year or two before they enter school, we are told, the large differences in scholastic aptitude would be minimized and the schools could go on thereafter treating all children very much alike and expect nearly all to perform as "average children" for their grade in school.

It hasn't worked. And educators are now beginning to say: "Let's really look at individual differences and try to find a variety of instructional methods and differentiated programs that will accommodate these differences." Whatever their causes may be, it now seems certain that they are not so superficial as to be erased by a few months of "cultural enrichment," "verbal stimulation," and the like. I have pointed out that some small-scale experimental intervention programs, which gear specific instructional methods to developmental differences, have shown more promise of beneficial results than the large-scale programs based on a philosophy of general cultural enrichment and a multiplication of the resources in already existing programs for the "average child."

The Opportunities

One of the chief obstacles to providing differentiated educational programs for children with different patterns of abilities, aside from the lack of any detailed technical knowledge as to how to go about this most effectively, is the fact that children in different visibly identifiable sub-populations probably will be disproportionately represented in different instructional programs. This highly probable consequence of taking individual differences really seriously is misconstrued by some critics as inequality of opportunity. But actually, one child's opportunity can be another's defeat. To me, equality of opportunity does not mean uniform treatment of all children, but equality of opportunity for a diversity of educational experiences and services. If we fail to take account either of innate or acquired differences in abilities and traits, the ideal of equality of educational opportunity can be interpreted so literally as to be actually harmful, just as it would be harmful for a physician to give all his patients the same medicine.

I know personally of many instances in which children with educational problems were denied the school's special facilities for dealing with such problems (small classes, specialist teachers, tutorial help, diagnostic services, etc.), not because the children did not need this special attention or because the services were not available to the school, but simply because the children were black and no one wanted to single them out as being different or in need of special attention. So instead, white middle-class children with similar educational problems get nearly all the attention and special treatment, and most of them benefit from it. No one objects, because this is not viewed by anyone as "discrimination." But some school districts have been dragged into court for trying to provide similar facilities for minority children with educational problems. In these actions the well-intentioned plaintiffs undoubtedly viewed themselves as the "good guys." Many children, I fear, by being forced into the educational mold of the "average child" from Grade 1 on, are soon "turned off" on school learning and have to pay the consequences in frustration and defeat, both in school and in the world of work for which their schooling has not prepared them.

I do not advocate abandoning efforts to improve the education of the disadvantaged. I urge increased emphasis on these efforts, in the spirit of experimentation, expanding the diversity of approaches and improving the rigor of evaluation in order to boost our chances of discovering the methods that will work best.

Learning and IQ

My article also dealt with my theory of two broad categories of mental abilities, which I call intelligence (or abstract reasoning ability) and associative learning ability. These types of ability appear to be distributed differently in various social classes and racial groups. While large racial and social class differences are found for intelligence, there are practically negligible differences among these groups in associative learning abilities, such as memory span and serial and paired-associate rote learning.

Research should be directed at delineating still other types of abilities and at discovering how the particular strengths of each individual's pattern of abilities can be most effectively brought to bear on school learning and on the attainment of occupational skills. By pursuing this path, I believe we can discover the means by which the reality of individual differences need not mean educational rewards for some children and utter frustration and defeat for others.

Intelligence

I pointed out that IQ tests evolved to predict scholastic performance in largely European and North American middle-class populations around the turn of the century. They evolved to measure those abilities most relevant to the curriculum and type of instruction, which in turn were shaped by the pattern of abilities of the children the schools were then intended to serve.

IQ or abstract reasoning ability is thus a selection of just one portion of the total spectrum of human mental abilities. This aspect of mental abilities measured by IQ tests is important to our society, but is obviously not the only set of educationally or occupationally relevant abilities. Other mental abilities have not yet been adequately measured; their distributions in various segments of the population have not been adequately determined; and their educational relevance has not been fully explored.

I believe a much broader assessment of the spectrum of abilities and potentials, and the investigation of their utilization for educational achievement, will be an essential aspect of improving the education of children regarded as disadvantaged.

Inheritance

Much of my paper was a review of the methods and evidence that led me to the conclusion that individual differences in intelligence—that is, IQ—are predominantly attributable to genetic differences, with environmental factors contributing a minor portion of the variance among individuals. The heritability of the IQ—that is, the percentage of individual differences variance attributable to genetic factors—comes out to about 80 per cent, the average value obtained from all relevant studies now reported.

These estimates of heritability are based on tests administered to European and North American populations and cannot properly be generalized to other populations. I believe we need similar heritability studies in minority populations if we are to increase our understanding of what our tests measure in these populations and how these abilities can be most effectively used in the educational process.

Class Differences

Although the full range of IQ and other abilities is found among children in every socioeconomic stratum in our population, it is well established that IQ differs, on the average, among children from different social class backgrounds. The evidence, some of which I referred to in my article, indicates to me that some of this IQ difference is attributable to environmental differences and some of it is attributable to genetic differences among social classes—largely as a result of differential selection of the parent generations for different patterns of ability.

I have not yet met or read a modern geneticist who disputes this interpretation of the evidence. In the view of geneticist C. O. Carter: "Sociologists who doubt this show more ingenuity than judgment." At least three sociologists who are students of this problem—Pitirim Sorokin, Bruce Eckland and Otis Dudley Duncan

—all agree that selective factors in social mobility and assortative mating have resulted in a genetic component in social class intelligence differences. As Eckland points out, this conclusion holds within socially defined racial groups but cannot properly be generalized among racial groups, since barriers to upward mobility have undoubtedly been quite different for various racial groups.

Race Differences

I have always advocated dealing with persons as individuals, each in terms of his own merits and characteristics and I am opposed to according treatment to persons solely on the basis of their race, color, national origin or social class background. But I am also opposed to ignoring or refusing to investigate the causes of the well-established differences among racial groups in the distribution of educationally relevant traits, particularly IQ.

I believe that the causes of observed differences in IQ and scholastic performance among different ethnic groups is, scientifically, still an open question, an important question and a researchable one. I believe that official statements such as: "It is a demonstrable fact that the talent pool in any one ethnic group is substantially the same as in any other ethnic group" (U.S. Office of Education, 1966), and "Intelligence potential is distributed among Negro infants in the same proportion and pattern as among Icelanders or Chinese, or any other group" (U.S. Dept. of Labor, 1965) are without scientific merit. They lack any factual basis and must be regarded only as hypotheses.

The fact that different racial groups in this country have widely separated geographic origins and have had quite different histories which have subjected them to different selective social and economic pressures make it highly likely that their gene pools differ for some genetically conditioned behavioral characteristics, including intelligence or abstract reasoning ability. Nearly every anatomical, physiological and biochemical system investigated shows racial differences. Why should the brain be any exception? The reasonableness of the hypothesis that there are racial differences in genetically conditioned behavioral characteristics, including mental abilities, is not confined to the poorly informed, but has been expressed in writings and public statements by geneticists such as K. Mather, C. D. Darlington, R. A. Fisher and Francis Crick, to name a few.

In my article, I indicated several lines of evidence which support my assertion that a genetic hypothesis is not unwarranted. The fact that we still have only inconclusive results with respect to this hypothesis does not mean that the opposite of the hypothesis is true. Yet some social scientists speak as if this were the case and have even publicly censured me for suggesting an alternative to purely environmental hypotheses of intelligence differences. Scientific investigation proceeds most effectively by means of what Platt has called "strong inference," pitting alternative hypotheses that lead to different predictions against one another and then putting the predictions to an empirical test.

Most environmentalist theories are so inadequate that they often fail to explain even the facts they were devised to account for. In this area, psychologists, sociologists, and anthropologists have not followed the usual methods of scientific investigation, which consist in part in testing rival hypotheses in such a way that empirical evidence can disconfirm either one or the other, or both. There has been only one acceptable hypothesis—the environmentalists'—and research has consisted largely of endless enumeration of subtler and subtler environmental differences among subpopulations and of showing their psychological, educational and sociological correlates, without even asking if genetic factors are in any way implicated at any point in the correlational network. Social scientists for the most part simply decree, on purely ideological grounds, that all races are identical in the genetic factors that condition various behavioral traits, including intelligence. Most environmental hypotheses proposed to account for intelligence differences among racial groups, therefore, have not had to stand up to scientific tests of the kind that other sciences have depended upon for the advancement of knowledge. Until genetic, as well as environmental, hypotheses are seriously considered in our search for causes, it is virtually certain that we will never achieve a scientifically acceptable answer to the question of racial differences in intellectual performance.

Dysgenic Trends

Lewontin does not comment on my article's pointing to a problem which is socially more important than the question of racial differences per se, namely, the high probability of dysgenic trends in our urban slums. At least 16 per cent of black children (as compared with less than two per cent of white children) in our nation's schools are mentally retarded by the criterion of IQs under 70 and scholastic performance commensurate with this level of ability. The figure is much higher in " inner city " schools, and these children come from the largest families. How much of this retardation is attributable to genetic factors and how much to environmental influences, we do not know. It is my position that we should try to find out. What hope is there for improving this condition, and for ameliorating the frustration and suffering obviously implied by these facts, if we do not discover the causes? Some of the causes are undoubtedly environmental, nutritional, pre- and perinatal, and cultural, and my article includes sections on all these factors. But I also suggest that genetic hypotheses (which, of course, do not exclude the effects of environment) be considered in our efforts to understand these conditions.

Census data show markedly higher birth rates among the poorest segments of the Negro population than among successful, middle-class Negroes. This social class differential in birth rate appears to be much greater in the Negro than in the white population. That is, the educationally and occupationally least able among Negroes have a higher reproductive rate than their white counterparts, and the most able segment of the Negro population has a lower reproductive rate than its white counterpart.

If social class intelligence differences within the Negro population have a genetic component, as in the white population, the condition I have described could create and widen the genetic intelligence differences between Negroes and whites. The social and educational implications of this trend, if it exists and persists, are enormous. The problem obviously deserves thorough investigation by social scientists and geneticists and should not be ignored or superficially dismissed as a result of well-meaning wishful thinking. The possible consequences of our failure seriously to study these questions may well be viewed by future generations as our society's greatest injustice to Negro Americans.

Specific Comments

I agree with Lewontin that much of educational psychology and educational practices are still in the seventeenth century, especially as regards recognition of individual and group differences. Just as the seventeenth-century alchemists tried to transmute base metals into gold, the twentieth-century alchemists in our schools would like to make all children conform to their concept of the average child, so that all can be taught the same things in the same way at the same pace.

Lewontin seems to believe that anything is possible, given sufficient technological implementation. But reality does not bow to technology. Technology depends upon a correct assessment of reality. With all our technological progress in the physical sciences since the seventeenth century, we have not yet produced the philosopher's stone that can change base metals into gold. Though this was the most highly sought goal of the forerunners of modern chemistry, it was abandoned as soon as scientists discovered the actual nature of matter. Scientific inquiry took the place of wishful thinking. So tremendous technological capabilities were never brought to bear on this pre-scientific goal of discovering the philosopher's stone. Yet men have found other ways to create wealth, ways compatible with reality.

Lewontin points out that "to say that children do not change their IQ is not the same as saying they cannot." I have never said anything to the contrary, but I would point out that no one knows how to change IQs appreciably, and in those few children in whom true large shifts in IQ are found, either there is no explanation or the explanation involves changes in physiological and biochemical factors. Except in the case of children reared in almost total social isolation, there is no known psychological or educational treatment that systematically will boost IQs more than the few points' gain that comes from direct practice in taking the tests. In writing about the high heritability of intelligence, I have stated: "This is not to say, however, that as yet undiscovered biological, chemical, or psychological forms of intervention in the genetic or developmental processes could not diminish the relative importance of heredity as a determinant of intellectual differences."

Although Lewontin dislikes E. L. Thorndike's statement ("In the actual race of life, which is not to get ahead but to get ahead of somebody, the chief determining factor is heredity"), it should be noted that the statement is found in an empirical paper by Thorndike based on twin correlations. The statement thus was not made out of "utter ignorance," and in fact it still emphasizes a most important point about heritability—that the genes do not fix an absolute level of performance but determine differences among individuals given equal opportunity.

Lewontin states that "one can find a number of faults with Jensen's treatment of the published data" pertaining to the heritability of IQ. I assume they are not very important faults, if existent at all, or Lewontin surely would have enumerated them. (A number of highly qualified geneticists have reviewed my treatment of quantitative genetics in the article and have found no faults with it.) I point out that heritability estimates for IQ range between about 0.6 and 0.9. Lewontin thinks I prefer the "higher" estimate of 0.8. I don't prefer it; I simply find that 0.8 turns out to be the average heritability value based on all the data which has been reported in the literature, and I have made a most thorough survey. Surely no one at all familiar with the relevant literature could reasonably argue that the evidence leads to conclusions significantly at variance with those in my article: that heredity is about twice as important as environment in accounting for IQ differences in the populations on which the heritability of IQ has been investigated.

The main thrust of Lewontin's argument, as he sees it, actually attacks only a straw man set up by himself: the notion that heritability of a trait within a population does not prove that genetic factors are involved in the mean difference between two different populations on the same trait. I agree. But nowhere in my "Harvard Educational Review" discussion of race differences do I propose this line of reasoning, nor have I done so in any other writings. I do, however, discuss many other lines of evidence which I believe are more consistent with a hypothesis that genetic factors are involved in the average Negro-white IQ differences than with purely environmental theories.

But let us further consider Lewontin's statement that heritability (i.e., proportion of variance attributable to genetic factors) within populations is irrelevant to the question of genetic differences between populations. Theoretically, this is true: it is possible to have genetic differences within populations and no genetic differences between populations which differ phenotypically; conversely, it is possible to have zero heritability within populations and complete genetic determination of the mean difference between populations. Therefore, heritability coefficients obtained within populations, no matter how high, cannot prove the existence of a genetic difference between populations. All this follows strictly from the quantitative logic of estimating heritability, and Lewontin has given some good concrete examples of this logic in the case of plant physiology. But it is necessary to distinguish between the possible and the probable, and between proof in the sense of mathematical tautology and the probabilistic statements that result from hypothesis testing in empirical science. The real question is not whether a heritability estimate, by its mathematical logic, can prove the existence of a genetic difference between two groups, but whether there is any probabilistic connection between the magnitude of the heritability and the magnitude of group differences. Given two populations (A and B) whose means on a particular characteristic differ by x amount, and given the heritability (h_A^2 and h_B^2) of the characteristic in each of the two populations, the probability that the two populations differ from one another genotypically as well as phenotypically is some monotonically increasing function of the magnitudes of h_A^2 and h_B^2. Such probabilistic statements are commonplace in all branches of science. It seems that only when we approach the question of genetic race differences do some scientists talk as though only one of two probability values is possible, either 0 or 1. The possibility for scientific advancement in any field would be in a sorry state if this restriction were a universal rule. Would Lewontin maintain, for example, that there would be no difference in the probability that two groups differ genetically where h^2 for the trait in question is 0.9 in each group as against the case where h^2 is 0.1? Pygmies average under five feet in height; the Watusis average over six feet. The fact that the heritability of physical stature is close to 0.9 does not prove that all the difference is not caused by environmental factors, but it is more probable that genetic factors may be involved in the difference than would be the probability in the case of a group difference in the amount of scarification (body markings) which very likely has a heritability close to zero. Since pygmies and Watusis live in very different environments, why should we not bet on the proposition that their difference in mean height is attributable entirely to environment? In short, the high heritability of height suggests a reasonable hypothesis. We would then look for other lines of evidence to test the hypothesis—for example, comparing the heights of pygmy orphans from birth in the Watusis tribe and vice versa; of pygmies and Watusis living in highly similar environments and eating the same foods; of the offspring of pygmy and Watusis matings, and so on. We can proceed similarly in studying group differences in behavioral characteristics. Within-group heritability estimates thus can give us probabilistic clues as to which characteristics are most likely to show genetic differences between groups when investigated through all other available lines of evidence. If a genetic hypothesis of Negro-white differences in intelligence is not plausible to Lewontin, he does not tell us why, nor does he offer a more plausible hypothesis. Lewontin merely shows his bias when he repeatedly says I am "wrong" and "in error," instead of saying why he disagrees with the tenability of the hypothesis I have proposed to account for the data.

Negro and Indian

The comparison I drew between the Negro and American Indian children in IQ and scholastic performance was perfectly valid. It shows that despite greater environmental disadvantage, as assessed by 12 different indices, the Indian children, on the average, exceeded the Negro in IQ and achievement. But I did not pick the environmental indices. The sociologists picked them. They are those environmental factors most often cited by social scientists as the cause of the Negroes' poor performance on IQ tests and in school work. Does not the fact that another group which rates even lower than the Negro on these environmental indices (Indians are as far below Negroes as Negroes are below whites), yet displays better intellectual performance, bring into question the major importance attributed to these environmental factors by sociologists? Or should we grant immunity from empirical tests to sociological theories when they are devised to explain racial differences?

There is an understandable reluctance to come to grips scientifically with the problem of race differences in intelligence—to come to grips with it, that is to say, in the same way that scientists would approach the investigation of any other phenomenon. This reluctance is manifested in a variety of "symptoms" found in most writings and discussions of the psychology of race differences, particularly differences in mental ability. These include a tendency to remain on the remotest fringes of the subject; to sidestep central questions; to blur the issues and tolerate a degree of vagueness in definitions, concepts and inferences that would be unseemly in any other realm of scientific discourse. The writings express an unwarranted degree of skepticism about reasonably well-established quantitative methods and measurements. They deny or belittle already generally accepted facts —accepted, that is, when brought to bear on inferences outside the realm of race differences—and demand practically impossible criteria of certainty before even seriously proposing or investigating genetic hypotheses, as contrasted with extremely uncritical attitudes toward purely environmental hypotheses. There is a failure to distinguish clearly between scientifically answerable aspects of the question and the moral, political, and social policy issues; a tendency to beat dead horses and to set up straw men on what is represented as the genetic side of the argument. We see appeals to the notion that the topic is either really too unimportant to be worthy of scientific curiosity or too complex, or too difficult, or that it is forever impossible for any kind of research to be feasible, or that answers to key questions are fundamentally "unknowable" in any scientifically acceptable sense. Finally, there is complete denial of intelligence and race as realities, or as quantifiable attributes, or as variables capable of being related to one another and there follows, ostrichlike, dismissal of the subject altogether.

These tendencies will be increasingly overcome the more widely and openly the subject is discussed among scientists and scholars. As some of the taboos against the public discussion of the topic fall away, the issues will become clarified on a rational basis. We will come to know better just what we do and do not yet know about the subject, and we will be in a better position to deal with it objectively and constructively. I believe my article has made a substantial contribution toward this goal. It has provoked serious thought and discussion among leaders in genetics, psychology, sociology and education concerned with these important fundamental issues and their implications for public education. I expect that my work will stimulate further relevant research as well as efforts to apply the knowledge gained thereby to educationally and socially beneficial purposes.

In my view, society will benefit most if scientists and educators treat these problems in the spirit of scientific inquiry rather than as a battlefield upon which one or another preordained ideology may seemingly triumph.

39   Race and Mental Ability

Richard A. Goldsby

Those who favor an explanation that considers heredity the sole or an important factor in determining the difference in the average IQ scores of Blacks and Whites offer these arguments in support of their point of view:

1. The average distribution of IQ scores in American Blacks is not only lower than that of Whites but also lower than that of an even more economically disadvantaged minority, the American Indian. Furthermore on tests of scholastic achievement that are known to correlate with IQ tests, Blacks receive lower scores than Whites, Oriental Americans, Mexican Americans, Indian Americans, and Puerto Ricans.
2. When American Black and White children are grouped into five socioeconomic levels ranging from high to low and their average IQs are compared, Black children average lower than White children in their own or even lower socioeconomic levels.
3. The strongly heritable nature of IQ.

Our imperfect understanding of the factors operating to produce the observed average IQ differences between the White and Black populations is highlighted by the paradox of the foregoing arguments. All of them are true, yet individually or collectively they cannot demonstrate that genetics is to any significant degree responsible for the average differences in the IQs of Black and White populations. This is because the observations mentioned here were not made under conditions where the only difference between the populations tested was race. This is a consequence of the fact that these populations, the Black and the White, differ not only in race but in numerous aspects of their present and past environments. It is well known that historically and currently the American White and the American Black populations are divided one from the other. By and large these populations learn in different schools, from different teachers, and with different classmates. They live in different neighbourhoods, earn different salaries and, we are beginning to realize, think different thoughts and honor different values. Because of the sharp separation between Blacks and Whites or even Blacks and other minority groups, the comparative performance of other disadvantaged minority groups on IQ tests is not very helpful in interpreting the average difference in Black and White IQ scores.

It is certainly true that the Puerto Rican, the Mexican American, and most certainly the American Indian have been the victims of disadvantage and discrimination. These populations share an equal or, in the case of the Indian, an even greater state of economic disadvantage with the Black. Why then do these groups not provide the comparisons necessary to adjust for whatever cultural and economic disadvantages exist between White and Black populations? Ignoring for the moment the fact that none of these groups shares a slave heritage with the Black, we cannot overlook the profound difference between them and the Blacks in school environments. A glance at Table 1 demonstrates that other minority groups attended schools with predominantly White student bodies and faculties. Interestingly, some studies show that Blacks enrolled in predominantly White schools, like Indians and other minority groups, score higher on achievement tests than Blacks in predominantly Black schools. When one recalls the use of White reference populations in the construction and standardization of IQ tests, the difference in the school experience that most Blacks have had and the one that the other minorities have had is an environmental factor one can hardly ignore.

Comparisons of Black with White children of apparently similar socioeconomic level also include environmental variables that often go unrecognized. The living patterns of Whites are much more stratified according to socioeconomic level than those of Blacks. Given the more homogeneous composition of White neighborhoods and schools, the White child often finds that his associates and schoolmates are from a socioeconomic range similar to his own. Not so with the Black child of

Table 1. Racial Composition of Schools Attended by Various Groups.

	Nationwide (percentage of indicated population in schools with indicated characteristics)					
	Mexican American	Puerto Rican	American Indian	Oriental American	American Black	American White
Elementary School						
Mostly White students	59	52	66	63	19	89
All White teachers	75	68	77	74	53	88
Secondary School						
Mostly White students	72	56	72	57	10	91
All White teachers	73	57	75	57	25	89

	Urban South (percentage of indicated population in schools with indicated characteristics)	
	American Black	American White
Elementary School		
Mostly White students	7	91
All White teachers	49	89
Secondary Schools		
Mostly White students	4	95
All White teachers	3	92

From "Equality of Educational Opportunity," a report to the United States Department of Health, Education and Welfare, by James F. Coleman, et al., 1966.

higher socioeconomic status. Because of the traditionally restricted pool of housing available for Blacks, the neighborhoods and the schools that serve them tend to be much more socioeconomically mixed. Remembering that the majority of the Black population is of a low socioeconomic level, it can be predicted that higher socioeconomic level Blacks, as opposed to Whites, will more often find their classmates and associates are of lower socioeconomic level. Finally, we need to inquire into the uniformity of socioeconomic level in Black as opposed to White families. Is it not likely that the Black of higher socioeconomic level is the only one in the family to have "made it"? On the other hand, is it not more usual for Whites, even of low socioeconomic level, to be able to point to a relative of high socioeconomic level than it is for Blacks? Clearly, one does not adequately match the socioeconomic level of Blacks and Whites by merely matching incomes and job titles. The problem has many more levels than these two surface parameters.

This brings us to the consideration that IQ, and by extension, intelligence, has a very high component of heritability (0.80). However, height, which has an even higher heritability (0.95), can be influenced by environmental factors. It is known than an environmental factor such as diet can affect the distribution of height in whole populations. An increase in average height is seen when one compares Japanese born since World War II with those born prior to the war. The essential cause has not been a change in the genetic make-up of the Japanese population, but rather an improvement in the quality of a crucial environmental factor, diet, that has allowed the population to approach its full genetic potential. Studies on identical twins has demonstrated that environmental factors, like schooling, have a significant effect on IQ. Considering the environmental plasticity of the IQ and the sharp qualitative and quantitative environmental differences in the experience of Black and White populations, the fifteen-point difference in their average IQs is clearly within the reach of so multidimensional a variable as environment. Indeed, considering the nature and intensity of the environmental differences that have existed between the Black and White populations of the United States, Dr. S. L. Washburn said in his presidential address to the American Anthropological Association:

> I am sometimes surprised to hear it stated that if Negroes were given an equal opportunity, their IQ would be the same as the Whites. If one looks at the degree of social discrimination against Negroes and their lack of education, and also takes into account the tremendous overlapping between the observed IQs of both, one can make an equally good case that, given a comparable chance to that of the White, their IQs might be higher.

Although there are no data available to refute or justify such a conclusion, the statement makes the point forcefully that these two populations experience markedly different environments.

40   Genes and People

Curt Stern

A skeptical attitude in regard to the possible involvement of genes in mental illness is largely based on a still lingering fear that proof of genic causation may interfere with attempts at nongenic prevention or healing. A similar unjustified fear also distorts discussions of the variations in intellectual performance among individuals in populations at large and of subpopulations. It is, of course, a fact that people vary greatly in the scores which they achieve in the so-called intelligence tests, be it a general IQ determination or a battery of tests for presumably independent components of intelligent performance. It is equally true that the mean scores obtained by children from different socioeconomic layers are not the same but are positively correlated with the level of their group. It is also true that the IQ of an individual is not a fixed quantity but can decrease or increase parallel with environmental deprivation or enrichment. It follows that the better environment which many homes of the upper socioeconomic groups provide for the intellectual stimulation and motivation of their children must lead to better average test performance as compared to that of children from lower layers. It does not follow, however, that the differences observed between individuals and between subpopulations are due exclusively to nongenetic circumstances. On the contrary, there has long been evidence from a variety of approaches for genetic components in the variability of intelligence. This is particularly clear in some specific instances of low mental performance. Take, for instance, the rare but widely known disease "PKU," phenylketonuria. This mental deficiency syndrome is the result of a single recessive gene. Its biochemical mode of action is known. Persons homozygous for it are unable to synthesize a specific liver enzyme which in normal individuals prevents accumulation of a specific amino acid. Lack of the enzyme results in a high level of the substance in the blood and this in turn causes damage to brain function.

PKU is present in approximately 1 per cent of the patients in institutions for severe mental defectives. What accounts for the 99 per cent majority? May we expect many other single genes to be responsible for mental defect, each one in its different way disrupting the normal biochemical machinery so as to result in damage to the brain? We have only partial answers to these questions. Some mental defect in addition to PKU is indeed caused by one or another of a whole array of rare single genes. Other severe mental defects are, as in Down's syndrome, the consequences of chromosomal imbalance. Still other individuals with mental defect, particularly those with relatively small impediments, almost certainly do not differ from normal persons in a single major gene. Rather, normal brain function depends on the collaboration of many genes each of which may exist in a variety of types.

Some will act toward high degrees of functioning, others will tend toward interference with normal function, and still others may be relatively neutral in their effect. The average person may have many such neutral genes or a more or less equal number of genes with positive and negative effect on mental function. If the shuffling of the genes in the egg and sperm which enter into the conception of a child happens to assign to it a majority of genes with negative effects subnormality may result.

There are thus single and multiple gene defectives. Still other defectives owe their status to external injury suffered before birth, during birth, or later. We do not know precisely what fractions of the mentally deficient belong to the different classes of causation, but it seems that the sum of the various genetic types far outweighs the nongenetic ones. This statement applies particularly to the medium and low grades of feeblemindedness. Undoubtedly the upper grades include a considerable number of people whose innately normal mind has been stunted by excessive cultural deprivation.

It has sometimes been suggested that there is a fundamental difference in the causation of truly low grade defect and the variability of mental performance in the range from somewhat below average to very high. The genetic nature of severe defect is granted but a genetic predisposition to more or less normal mental performance is denied. This judgment is based on the higher quality of direct evidence for a genetic basis of severe defect as compared to that for normal variability. The difference in the quality of the evidence, however, depends on the ease with which it can be obtained rather than on a basic difference in causation. It is intrinsically easier to recognize single gene substitutions, as in PKU, which block normal biochemical pathways so effectively as to lead to disaster, than to unravel a whole constellation of genes whose joint effects result in a performance potential somewhat below or above normal.

Research with experimental organisms demonstrates that all genes exist in a variety of forms which can be ordered in their effectiveness concerning a specific chemical reaction. Not only will one gene variety control the production of a highly efficient enzyme and another variety lead to complete absence of the enzyme but further varieties will make enzymes with a wide array of intermediate and even negative efficiencies. It would be strange indeed if the genetic control of brain function in man were fundamentally different from the genetic control of innumerable functions in other organisms, including brain function in nonhuman mammals.

It is not necessary, however, to argue from analogy only. We are now acquainted with a number of phenomena which provide unequivocal evidence for genetic determination of mental performance within the more or less normal range. For one, deviations from the normal sex-chromosome number result in some lowering of mental function. Many persons with such unbalanced chromosomal constitutions are normal and only statistical comparisons show that on the average they are somewhat inferior in mental ability. Among these kinds of individuals are women with only one X-chromosome and women with three X-chromosomes, as well as men with XXY constitution. It would have been difficult to prove the genetic basis of some of these slight impairments of mental performance had they not occurred as by-products of readily recognizable chromosome aberrations. The discovery of the genetic basis under these circumstances justifies the conviction that other genotypes more difficult to analyze are involved in similar slight variations in mental function.

We know of no chromosome aberration responsible for increased mental performance but very recently a suggestive relation has been reported between a genetically conditioned abnormally high production of a hormone and higher than average IQ-test intelligence. A rare disease, the adrenogenital syndrome, results from a recessive gene. This induces a metabolic error inhibiting the production of cortisone. As a result the adrenals produce excessive amounts of male sex hormone. In a sample of seventy patients affected with the syndrome a variety of intelligence tests yielded a mean ten points above the average. Moreover, 60 per cent of the individuals scored above 110 instead of the expected 25 per cent. It seems that the high mean performance is causally connected with the special genetically controlled hormonal situation but, it must be stressed, further evidence is required.

The existence of genetic factors influencing test intelligence is also proven by the performance of children from consanguineous marriages as compared to children from unrelated parents. Both in Sweden and in Japan where relevant studies have been made, formal mental tests showed a small but significant depression in the average performance by the inbred children. Where different degrees of inbreeding could be distinguished, increased inbreeding led to increasingly poorer ratings. Inbred children also performed slightly worse at school. These inbreeding effects were the residue of greater differences after socioeconomic factors had been taken into account. The data demonstrate the existence of genes which influence mental performance in the normal range and which have been made homozygous as a consequence of consanguinity.

Test intelligence has long been the object of family studies and of investigations of twins both reared together and apart. Again the results fit genetic interpretations, after giving due weight to the presence of clearly apparent, powerful environmental components. In these types of studies, however, proof of the presence of genetic components is less unassailable than in those reported above.

It has often been stressed that test intelligence is strongly influenced by personality traits, such as drive and motivation, and that such traits are dependent on the cultural milieu which varies among families and among various subpopulations, such as ethnic groups and socioeconomic classes. Simultaneously it is frequently implied that individual and mean differences in personality traits are the results of nongenetic factors. Recent studies, however, show this not to be true. On the average, genetically identical twin partners score much more similarly in batteries of personality tests than nonidentical partners, and observations on behavior of twin infants agree with the test results. This points to genetic components in the variance of personality traits. As in intelligence scores the genetic basis of personality variance consists presumably of a multitude of genes and is therefore not easily broken down into individual genetic units. There are, however, some special types of personality in which specific genetic influences are discernible. Thus, the extra chromosome in Down's syndrome is responsible not only for physical abnormalities and mental defect but typically provides its bearers with cheerful and friendly per-

411

sonalities. In contrast, the gene for Huntington's chorea seems often to lead to personality traits of less desirable character. Perhaps some cases of the specific personality trait "aggressiveness" represent an unusually striking example of genetic co-determination. If preliminary findings should be confirmed, inmates in institutions for particularly aggressive mental defectives seem to include a significantly higher proportion of males with two Y-chromosomes than found in the general population and in less aggressive mental defectives. These observations suggest that the double-Y condition predisposes to aggressiveness. It is tempting to extrapolate from here and to speculate that the female sex owes its gentleness to the absence of a Y-chromosome and the normal male his moderate, socially approved aggressiveness to his single-Y!

The nature-nurture discussion of mental attributes has been called futile. It is futile only if in compassion for our underprivileged fellow men we close our minds to the facts of life. It seems to me that Galton's dictum of 1869 still stands. " I have no patience with the hypothesis occasionally expressed, and often implied, especially in tales written to teach children to be good, that babies are born pretty much alike, and that the whole agencies in creating differences between boy and boy, and man and man, are steady application and moral effect. It is in the most unqualified manner that I object to pretentions of natural equality."

A genetic interpretation of the fact that the mean test intelligence of different socioeconomic layers varies is primarily based on studies of adopted and of orphanage children. On the average, children born in the upper social layers score considerably higher than those of the lower layers, but children adopted into the upper groups score only moderately higher than those adopted into the lower groups. Apparently the children do not experience only the positive or negative influences of their homes on test intelligence but also to some measure the positively or negatively acting genes of their parental classes. On the other hand, the adopted children are genetically not correlated with their adoptive parents and experience only environmental differences.

Limitations of the power of nongenetic factors are also apparent from a study of IQ scores of children raised from infancy in an orphanage under relatively uniform conditions. In spite of the nongenetic uniformity, the children of parents from upper occupational groups on the average scored higher than those from lower groups. Methodologically none of these studies on adopted and orphanage children are beyond criticism but in the aggregate they leave little doubt as to the existence of genetic differences in the mean endowments of different socioeconomic subpopulations.

The inequality of the mean test performance of members of different socioeconomic layers is important in several ways. Its genetic aspects are related to the differential fertility of the different layers. During the last hundred years in many different countries the average number of children per family was low in the top layers and increasingly higher the more one descended in the social scale. The consequence was that proportionally more of the people comprising a later generation were offspring from parents of the lower than the higher groups. This would be of no genetic consequence if the mean endowments of the various groups were alike. Since, however, they are not, the inverse relation between fertility and test intelligence of the different layers would, so it seemed to many, result in "gene erosion" of the population.

Gene erosion is one of the possible changes in genic content in successive generations of groups of mankind. Evolutionarily, there has been a striking rise in intelligence endowment between the times when there were only prehuman primate ancestors and the emergence of *Homo sapiens*. What has happened since the appearance of modern man, specifically in the last 50,000 years, we do not know. Has the rise been continued, though perhaps at a lesser pace, or has it levelled off completely? Could the genetic endowment even have decreased? What may be the trends for the next hundred years, or the next thousand? Few attempts have been made to get direct information on these important matters. The most extensive study was the Scottish Survey of the test intelligence of two complete cohorts of Scottish eleven-year-old school children, one consisting of those born in 1921, the other of those born in 1936. The advocates of the gene erosion theory who predicted a fall in performance during the fifteen-year interval were not sustained by the results. The mean scores of the second group, tested in 1947, were significantly higher than those of the first group tested in 1932. Did then the Scottish national intelligence rise intrinsically? A very careful analysis of the data did not lead to a decision. We do not know whether the genetic endowment had improved or whether an unchanged endowment resulted in better performance due possibly to such nongenetic factors as improved educational background, test sophistication, accelerated physical and mental maturation of the children, or still other environmental forces. The results do not even exclude the possibility that the better performance in 1947 was due to a combination of decreased endowment and increased environmental quality.

During the last decades the gaps between the fertilities of different socioeconomic layers have diminished, partly due to the spread of contraception to the lower layers which earlier had practiced it least, partly due to volitional raising of larger families in upper layers. Moreover, data on the fertility and mean test performance of members of different population layers give only a coarse and possibly misleading measure of the situation. Within each layer there is a wide and greatly overlapping range of endowments. It is very likely that fertility differentials exist between differently endowed persons within the same layer, and it is conceivable that in part the differentials are in opposite direction to those between layers. Some recent studies, limited in extent but nonetheless highly noteworthy, have been concerned with the fertility of individuals classified by intelligence test performance independently of their socioeconomic status. By using data on the completed fertility of women whose IQ scores were available from the time they had been at school it has become apparent that not only the fertility of the poorest test performers is high but also that of the best performers. In other words, the relation is not inverse throughout but shows a bimodal distribution. Calculations indicate that in these sample populations the mean IQ range holds its own reproductively. Much more information is needed to judge the situation in general in order to know what happens to our genes in the course of generations. It is unlikely that a permanent

413

trend exists. A continuous watch for fertility changes is necessary, just as general census data have to be collected on a continuing basis. It may be true that at the present time no striking change in the pool of our intelligence genes is taking place, and this may remain so in the foreseeable future. Even then the question is bound to come up whether society should be acquiescent in the absence of gene erosion or should institute measures of improving its endowment.

I have not yet touched on the question whether different racial groups are differently endowed in intelligence. The answer is: we do not know. Our tests are worthless if used to compare genetic components of groups in which social and historical differences alone account for an overwhelming part of differences in performance. This state of affairs is neither compatible with pronouncements alleging inferiority of some races nor with pronouncements postulating strict equality. Examples of the former are well known. One of the latter may be quoted here from an impressive study, by the Office of Policy Planning and Research of a United States federal agency, on "The Negro Family." The study demonstrates convincingly certain social nongenetic circumstances which act to the detriment of Negro youth in the United States in the performance of mental tests. And it then extrapolates as follows: "There is absolutely no question of any genetic differential. Intelligence potential is distributed among Negro infants in the same proportion and pattern as among Icelanders or Chinese or any other group." Such statements lack a factual basis. Indeed it seems very improbable that the genetic endowments of groups which have been relatively isolated from one another for thousands of years are exactly alike. The prediction, however, that two racial groups will be found to be different does not include a specific guess concerning which of the groups is endowed more highly than the other with reference to a given quality.

Genes and people! We are different in the genes we received at conception, different in health and disease, intelligence endowment and personality traits. Like all other organisms we belong to a polymorphic species. We must educate ourselves to accept the many-faceted inequalities of man. We must not forget also that an individual person who ranks high in some respects may rank average or below in others. The same multidimensional aspects would apply to subpopulations such as socioeconomic layers or racial groups. If there are somewhat more persons in Group A endowed with one desirable quality, there may be more persons in Group B endowed with a different desirable one.

There is little new in what I have told you about the genetic diversity of people. To paraphrase Goethe: Who can say something clever or something stupid that has not been said before? Why then did I choose to speak to you as I did? Because truth and the search for it can be suppressed not only by ill will but also by good will. In our justified fear of mankind's misuse of its powers the cry for a moratorium on research has often been sounded. Those who plead for such a moratorium do not make entries on both sides of the balance sheet. From the discovery of fire to that of nuclear power the destructive consequences have been accompanied by beneficial ones. It is the moral tragedy of man that though the extinction of one life cannot be compensated by the preservation of even many lives he cannot avoid making choices. Moreover, it is impossible to wait with new explorations until man is a more moral being. It is not likely that he will soon attain the status of an angel, either by changing his basic nature or by being born into a perfect society. Rather, for a long time the inhuman nature in each of us will have to be overcome by slow individual effort. While this is the case, can we deny man the benefits of possible new discoveries on his genetic variability because there is also the possibility of harm arising from such discoveries?

We must learn to realize that changes in our inequalities are going on incessantly, often independent of our conscious actions and dependent on the social system under which we live. As facts become known we or our children must become willing to use them in our planning for the future. There is no urgency about this. For a long time the immense gene pools of human populations will include a reservoir of genes from which selection in any desired direction will be possible.

I have abstained from proposing an eugenic program. The need for more knowledge, the danger of rash political action, and the slowness with which genetic changes impinge on large populations justify a waiting attitude. Nevertheless, responsible thinking on problems of eugenics should be encouraged. Past errors of proponents of eugenic measures and the crimes committed under the pretense of eugenics should not stand in the way of new approaches. Culture and social organization are not the ultimate forces which form us. They themselves are made possible by our genes.

Some References

Bajema, C. Relation of fertility to educational attainment in a Kalamazoo public school population: A follow-up study. *Eugen. Quart.*, **13** : 306, 1966.

Book, J. A. Genetical investigations in a north Swedish population: the offspring of first-cousin marriages. *Ann. Hum. Genet.*, **21** : 191–221, 1957.

Higgins, J., E. Reed, and S. Reed. Intelligence and family size: A paradox resolved. *Eugen. Quart.*, **9** : 84, 1962.

Money, J., and V. Lewis. IQ, genetics and accelerated growth: Adrenogenital syndrome. *Johns Hopkins Hosp. Bull.*, **118(5)** : 365, 1966.

Schull, W. J., and J. V. Neel. *The Effects of Inbreeding on Japanese Children*. New York: Harper & Row, Inc., 1965.

Scottish Council for Research in Education. *Social Implications of the 1947 Scottish Mental Survey*. London: University of London Press, 1953.

Thompson, G. H. *The Trend of Scottish Intelligence*. London: University of London Press, 1949.

nine Gene Therapy and the Future

Wallmeyer cartoon reprinted by permission of The Register and Tribune Syndicate and the Long Beach Independent, Press-Telegram.

"A species which cannot as yet even control its own sheer numbers is obviously not likely to control its own genetic constitution" (Davis, 1966). Nevertheless, the topic of gene therapy engages the minds of the geneticist and science fiction writer alike (Freedman, 1974; Herbert, 1966)—not to mention sociologists and journalists.

Etzioni provides us with a scenario involving the slightest of soon-to-be-possible genetic changes, a shift in the sex ratio. And who would object to the chance to choose the sex of their children? But, as Etzioni points out, the ramifications of this choice on our social structure would be spectacular. One saving grace of the sex choice of one's children, however, which Etzioni does not mention, would be that if parents chose to avoid having sons the frequency of sex-linked recessive defects would decline—a eugenic boon. But several studies have found that if sex choice were possible, the number of males would increase, especially among the first born (Westoff and Rindfuss, 1974; Largey, 1972; Timson, 1974). Other reports bearing on sex selection include Bennett and Boyse (1973), Erickson et al. (1973), and Rhine et al. (1975). Attempted subversion of amniocentesis and karyotyping for sex selection has been reported (Stenchever, 1972).

The paper by Crow in Section Seven introduced the theme of "spicing up" human evolution by genetic engineering—a topic that dominates the subsequent selections in this section.

Is DNA the cure-all drug? Sinsheimer in his first paper here, published in 1969, provides us with a portfolio of rather complicated tricks to cure a disabling disease, diabetes, by one-shot gene therapy rather than the conventional way, by insulin treatment. Recent ideas on genetic improvement of the human species, such as those of Sinsheimer, were foreshadowed by the ideas of Galton (see Dunn's paper in Section One). To be sure, few geneticists have the tolerable optimism about gene therapy that Sinsheimer evinced in 1969, either in terms of efficacy or social approval. Among the pessimists are Davis (1970) and Fuhrmann (this section). Background for experiments discussed by Fuhrmann is given in Rabovsky (1971). As Murray Kempton and others have pointed out, an age of science is not necessarily an age of reason (but see Friedmann and Roblin, 1972).

What, then, of the ultimate in gene therapy, the deliberate fabrication of a genetically "desirable" human being? There are several methodologies for accomplishing, or at least approximating, such a person. The first requires little technology according to the ideas of H. J. Muller (1965); it is simply a free choice of sperm for artificial insemination of women from a frozen sperm bank to which select donors contribute and get listed in some "Who's Who Among Sperm Donors." To some, such facilities are anathema. To others, including Muller, the important thing is to make use of this opportunity to improve the human species. The second methodology is that of cloning (making numerous identical twins) of some desirable existing genotype (Lederberg, 1966; Watson, 1971). The secularity on which Lederberg's cloning vision rests has been indignantly challenged by Ramsey (1970). Ramsey says that no one should "be frightened out of his ethical wits by grand eugenic designs."

419

In 1974 reports occurred in the public press that human eggs had been fertilized *in vitro*, implanted in the uterus of women with defective oviducts, and subsequently developed into newborn infants (see Anon., 1974). The rationale given was permitting women who were sterile because of oviduct defects to have wanted children. But many people were alarmed at this preliminary step toward "test-tube babies." In 1976 a report was further made on this method of initiating pregnancy (Steptoe and Edwards, 1976).

The foundations for these developments come from the work on human embryogenesis of Edwards and his co-workers (Edwards, 1974; Edwards and Sharpe, 1971). It has been discussed pro and con by a variety of commentators (Austin, 1973; Kass, 1971; Marx, 1973). If it ever leads to egg-donation catalogs with "prenatal adoption," it will be the female counterpart of Muller's sperm-choice idea.

A more recent development in the possibilities for genetic engineering has caused another furor (Cohen, 1975; Wade, 1975; Anon., 1975a). Enzymes are now available that can snip DNA in gene-size pieces. The pieces can then be joined experimentally to a chromosome in a cell of another species. Thus, DNA pieces from frogs, viruses, and other microorganisms have been inserted into cells of the gut bacterium *Escherichia coli*. If gene insertion to rectify human mutational defects could be accomplished, it would be a strong stimulus to gene-based medical therapy. However, this research so far has been confined to bacterial cells and the danger hotly debated is not "shotgun" genetic engineering but the health hazard of DNA-manipulated gut bacteria escaping from the laboratory into the public domain and causing some bizarre epidemic. Another acknowledged hazard of this research is the possible engineering of novel organisms for biological warfare. These recognized dangers have prompted the researchers involved to self-restraint in experimentation in this area. Senator Edward Kennedy and technicians at a French research institute (Anon., 1975b), among others, have expressed doubts about the effectiveness of self-restraint, however. The 1975 paper by Sinsheimer in this section cautiously examines the issues and people involved in the decision making on DNA manipulation (see also Berg et al., 1975; Rogers, 1975; Marx, 1976).

Bibliography

Anonymous. Amber light for genetic manipulation. *Nature,* **253** : 295, 1975a.

Anonymous. Asilomar and the Pasteur Institute. *Nature,* **256** : 5, 1975b.

Anonymous. The baby maker. *Time Magazine,* July 29, 1974, pp. 58–59.

Ausabel, F., J. Beckwith, and K. Janssen. The politics of genetic engineering: Who decides who's defective? *Psychology Today,* June 1974, pp. 30–43.

Austin, C. R. Embryo transfer and sensitivity to teratogenesis. *Nature,* **244** : 333–334, 1973. Discusses the hazards of *in vitro* fertilization for human life.

Batt, J. They shoot horses, don't they?: An essay on the scotoma of one-eyed kings. *UCLA Law Review,* **15** : 510–550, 1968.

Bennett, D., and E. Boyse. Sex ratio in progeny of mice inseminated with serum treated with H–Y antiserum. *Nature,* **246** : 308–309, 1973. Mouse sex ratio can be somewhat controlled by antibody treatment of sperm.

Berg, P., D. Baltimore, S. Brenner, R. O. Roblin, and M. F. Singer. Asilomar conference on recombinant DNA molecules. *Science,* **188** : 991–994, 1975.

Berns, M. W. Directed chromosome loss by laser micro-irradiation. *Science,* **186** : 700–705, 1974. This technique may permit chromosome engineering.

Cohen, S. N. The manipulation of genes. *Sci. Amer.,* **233** : 25–33, 1975.

Davis, B. D. Prospects for genetic intervention in man. *Science,* **170** : 1279–1283, 1970.

Davis, K. Sociological aspects of genetic control. In *Genetics and the Future of Man,* J. D. Roslansky, ed. New York: Appleton, 1966, pp. 173–204.

Djerassi, C. Probabilities and practicalities. *Bull. Atomic Sci.,* **28** : 25–28, 1972. Part of a symposium on genetic engineering.

Edwards, R. G. Fertilization of human eggs *in vitro*: Morals, ethics, and the law. *Quart. Rev. Biol.,* **49** : 3–26, 1974.

Edwards, R. G., and D. J. Sharpe. Social values and research in human embryology. *Nature,* **231** : 87–91, 1971.

Ericsson, R. J., C. N. Langevin, and M. Nishino. Isolation of fractions rich in human Y sperm. *Nature,* **246** : 421–424, 1973.

Freedman, N. *Joshua, Son of None.* New York: Dell, 1974. Science fiction about a clonal man.

Friedmann, T., and R. Roblin. Gene therapy for human genetic disease? *Science,* **175** : 949–955, 1972.

Glass, B. Eugenic implications of the new reproductive

technologies. *Soc. Biol.*, **19** : 326–336, 1972. Updates H. J. Muller's ideas on genetic improvement.

Golding, M. P. Ethical issues in biological engineering. *UCLA Law Review*, **15** : 443–479, 1968.

Hamilton, M., ed. *The New Genetics and the Future of Man.* Grand Rapids, Mich.: Eerdmanns, 1972.

Herbert, F. *The Eye of Heisenberg.* New York: Berkley Medallion, 1966. *In vitro* embryogenesis and mutational repair in science fiction.

Hudock, G. A. Gene therapy and genetic engineering: Frankenstein is still a myth, but it should be reread periodically. *Indiana Law J.*, **48** : 531–580, 1973.

Jones, A., and W. F. Bodmer. *Our Future Inheritance: Choice or Chance?* New York: Oxford University Press, 1974. A nontechnical report from a British group.

Kass, L. R. Babies by means of *in vitro* fertilization: Unethical experiments on the unborn? *New Eng. J. Med.*, **285** : 1174–1179, 1971.

Kaufman, M. H., E. Huberman, and L. Sachs. Genetic control of haploid parthenogenetic development in mammalian embryos. *Nature*, **254** : 694–695, 1975. Haploid hamster embryos have developed *in vitro* for several days.

Largey, G. Sex control, sex preferences, and the future of the family. *Soc. Biol.*, **19** : 379–392, 1972.

Lederberg, J. Experimental genetics and human evolution. *Amer. Naturalist*, **100** : 519–531, 1966.

Lederberg, J. Genetic engineering and the amelioration of genetic defect. *BioScience*, **20** : 1307–1310, 1970.

Lerner, I. M. Ethics and the new biology. In *Genetics and the Quality of Life*, C. Birch and P. Abrecht, eds. New York: Pergamon, 1975, pp. 20–35.

Marx, J. L. Molecular cloning: Powerful tool for studying genes. *Science*, **191** : 1160–1162, 1976.

Marx, J. L. Out of the womb—into the test tube. *Science*, **182** : 811–814, 1973.

Muller, H. J. Man's future birthright. In *Essays on Science and Humanity*, E. A. Carlson, ed. Albany: State University of New York Press, 1973.

Rabovsky, D. Molecular biology: Gene insertion into mammalian cells. *Science*, **174** : 933–934, 1971.

Ramsey, P. *Fabricated Man.* New Haven, Conn.: Yale University Press, 1970.

Rhine, S. A., J. L. Cain, R. E. Cleary, C. G. Palmer, and J. F. Thompson. Prenatal sex detection with endocervical smears: Successful results utilizing Y-body fluorescence. *Amer. J. Obstet. Gyn.*, **122** : 155–160, 1975.

Rivers, C. Grave new world. *Saturday Review*, Apr. 8, 1972, pp. 23–27.

Rogers, M. The Pandora's box congress. *Rolling Stone,* June 19, 1975, pp. 37–82 ff.

Rogers, S. Skills for genetic engineers. *New Scientist,* Jan. 29, 1970, pp. 194–196.

Roslansky, J. D., ed. *Genetics and the Future of Man.* New York: Appleton, 1966.

Sonneborn, T. M., ed. *The Control of Human Heredity and Evolution.* New York: Macmillan, 1965.

Stenchever, M. A. An abuse of prenatal diagnosis. *J. Amer. Med. Assoc.,* **221** : 408, 1972. One of many reports of requests for abortion of a fetus of unwanted sex following amniocentesis and karyotyping for other reasons.

Steptoe, P. C., and R. G. Edwards. Reimplantation of a human embryo with subsequent tubal pregnancy. *Lancet,* **1** : 880–882, 1976.

Timson, J. The preselection of sex. *Eugenics Soc. Bull.,* **6**: 7–11, 1974.

Vogel, F. Eugenic aspects of genetic engineering. *Adv. Biosciences,* **8**: 397–406, 1972. Part of the same symposium as Fuhrmann's paper.

Wade, N. Genetics: Conference sets strict controls to replace moratorium. *Science,* **187** : 931–935, 1975.

Wade, N. Recombinant DNA. *Science,* **190** : 1175–1179, 1975, and **191** : 834–836, 1976.

Wallace, B. Man's humanity. *Sat. Rev. Sciences,* **1** : 48–49, 1973.

Watson, J. D. Moving toward the clonal man. *Atlantic Monthly,* **227** : 50–53, 1971.

Westoff, C. F., and R. R. Rindfuss. Sex preselection in the United States. *Science,* **184** : 633–636, 1974.

41   Sex Control, Science, and Society

Amitai Etzioni

Using various techniques developed as a result of fertility research, scientists are experimenting with the possibility of sex control, the ability to determine whether a newborn infant will be a male or a female. So far, they have reported considerable success in their experiments with frogs and rabbits, whereas the success of experiments with human sperm appears to be quite limited, and the few optimistic reports seem to be unconfirmed. Before this new scientific potentiality becomes a reality, several important questions must be considered. What would be the societal consequences of sex control? If they are, on balance, undesirable, can sex control be prevented without curbing the freedoms essential for scientific work? The scientific ethics already impose some restraints on research to safeguard the welfare and privacy of the researched population. Sex control, however, might affect the whole society. Are there any circumstances under which the societal well-being justifies some limitation on the freedom of research? These questions apply, of course, to many other areas of scientific inquiry, such as work on the biological code and the experimental use of behavior and thought-modifying drugs. Sex control provides a useful opportunity for discussion of these issues because it presents a relatively "low-key" problem. Success seems fairly remote, and, as we shall see, the deleterious effects of widespread sex control would probably not be very great. Before dealing with the possible societal effects of sex control, and the ways they may be curbed, I describe briefly the work that has already been done in this area.

The State of the Art

Differential centrifugation provided one major approach to sex control. It was supposed that since X and Y chromosomes differ in size (Y is considerably smaller), the sperm carrying the two different types would also be of two different weights; the Y-carrying sperm would be smaller and lighter, and the X-carrying sperm would be larger and heavier. Thus, the two kinds could be separated by centrifugation and then be used in artificial insemination. Early experiments, however, did not bear out this theory. And, Witschi pointed out that, in all likelihood, the force to be used in centrifugation would have to be of such magnitude that the sperm might well be damaged (1).

In the 1950's a Swedish investigator, Lindahl (2), published accounts of his results with the use of counter-streaming techniques of centrifugation. He found that by using the more readily sedimenting portion of bull spermatozoa that had undergone centrifugation, fertility was decreased but the number of male calves among the offspring was relatively high. His conclusion was that the female-

From *Science*, **161** (Sept. 13, 1968), 1107–1112. Copyright 1968 by the American Association for the Advancement of Science. Reprinted with permission of the author and the publisher.

determining spermatozoa are more sensitive than the male and are damaged due to mechanical stress in the centrifuging process.

Electrophoresis of spermatozoa is reported to have been successfully carried out by a Soviet biochemist, V. N. Schröder, in 1932 (3). She placed the cells in a solution in which the pH could be controlled. As the pH of the solution changed, the sperm moved with different speeds and separated into three groups: some concentrated next to the anode, some next to the cathode, and some were bunched in the middle. In tests conducted by Schröder and N. K. Kolstov (3), sperm which collected next to the anode produced six offspring, all females; those next to the cathode—four males and one female; and those which bunched in the center—two males and two females. Experiments with rabbits over the subsequent 10 years were reported as successful in controlling the sex of the offspring in 80 percent of the cases. Similar success with other mammals is reported.

At the Animal Reproduction Laboratory of Michigan State University, Gordon replicated these findings, although with a lower rate of success (4). Of 167 births studied, in 31 litters, he predicted correctly the sex of 113 offspring, for an average of 67.7 percent. Success was higher for females (62 out of 87, or 71.3 percent) than for males (51 out of 80, or 63.7 percent).

From 1932 to 1942, emphasis in sex control was on the acid-alkali method. In Germany, Unterberger reported in 1932 that in treating women with highly acidic vaginal secretions for sterility by use of alkaline douches, he had observed a high correlation between alkalinity and male offspring. Specifically, over a 10-year period, 53 out of 54 treated females are reported to have had babies, and all of the babies were male. In the one exception, the woman did not follow the doctor's prescription, Unterberger reported (5). In 1942, after repeated tests and experiments had not borne out the earlier results, interest in the acid-alkali method faded (6).

It is difficult to determine the length of time it will take to establish routine control of the sex of animals (of great interest, for instance, to cattle breeders); it is even more difficult to make such an estimate with regard to the sex control of human beings. In interviewing scientists who work on this matter, we heard conflicting reports about how close such a breakthrough was. It appeared that both optimistic and pessimistic estimates were vague—"between 7 to 15 years"—and were not based on any hard evidence but were the researchers' way of saying, "don't know" and "probably not very soon." No specific road blocks which seemed unusually difficult were cited, nor did they indicate that we have to await other developments before current obstacles can be removed. Fertility is a study area in which large funds are invested these days, and we know there is a correlation between increased investment and findings (7). Although most of the money is allocated to birth-control rather than sex-control studies, information needed for sex-control research has been in the past a by-product of the originally sponsored work. Schröder's findings, for example, were an accidental result of a fertility study she was conducting (4, p. 90). Nothing we heard from scientists working in this area would lead one to conclude that there is any specific reason we could not have sex control 5 years from now or sooner.

In addition to our uncertainty about when sex control might be possible, the question of how it would be effected is significant and also one on which there are differences of opinion. The mechanism for practicing sex control is important because certain techniques have greater psychic costs than others. We can see today, for example, that some methods of contraception are preferred by some classes of people because they involve less psychic "discomfort" for them; for example, the intrauterine device is preferred over sterilization by most women. In the same way, although electrophoresis now seems to offer a promising approach to sex control, its use would entail artificial insemination. And, whereas the objections to artificial insemination are probably decreasing, the resistance to it is still considerable (8). (Possibly, the opposition to artificial insemination would not be as great in a sex-control situation because the husband's own sperm could be used.) If drugs taken orally or douches could be relied upon, sex control would probably be much less expensive (artificial insemination requires a doctor's help), much less objectionable emotionally, and significantly more widely used.

In any event both professional forecasters of the future and leading scientists see sex control as a mass practice in the foreseeable future. Kahn and Wiener, in their discussion of the year 2000, suggest that one of the "one hundred technical innovations likely in the next thirty-three years" is the "capability to choose the sex of unborn children" (9). Muller takes a similar position about gene control in general (10).

Societal Use of Sex Control

If a simple and safe method of sex control were available, there would probably be no difficulty in finding the investors to promote it because there is a mass-market potential. The demand for the new freedom to choose seems well established. Couples have preferences on whether they want boys or girls. In many cultures boys provide an economic advantage (as workhorses) or a form of old-age insurance (where the state has not established it). Girls in many cultures are a liability; a dowry which may be a sizable economic burden must be provided to marry them off. (A working-class American who has to provide for the weddings of three or four daughters may appreciate the problem.) In other cultures, girls are profitably sold. In our own culture, prestige differences are attached to the sex of one's children, which seem to vary among ethnic groups and classes (11, pp. 6–7).

Our expectations as to what use sex control might be put in our society are not a matter of idle speculation. Findings on sex preferences are based on both direct "soft" and indirect "hard" evidence. For soft evidence, we have data on preferences parents expressed in terms of the number of boys and girls to be conceived in a hypothetical situation in which parents would have a choice in the matter. Winston studied 55 upperclassmen, recording anonymously their desire for marriage and children. Fifty-two expected to be married some day; all but one of these desired children; expectations of two or three children were common. In total, 86 boys were desired as compared to 52 girls, which amounts to a 65 percent greater demand for males than for females (12).

A second study of attitudes, this one conducted on an Indianapolis sample, in 1941, found similar preferences for boys. Here, while about half of the parents had no preferences (52.8 percent of the wives and 42.3 percent of the husbands), and whereas the wives with a preference tended to favor having about as many boys as girls (21.8 percent to 25.4 percent), many more husbands wished for boys (47.7 percent as compared to 9.9 percent) (13).

Such expressions of preference are not necessarily good indicators of actual behavior. Hence of particular interest is "hard" evidence, of what parents actually did—in the limited area of choice they already have: the sex composition of the family at the point they decided to stop having children. Many other and more powerful factors affect a couple's decision to curb further births, and the sex composition of their children is one of them. That is, if a couple has three girls and it strongly desires a boy, this is one reason it will try "once more." By comparing the number of families which had only or mainly girls and "tried once more" to those which had only or mainly boys, we gain some data as to which is considered a less desirable condition. A somewhat different line was followed in an early study. Winston studied 5466 completed families and found that there were 8329 males born alive as compared to 7434 females, which gives a sex ratio at birth of 112.0. The sex ratio of the last child, which is of course much more indicative, was 117.4 (2952 males to 2514 females). That is, significantly more families stopped having children after they had a boy than after they had a girl.

The actual preference for boys, once sex control is available, is likely to be larger than these studies suggest for the following reasons. Attitudes, especially where there is no actual choice, reflect what people believe they ought to believe in, which, in our culture, is equality of the sexes. To prefer to produce boys is lower class and discriminatory. Many middle-class parents might entertain such preferences but be either unaware of them or unwilling to express them to an interviewer, especially since at present there is no possibility of determining whether a child will be a boy or a girl.

Also, in the situations studied so far, attempts to change the sex composition of a family involved having more children than the couple wanted, and the chances of achieving the desired composition were 50 percent or lower. Thus, for instance, if parents wanted, let us say, three children including at least one boy, and they had tried three times and were blessed with girls, they would now desire a boy strongly enough to overcome whatever resistance they had to have additional children before they would try again. This is much less practical than taking a medication which is, let us say, 99.8 percent effective and having the number of children you actually want and are able to support. That is, sex control by a medication is to be expected to be significantly more widely practiced than conceiving more children and gambling on what their sex will be.

Finally, and most importantly, such decisions are not made in the abstract, but affected by the social milieu. For instance, in small *kibbutzim* many more children used to be born in October and November each year than any other months because the community used to consider it undesirable for the children to enter classes in the middle of the school year, which in Israel begins after the high holidays, in

427

October. Similarly, sex control—even if it were taboo or unpopular at first—could become quite widely practiced once it became fashionable.

In the following discussion we bend over backward by assuming that actual behavior would reveal a smaller preference than the existing data and preceding analysis would lead one to expect. We shall assume only a 7 percent difference between the number of boys and girls to be born alive due to sex control, coming on top of the 51.25 to 48.75 existing biological pattern, thus making for 54.75 boys to 45.25 girls, or a surplus of 9.5 boys out of every hundred. This would amount to a surplus of 357,234 in the United States, if sex control were practiced in a 1965-like population (14).

The extent to which such a sex imbalance will cause societal dislocations is in part a matter of the degree to which the effect will be cumulative. It is one thing to have an unbalanced baby crop one year, and quite another to produce such a crop several years in a row. Accumulation would reduce the extent to which girl shortages can be overcome by one age group raiding older and younger ones.

Some demographers seem to believe in an invisible hand (as it once was popular to expect in economics), and suggest that overproduction of boys will increase the value of girls and hence increase their production, until a balance is attained under controlled conditions which will be similar to the natural one. We need not repeat here the reasons such invisible arrangements frequently do not work; the fact is they simply cannot be relied upon, as recurrent economic crises in pre-Keynesian days or overpopulation show.

Second, one ought to note the deep-seated roots of the boy-favoring factors. Although there is no complete agreement on what these factors are, and there is little research, we do know that they are difficult and slow to change. For instance, Winston argued that mothers prefer boys as a substitute for their own fathers, out of a search for security or Freudian considerations. Fathers prefer boys because boys can more readily achieve success in our society (and in most others). Neither of these factors is likely to change rapidly if the percentage of boys born increases a few percentage points. We do not need to turn to alarmist conclusions, but we ought to consider what the societal effects of sex control might be under conditions of relatively small imbalance which, as we see it, will cause a significant (although not necessarily very high) male surplus, and a surplus which will be cumulative.

Societal Consequences

In exploring what the societal consequences may be, we again need not rely on the speculation of what such a society would be like; we have much experience and some data on societies whose sex ratio was thrown off balance by war or immigration. For example, in 1960 New York City had 343,470 more females than males, a surplus of 68,366 in the 20- to 34-age category alone (15).

We note, first, that most forms of social behavior are sex correlated, and hence that changes in sex composition are very likely to affect most aspects of social life. For instance, women read more books, see more plays, and in general consume more culture than men in the contemporary United States. Also, women attend church more often and are typically charged with the moral education of children. Males, by contrast, account for a much higher proportion of crime than females. A significant and cumulative male surplus will thus produce a society with some of the rougher features of a frontier town. And, it should be noted, the diminution of the number of agents of moral education and the increase in the number of criminals would accentuate already existing tendencies which point in these directions, thus magnifying social problems which are already overburdening our society.

Interracial and interclass tensions are likely to be intensified because some groups, lower classes and minorities specifically (16), seem to be more male oriented than the rest of the society. Hence while the sex imbalance in a society-wide average may be only a few percentage points, that of some groups is likely to be much higher. This may produce an especially high boy surplus in lower status groups. These extra boys would seek girls in higher status groups (or in some other religious group than their own) (11)—in which they also will be scarce.

On the lighter side, men vote systematically and significantly more Democratic than women; as the Republican party has been losing consistently in the number of supporters over the last generation anyhow, another 5-point loss could undermine the two-party system to a point where Democratic control would be uninterrupted. (It is already the norm, with Republicans having occupied the White House for 8 years over the last 36.) Other forms of imbalance which cannot be predicted are to be expected. "All social life is affected by the proportions of the sexes. Wherever there exists a considerable predominance of one sex over the other, in point of numbers, there is less prospect of a well-ordered social life." " Unbalanced numbers inexorably produce unbalanced behavior" (17).

Society would be very unlikely to collapse even if the sex ratio were to be much more seriously imbalanced than we expect. Societies are surprisingly flexible and adaptive entities. When asked what would be expected to happen if sex control were available on a mass basis, Davis, the well-known demographer, stated that some delay in the age of marriage of the male, some rise in prostitution and in homosexuality, and some increase in the number of males who will never marry are likely to result. Thus, all of the " costs " that would be generated by sex control will probably not be charged against one societal sector, that is, would not entail only, let us say, a sharp rise in prostitution, but would be distributed among several sectors and would therefore be more readily absorbed. An informal examination of the situation in the U.S.S.R. and Germany after World War II (sex ratio was 77.7 in the latter) as well as Israel in early immigration periods, support Davis' non-alarmist position. We must ask, though, are the costs justified? The dangers are not apocalyptical; but are they worth the gains to be made?

A Balance of Values

We deliberately chose a low-key example of the effects of science on society. One can provide much more dramatic ones; for example, the invention of new " psychedelic" drugs whose damage to genes will become known only much later (LSD was reported to have such effects), drugs which cripple the fetus (which has already occurred with the marketing of thalidomide), and the attempts to control birth with devices which may produce cancer (early versions of the intrauterine device were held to have such an effect). But let us stay with a finding which generates only relatively small amounts of human misery, relatively well distributed among various sectors, so as not to severely undermine society but only add, maybe only marginally, to the considerable social problems we already face. Let us assume that we only add to the unhappiness of seven out of every 100 born (what we consider minimum imbalance to be generated), who will not find mates and will have to avail themselves of prostitution, homosexuality, or be condemned to enforced bachelorhood. (If you know someone who is desperate to be married but cannot find a mate, this discussion will be less abstract for you; now multiply this by 357,234 per annum.) Actually, to be fair, one must subtract from the unhappiness that sex control almost surely will produce, the joy it will bring to parents who will be able to order the sex of their children; but as of now, this is for most, not an intensely felt need, and it seems a much smaller joy compared to the sorrows of the unmatable mates.

We already recognize some rights of human guinea pigs. Their safety and privacy are not to be violated even if this means delaying the progress of science. The "rest" of the society, those who are not the subjects of research, and who are nowadays as much affected as those in the laboratory, have been accorded fewer rights. Theoretically, new knowledge, the basis of new devices and drugs, is not supposed to leave the inner circles of science before its safety has been tested on animals or volunteers, and in some instances approved by a government agency, mainly the Federal Drug Administration. But as the case of lysergic acid diethylamide (LSD) shows, the trip from reporting of a finding in a scientific journal to the bloodstream of thousands of citizens may be an extremely short one. The transition did take quite a number of years, from the days in 1943 when Hoffman, one of the two men who synthesized LSD-25 at Sandoz Research Laboratories, first felt its hallucinogenic effect, until the early 1960's, when it "spilled" into illicit campus use. (The trip from legitimate research, its use at Harvard, to illicit unsupervised use was much shorter.) The point is that no additional technologies had to be developed; the distance from the chemical formula to illicit composition required in effect no additional steps.

More generally, Western civilization, ever since the invention of the steam engine, has proceeded on the assumption that society must adjust to new technologies. This is a central meaning of what we refer to when we speak about an industrial revolution; we think about a society being transformed and not just a new technology being introduced into a society which continues to sustain its prior values and institutions. Although the results are not an unmixed blessing (for instance, pollution and traffic casualties), on balance the benefits in terms of gains in standards of living and life expectancy much outweight the costs, [whether the same gains could be made with fewer costs if society would more effectively guide its transformation and technology inputs is a question less often discussed (18)]. Nevertheless we must ask, especially with the advent of nuclear arms, if we can expect such a favorable balance in the future. We are aware that single innovations may literally blow up societies or civilization; we must also realize that the rate of social changes required by the accelerating stream of technological innovations, each less dramatic by itself, may supersede the rate at which society can absorb. Could we not regulate to some extent the pace and impact of the technological inputs and select among them without, by every such act, killing the goose that lays the golden eggs?

Scientists often retort with two arguments. Science is in the business of searching for truths, not that of manufacturing technologies. The applications of scientific findings are not determined by the scientists, but by society, politicians, corporations, and the citizens. Two scientists discovered the formula which led to the composition of LSD, but chemists do not determine whether it is used to accelerate psychotherapy or to create psychoses or, indeed, whether it is used at all, or whether, like thousands of other studies and formulas, it is ignored. Scientists split the atom, but they did not decide whether particles would be used to produce energy to water deserts or to fuel superbombs.

Second, the course of science is unpredictable, and any new lead, if followed, may produce unexpected bounties; to curb some lines of inquiry—because they may have dangerous outcomes—may well force us to forego major payoffs; for example, if one were to forbid the study of sex control one might retard the study of birth control. Moreover, leads which seem " safe " may have dangerous outcomes. Hence, ultimately, only if science were stopped altogether, might findings which are potentially dangerous be avoided.

These arguments are often presented as if they themselves were empirically verified or logically true statements. Actually they are a formula which enables the scientific community to protect itself from external intervention and control. An empirical study of the matter may well show that science does thrive in societies where scientists are given less freedom than the preceding model implies science must have, for example, in the Soviet Union. Even in the West in science some limitations on work are recognized and the freedom to study is not always seen as the ultimate value. Whereas some scientists are irritated when the health or privacy of their subject curbs the progress of their work, most scientists seem to recognize the priority of these other considerations. (Normative considerations also much affect the areas studied; compare, for instance, the high concern with a cancer cure to the almost unwillingness of sociologists, since 1954, to retest the finding that separate but equal education is not feasible.)

One may suggest that the society at large deserves the same protection as human subjects do from research. That is, the scientific community cannot be excused from the responsibility of asking what effects its endeavours have on the community. On the contrary, only an extension of the existing codes and mechanisms of self-control will ultimately protect science from a societal backlash and the heavy hands of external regulation. The intensification of the debate over the scientists' responsibilities with regard to the impacts of their findings is by itself one way of exercising it, because it alerts more scientists to the fact that the areas they choose to study, the ways they communicate their findings (to each other and to the community), the alliances they form or avoid with corporate and governmental interests—all these affect the use to which their work is put. It is simply not true that a scientist working on cancer research and one working on biological warfare are equally likely to come up with a new weapon and a new vaccine. Leads are not that random, and applications are not that readily transferable from one area of application to another.

Additional research on the societal impact of various kinds of research may help to clarify the issues. Such research even has some regulatory impact. For instance, frequently when a drug is shown to have been released prematurely, standards governing release of experimental drugs to mass production are tightened (*19*), which in effect means fewer, more carefully supervised technological inputs into society; at least society does not have to cope with dubious findings. Additional progress may be achieved by studying empirically the effects that various mechanisms of self-regulation actually have on the work of scientists. For example, urging the scientific community to limit its study of some topics and focus on others may not retard science; for instance, sociology is unlikely to suffer from being now much more reluctant to concern itself with how the U.S. Army may stabilize or undermine foreign governments than it was before the blowup of Project Camelot (*20*).

In this context, it may be noted that the systematic attempt to bridge the "two cultures" and to popularize science has undesirable side effects which aggravate the problem at hand. Mathematical formulas, Greek or Latin terminology, and jargon were major filters which allowed scientists in the past to discuss findings with each other without the nonprofessionals listening in. Now, often even preliminary findings are reported in the mass media and lead to policy adaptations, mass use, even legislation (*21*), long before scientists have had a chance to double-check the findings themselves and their implications. True, even in the days when science was much more esoteric, one could find someone who could translate its findings into lay language and abuse it; but the process is much accelerated by well-meaning men (and foundations) who feel that although science ought to be isolated from society, society should keep up with science as much as possible. Perhaps the public relations efforts on behalf of science ought to be reviewed and regulated so that science may remain free.

A system of regulation which builds on the difference between science and technology, with some kind of limitations on the technocrats serving to protect societies coupled with little curbing of scientists themselves, may turn out to be much more crucial. The societal application of most new scientific findings and principles advances through a sequence of steps, sometimes referred to as the R & D process. An abstract finding or insight frequently must be translated into a technique, procedure, or hardware, which in turn must be developed, tested, and mass-produced, before it affects society. While in some instances, like that of LSD, the process is extremely short in that it requires few if any steps in terms of further development of the idea, tools, and procedures, in most instances the process is long and expensive. It took, for instance, about $2 billion and several thousand applied scientists and technicians to make the first atomic weapons after the basic principles of atomic fission were discovered. Moreover, technologies often have a life of their own; for example, the intrauterine device did not spring out of any application of a new finding in fertility research but grew out of the evolution of earlier technologies.

The significance of the distinction between the basic research ("real" science) and later stages of research is that, first, the damage caused (if any) seems usually to be caused by the technologies and not by the science applied in their development. Hence, if there were ways to curb damaging technologies, scientific research could maintain its almost absolute, follow-any-lead autonomy and society would be protected.

Second, and most important, the norms to which applied researchers and technicians subscribe and the supervisory practices, which already prevail, are very different than those which guide basic research. Applied research and technological work are already intensively guided by societal, even political, preferences. Thus, while about $2 billion a year of R & D money are spent on basic research more or less in ways the scientists see fit, the other $13 billion or so are spent on projects specifically ordered, often in great detail, by government authorities, for example, the development of a later version of a missile or a "spiced-up" tear gas. Studies of R & D corporations—in which much of this work is carried out, using thousands of professionals organized in supervised teams which are given specific assignments —pointed out that wide freedom of research simply does not exist here. A team assigned to cover a nose cone with many different alloys and to test which is the most heat-resistant is currently unlikely to stumble upon, let us say, a new heart pump, and if it were to come upon almost any other lead, the boss would refuse to allow the team to pursue the lead, using the corporation's time and funds specifically contracted for other purposes.

Not only are applied research and technological developments guided by economic and political considerations but also there is no evidence that they suffer from such guidance. Of course, one can overdirect any human activity, even the carrying of logs, and thus undermine morale, satisfaction of the workers, and their productivity, but such tight direction is usually not exercised in R & D work nor is it required for our purposes. So far guidance has been largely to direct efforts toward specific goals, and it has been largely corporate, in the sense that the goals have been chiefly set by the industry (for example, building flatter TV sets) or mission-oriented government agencies (for instance, hit the moon before the Russians). Some "preventive" control, like the suppression of run-proof nylon stockings, is believed to have taken place and to have been quite effective.

I am not suggesting that the directive given to technology by society has been a wise one. Frankly, I would like to see much less concern with military hardware and outer space and much more investment in domestic matters; less in developing new consumer gadgets and more in advancing the technologies of the public sector (education, welfare, and health); less concern with nature and more with society. The point though is that, for good or bad, technology is largely already socially guided, and hence the argument that its undesirable effects cannot be curbed because it cannot take guidance and survive is a false one.

What may have to be considered now is a more preventive and more national effective guidance, one that would discourage the development of those technologies which, studies would suggest, are likely to cause significantly more damage than payoffs. Special bodies, preferably to be set up and controlled by the scientific

433

community itself, could be charged with such regulation, although their decrees might have to be as enforceable as those of the Federal Drug Administration. (The Federal Drug Administration, which itself is overworked and understaffed, deals mainly with medical and not societal effects of new technologies.) Such bodies could rule, for instance, that whereas fertility research ought to go on uncurbed, sex-control procedures for human beings are not to be developed.

One cannot be sure that such bodies would come up with the right decisions. But they would have several features which make it likely that they would come up with better decisions than the present system for the following reasons: (i) they would be responsible for protecting society, a responsibility which so far is not institutionalized; (ii) if they act irresponsibly, the staff might be replaced, let us say by a vote of the appropriate scientific associations; and (iii) they would draw on data as to the societal effects of new (or anticipated) technologies, in part to be generated at their initiative, while at present—to the extent such supervisory decisions are made at all—they are frequently based on folk knowledge.

Most of us recoil at any such notion of regulating science, if only at the implementation (or technological) end of it, which actually is not science at all. We are inclined to see in such control an opening wedge which may lead to deeper and deeper penetration of society into the scientific activity. Actually, one may hold the opposite view—that unless societal costs are diminished by some acts of self-regulation at the stage in the R & D process where it hurts least, the society may "backlash" and with a much heavier hand slap on much more encompassing and throttling controls.

The efficacy of increased education of scientists to their responsibilities, of strengthening the barriers between intrascientific communications and the community at large, and of self-imposed, late-phase controls may not suffice. Full solution requires considerable international cooperation, at least among the top technology-producing countries. The various lines of approach to protecting society discussed here may be unacceptable to the reader. The problem though must be faced, and it requires greater attention as we are affected by an accelerating technological output with ever-increasing societal ramifications, which jointly may overload society's capacity to adapt and individually cause more unhappiness than any group of men has a right to inflict on others, however noble their intentions.

References and Notes

1. E. Witschi, personal communication.
2. P. E. Lindahl. *Nature* 181, 784 (1958).
3. V. N. Schröder and N. K. Koltsov. *Ibid.* **131**, 329 (1933).
4. M. J. Gordon. *Sci. Amer.* **199**, 87–94 (1958).
5. F. Unterberger. *Deutsche Med. Wochenschr.* **56**, 304 (1931).
6. R. C. Cook, *J. Hered.* **31**, 270 (1940).
7. J. Schmookler. *Invention and Economic Growth* (Harvard Univ. Press, Cambridge, Mass., 1966).
8. Many people prefer adoption to artificial insemination. See G. M. Vernon and J. A. Boadway, *Marriage Family Liv.* **21**, 43 (1959).

9. H. Kahn and A. J. Wiener. *The Year 2000: A Framework for Speculation on the Next Thirty-Three Years* (Macmillan, New York, 1967), p. 53.

10. H. J. Muller. *Science* **134,** 643 (1961).

11. C. F. Westoff. "The social-psychological structure of fertility," in *International Population Conference* (International Union for Scientific Study of Population, Vienna, 1959).

12. S. Winston. *Amer. J. Sociol.* **38,** 226 (1932). For a critical comment which does not affect the point made above, see H. Weiler, *Ibid.* **65,** 298 (1959).

13. J. E. Clare and C. V. Kiser. *Milbank Mem. Fund Quart.* **29,** 441 (1951). See also D. S. Freedman, R. Freedman, P. K. Whelpton. *Amer. J. Sociol.* **66,** 141 (1960).

14. Based on the figure for 1965 registered births (adjusted for those unreported) of 3,760,358 from *Vital Statistics of the United States 1965* (U.S. Government Printing Office, Washington, D.C., 1965), vol. 1, pp. 1–4, section 1, table 1–2. If there is a "surplus" of 9.5 boys out of every hundred, there would have been 3,760,358/100 × 9.5 = 357,234 surplus in 1965.

15. Calculated from C. Winkler, Ed. *Statistical Guide 1965 for New York City* (Department of Commerce and Industrial Development, New York, 1965), p. 17.

16. Winston suggests the opposite but he refers to sex control produced through birth control which is more widely practiced in higher classes, especially in the period in which his study was conducted, more than a generation ago.

17. Quoted in J. H. Greenberg. *Numerical Sex Disproportion: A Study in Demographic Determinism* (Univ. of Colorado Press, Boulder, 1950), p. 1. The sources indicated are A. F. Weber, *The Growth of Cities in the Nineteenth Century*, Studies in History, Economics, and Public Law, vol. 11, p. 85, and H. von Hentig, *Crime: Causes and Conditions* (McGraw-Hill, New York, 1947), p. 121.

18. For one of the best discussions, see E. E. Morison, *Men, Machines, and Modern Times* (M.I.T. Press, Cambridge, Mass., 1966). See also A. Etzioni, *The Active Society: A Theory of Societal and Political Processes* (Free Press, New York, 1968), chaps. 1 and 21.

19. See reports in the *New York Times*: "Tranquilizer is put under U.S. curbs; side effects noted," 6 December 1967; "F.D.A. is studying reported reactions to arthritis drug," 19 March 1967; "F.D.A. adds 2 drugs to birth defect list," 3 January 1967. On 24 May 1966, Dr. S. F. Yolles, director of the National Institute of Mental Health, predicted in testimony before a Senate subcommittee: "The next 5 to 10 years ... will see a hundredfold increase in the number and types of drugs capable of affecting the mind."

20. I. L. Horowitz. *The Rise and Fall of Project Camelot* (M.I.T. Press, Cambridge, Mass., 1967).

21. For a detailed report, see testimony by J. D. Cooper, on 28 Febuary 1967, before the subcommittee on government research of the committee on government operations, United States Senate, 90th Congress (First session on Biomedical Development, Evaluation of Existing Federal Institutions), pp. 46–61.

42   The Prospect of Designed Genetic Change

Robert L. Sinsheimer

" It has now become a serious necessity to better the breed of the human race. The average citizen is too base for the everyday work of modern civilization. Civilized man has become possessed of vaster powers than in old times for good or ill but has made no corresponding advance in wits and goodness to enable him to direct his conduct rightly." This was written in 1894 by Sir Francis Galton. The concerns of the present are clearly not new.

It has long been apparent that you and I do not enter this world as unformed clay compliant to any mold; rather, we have in our beginnings some bent of mind, some shade of character. The origin of this structure—of the fiber of this clay— was for centuries mysterious. In earlier times men sought its trace in the conjunction of the stars or perhaps in the momentary combination of the elements at nativity. Today, instead, we know to look within. We seek not in the stars but in our genes for the herald of our fate.

Today there is much talk about the possibility of human genetic modification— of designed genetic change, specifically of mankind. A new eugenics has arisen, based upon the dramatic increase in our understanding of the biochemistry of heredity and our comprehension of the craft and means of evolution. I think this possibility, which we now glimpse only in fragmented outline, is potentially one of the most important concepts to arise in the history of mankind. I can think of none with greater long-range implications for the future of our species. Indeed this concept marks a turning point in the whole evolution of life. For the first time in all time a living creature understands its origin and can undertake to design its future. Even in the ancient myths man was constrained by his essence. He could not rise above his nature to chart his destiny. Today we can envision that chance—and its dark companion of awesome choice and responsibility.

It is all too easy, albeit useful, to let our imagination in these matters roam far beyond our technical base. It is easy, even for modest men given to cautious projection, because in truth all that seems needed is the technology and the resolution to transfer to man what we already know to be feasible in bacteria or carrot cells or frogs. It is easy because there are no known natural laws to repeal or contravene. None of the time warps or hyper-drives or teleportation of science fiction are needed to envision vegetative reproduction, organ regeneration, genetic therapy, or eugenic transformation of our species.

I would like, however, to consider a very specific and possible use of our newer knowledge, relating to a major biomedical problem. This application may well seem of small dimensions as compared to some of the more sweeping prospects,

From *Engineering and Science* (Apr. 1969), pp. 8–13. © 1969 Alumni Association, California Institute of Technology. Reprinted with permission of the author and the publisher.

but I believe it will illuminate the state of our knowledge and our technology and will thereby reveal the shape of things to come.

I want to use the phrase "genetic change" in a broad sense, in the sense of altering some physiological or psychological process which at present we believe has been programmed into us through our inheritance. And I will assume that such change might be achieved either in a strictly genetic mode through a change in our inherited characteristics, or in a somatic (non-inheritable) mode—possibly through a change in the time or place or degree of action of our inherited genetic components, or possibly through the somatic addition of genetic components. Obviously changes of the former—the truly genetic type—have the greater ultimate potential; for the very nature of the species seems potentially susceptible to change. Changes of the latter type—somatic genetic modifications—are more limited. Their scope and function are the more restricted, but they are also undoubtedly the more accessible possibilities which we will first achieve.

There are in the United States today some 4,000,000 clinical diabetics. Many of these people are kept alive only by repeated, frequent injections of the hormone insulin. It is believed that there are several million more cases with marginal symptoms. Without recurrent injections of insulin many of these people would perish. While it keeps them alive, the injection of insulin is not the full equivalent of a normal physiological function; diabetics are known to be more susceptible to disease, to heart and circulatory illnesses, and other physical limitations than non-diabetics.

I propose that genetic therapy offers the promise of a much more elegant, and indeed more satisfactory, physiological solution to this ailment. And there are various possible genetic approaches.

To begin we must understand the normal process of insulin formation. Insulin is a protein, composed of two polypeptide chains—one of 21 amino acids and one of 30—joined by two disulphide bonds. There is recent evidence that indicates strongly that the insulin molecule is initially formed as a single polypeptide chain, and an internal segment is subsequently excised by the action of a specific proteolytic enzyme (Figure 42–1).

The synthesis of this protein, the proinsulin, is accomplished in the usual manner: The hereditary instructions specifying the sequence of amino acids for insulin are

Figure 42-1. The chemical structure of human insulin consists of two polypeptide chains, one of 21 amino acids and the other of 30, which are joined by two disulphide bonds.

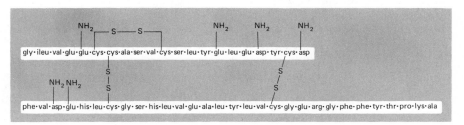

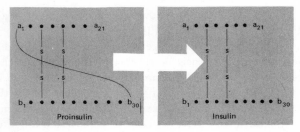

Figure 42-2. This schematic drawing of the conversion of proinsulin to insulin illustrates the recent evidence that the insulin molecule is formed as a single polypeptide chain and that an internal segment is subsequently excised by the action of a specific proteolytic enzyme.

encoded in a segment of the DNA from the cell nucleus (Figure 42-2). The instructions are copied and transcribed into a messenger RNA molecule which then is transported out of the nucleus to the cytoplasm (Figure 42-3). There protein synthesis takes place upon the ribosomes. In this process the sequence of nucleotides in the RNA is translated into the corresponding amino acid sequence with the help of the transfer RNA molecules, the activating enzymes, the initiators, the coupling factors, and all the rest of a very complex machinery.

It is well known that this synthesis of insulin normally takes place only in the beta cells in the Islands of Langerhans in the human pancreas. In the diabetic these cells fail to produce an adequate amount of insulin. Now it is believed, and there is good reason for this belief from studies of lower animals, that the *full* DNA content of the genome is present in every somatic cell. And thus we believe that the genetic instructions specifying the sequence of proinsulin are present in all the cells of the body and not only the beta cells of the Islands of Langerhans. Evidently these instructions, though present in other cells, are not in use. Either they are not activated, or, as it is more fashionable to assume these days, they are repressed. Repression could take place at any of several levels.

A typical somatic cell is only called upon to use a small fraction of its genome. There is good evidence that in a liver or a muscle cell no more than 5 percent of the DNA is ever transcribed into RNA, so there is repression at the chromosomal level. Further, it is clear that perhaps half or more of that which *is* transcribed never reaches the cytoplasm to be translated. And even if the RNA reaches the cytoplasm, there is evidence for specific blocks at the translational level. There are clearly many opportunities for the restriction of expression of the inherited genetic instructions.

Figure 42-3. Steps in the biosynthesis of insulin. Repression of insulin synthesis could take place at any of these stages.

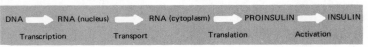

In the case of insulin we do not know by what means the expression of this gene is limited to a few islands of cells. We do not know at what level the restriction is imposed. However, one approach to the problem of diabetes would be to attempt to turn on the synthesis of insulin in another set of cells.

We do know that genes can be turned on by external influence. Hormones do this every day. For example, under the influence of cortisone, liver cells initiate the synthesis of a variety of enzymes including tryptophan pyrrolase and tyrosine-alpha-ketoglutarate transaminase.

In some instances the prior repression appears to be lifted hormonally at the chromosomal level of transcription, in others, at the translational level. We do not now know how we might do this for insulin, but we can see a clear model. And in fact just such an activation or derepression for insulin *must* have occurred through some chain of ontogenetic events during embryonic development to activate—to turn on—the appropriate genes in the beta cells of the Islands of Langerhans.

We should not oversimplify this problem. Obviously if we want a new group of cells to synthesize insulin, we must not only activate the gene for proinsulin but also arrange for its conversion to insulin and for its release from the cells. But, if we were fortunate, these functions might all come as a genetic package.

There is a radically different genetic approach that we might take alternatively. Instead of an attempt to lift this profound repression of the expression of the gene for insulin, we might, in principle, supply to a group of cells a wholly new gene or set of genes which would code for the synthesis of insulin and which might *not* be subject to the normal somatic pattern of repression.

How might we add, in such a specific manner, to the genetic components of a cell? Our models come from studies with bacterial cells. In these organisms a variety of means exist to permit exchange and in so doing to provide small increments of genetic material. These include transformation, contact transfer both chromosomal and episomal, and transduction both general and specific. Organelles for contact transfer are not known among mammalian cells, and transformation as such has not yet been convincingly demonstrated in mammalian cells. Therefore, the possible use of transduction as a means to genetic modification of cells of higher organisms should be specifically considered.

Transduction among bacteria involves the transfer of genetic material, DNA, from one cell to another through viral mediation. I would like to present two particular cases.

The first case is that of the bacterial virus P1, which contains one molecule of DNA of about 60,000,000 in molecular weight. Upon infection of the cell by certain types of P1, the cell is lysed (broken down) after half an hour to produce a few hundred progeny virus particles. Most of these will contain a DNA identical to that of the virus that initiated the infection. However, a little less than 1 percent of the particles will contain *instead* a piece of the DNA of the chromosome of the host bacterium, a piece also about 60,000,000 in molecular weight. Which piece of DNA—which particular 60,000,000 out of the 3 billion molecular weight of host DNA—is random. The particular virus will contain the piece carrying, say, genes D and E, while another carries a piece with the genes P and Q, etc.

439

By appropriate means these particles carrying host DNA, called transducing particles, can be separated from those carrying the normal viral DNA. When such transducing particles are added to susceptible bacteria, the DNA inside the virus particle is, in the usual way of bacterial viruses, injected into the cell. But *now* we have added to the cell not a destructive virus genome, but a piece of bacterial DNA which may well carry genetic markers not present in this particular host. This DNA may be transcribed at once to yield new protein.

For this piece of DNA to perpetuate itself, however, it must, in general, become incorporated *into* the host chromosome by a process of genetic recombination. Normal bacterial cells have the enzymatic machinery to do this, and, in the case of the transducing particles of phage P1, there is about one chance in ten that the particular piece of DNA will be so incorporated and perpetuated.

In bacterial cells there are often small secondary chromosomes—episomes usually containing 1 or 2 percent as much DNA as the principal chromosome. These are physically separate from the principal chromosome, but usually replicate in synchrony with it. It is possible for a P1 phage to pick up and transfer an entire episome as well as a piece of bacterial chromosome.

A second case of transduction concerns the temperate (frequently non-lethal) bacteriophage lambda. Upon infection with the bacteriophage lambda, the result in an appreciable percentage of the cells (it can be the majority) is the physical incorporation of either the viral DNA or of one of its descendents into the chromosomes of the host. Following this, the *virus-like* tendencies of this DNA are suppressed. The cell survives and multiplies, and the incorporated viral DNA is replicated into each daughter cell along with the rest of the bacterial chromosome. Such a virus-carrying cell is said to be a lysogen.

An important feature is that the point of insertion for the lambda DNA into the host DNA is specific, and it is determined by the particular virus which in turn specifies an enzyme—an integrase—which brings about its incorporation at that site. Related strains of lambda-type viruses are known which integrate into *other* chromosomal sites because they have different integrases.

It is possible, however, to induce an activation of this carried viral DNA in the lysogenic cells—to cause it to remember that it really is a virus, to cause it to break out of the bacterial chromosome, to begin to multiply, to produce progeny, and to lyse the cell, producing new virus particles.

Occasionally in such an activation, which is called induction, the piece of DNA which splits out of the chromosome is not strictly the viral genome but may incorporate a piece of the neighboring bacterial chromosome with its genetic material in lieu of a piece of the viral genome. Under certain circumstances viral development can proceed anyway. Such pieces of DNA, partially viral and partially host, can multiply and can be incorporated into virus particles (Figure 42-4). Particles with this mixed DNA can be isolated from the bulk of progeny. If they are now added to susceptible cells, this DNA can still integrate into the host chromosome at the same locus but now adding along with the viral genes a specific piece of DNA from the former host—which may carry specific novel genetic traits into the new host.

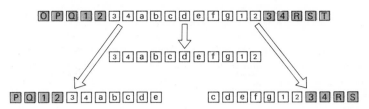

Figure 42-4. In excision of a lambda DNA from the host chromosome, normal excision (center) yields one complete viral genome, while abnormal excision (left and right) yields mixed genomes, part viral and part host.

In both of these cases, then—P1 and lambda—the net result is the introduction, via a particle normally indistinguishable from a virus, of new genetic material into the host cell. In the first instance the new factors added are random relative to the host genome. In the second, they are factors found at specific sites near the normal region of integration of the virus. The region varies in different viruses.

Could a similar transfer be accomplished with a virus in the cells of higher organisms? We have every reason to think that it does occur. Upon infection of mammalian cells with the simian virus 40, or with polyoma virus, or with some strains of adenovirus, in a fraction of such infected cells the viral DNA becomes established within the cell. Whether it is integrated into the chromosomal DNA or is an episome is not known. It is then perpetuated within the clone of cells descended from the original infected cell, as in a lysogenic bacterium. The information carried in the viral DNA is certainly expressed; cells carrying such DNA have altered properties; that messenger RNA derives from this viral DNA can be demonstrated; new protein antigens have been detected within such cells; and in special circumstances the entire genome of the virus can be recovered (and hence must have been present) from remote descendents of such altered cells.

Technically and literally the stage is set. If we could obtain a virus analogous to simian virus 40—able to persist within altered cells and carrying an expressible gene for proinsulin in lieu of a normal viral gene—we might indeed be able to provide a genetic alternative to the daily injection of insulin.

The problem then is, where we are to find this virus so propitiously carrying a gene to provide insulin? Such a virus might exist in nature, but I propose that we should quite literally, in time, be able to make it to order. We will have the ability in the not distant future to synthesize a polynucleotide chain capable of coding for insulin and for the other genes necessary to integrate the DNA into a chromosome, or to maintain it as an episome, or whatever. And we will then also be able to package this *de novo* DNA into an appropriate virus coat.

Is this pure fantasy? No, not really. The DNA of simian virus 40 consists of a chain of 5,000 nucleotides. The art of specific polynucleotide synthesis is young but thriving. It is now feasible to construct a specific sequence of 50 deoxyribonucleotides. A sequence of up to 100 seems close at hand, and a thousand or a few thousand is by no means inconceivable.

Furthermore, such a synthesis needs to be done only once. Once the DNA is available, nature provides the means to copy it with the highest fidelity.

Similarly, our understanding of the process of viral self-assembly is growing swiftly, and but a small step behind is the art of viral assembly *in vitro*. The technology needed for such a radically different approach to a major clinical problem is almost in reach.

Though the analogy is not perfect, in describing these prospects I feel strangely akin to the physicists who pointed out in the 1930's that the principles required for the release of the energy locked in the atomic nucleus were understood. All that was needed was a practical breakthrough and the requisite technology. Here, too, the principles seem in hand. All that seems really needed is optimism, sustained effort, and support commensurate with the importance of the problem.

The larger and the deeper challenges—those concerned with the defined genetic improvement of man—perhaps fortunately are not yet in our grasp, but they are etched clear upon the horizon. We should begin to prepare now for their reality.

It is worthwhile to consider specifically wherein the potential of the new genetics exceeds that of the old. To implement the older eugenics of Galton and his successors would have required a massive social program carried out over many generations. Such a program could not have been initiated without the consent and cooperation of a major fraction of the population, and would have been continuously subject to social control. In contrast, the new eugenics could, at least in principle, be implemented on a quite individual basis, in one generation, and subject to no existing social restrictions.

The old eugenics would have required a continual selection for breeding of the fit, and a culling of the unfit. The new eugenics would permit in principle the conversion of all of the unfit to the highest genetic level.

The old eugenics was limited to a numerical enhancement of the best of our existing gene pool. The horizons of the new eugenics are in principle boundless—for we should have the potential to create new genes and new qualities yet undreamed. But of course the ethical dilemma remains. What are the best qualities, and who shall choose?

It is a new horizon in the history of man. Some may smile and may feel that this is but a new version of the old dream of the perfection of man. It is that, but it is something more. The old dreams of the cultural perfection of man were always sharply constrained by his inherent, inherited imperfections and limitations. Man is all too clearly an imperfect and flawed creature. Considering his evolution, it is hardly likely that he could be otherwise. To foster his better traits and to curb his worse by cultural means alone has always been, while clearly not impossible, in many instances most difficult. It has been an Archimedian attempt to move the world, but with the short arm of a lever. We now glimpse another route—the chance to ease the internal strains and heal the internal flaws directly, to carry on and consciously perfect far beyond our present vision this remarkable product of two billion years of evolution.

I know there are those who find this concept and this prospect repugnant—who fear, with reason, that we may unleash forces beyond human scale and who recoil from this responsibility. I would suggest to them that they do not see our present situation whole. They are not among the losers in that chromosomal lottery that so firmly channels our human destinies. This response does not come from the 250,000 children born each year in this country with structural or functional defects, of which an estimated 80 percent involve a genetic component. And this figure counts only those with gross evident defects outside those ranges we choose to call natural. It does not include the 50,000,000 "normal" Americans with an IQ of less than 90.

We are among those who were favored in the chromosomal lottery, and, in the nature of things, it will be our very conscious choice whether as a species we will continue to accept the innumerable individual tragedies inherent in the outcome of this mindless, age-old throw of dice, or instead will shoulder the responsibility for intelligent genetic intervention.

As we enlarge man's freedom, we diminish his constraints and that which he must accept as given. Equality of opportunity is a noble aim given the currently inescapable genetic diversity of man. But what does equality of opportunity mean to the child born with an IQ of 50?

The application of knowledge requires technology, but the impact of knowledge can precede its application. Knowledge brings understanding, and the consequences of understanding can overflow the mind into the heart. It may be that in the near future the most important consequence of our new knowledge of ourselves will be a new sense of the power and responsibility—of the pivotal role—of man in this universe. Copernicus and Darwin demoted man from his bright glory at the focal point of the universe to be merely the current head of the animal line on an insignificant planet. In the mirror of our newer knowledge we can begin to see that in truth we are far more than another ephemeral form in the chain of evolution. Rather we are an historic innovation. We can be the agent of transition to a wholly new path of evolution. This is a cosmic event.

43   Therapy of Genetic Diseases in Man and the Possible Place of Genetic Engineering

Walter Fuhrmann

SUMMARY: *Conventional therapy of genetic diseases has been very successful in selected disorders. It can be applied on a large scale and a great number of patients may benefit. Treatment, however, in most cases has to be continued for life and may be costly and strenuous. Moreover, at best it succeeds in correcting the phenotype but it does not alter*

From *Advances in the Bioscience*, **8** : 387–395, 1972. Copyright © 1972, Friedr. Vieweg & Sohn, Braunschweig, Germany. Reprinted with permission of the author and the publisher.

the genotype. In many other cases even correction of the phenotype cannot be achieved. The search for early and lasting correction of the primary defect, therefore, is urgent but has been unsuccessful so far. Attempts and speculations on practical application of genetic engineering in human disease states are reviewed and difficulties pointed out. It is concluded that for some time to come research on the exact definition of genetic diseases and elucidation of the primary defect must have priority in medical genetics. This will constitute the basis for more effective palliative therapy as well as for possible correction of the primary defect in the future. Genetic exchange is a valuable research tool in this respect as well as in mapping of the human genome; its direct application as a therapeutic principle, when at all possible, will be restricted to a small number of selected cases.

When I was asked to review the present scene of clinical genetics and therapy and its outlook for the future in regards to genetic engineering, I could only speculate on what would be said by the preceding speakers involved in the experimental work. I, therefore, thought it appropriate to add a word of caution to those in the audience who might be too ready to assume that therapy of genetic diseases will be revolutionized in the near future by direct manipulation of the genetic material, an impression that could arise easily if one were not familiar with the achievements and problems of medical genetics. With great satisfaction I then noted the sober and cautious view taken by almost all speakers. To set the scale right, one should not simply list the difficulties still unsolved, but also try to balance the possible benefits of genetic engineering against the achievements of conventional therapy.

Conventional Therapy of Genetic Disease

All presently available treatment of inherited disease in man is symptomatic. It aims at a change of the phenotype, without alteration of the genetic information. The place of action is somewhere between the altered primary gene product and the phenotypic manifestation. Such treatment could frequently be instituted without exact knowledge of the primary genetic defect and even before inheritance was recognized. Methods of intervention include surgical repair, prostheses, and drug therapy. More elegantly, the environment has been adapted to the patient's needs, for example by diet therapy which had dramatic results in certain inborn errors of metabolism, as for example in phenylketonuria, galactosemia, and fructose intolerance.

A somewhat more convincing approach is substitution of a missing or functionally defective gene product. With few exceptions this has been successful only in cases of humoral substances acting when circulating in the blood stream.

Closely related in principle are procedures which aim at an influence on the regulation of hormone or enzyme synthesis and action. A well established case is the cortisol treatment of patients with the adrenogenital syndrome. These patients have a defective cortisol synthesis, but the major symptoms derive from the resulting dysregulation of adrenocortical hormones, some of them with virilizing effect. Cortisol substitution restores the regulatory cycle.

444

Such measures can be applied on a large-scale basis; a great number of patients may benefit, but continuous treatment is required. Such treatment may be costly and strenuous, and results usually are far from perfect. Nevertheless, the number of patients is increasing who are enabled to lead an almost normal life and to propagate their defective genes. For the majority of genetic diseases, however, no treatment is available as yet. Here any possibility of betterment will be appreciated.

Attempts Towards Causal Therapy

What is wanted ideally is a therapeutic approach to correction of the primary defect, which could be instituted early enough to prevent irreversible damage and, if possible, do away with the need for continuous treatment. What has been achieved along this line? An early idea was the transplantation of glands in order to give a patient a permanent internal supply of a hormone. This usually failed due to lack of specificity and, among other reasons, due to immunologic intolerance and degeneration of the transplant. As a logical extension one may view the attempts recently undertaken with variable success to transplant bone marrow cells from healthy donors to patients with specific defects, or to transplant thymus to patients with congenital absence of the organ. Success of bone marrow transplantation has been reported particularly in immunodeficiency diseases and will be discussed later by Dr. Good [7]. It could offer a chance also in certain hemoglobinopathies, if one succeeded to establish sufficiently large cell clones able to produce normal hemoglobin, or in the mucopolysaccharidoses, if such clones could be shown to supply to the recipient's cells the correction factors found by Neufeld and her group [5, 15].

In hemophilia A, it has been shown by transplantation experiments in dogs and by organ perfusion that the antihemophilic factor, factor VIII, or its precursor is produced mainly in the liver and is activated especially by the spleen. From these experiments a therapy appears possible by either transplantation of liver and spleen to a patient or perhaps by the infusion of cell suspensions from these organs [3, 10, 11, 14, 29].

Use of in Vitro Mutation or Gene Transfer?

Major problems in all such therapeutic attempts remain the graft versus host reactions and the immunosuppression often required to facilitate the acceptance of the graft. This could be avoided if the old suggestion of Tatum [23, 24] could be realized. This was to culture cells of the patient himself and, by induced mutation and selective cloning, produce a clone of reverted cells which could be reimplanted into the patient to supply the missing (or defective) gene product.

As Professor Henry Harris and his colleagues at Oxford have shown, genetic repair could also be achieved in cultured mammalian cells by hybridization procedures without necessarily introducing or retaining genes specifying new antigens, which would cause homograft rejection on reimplantation or reinjection. Such cells, therefore, should be accepted and be able to propagate. A drawback at present is

445

that reverted cells produced by the latter method don't have sufficient genetic stability [21]. They tend to lose the newly introduced genetic material. If these obstacles could be overcome, this then would constitute the use of genetic engineering and exchange in practical therapy but only for a small fraction of cells. The obvious difficulties of such an experiment have prevented its realization until now. Besides, this method could be applied only to diseases correctable by factors secreted by these cells into their environment or perhaps again in certain hemoglobinopathies.

Specific Mutagenesis Not Available

An old dream in correction of genetic defects is specific mutation in vivo. Although particular bases may be attacked preferentially by certain mutagenic substances, and particular regions within genes may be altered in preference by different mutagens, no specificity of mutagenesis is known in the sense needed for therapeutic use in vivo or even in vitro. Even a refined technique, which may serve to bring a chosen mutagen in close contact with a specified region of the genetic material, would certainly not give the desired effect, but would result in various mutations at this site and be accompanied by so much "noise" elsewhere as to do more harm than good. No one has suggested such methods during this meeting.

Gene Transfer in Vivo as a Therapeutic Principle

Much more promising for in vivo treatment and more easily accessible was the attempt to use viruses to introduce new information into host cells. In this way one could hope to infect a great number of cells and supply them with useful additional genetic information. A possibility to realize this concept was suggested by three observations: In 1959, Rogers [18] found that rabbit cells infected with Shope rabbit papilloma virus contain an enzyme that degrades arginine. This arginase induced by the virus can be distinguished by several properties from the enzyme naturally occurring in organs of healthy rabbits. Further studies of the same author in 1966 [19] showed that persons who were in close contact with the virus and the infected rabbits had a diminished arginine concentration in blood. The decrease reached up to about 50% of normal values and persisted in some individuals for as long as 30 years. The persons studied had never shown any ill effects due to the apparent infection. Finally, two sibs were observed who suffered from a hyperargininemia due to a lack of arginase [26, 27].

The occurrence in sibs suggested autosomal-recessive inheritance of the defect in the urea cycle at the point of degradation of arginine to ornithine and urea. Clinical features include mental and motor retardation, seizures, electro-encephalographic anomalies, spastic paresis, periodic spells of vomiting, and liver enlargement. Use of a diet low in proteins were unsuccessful. In view of the experience with people accidentally infected and in the absence of other effective therapy, it was tempting to try the effect of an infection with the Shope virus. This has been undertaken by Terheggen in 1970 at Cologne, Germany, in cooperation with Drs. Lowenthal,

446

Antwerp, Belgium; Colombo, Berne, Switzerland; and Rogers, Oak Ridge, USA. (I quote this unpublished work with permission of Dr. Terheggen [25.])

> A virus suspension was injected intravenously to the younger child, but no change could be observed in the serum arginine concentration. A higher dose of virus given to the elder sib likewise remained without notable result. Meanwhile, another child had been born to this family in whom the diagnosis of the same metabolic error could be established from cord blood. Since experimental infection of rabbits, rats, and donkeys consistently led to decreased arginine levels [20], this youngest child again was inoculated with the virus, now using an again higher dose of virus corresponding to the dose being used in the animal experiments. Results of this attempt cannot be evaluated as yet.

Certainly one could have had little hope for clinical improvement in the older children since therapy of metabolically caused mental retardation is usually successful only if instituted very early in life. Chances for the youngest child could be much better. It must be noted, though, that only about one-half to one-third of all persons exposed to the virus and studied by Rogers had a diminished serum arginine concentration, and then the decrease amounted to only about 50%, whereas serum arginine concentration in the affected children was increased 7 to 13 times.

It is unlikely that we will find many naturally occurring viruses, which specify enzymes able to correct genetic defects in man, and that also satisfy the requirement of doing no harm to the cell—particularly if one wants to be sure about late effects. Most often the viral enzyme will also be different from the missing enzyme in certain of its properties and kinetics.

The prospects of *de novo* synthesis of genes opened by Professor Khorana's work [1, 8] immediately led to speculation that eventually it will become possible to produce pieces of DNA exactly resembling the missing information and to introduce them into the cells of patients. To many popular writers this problem seems close to solution, but I don't believe it is. There is very little hope that injected naked DNA could effectively reach its proper place of action within the cell. I will not discuss here the problems concerning the possible means of transferring such artificial or elsewhere-derived genes to the cell by the use of viruses or pseudovirions, because this has been done by others much more competent than I. I may only point to the work of Osterman and associates [17] who have shown that pseudovirions used for transport of DNA fragments indeed can be uncoated by animal cells, and to the work of Merril and coworkers [12, 13] who, as you have just heard, succeeded in transducing human fibroblasts in culture from a patient with galactosemia by infection with Lambda phages carrying the *E. coli* galactose operon including the missing enzyme alpha-D-galactose-1-phosphate uridyl (G. P. U.) transferase. They could demonstrate that the phage DNA under these conditions was transcribed in human cells.

It remains an open question, of course, whether a phage could also be used to transport selected bacterial genes or, by some technical modification, artificially synthesized DNA stretches into human cells in vivo. A special problem is raised by the fact that in order to produce pseudovirions in the usual fashion, the artificial

DNA first would have to be incorporated into the nuclei of the host cells of the virus. Strauss [22], and with some variation Klingmüller [9], therefore, recently pictured the possibility of going one step further and producing artificial pseudo-virions by coating the synthesized DNA pieces with proteins also derived synthetically or isolated from virus particles, thus facilitating their entrance into living cells by simulating the steps known in viral infection.

No less important an obstacle is that, with a few exceptions, we do not know the primary gene defect in most human inherited diseases. In some we can pinpoint at least a certain enzyme activity that seems missing; in most of the over 1500 monogenic human diseases known not even this can be done. A prominent example of popular misquotation is diabetes mellitus. Since insulin is used successfully in therapy and since the amino-acid sequence of insulin has been known for some time, it was concluded that within the near future it will become possible to assemble a DNA sequence coding for this enzyme. Such an artificial gene could then be introduced into the cell. Notwithstanding the difficulties for such transfer, already discussed, this speculation also does not account for the fact that the insulin molecule consisting of 51 amino acids is derived from a much larger proinsulin with 84 amino acids, and that from a given amino-acid sequence it is not possible to unequivocally deduce the natural sequence of DNA triplets. Moreover, from the fact that externally supplied insulin is effective in therapy, it does not follow that the primary defect in diabetes is a defect of the structural gene for insulin. Many studies indeed have disproved this assumption. Quite likely diabetes mellitus includes several genetically distinct entities with different causes on various levels of regulation, synthesis, and transport [30].

On the long road from the transduction of some fibroblasts in culture to therapeutically useful transduction in an intact mammalian organism or in man, more profound difficulties will be encountered. It will be necessary to introduce the new DNA parts into the genome of the cell so that successive cell generations will preserve the corrected information. Such preservation apparently took place in the experiments of Merril and coworkers, but it could not be said whether this was due to integration of the new information into the genome, due to plasmid-like existence in the cytoplasm, to interaction with mitochondria, or some other as yet unknown mechanisms. Cells supplied with foreign genes by the technique of Harris and coworkers [21] tend to shed the introduced genetic material.

Clinical success may come first in a case where the correction of the defect in a considerable number of cells could perhaps suffice to supply a factor to the blood stream. It will be much harder to correct a defect in all or nearly all cells of the body in similar fashion.

Problems of Gene Regulation and Control

Will it be necessary and will it be possible to place the newly introduced gene under the genetic control at the correct locus? Little more than nothing is known about the exact position of the probably not less than 40,000, and possibly more, human structural genes. On the other hand, if, as Ohno assumed [16], most of the human DNA is noninformational, would the exact position of a foreign gene matter? Will the defective DNA unit, retained in the genome, interfere with effective control of the new gene? A defective gene product coded by the original DNA may also compete with the correct protein for its binding sites.

It is impossible to answer these questions at present. The answers, moreover, will most probably be different for various loci and genes. However, we know very well of cases in which part of the normal genome is present in excess, and this usually leads to severe developmental disturbance and often death of the organism. Such examples are supplied by the trisomies or partial trisomies—best known of the group is mongolism. It may well be that the actual defect here derives mainly from disturbed regulation; but could we be sure that transduction or other means of gene transfer would not also alter regulation? Admittedly the amount of genetic material present in excess in mongolism is much greater than the one probably used for therapeutic gene transfer, but much smaller additional parts of a chromosome than a chromosome 21 have been associated with human maldevelopment. On the other hand, some marker chromosomes with little excess of material obviously are tolerated without apparent ill effect.

If the newly introduced gene does not come under proper regulatory control, could it not exhaust the protein synthesizing capacity of a cell at least to the expense of other functions?

There is a real danger in genetic engineering with the help of viruses that infection may turn out to be oncogenic because many viruses used in these experiments can transform cells to potentially neoplastic cells and cause neoplasms in animals [4].

Very little is known about gene regulation in higher organisms. Most researchers favor the view that the vast majority of DNA in higher organisms and man is used for control purposes and that control may be extremely complicated (an opinion that recently has been put forward again by Crick [2]). Only a few weeks ago Ohno [16] suggested that mammalian regulatory systems will turn out to be much less complicated than generally anticipated. According to him, they should be as simple or even simpler than the lac-operon system of *E. coli*. In the latter case, some hope may be justified that one could introduce not only defined genes but also the appropriate control regions attached to them.

At present very little is known about possibilities to influence gene expression directly and specifically by influencing gene control. One may, of course, view hormone therapy as a first step on this road. Further progress will have to await the analysis of genetic regulatory systems in mammals.

Conclusions

We must conclude that there is little hope that genetic exchange or other forms of genetic engineering can be applied effectively in the therapy of human inherited disease in the near future. The achievement of genetic repair involving not only a limited number of body cells but all cells including those of the gonads appears highly improbable indeed. This ultimate goal alone could be named "radical therapy." It is questionable whether it ever can be reached—certainly not in our generation's lifetime, and equally certain not on a large-scale basis for a much longer time.

Considering the economy of means and man power, if therapy of inherited disease is the matter in question, at the present time priority should be given in medical genetics to research on the exact definition of entities of genetic disease and elucidation of the basic defects and mechanisms involved. For this, the techniques of gene transfer and exchange may contribute very important information. This analysis will constitute the basis for the second step: search for effective, though palliative therapy. It also is the prerequisite for any future therapy aimed at the correction of the primary defect.

Genetic exchange has proved to be an excellent research tool; it will help to map the human genome, and in the future it also may have an application as a therapeutic principle in selected cases. It, therefore, should be explored thoroughly. However, it would be unwise, even unethical, to seek publicity with precocious promises and speculations regarding clinical therapy which will not be fulfilled and again may bring genetic research into disrepute. For some time to come, prevention will remain the paramount task in human genetics. Genetic counseling will be more important than genetic engineering.

References

1. Agarwal, K. L., Büchi, H., Caruthers, M. H., Gupta, N., Khorana, H. G., Kleppe, K., Kumar, A., Ohtsuka, E., Rajbhandara, U. L., Van de Sande, J. H., Sgaramella, V., Weber, H., and Yamada, T. 1970. Total synthesis of the gene for an alanine transfer ribonucleic acid from yeast. Nature 227: 27.
2. Crick, F. 1971. General model for the chromosomes of higher organisms. Nature 234: 25.
3. Dodds, W. J. 1969. Hepatic influence on splenic synthesis and release of coagulation activities. Science 166: 882.
4. Dulbecco, R. 1969. Cell transformation by viruses. Science 166: 962.
5. Fratantoni, J. C., Hall, C. W., and Neufeld, E. F. 1969. The defect in hurler and hunter syndromes. II. Deficiency of specific factors involved in mucopolysaccharide degradation. Proc. Nat. Acad. Sci., USA 64: 360.
6. Gatt, R. A., Meuwissen, H. J., Allen, H. D., Hong, R., and Good, R. A. 1968. Immunological reconstitution of sex-linked lymphopenic immunological deficiency. Lancet 2: 1366.
7. Good, R. A. 1972. Cellular engineering—an approach to treatment of genetically determined disease. This volume, p. 411 ff.
8. Khorana, H. G. 1972. The synthesis of transfer RNA genes. This volume, p. 89 ff.
9. Klingmüller, W. 1971. Heilung von Erbkrankheiten durch gezielte Eingriffe in das Erbgut. Biologie in unserer Zeit 1: 87.
10. Marchioro, T. L., Hougie, C., Radge, H., Epstein, R. B., and Thomas, E. D. 1969. Hemophilia: Role of organ homografts. Science 163: 188.
11. Marchioro, T. L., Hougie, C., Radge, H., Epstein, R. B., and Thomas, E. D. 1969. Organ homografts for hemophilia. Transplant. Proc. 1: 316.

12. Merrill, C. R. et al. 1972. Bacterial gene expression in mammalian cells. This volume, p. 329 ff.
13. Merrill, C. R., Geier, M. R., and Petricianni, J. C. 1971. Bacterial virus gene expression in human cells. Nature **234**: 398.
14. Müller-Berghaus, G. 1971. Zur Pathophysiologie der Hämophilia A. Dtsche. Med. Wschr. **96**: 1723.
15. Neufeld, E. F., and Fratantoni, J. C. 1970. Inborn errors of mucopolysaccharide metabolism. Science **169**: 141.
16. Ohno, S. 1971. Simplicity of mammalian regulatory systems inferred by single gene determination of sex phenotypes. Nature **234**: 134.
17. Osterman, J. V., Waddell, A., and Aposhian, H. V. 1970. DNA end gene therapy: Uncoating of polyoma pseudovirus in mouse embryo cells. Proc. Nat. Acad. Sci. USA **67**: 37.
18. Rogers, S. 1959. Induction of arginase in rabbit epithelium by the shope rabbit papilloma virus. Nature **183**: 1815.
19. Rogers, S. 1966. Shope papilloma virus: A passenger in man and its significance to the potential control of the host genome. Nature **212**: 1220.
20. Rogers, S. 1971. Genetic engineering. 4th International Congress of Human Genetics, Paris 1971. Paper presented and abstract published in: International Congress Series No. 233, p. 9 by Amsterdam, Excerpta Medica.
21. Schwartz, A. G., Cook, P. R., and Harris H. 1972. Correction of a genetic defect in a mammalian cell. Nature New Biology **230**: 5.
22. Strauss, B. S. 1971. Clinical application of genetic engineering. Prospects for the future. *In* Lipid storage diseases, eds. *Bernsohn* and *Grossman*. New York–London. Academic Press.
23. Tatum, E. L. 1965. The nature of the revolutionary new biology. *In* The control of human heredity and evolution, ed. *Sonneborn*, New York. Macmillan.
24. Tatum, E. L. 1966. The possibility of manipulating genetic change. *In* Genetics and the future of man, ed. *Roslansky*. Amsterdam: North-Holland Publishing Comp.
25. Terheggen, H. G. 1971. Personal communication.
26. Terheggen, H. G., Schwenk, A., Lowenthal, A., von Sande, M., and Colombo, J. P. 1969. Argininaemia with arginase deficiency. Lancet **2**: 748.
27. Terheggen, H. G., Schwenk, A., Lowenthal, A., van Sande, M., and Colombo, J. P. 1970. Hyperargininämie mit Arginasedefekt. Eine neue familiäre Stoffwechselstörung. Z. Kinderheilkunde **107**: 298, und **107**: 312.
28. Warkany, J. 1959. Congenital malformations in the past. J. Chron. Dis. St. Louis **10**: 83.
29. Webster, W. P., Zukoski, C. F., Hutchin, P., and Penick, G. D. 1970. Hepatic and extrahepatic synthesis of factor VIII (abstract). XIIIth Int. Congr. Hematol., München 1970, p. 222.
30. Editorial: Diabetes mellitus: Disease or syndrome. 1972. Lancet **1**: 583.

44   Troubled Dawn for Genetic Engineering

Robert L. Sinsheimer

The essence of engineering is design and, thus, the essence of genetic engineering, as distinct from applied genetics, is the introduction of human design into the formulation of new genes and new genetic combinations. These methods thus supplement the older methods which rely upon the intelligent selection and perpetuation of those chance genetic combinations which arise in the natural breeding process.

The possibility of genetic engineering derives from major advances in DNA technology—in the means of synthesising, analysing, transposing and generally manipulating the basic genetic substance of life. Three major advances have all neatly combined to permit this striking accomplishment: these are, 1, the discovery of means for the cleavage of DNA at highly specific sites; 2, the development of simple and generally applicable methods for the joining of DNA molecules; and 3, the discovery of effective techniques for the introduction of DNA into previously refractory organisms.

The art of DNA cleavage and degradation languished in a crude and unsatisfactory state until the discovery and more recent application of enzymes known as restriction endonucleases. These enzymes protect the host cells against invasion by foreign genomes by specifically severing the intruding DNA strands. For the purposes of genetic engineering, restriction enzymes provide a reservoir of means to cleave DNA molecules reproducibly at a limited number of sites by recognising specific tracts of DNA ranging for four to eight nucleotides in length. These sites may be deliberately varied by the choice of the restriction enzyme.

The enzymes cut both strands of the DNA double helix, and the break may be at the same base pair or staggered by several bases (Figure 44–1). In the latter case the two fragments of DNA are each left with a terminal unpaired strand—a so-called cohesive or "sticky" end. This is particularly valuable in joining together two pieces of DNA end to end.

The number of susceptible tracts in a DNA obviously depends on the particular DNA and the particular enzyme. In some important instances there is only one such tract. For instance, the restriction enzyme coded by the *E. coli* drug resistance transfer factor I—Eco RI—cleaves the DNA of the simian virus 40 at only one site. Similarly it cleaves the circular DNA of the plasmid PSC 101 at only one site. The DNA of bacteriophage lambda is, however, severed at five sites. It is possible to produce mutants of lambda with progressively fewer sites, until lambda strains are now available with just one or two sites.

For some purposes more numerous cleavage sites are useful. In a number of

This article is based on a lecture given to the Genetics Society of America.

From *New Scientist*, **68** (Oct. 16, 1975), 148–151. Reprinted with permission of the author and the publisher.

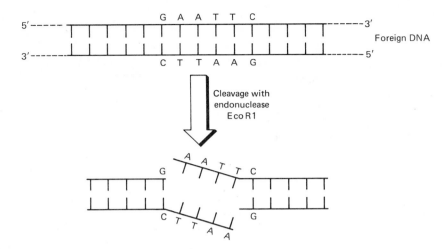

Cleaved DNA with specific sticky end

Figure 44-1. Endonucleases can cleave double-stranded DNA at one point, or staggered as shown in the diagram. The staggered cut produces sticky ends which can join with sections of DNA severed by the same enzyme.

laboratories, including my own, the ϕX virus replicating form can be cut at up to 13 sites using selected restriction enzymes. Because these enzymes yield overlapping fragments, a physical map of the DNA can be formed and correlated with the viral genetic map.

Restriction enzymes thus permit us to obtain specific fragments of DNA. For genetic engineering one would like to be able to rejoin such fragments in arbitrary ways. Two general methods have been developed to achieve this, both of which depend on the " sticky end" principle in which complementary single strand ends combine (Figures 44–1 and 44–2). Restriction enzymes which inflict staggered cuts automatically produce "sticky ends" in the DNA chain severed. Alternatively, a combination of enzymic and chemical manipulation can create a "sticky end."

Modified Plasmids in E. coli

By these means, then, any arbitrarily selected piece of DNA from any source can be inserted into the DNA of an appropriately chosen plasmid or virus. The new combination must then, for most purposes, be reintroduced into an appropriate host cell. This was achieved just a few years ago when Stanley Cohen, at Stanford, discovered that plasmid DNA could be reintroduced, albeit with low efficiency, into appropriately treated *E. coli* cells and that these could then subsequently grow and propagate the plasmid. Foreign genes can therefore be introduced into *E. coli* plasmids which can be propagated indefinitely in ordinary bacterial cultures. As one instance, the ribosomal RNA genes of *Xenopus laevis* (the African clawed toad)

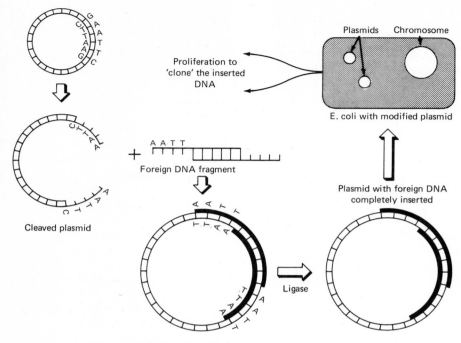

Proliferation to
'clone' the inserted
DNA

Plasmids   Chromosome

E. coli with modified plasmid

Foreign DNA fragment

Cleaved plasmid

Plasmid with foreign DNA
completely inserted

Ligase

Figure 44-2. Cloning a gene: A nick is made in the circular DNA of a plasmid; the required DNA sequence, excised with the same restriction enzyme, is inserted into the gap; the DNA chains are repaired by a ligase enzyme; the plasmids are reintroduced into E. coli; when the coli culture multiplies, the plasmids, and the foreign genes with them, are multiplied (cloned) too.

have been introduced into an *E. coli* plasmid and propagated for over 100 cell generations. And these genes are transcribed in their new host (Figure 44–2).

A similar result can, in principle, be achieved with the bacteriophage lambda. A foreign gene can be inserted into lambda DNA; spheroplasts or treated cells infected with this DNA will yield virus which can then be used to infect normal cells. By clever manipulation a recombinant DNA can be obtained which can subsequently be integrated into the host chromosome and propagated thereafter with the host.

To what purposes may these novel genetic combinations be put? One can conceive of a variety of benign purposes. Unfortunately one can also conceive of malign purposes, and of major, if unintended, hazards.

The first purposes that come to mind are of a purely scientific character. The structure and organisation of the eukaryotic (higher organism) genome is currently being studied intensively. This research has been grossly impeded by the complexity of these genomes and the lack of means to isolate particular portions in adequate quantities for experimental analysis. The insertion of fragments of eukaryotic DNA into plasmids, followed by cloning (cellular multiplication), permits one to grow cultures of any size containing just one particular fragment. At present the choice of fragments to be inserted cannot in general be precisely defined, although some prior selection can be introduced. However, ingenious methods are being devised to permit subsequent selection of those bacterial clones carrying fragments of particular interest.

Clones of bacteria bearing, say, histone genes, ribosomal RNA genes, genes from individual bands of *Drosophila* DNA, DNA of a certain degree of repetition in the sea urchin genome, and so forth, are currently being investigated. There are numerous questions to ask and numerous matters of interest concerning the transcription and translation of such genes in the bacterial host: for instance, the rates at which they may mutate, and the use of such cloned genes as probes of the eukaryotic genome.

It is very probable that in time the appropriate genes can be introduced into bacteria to convert them into biochemical factories for producing complex substances of medical importance: for example, insulin (for which a shortage seems imminent), growth hormone, specific antibodies, and clotting factor VIII which is defective in hemophiliacs. Even if these specific genes cannot be isolated from the appropriate organisms, the chances of synthesising them from scratch are now significant.

Other more grandiose applications of microbial genetic engineering can be envisaged. The transfer of genes for nitrogen fixation into presently inept species might have very significant agricultural applications. Appropriate design might permit appreciable modifications of the normal bacterial flora of the human mouth with a significant impact upon the incidence of dental caries. Even major industrial processes might be carried out by appropriately planned microorganisms.

However, we must remember that we are creating here novel, self-propagating organisms. And with that reminder, another darker side appears on this scene of brilliant scientific enterprise. For instance, for scientific purposes there is great interest in the insertion of particular regions of viral DNA into plasmids—particularly, portions of oncogenic (cancer-inducing) viral DNA—so as to be able to obtain such portions and their gene products in quantity and subsequently to study the effects of these substances on their normal host cells. Abruptly we come to the potential hazard of research in this field, in fact the specific hazard which inspired the widely known "moratorium" proposed last year by a committee of the US National Academy, chaired by Paul Berg.

This moratorium and its related issues deserve very considerable discussion. Briefly, it became apparent to the scientists involved—at almost the last hour when all of the techniques were really at hand—that they were about to create novel forms of self-propagating organisms—derivatives of strains known to be normal components of the human intestinal flora—with almost completely unknown potential for biological havoc. Could an *Escherichia coli* strain carrying all or part of an oncogenic virus become resident in the human intestine? Could it thereby become a possible source of malignancy? Could such a strain spread throughout a human population?

What would be the consequence if even an insulin-secreting strain became an intestinal resident? Not to mention the more malign or just plain stupid scenarios such as those which depict the insertion of the gene for botulinus toxin into *Escherichia coli*.

Unknown Probabilities

Unfortunately the answers to these questions in terms of probabilities that some of these strains could persist in the intestines, the probabilities that the modified plasmids might be transferred to other strains better adapted to intestinal life, the probabilities that the genome of an oncogenic virus could escape, could be taken up, could transform a host cell, are all largely unknown.

Following the call for a moratorium a conference was held at Asilomar at the end of last February to assess these problems. While it proved possible to rank various types of proposed experiments with respect to potential hazard, for the reasons already stated it proved impossible to establish, on any secure basis, the absolute magnitude of hazard. Various distinguished scientists differed very widely, but sincerely, in their estimates. Historical experience indicated that simple reliance upon the physical containment of these new organisms could not be completely effective.

In the end a broad, but not universal, consensus was reached which recommended that the seemingly more dangerous experiments be deferred until means of "biological containment" could be developed to supplement physical containment. By biological containment is meant the crippling of all vehicles—cells or viruses—intended to carry the recombinant genomes through the insertion of a variety of genetic defects so as to reduce very greatly the likelihood that the organisms could survive outside of a protective, carefully supplemented laboratory culture.

This seems a sensible and responsible compromise. However, several of the less prominent aspects of the Asilomar conference also deserve much thought. The lens of Asilomar was focused sharply upon the potential biological and medical hazard of this new research, but other issues drifted in and out of the field of discussion. There was, for instance, no specific consideration of the wisdom of diverting appreciable research funds and talent to this field, in lieu of others. An indirect discussion of this question was perhaps implicit in the description of the significance and scientific potential of research in this field presented by those who were impatient of any delay.

Indeed the eagerness of the researchers to get on with the work in this field was most evident. To a scientist this was exhilarating. Obviously these new techniques open many previously closed doorways leading to the potential resolution of long-standing and important problems. I think also there is a certain romance in this joining together of DNA molecules that diverged billions of years ago and have pursued separate paths through all of these millenia. Personally I feel confident one could easily justify this new research direction. But a sociologist of science might see other under-currents in this impetuous eagerness, and the bright scientific promise should not blind us to the realities of other concerns.

Nor was there any sustained discussion at Asilomar of ancillary issues such as the absolute right of free inquiry claimed quite vigorously by some of the participants. Here, I think, we have come to recognise that there are limits to the practice of any human activity. To impose any limit upon freedom of inquiry is especially bitter for the scientist whose life is one of inquiry; but science has become too potent. It is no longer enough to wave the flag of Galileo.

Rights are not found in nature. Rights are conferred within a human society and for each there is expected a corresponding responsibility. Inevitably at some boundaries different rights come into conflict and the exercise of a right should not destroy the society that conferred it. We recognise this in other fields. Freedom of the press is a right but it is subject to restraints, such as libel and obscenity and, perhaps more dubiously, national security. The right to experiment on human beings is obviously constrained. Similarly, would we wish to claim the right of individual scientists to be free to create novel self-perpetuating organisms likely to spread about the planet in an uncontrollable manner for better or worse? I think not.

This does not mean we cannot advance our science or that we must doubt its ultimate beneficence. It simply means that we must be able to look at what we do in a mature way.

There was, at Asilomar, no explicit consideration of the potential broader social or ethical implications of initiating this line of research—of its role, as a possible prelude to longer-range, broader-scale genetic engineering of the flora and fauna of the planet, including, ultimately, man. It is not yet clear how these techniques may be applied to higher organisms but we should not underestimate scientific ingenuity. Indeed the oncogenic viruses may provide a key; and mitochondria may serve as analogues for plasmids.

Controlled Evolution?

How far will we want to develop genetic engineering? Do we want to assume the basic responsibility for life on this planet—to develop new living forms for our own purpose? Shall we take into our own hands our own future evolution? These are profound issues which involve science but also transcend science. They deserve our most serious and continuing thought. I can here mention only a very few of the more salient considerations.

Clearly the advent of genetic engineering, even merely in the microbial world, brings new responsibilities to accompany the new potentials. It is always thus when we introduce the element of human design. The distant, yet much discussed application of genetic engineering to mankind would place this equation at the centre of all future human history. It would in the end make human design responsible for human nature. It is a responsibility to give pause, especially if one recognises that the prerequisite for responsibility is the ability to forecast, to make reliable estimates of the consequence.

Can we really forecast the consequence for mankind, for human society, of any major change in the human gene pool? The more I have reflected on this the more

I have come to doubt it. I do not refer here to the alleviation of individual genetic defects—or, if you will, to the occasional introduction of a genetic clone—but more broadly to the genetic redefinition of man. Our social structures have evolved so as to be more or less well adapted to the array of talents and personalities emergent by chance from the existing gene pool and developed through our cultural agencies. In our social endeavours we have, biologically, remained cradled in that web of evolutionary nature which bore us and which has undoubtedly provided a most valuable safety net as we have in our fumbling way created and tried out varied cultural forms.

To introduce a sudden major discontinuity in the human gene pool might well create a major mismatch between our social order and our individual capacities. Even a minor perturbation such as a marked change in the sex ratio from its present near equality could shake our social structures—or consider the impact of a major change in the human life span. Can we really predict the results of such a perturbation? And if we cannot foresee the consequence, do we go ahead?

It is difficult for a scientist to conceive that there are certain matters best left unknown, at least for a time. But science is the major organ of inquiry for a society —and perhaps a society, like an organism, must follow a developmental programme in which the genetic information is revealed in an orderly sequence.

The dawn of genetic engineering is troubled. In part this is the spirit of the time —the very idea of progress through science is in question. People seriously wonder if through our cleverness we may not blunder into worse dilemmas than we seek to solve. They are concerned not only for the vagrant lethal virus or the escaped mutant deadly microbe, but also for the awful potential that we might inadvertently so arm the anarchic in our society as to shatter its bonds or conversely so arm the tyrannical in our society as to forever imprison liberty.

It is grievous that the elan of science must be tempered, that the glowing conviction that knowledge is good and that man can with knowledge lift himself out of hapless impotence must now be shaded with doubt and caution. But in this we join a long tradition. The fetters that are part of the human condition are not so easily struck.

We confront again the enduring paradox of emergence. We are each a unit, each alone. Yet, bonded together, we are so much more. As individuals men will have always to accept their genetic constraints, but as a species we can transcend our inheritance and mould it to our purpose—if we can trust ourselves with such powers. As geneticists we can continue to evolve possibilities and take the long view.

Clone Order Form

CLONE ORDER FORM	**PACIFIC INSTITUTE OF TECHNOLOGY** **MEDICAL SCHOOL** 8444 Wilshire Blvd., Beverly Hills, California 90211

April 22, 1983

TO PROSPECTIVE PARENTS:

Congratulations on your decision to become a parent. We are sending this order form for your clone in response to your inquiry.

We are indeed fortunate to live in the Nineteen Eighties for, through science, we are no longer at the mercy of mother nature when it comes to acquiring children. There was a time—many of you may remember—when to produce an offspring we were limited to sexual reproduction.

In ordering your child (clone) fill in the spaces below with careful consideration. Submit this form in triplicate to the biogenetics culture laboratories of Pacific Institute of Technology's medical school along with (a) The U.S. Department of Genetic and Cloning Control Certificate of Permission for Parenthood, and (b) a sample of epidermal cells of each prospective parent. (If there is to be one parent, only one sample of cells is necessary. Of course, if there are to be two, three, or more parents of the clone, a sample from each for the proper genetic combination is necessary.) Your family physician can take these samples in a simple office consultation.

We will endeavor to provide you with the combination in genetic make-up most closely matching your specifications as set forth below. You should receive your clone after the normal nine month in-vitro fertilization and extra-utero gestation period.

James R. Kemp

James R. Kemp, M.D.
Institute Director

UNAUTHORIZED CLONING IS A FEDERAL OFFENSE

For Government Use Only (43–67–J8)
Approval: Dept. A ☐ Dept. B ☐ Dept. C ☐
Suggested Alterations:
A. XM–21 ☐   B. J5R–VB ☐   C. U90–1 ☐
Statistical   Statistical

From *Genetic Engineering, Science and Society*, Series no. 3, Science for the People, 1973.

[1]	Parent(s) Name(s)			Sex		other
	(last)	(first)	(initial)	M	F	
1st						
2nd						
3rd						

The following are characteristics you desire in your child (clone). Check appropriate boxes. Please type or use pen.

[2]					Rate
Sex	Male ☐	Female ☐	Comb. ☐		$73.00

[3]					
Ht. in Adulthood: ☐ Tall	☐ Med	☐ Short			$61.00

[4]				
Wt. in Adulthood: ☐ Obese	☐ Norm	☐ Slender		$52.00

[5] Color of Hair: Blk ☐ Brn ☐ Blond ☐ Red ☐ Comb. ☐ — $71.00

[6] Color of Eyes: Brn ☐ Blue ☐ Hazel ☐ Blk ☐ Comb. ☐ — $72.00

[7] Intelligence Quotient:

100–110	111–120	121–130	131–140	Above 140
$320	$400	$480	$560	$620 ◄

[8] Exceptional aptitude in selected profession. Please specify:

[9] Geno-Physical Type:

If desired, physical appearance can be cloned from cells in storage of the individuals listed below. Check one.

		a, b, f, h, i, l — $415
☐ a. Mick Jagger	☐ b. Johnny Carson	
☐ c. Orson Welles	☐ d. Burt Reynolds	
☐ e. Henry Kissinger	☐ f. Richard Nixon	
☐ g. Wilt Chamberlin	☐ h. Raquel Welch	c, d, e, g, j, k — $375
☐ i. Jacqueline Kennedy	☐ j. Kate Smith	
☐ k. Phyllis Diller	☐ l. Angela Davis	

[10] Geno-Cerebral Type: If desired, special talent and professional characteristics are available from cells in storage of the following individuals. Check one. Prices available on request.[*]

☐ Albert Einstein	☐ Mohammed Ali
☐ Emilia Earhart	☐ Leonard Bernstein
☐ William Shakespeare	☐ Francis Ford Coppola
☐ Pablo Picasso	☐ Sigmund Freud
☐ Plato	☐ Adolph Hitler
☐ Howard Hughes	☐ Golda Meir
☐ Marilyn Monroe	☐ Marie Curie
☐ Tokyo Rose	☐ Betsy Ross
☐ Amenhotep II	☐ Pope Pius (I–XII)
☐ Mozart	☐ Yoko Ono
☐ Noble "Kid" Chisel	☐ Doodles Weaver

[*]Inquire for additional selections

[11] Personality Preferences: Check one position for each scale

	1	2	3	4	5	6	
Passive							Aggressive
Dominant							Submissive
Introvert							Extrovert
Hyperkinetic							Hypokinetic
Affectionate							Nonaffectionate
Sensitive							Insensitive